MW01323318

Quantum Mechanics

Springer
Berlin
Heidelberg
New York
Barcelona
Hong Kong
London
Milan
Paris
Tokyo

Advanced Texts in Physics

This program of advanced texts covers a broad spectrum of topics which are of current and emerging interest in physics. Each book provides a comprehensive and yet accessible introduction to a field at the forefront of modern research. As such, these texts are intended for senior undergraduate and graduate students at the MS and PhD level; however, research scientists seeking an introduction to particular areas of physics will also benefit from the titles in this collection.

The École Polytechnique, one of France's top academic institutions, has a long-standing tradition of producing exceptional scientific textbooks for its students. The original lecture notes, the *Cours de l'École Polytechnique*, which were written by Cauchy and Jordan in the nineteenth century, are considered to be landmarks in the development of mathematics.

The present series of textbooks is remarkable in that the texts incorporate the most recent scientific advances in courses designed to provide undergraduate students with the foundations of a scientific discipline. An outstanding level of quality is achieved in each of the scientific fields taught at the *École*: pure and applied mathematics, computer sciences, mechanics, physics, chemistry, biology, and economics. The uniform level of excellence is the result of the unique selection of academic staff there which includes, in addition to the best researchers in its own renowned laboratories, a large number of world-famous scientists, appointed as part-time professors or associate professors, who work in the most advanced research centers France has in each field.

Another distinctive characteristic of these courses is their overall consistency; each course makes appropriate use of relevant concepts introduced in the other textbooks. This is because each student at the École Polytechnique has to acquire basic knowledge in the various scientific fields taught there, so a substantial link between departments is necessary. The distribution of these courses used to be restricted to the students at the École. Some years ago we were very successful in making these courses available to a larger French-reading audience. We now build on this success by making these textbooks also available in English.

Jean-Louis Basdevant Jean Dalibard

Quantum Mechanics

Including a CD-ROM by Manuel Joffre

With 84 Figures and 92 Exercises with Solutions

Professor Jean-Louis Basdevant
Professor Manuel Joffre

École Polytechnique
Département de Physique
91128 Palaiseau Cedex, France

E-mails:
basdevan@poly.polytechnique.fr
manuel.joffre@polytechnique.fr

Professor Jean Dalibard

École Normale Supérieure
Département de Physique
Laboratoire Kastler Brossel
24, rue Lhomond
75231 Paris Cedex 05, France

E-mail: jean.dalibard@lkb.ens.fr

Cover picture: Shows a schematic drawing of a Young double slit interference experiment performed with ultracold atoms (drawing by the authors); see also Chap. 1, Figs. 1.3 and 1.4.

ISSN 1439-2674

ISBN 3-540-42739-2 Springer-Verlag Berlin Heidelberg New York

Library of Congress Cataloging-in-Publication Data applied for.

Die Deutsche Bibliothek - CIP-Einheitsaufnahme
Basdevant, Jean-Louis:
Quantum mechanics : including a CD-ROM ; with 84 figures and 92 exercises with solutions/ Jean-Louis Basdevant ; Jean Dalibard. - Berlin ; Heidelberg ; New York ; Barcelona ; Hong Kong ; London ; Milan ; Paris ; Tokyo : Springer, 2002
(Physics and astronomy online library)
(Advanced texts in physics)
ISBN 3-540-42739-2

Springer-Verlag Berlin Heidelberg New York
a member of BertelsmannSpringer Science+Business Media GmbH

http://www.springer.de

Typesetting by the authors using a Springer LaTEX macro
Data conversion by EDV-Beratung F. Herweg, Hirschberg
Cover design: *design & production* GmbH, Heidelberg

Printed on acid-free paper SPIN 10851005 56/3141/di 5 4 3 2 1 0

Preface

Felix qui potuit rerum cognoscere causas
(Lucky are those who have been able
to understand the causes of things.)
R. Goscinny and A. Uderzo, Asterix in Corsica, 1973, page 22;
see also: Virgil, Georgics II

Quantum mechanics has the unexpected feature that there is as yet no empirical evidence that it has limited applicability. The only hypothetical indication that some "new physics" might exist comes from cosmology, and concerns the first 10^{-43} s of the universe. This is quite unlike the situation for other physical theories. Quantum physics was born at the beginning of the 20th century from the questioning of physicists faced with an incredible variety of experimental facts which were steadily accumulating without any global explanation. This questioning was amazingly ambitious and fruitful. In fact, quantum theory is undoubtedly one of the greatest intellectual endeavors of mankind, perhaps the greatest of the 20th century.

It was born in an unexpected way. At the beginning of the 19th century, the sagacious French philosopher Auguste Comte claimed that one could never know the chemical composition of stars since it was impossible for us to visit them. Had he thought that the same remark could apply just as well to a hot oven, he would have described unintentionally, and by pure reasoning, the cradle of quantum physics.

Quantum physics appeared fortuitously in an idea of Planck about the black-body radiation spectrum, which was acknowledged to be a fundamental problem. Quantum physics first developed by disentangling spectroscopic data. In that sense it owes much to astrophysics, which was developing at the same time, and revealed the complex spectra of elements. The phenomenological analysis of the regularities of spectra (by Balmer, Rydberg, Ritz and Rayleigh) had led to a set of efficient recipes. But there was no indication that this scrupulous classification would lead to such an upheaval of the foundations of physics.

In fact, the fate of quantum physics was unexpected. It started by explaining the laws of radiation, and no one could have imagined that it would end up giving a complete explanation of the structure of matter, of atoms and molecules. Atomic theory ceased to be a qualitative controversy. It became a fact, and this struck the minds of people. In an article published in 1948 entitled "2400 years of quantum mechanics", Schrödinger said that Democritus and the inventors of atomism were the "first quantum physicists".[1] He paid tribute to all those who had tried to understand the fundamental structure of matter. This had been difficult for many reasons. The Catholic Church, for instance, remained strongly opposed to the idea for a long time since atoms do not have souls. Even Leibniz thought he could disprove the existence of atoms.[2] Our first quantitative ideas about atoms came on one hand from the chemists of the 19th century, who discovered that they could reduce chemical reactions to an interplay of integers, and on the other hand from the initiators of statistical physics, Maxwell and Boltzmann, who showed that the thermodynamic properties of gases found natural explanations within the molecular hypothesis. Because it succeeded in describing quantitatively the structure of atoms, quantum mechanics consecrated their existence.

The range of its applications was also unexpected. Quite rapidly, all physics and all chemistry became quantum theories. The theory accounts not only for atoms and molecules, but also for the structure of nuclei, for particle physics and cosmology, for the electrical and mechanical properties of solid-state materials, etc. Astrophysics was well paid back and underwent spectacular developments because of quantum theory. These developments led to new observational means to probe the cosmos, and also to the explanation of truly macroscopic quantum objects such as white dwarfs and neutron stars.

Since its beginning, quantum theory has also generated considerable intellectual and philosophical turmoil. For the first time, not only pure reasoning but also what we think to be common sense appeared to be falsified by experimental facts. We needed a new way of thinking about reality, a new logic. It was necessary to develop a quantum intuition, which often seemed contrary to common intuition. As one can guess, an epistemological revolution took place. Philosophers such as Kirkegaard, Höffding, Husserl, Wittgenstein and many others had already discovered how treacherous common language may be. It is full of a priori conclusions on the nature of things, and any new experimental field can be analyzed only with new concepts and a new language. Quantum mechanics seems to have been invented to prove the philosophers were right. In some respects it goes against some aspects of rationalism. It is quite remarkable that, although at present everyone accepts the mathematical and operational framework of the theory, there are still bitter disputes about its interpretation and its philosophical content.[3]

---

[1] E. Schrödinger, "2400 Jahre Quantenmechanik", Ann. Phys. **3**, 43 (1948).

[2] G.W. Leibniz, *New Essays on Human Understanding*, Leibnitii Opera Omnia, L. Duten (ed.) Geneva (1768).

[3] See for instance *Quantum Theory and Measurement*, edited by J.A. Wheeler and W.H. Zurek, Princeton University Press, Princeton (1983).

What was really unexpected in quantum theory was that it would tackle so directly and so successfully the fundamental structure of matter. There is no experimental evidence at present that a more elaborate conceptual framework is necessary in order to understand the fundamental constituents of matter and their interactions. By its predictive power, quantum physics has been able to radically transform numerous technological sectors in the last 50 years. It has changed the orders of magnitude of what was conceivable. It is now possible to manufacture a material with a virtually unlimited range of thermal, optical, mechanical and electrical properties. It is more and more feasible to detect a deficiency in a biological function and to cure it in a planned and reasoned manner. The results of the development of semiconductor physics and of microelectronics fill our daily life. In the history of mankind, it is a true revolution which multiplies the power of man's mind, just as the industrial revolution multiplied our strength. This gigantic technological progress is modifying deeply the structure of social, economic and political life, and the mere question of how to adapt our societies to these developments has become a major problem.

Obviously, the number of problems to be solved increases faster than those which have been solved. For instance, in order to go from elementary processes to macroscopic phenomena, one needs the concepts of statistical physics. It is one of the great discoveries of the past decades that it is impossible to reduce everything to microscopic processes. However, one cannot deny that the dimensions and perspectives of physics have changed radically since it has entered the quantum era.

Let us recall that the construction of quantum mechanics benefited considerably from the collaboration of mathematicians. The mathematical framework of the theory was discovered very soon by Hilbert and Von Neumann. The mathematical structure of quantum mechanics and of quantum field theory has always been a fruitful field of research for mathematicians.

Conversely, one must admit that one of the difficulties one meets in apprehending the theory lies in the fact that the experimental reality of the quantum world is quite far from what is directly accessible. Many intermediate steps are necessary in order to build one's own representation of a phenomenon. This is of course reflected by the mathematical structure of the theory, which certainly deserves the criticism of being abstract. In the epigraph of his book *An Introduction to the Meaning and Structure of Physics* (Leon N. Cooper, Harper & Row, New York (1968)), Leon Cooper writes, in beautiful French, *S'il est vrai qu'on construit des cathédrales aujourd'hui dans la Science, il est bien dommage que les gens n'y puissent entrer, ne puissent pas toucher les pierres elles-mêmes.*[4]

[4] "While it is true that we build cathedrals nowadays in Science, it is a great pity that people cannot enter them, and touch the stones they are made of." *(Free translation).*

How to teach quantum mechanics has been a source of discussion perhaps as rich as that of its foundations. Many of the first textbooks were oriented along one of the two following lines. The first consisted in explaining at length the failure of classical conceptions and in using similarities which were often as long as they were obscure. The other method, which was more radical, consisted in expounding first the mathematical beauty and virtues of the theory, and in mentioning briefly some restricted set of assertions or experimental facts. A third approach appeared in the 1960s. It consisted in first describing quantum phenomena and then in introducing, or sometimes inventing, the mathematical structures as they became necessary.

In the last twenty years or so, the situation has evolved considerably for three main reasons.

The first reason is experimental. Many fundamental experiments which are easy to discuss but difficult to achieve technically have become possible. A first example is the Young double-slit interference experiment with atoms, which was performed in the 1990s and which we present in Chap 1. This experiment allows us to discuss in a clear and concrete way what was a gedanken experiment before that. A second example is the neutron interference experiments which we mention in Chap 12. Such experiments were carried out in the early 1980s near high-flux nuclear reactors. They put an end to a 50 year old dispute about the measurability of the phase of the wave function, be it in a magnetic field or in a gravitational field.

The second reason comes from what we may call the breakdown of paradoxes. The formulation of Bell's inequalities and their experimental study are undoubtedly major intellectual steps in the history of quantum mechanics. We now possess quantitative experimental answers to questions which used to come within a hair's breadth of metaphysics. These experiments, together with other experiments on entangled states which we refer to in Chap 14, have changed our way of thinking. In some sense, one discovers that Einstein was right when he claimed that the interpretation of quantum mechanics causes genuine physical problems, even though the solution he had in mind was apparently not the correct one. More recently, the development of the theory of decoherence and its verification on mesoscopic systems have constituted a major step forward in the understanding of the foundations of quantum mechanics.

The third reason stems from the remarkable development of numerical simulation methods and imaging with modern computers. We are now able to perform true visual representations of processes on very small space–time scales. This allows a direct intuitive visualization of the theory and of its consequences which is radically different from what could be done previously.

In this book, whose origin lies in the 25 year teaching experience of one of us with third year undergraduates at the Ecole Polytechnique, we have made use of these three aspects. Perhaps more of the first and third, even though the second has played a major role psychologically, as illustrated in Chap. 14.

We have followed a rather traditional path, starting from wave mechanics in order to become familiarized with the relevant mathematical notions. We have done our best to introduce the mathematical tools of the theory starting, as much as possible, from the structure of observed phenomena. This textbook contains a set of 90 (rather simple) exercises and their solutions. Its natural complement is *The Quantum Mechanics Solver*,[5] which we published ahead of time, a year ago, and where applications to genuine, recent physical phenomena can be found, such as neutrino oscillations, entangled states, quantum cryptography, Bell's inequalities, laser cooling, Bose–Einstein condensates, etc. In addition, the book comes with a CD-ROM, due basically to Manuel Joffre, which contains examples, applications and web links which, we hope, will help the user to become familiar both with the theory and with its present applications.

But we cannot avoid two obstacles. First, an axiomatic presentation is more economical and easier for the person who teaches. Secondly, the relation between physical concepts and their mathematical representations is not as direct as in classical physics. If this book seems too abstract or theoretical, we cannot avoid pleading guilty. Building one's own representation of quantum phenomena is a personal matter which can only result from practicing with the theory and from experimental results, including all the unexpected features they reveal.

We wish to thank all our colleagues who contributed to this book, in the exceptional teaching team which was created around us. We pay tribute to the memory of Eric Paré and Dominique Vautherin. Eric, who was a remarkable particle physicist and a marvelous friend, died accidentally in July 1998 at the age of 39. Dominique, a theorist who made decisive contributions to nuclear physics and the many-body problem, kept all his sense of humor, his generosity and his beautiful intelligence during a one-year fight against a disease which defeated him in December 2000 at the age of 59. Both of them made important contributions to this text.

We thank Florence Albenque, Hervé Arribart, Alain Aspect, Gérald Bastard, Adel Bilal, Alain Blondel, Jean-Noël Chazalviel, Jean-Yves Courtois, Nathalie Deruelle, Henri-Jean Drouhin, Claude Fabre, Hubert Flocard, Philippe Grangier, Denis Gratias, Gilbert Grynberg, François Jacquet, Thierry Jolicoeur, David Langlois, Rémy Mosseri, Pierre Pillet, Daniel Ricard, Jim Rich, André Rougé, Emmanuel Rosencher, Michel Spiro, Alfred Vidal-Madjar and Henri Videau. They all contributed significantly to what is interesting in this book.

One of us (JLB) expresses his gratitude to Yves Quéré, Bernard Sapoval, Ionnel Solomon and Roland Omnès for their interest and help when this course started. He thanks his mathematician colleagues Laurent Schwartz, Alain Guichardet, Yves Meyer, Jean-Pierre Bourguignon and Jean-Michel

[5] J.-L. Basdevant and J. Dalibard, *The Quantum Mechanics Solver: How to Apply Quantum Theory to Modern Physics*, Springer, Berlin Heidelberg (2000).

Bony, and the chemist Marcel Fétizon, for useful interdisciplinary collaborations. He also pays tribute to the memory of Bernard Gregory and to that of Michel Métivier.

Palaiseau, Paris,
January 2002

Jean-Louis Basdevant
Jean Dalibard

Contents

Physical Constants

Units

Angström	1 Å$= 10^{-10}$ m ($\sim$ size of an atom)
Femtometer*	1 fm $= 10^{-15}$ m ($\sim$ size of a nucleus)
Electron-volt	1 eV $= 1.60218\ 10^{-19}$ J

Fundamental constants

Planck's constant	$h = 6.6261 \times 10^{-34}$ J s, $\hbar = h/2\pi = 1.05457 \times 10^{-34}$ J s $= 6.5821 \times 10^{-22}$ MeV s
Velocity of light	$c = 299\,792\,458\,\mathrm{m\,s^{-1}}$ $\hbar c = 197.327$ MeV fm $\simeq 1973$ eV Å
Vacuum permeability	$\mu_0 = 4\pi 10^{-7}$ H m^{-1}, $\quad \epsilon_0\mu_0 c^2 = 1$
Boltzmann's constant	$k_\mathrm{B} = 1.38066 \times 10^{-23}\,\mathrm{J\,K^{-1}} = 8.6174 \times 10^{-5}\,\mathrm{eV\,K^{-1}}$
Avogadro's number	$N_\mathrm{A} = 6.0221 \times 10^{23}$
Electron charge	$q_\mathrm{e} = -q = -1.60218 \times 10^{-19}$ C and $e^2 = q^2/(4\pi\epsilon_0)$
Electron mass	$m_\mathrm{e} = 9.1094 \times 10^{-31}$ kg, $\quad m_\mathrm{e}c^2 = 0.51100$ MeV
Proton mass	$m_\mathrm{p} = 1.67262 \times 10^{-27}$ kg, $\quad m_\mathrm{p}c^2 = 938.27$ MeV, $m_\mathrm{p}/m_\mathrm{e} = 1836.15$
Neutron mass	$m_\mathrm{n} = 1.67493 \times 10^{-27}$ kg, $\quad m_\mathrm{n}c^2 = 939.57$ MeV
Fine structure constant (dimensionless)	$\alpha = e^2/(\hbar c) = 1/137.036$
Classical radius of the electron	$r_\mathrm{e} = e^2/(m_\mathrm{e}c^2) = 2.818 \times 10^{-15}$ m
Compton wavelength of the electron	$\lambda_c = h/(m_\mathrm{e}c) = 2.426 \times 10^{-12}$ m
Bohr radius	$a_1 = \hbar^2/(m_\mathrm{e}e^2) = 0.52918\ 10^{-10}$ m
Ionization energy of hydrogen	$E_\mathrm{I} = m_\mathrm{e}e^4/(2\hbar^2) = \alpha^2 m_\mathrm{e}c^2/2 = 13.6057$ eV
Rydberg's constant	$R_\infty = E_\mathrm{I}/(hc) = 1.09737 \times 10^7\,\mathrm{m^{-1}}$
Bohr magneton	$\mu_\mathrm{B} = q_\mathrm{e}\hbar/(2m_\mathrm{e}) = -9.2740\ 10^{-24}\,\mathrm{J\,T^{-1}}$ $= -5.7884 \times 10^{-5}\,\mathrm{eV\,T^{-1}}$
Nuclear magneton	$\mu_\mathrm{N} = q\hbar/(2m_\mathrm{p}) = 5.0508\ 10^{-27}\,\mathrm{J\,T^{-1}}$ $= 3.1525 \times 10^{-8}\,\mathrm{eV\,T^{-1}}$

Updated values can be found at `http://wulff.mit.edu/constants.html`

---
* Nuclear physicists honor Enrico Fermi by also calling this unit the Fermi.

1. Quantum Phenomena

Any matter begins
with a great spiritual disturbance.
Antonin Artaud

The birth of quantum physics occurred on December, 14 1900, when Max Planck, at the German Physical Society, proposed a simple formula in excellent agreement with the observed black-body radiation spectrum. Planck had first obtained his result empirically, but he noticed that he could deduce the key point of his argument from Boltzmann's statistical thermodynamic theory by making the puzzling assumption that charged mechanical oscillators of frequency ν emit or absorb radiation only in discrete amounts, energy "quanta" which are integer multiples of $h\nu$. The *quantum of action* h is a fundamental constant, as Planck realized:

$$h \simeq 6.6261 \times 10^{-34}\ \mathrm{J\,s}\,. \tag{1.1}$$

Planck's quanta were mysterious, but his result was amazingly successful. Until 1905, neither the scientific community nor Planck himself fully appreciated this discovery. In that year, Einstein published his famous article, "On a heuristic point of view concerning the production and transformation of light",[1] where he analyzed Planck's argument. Einstein found some inconsistencies and corrected them. If one pushes Planck's argument a bit further, one must admit that light itself has "quantum" properties, and Einstein introduced the concept of a *quantum of radiation*, called the *photon* by G.N. Lewis in 1926. A quantum of light of frequency ν (or angular frequency ω) has an energy

$$E = h\nu = \hbar\omega\,, \qquad \text{where} \quad \hbar = \frac{h}{2\pi} = 1.0546 \times 10^{-34}\,\mathrm{J\,s}. \tag{1.2}$$

In the course of his work, Einstein realized that he could also explain the laws of the photoelectric effect, discovered in 1887 by Hertz and systematically studied by Lenard and Millikan. He also proposed that the photon carries a momentum

[1] Ann. Phys. **17**, 132 (1905); translated into English by A.B. Arons and M.B. Peppard, Am. J. Phys. **33**, 367 (1965).

$$\boldsymbol{p} = \hbar \boldsymbol{k}\,, \qquad |\boldsymbol{k}| = 2\pi/\lambda\,, \tag{1.3}$$

where $\boldsymbol{k}$ is the wave vector of the electromagnetic wave. This idea was confirmed by Compton's experiments (scattering of X rays by thin aluminum sheets) in 1923.

Planck's quanta were mysterious, although they were well accepted by the scientific community owing to the remarkable efficiency of his formula. Conversely, Einstein's quanta were quite controversial, and remained so for some time. Many people considered the idea of quanta was nonsense since it contradicted Maxwell's equations, which describe the energy of radiation as a continuous function in space and time. Einstein was aware of that, but he considered that measurements in optics involved only time averages, and that it was conceivable that Maxwell's theory could be insufficient whenever quasi-instantaneous processes take place. Einstein called his introduction of the light-quanta hypothesis a "revolutionary" step. He had the premonition that the manifestations of the properties of light could be *both* wave-like *and* particle-like. This can be considered as the true starting point of quantum theory.

The second step took place in the years 1912–1914. Niels Bohr, in trying to find a consistent model for atomic structure, managed to reconcile the Ritz combination principle of spectral lines, the atomic model of Rutherford (who had discovered the existence of the nucleus in 1911) and the quanta of Planck and Einstein. Bohr assumed that the energies of atomic and molecular systems can only take discrete values, and that the emission and the absorption of light by these systems take place only at fixed frequencies:

$$\nu_{\mathrm{if}} = |E_{\mathrm{i}} - E_{\mathrm{f}}|/h\,, \tag{1.4}$$

where E_{i} and E_{f} are the energies of the system before and after the emission (or the absorption). By chance, Bohr learned in 1913 of the existence of the empirical Balmer formula. In a few weeks, he guessed the rule for energy quantization, and developed his celebrated model of the hydrogen atom. The mechanism for absorption and emission of light remained obscure in Bohr's theory. It was only explained a few years later, in particular by Einstein. However, as soon as 1914, the experiments of Franck and Hertz gave a direct proof of the quantization of matter, i. e. of energies in atoms.

Thus the quantization of radiation had been discovered before the quantization of matter. This latter quantization seemed to imply a "discontinuous" aspect of the laws of nature. It was therefore with some relief and enthusiasm that many physicists, including Einstein, reacted to the idea Louis de Broglie proposed in 1923. In the same way as light manifests a particle-like behavior, de Broglie suggested that particles should have a wave-like behavior under appropriate circumstances. With a particle of velocity $\boldsymbol{v}$ and momentum $\boldsymbol{p} = m\boldsymbol{v}$, de Broglie "associated" a wave, of wavelength

$$\lambda = h/p\,. \tag{1.5}$$

This wave hypothesis could provide a means to understand energy quantization as a stationary wave problem, and to restore the continuity of nature.

Louis de Broglie's idea originated from a series of remarkable theoretical works by Marcel Brillouin (the father of Léon Brillouin). He also spent some time in the laboratory of his brother Maurice de Broglie, and he was surprised to hear physicists talk about the same physical concept both as an "electron" and as a "β ray" in radioactivity.

In its present formalism, the theory of quantum mechanics was developed quite rapidly between 1925 and 1927. It is the collective fruit of the work of an exceptional group of physicists and mathematicians such as Schrödinger, Heisenberg, Born, Bohr, Dirac, Pauli, Hilbert, von Neumann, etc. This remarkable synthetic work, which was followed by crucial experiments, was based on extensive theoretical and experimental investigations which had taken place in the first quarter of the century. We have mentioned above a few important steps.

Among all these experiments, we have chosen to discuss three which are particularly significant. In Sect. 1.1 we present the experiment of Franck and Hertz, which was the first experimental demonstration of the quantization of energy in atoms. In Sect. 1.2 we discuss an experiment which shows the wave behavior of particles, specifically atoms. In analyzing the results of this experiment, we shall demonstrate the probabilistic features of quantum physics and see in what sense they differ from the usual probabilistic effects. Finally, in Sect. 1.3 we sketch the first experimental proof of the wave behavior of electrons, due to Davisson and Germer, who observed the diffraction of an electron beam impinging on a crystal.

These experiments are significant insofar as they show how difficult it is at first to gain insight into the quantum world. Our usual intuition, "reason" and "common sense", which we have built by observing the world of classical phenomena, conceal unexpected traps. The experiments described here are also significant because they contain the essential basic concepts one needs in order to understand *quantum phenomena.*

1.1 The Franck and Hertz Experiment

The ideas of Niels Bohr received a spectacular and unexpected confirmation as soon as 1914. At that time, James Franck and Gustav Hertz were studying cathode rays and discharges in rarefied gases. This had important applications in the improvement of electron tubes, in particular for long-distance telephone technology.

In 1914 Franck and Hertz, who were shooting electrons with an energy of a few electron-volts at mercury atoms, discovered the following remarkable fact. As long as the energy of the electrons is smaller than some threshold energy, $E_{\text{th}} = 4.9\,\text{eV}$, the collision is elastic; the outgoing electrons have the same energy as the incident ones. Nothing surprising, since the mass of a mercury atom is $\sim 400\,000$ times larger than the electron mass and its recoil is negligible. The surprising fact was that when one reaches the threshold energy $E_{\rm th} = 4.9$ eV, the outgoing electrons lose practically all their energy in the process. Above that energy, a fraction of the outgoing electrons have an energy smaller than their initial energy by exactly the amount 4.9 eV, the other electrons keep all their energy in the collision process.

Furthermore, when the energy of the electrons is larger than the threshold $E_{\rm th}$, the mercury atoms emit ultraviolet radiation of wavelength $\lambda = 253.7\,$nm, and this does not occur if the energy of the incident electrons is smaller than the threshold. The spectral line of mercury at $\lambda = 253.7\,$nm was well known to spectroscopists; the corresponding frequency ν is indeed such that $h\nu = 4.9\,$eV!

This was a simple and spectacular confirmation of Niels Bohr's ideas on the structure of atoms, and of his explanation of spectroscopic measurements. The results of Franck and Hertz supported the idea that the energy of an atom can only take discrete (or quantized) values and that the lines observed in spectroscopy correspond to transitions between such energy levels. In the collision, an electron can transfer its energy to an atom and excite it to a higher-energy state, losing the corresponding energy difference in the process. This can only happen if the electron energy is equal to or larger than this energy difference. Once the atom is in its excited state, it can decay to its ground state by emitting radiation at the Bohr frequency.

The result of Franck and Hertz, who were awarded the Nobel prize for this discovery in 1925, was acknowledged by the scientific community. It is a direct mechanical proof of energy quantization in atomic and molecular systems. Franck and Hertz pursued their explorations systematically. They were able to observe other known spectral lines of mercury and of other elements such as helium. The experiment was also seminal work; many discoveries in nuclear and particle physics have been made by studying resonant effects in the collisions of electrons with nuclear and subnuclear targets.

A more recent example, shown in Fig. 1.1, concerns diatomic molecules. In such molecules, for instance carbon monoxide, CO, the two atoms can vibrate with respect to one another along the axis of the molecule. For small oscillations, the vibrational motion is harmonic. We shall study the quantum version of this problem in Chap. 4. The expression for the energy levels is particularly simple for a harmonic oscillator. If ν is the vibration frequency ($2\pi\nu = \sqrt{K/m}$, where K is the spring constant of the restoring force and m is the reduced mass) the energy levels, labeled by an integer n, are equally spaced:

$$E_n = \left(n + \frac{1}{2}\right) h\nu \,.$$

The energy difference between two levels is an integer multiple of $h\nu$.

The experimental verification consists in sending a beam of electrons of a given energy (typically 2 eV) into a molecular beam. A detector measures the

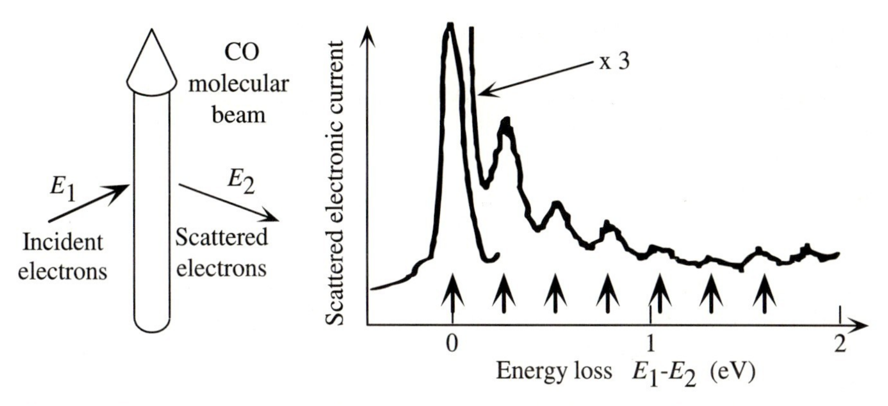

Fig. 1.1. Energy spectrum of electrons scattered by a beam of CO molecules, at an incident energy of $E_1 = 2.05\,\text{eV}$. The peaks in the signal correspond to excitation of the vibrational motion of the molecules after a collision with an electron. The curve corresponding to inelastic collisions (less probable than elastic ones) has been multiplied by 3

energy distribution of the outgoing electrons. The energy loss of an outgoing electron is transferred to the molecules; the theory predicts this loss to be $nh\nu$ ($n = 0$, elastic collision; $n = 1, 2, \ldots$, inelastic collisions). Therefore one expects a series of equally spaced peaks whose positions give the vibration frequency of the molecule.

The results[2] for the CO molecule are shown in Fig. 1.1. They indeed display the equal spacing of the vibrational levels. With these data one can determine the energy spacing $\Delta E = h\nu$ and the vibration frequency of the CO molecule:

$$\Delta E \sim 0.26\,\text{eV}\,, \qquad \nu \sim 6.5 \times 10^{13}\,\text{Hz}\,.$$

This type of experiment has been performed with many other molecular species and has provided information on their structure.

1.2 Interference of Matter Waves

The fundamental experiment which demonstrated the wave behavior of material particles, as predicted by Louis de Broglie, was performed in 1927. It was due to Davisson and Germer, who observed the diffraction of an electron beam by a nickel crystal. In order to analyze the fundamental quantum phenomena that this type of experiment reveals, we shall first refer to a two-slit interference experiment which is conceptually simpler although it is more difficult to achieve in practice. The discussion of the Davisson and Germer result will be done in the subsequent section.

[2] G.J. Schulz, Phys. Rev. **135**, 988 (1964).

1.2.1 The Young Double-Slit Experiment

The Young double-slit interference experiment, represented in Fig. 1.2, is usually quite simple to achieve with a monochromatic field associated with a given wave phenomenon. This can be a light wave, an acoustic wave, ripples on the surface of a liquid, etc. One starts with an initial plane wave of wavelength λ, which is sent at normal incidence onto an opaque plate. The plate is pierced with two parallel, identical slits S_1 and S_2, separated by a distance a. These two slits act as secondary sources oscillating in phase, and the intensity on a screen located downstream reveals the interference between the waves emerging from these two sources.

The amplitude A_c of the field at a point C of the screen is the sum of the two amplitudes A_1 and A_2 of the waves issuing from the two slits. The intensity I_c at point C is

$$I_c = |A_c|^2 = |A_1 + A_2|^2 . \tag{1.6}$$

This is the basic formula of the interference phenomenon. The intensity I_c is large if the two amplitudes A_1 and A_2 are in phase, and it vanishes if they are out of phase by π. We assume that both x and a are small compared with D, where x is the distance between C and the center of the screen and D is the distance between the slit plane and the screen (see Fig. 1.2). In this approximation, the difference between the two path lengths S_1C and S_2C is $\delta \simeq xa/D$. This difference of path lengths for the two waves issuing from S_1 and S_2 induces a phase shift ϕ between the corresponding amplitudes A_1 and A_2: $\phi = 2\pi\delta/\lambda = 2\pi xa/(\lambda D)$. Therefore the fringe spacing, defined as the distance x_s between two consecutive lines on the screen, is $x_s = \lambda D/a$.

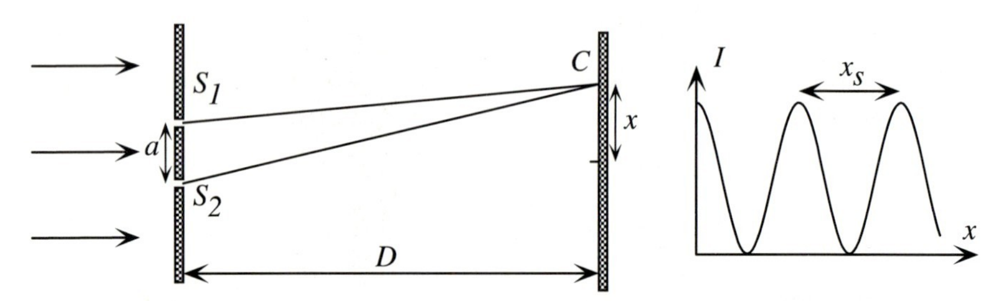

Fig. 1.2. Young double-slit experiment: a monochromatic wave with wavelength λ arrives at normal incidence at a plate pierced with two linear, parallel slits, separated by a distance a. The illumination $I(x)$ of a screen located at a distance $D \gg a$ reveals an interference pattern with a fringe spacing $x_s = \lambda D/a$

1.2.2 Interference of Atoms in a Double-Slit Experiment

The principle and the result of a Young double-slit interference experiment performed with atoms[3] are represented in Fig. 1.3. A small cloud of neon atoms was first captured in a laser trap, and cooled down to a temperature of the order of one millikelvin. It was then released and fell with zero initial velocity onto a plate pierced with two parallel slits of width 2 microns, separated by a distance of 6 microns. The plate was located 3.5 cm below the center of the laser trap. The atoms were detected when they reached a screen located 85 cm below the plane of the two slits. This screen registered the impacts of the atoms: each dot in Fig. 1.3 represents a single impact. The impacts are distributed in a system of fringes similar to the one obtained with light waves. On some lines, parallel to the slit direction, very few atoms are detected. On either side of such a line, a large atom flux is detected.

This type of experiment has also been performed with other particles: electrons, neutrons and molecules. In all cases the distribution of impacts on the detection screen reveals an interference-like pattern. The measured fringe spacing is $x_s = \lambda D/a$, where the wavelength λ and the momentum $\boldsymbol{p} = m\boldsymbol{v}$ of the particles of velocity $\boldsymbol{v}$ are related by the de Broglie relation

$$p = h/\lambda\,. \tag{1.7}$$

In the experiment shown in Fig. 1.3, an accurate calculation of the positions of the interference fringes must take into account the variation of the de Broglie wavelength $\lambda = h/p$ since the atoms are accelerated by gravity. The velocity of the atoms is $0.8\,\mathrm{m\,s^{-1}}$ in the plane of the slits and $4\,\mathrm{m\,s^{-1}}$ on the detecting screen.

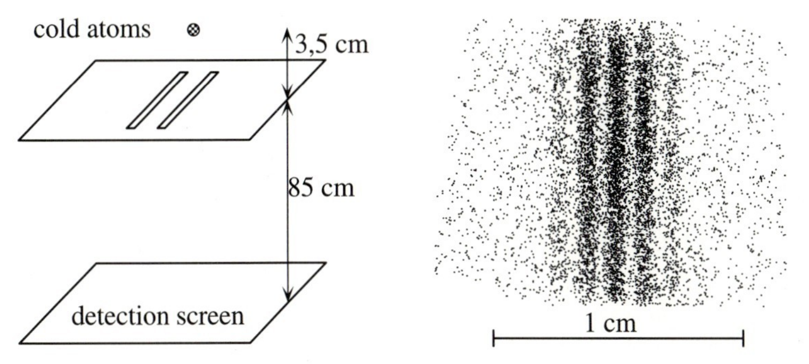

Fig. 1.3. *Left:* Young double-slit experiment with neon atoms cooled by lasers to a temperature $T \sim 1\,\mathrm{mK}$. *Right:* detected pattern, with clearly visible interference fringes. Each point of the figure corresponds to the impact of an atom on the detection screen

[3] F. Shimizu, K. Shimizu, H. Takuma, Phys. Rev. A **46**, R17 (1992) and private communication.

1.2.3 Probabilistic Aspect of Quantum Interference

The physical content of interference experiments with particles such as atoms or electrons extends much beyond the usual wave phenomena. As such, the result is extraordinary because the atoms are *point-like particles* in the experiment. Their dimensions are of the order of a fraction of a nanometer, which is much smaller than any of the length scales in the problem, i. e. the dimensions of the slits or the fringe spacing. Conversely, a wave *fills all space*! A ripple wave on a water surface, for instance, constitutes a field, defined as the set of liquid heights at *all* points of the surface.

In order to make progress, we shall now analyze what happens if one sends the atoms individually, one after the other, into the apparatus, each atom being prepared by the same experimental procedure. The relevant results are as follows:

1. Each atom is detected at a *single point* of the detector. This confirms the idea that the atoms or the electrons can be considered as point-like objects, whose position can be determined with a much better accuracy than the other distances, such as the slit separation or the fringe spacing.
2. As far as we can judge, the impact points are distributed at *random*. Two particles, prepared in what we think to be the same initial conditions, have different impact points.
3. When the opaque plate is pierced with only one slit, either S_1 or S_2, we observe a smooth distribution of impacts I_1 or I_2. These distributions are nearly flat on the length scale x_s if the width of the open slit is much smaller than a (Fig. 1.4a).
4. When the opaque plate is pierced with the two slits, the distribution of a large number of impacts reveals the interference-like pattern of Fig. 1.3 or Fig. 1.4b.

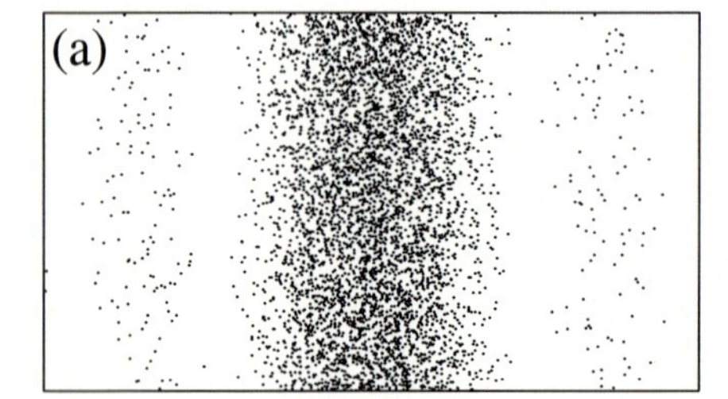

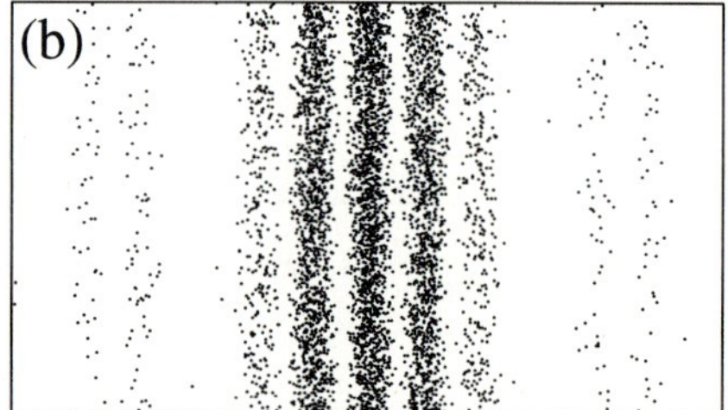

Fig. 1.4. Particles with momentum p arrive at normal incidence on a plate pierced with a single (**a**) or a double (**b**) slit. The slit width is $b = 10\,\lambda$ and the distance between the slits is $a = 30\,\lambda$ for the figure on the right (where $\lambda = h/p$). Each figure shows a typical (computer-simulated) distribution of impacts on a screen located at a distance $D = 10^4\,\lambda$ after the plate. The width of each rectangle is $4200\,\lambda$

Points 1, 2, 3 show that we are facing a phenomenon whose nature is *probabilistic.* Two experiments performed with what we believe to be the same initial conditions can lead to different results. This difficulty, in the absence of point 4, could be solved within the framework of classical physics. One could imagine that our control of the initial conditions is not as good as what we think it is, and that some hidden parameters, which remain to be discovered, fluctuate from one atom to the next and introduce the apparent randomness of the impact positions.

However, the addition of point 4 renders such a classical probabilistic account difficult if not impossible. In fact, the classical probabilistic account would lead to the following line of argument.

- We send the atoms one after the other. Therefore, the events are *independent.*
- Each atom necessarily goes through one of the slits.
- We can *measure* through which slit each atom has passed (with counters, or by shining light on the slits S_1 and S_2, etc.).
- If we perform this measurement, we can separate the events into two sets: those events for which the atoms passed through S_1, and those for which they passed through S_2.
- For all events in which the atoms passed through S_1, everything is such as if S_2 was blocked. Their impact distribution is the pattern shown in Fig. 1.4a (and a quasi-identical pattern is obtained for the events in which the atoms passed through S_2).

In the context of probabilities as encountered in everyday life, if we put together the two sets of events, the result obtained by opening the two slits should be the sum of these two distributions. However, this is not the case (see Figs. 1.3 and 1.4b). On the contrary, the act of opening a second slit, i. e. the act of providing an additional possible way to pass through the plate, has *prevented* the atoms from reaching the empty fringe regions, which they can very well reach if only one slit is open.

Fortunately we can still reach a consistent description. The key point is that in order to measure through *which slit* the atoms have passed, we must perform a different experiment from that of Fig. 1.3, and this second experiment indeed leads to the distribution of Fig. 1.4a. Conversely, if the experimental setup is such that we observe an interference-like pattern (Fig. 1.4b), we cannot determine physically through which slit each atom has passed.

From this we draw two fundamental conclusions:

1. If we do not measure through which slit the atoms pass, an interference-like pattern can be observed. If we perform such a measurement by any means, the interference pattern no longer shows up. In quantum physics, a measurement generally perturbs the system.
2. The atoms do not possess a trajectory in the classical sense. When we observe the impact of an atom in an interference experiment, we cannot

tell which path it followed before. The best we can say is that the atom passed through both slits, which is certainly a paradoxical statement for a classical particle. In quantum physics, the concept of a trajectory, which is one of the foundations of Newtonian mechanics, does not stand the test of experiment.

Naturally, such a phenomenon would presumably be very difficult to explain if it did not resemble so much an interference phenomenon of the usual kind. We are facing a phenomenon which is both wave-like *and* probabilistic, and we shall make use of what we know about waves in order to construct the theory.

In a way similar to that followed for explaining other interference phenomena, we shall introduce *probability amplitudes* $A_1(x)$ and $A_2(x)$ that an atom issuing from one slit (the other one being closed) reaches the detector at a point x. We shall assume that the probability amplitude $A(x)$ that the atom reaches this point when both slits are open is the sum $A(x) = A_1(x)+A_2(x)$, and that the probability for the atom to reach this point is, as in (1.6), the modulus squared of this sum:

$$P(x) = |A(x)|^2 = |A_1(x) + A_2(x)|^2 .$$

In the next chapter, we shall see how one can cast this into a quantitative form.

1.3 The Experiment of Davisson and Germer

In 1927, Davisson and Germer gave an experimental demonstration that electrons, which were known to be particles with a well-defined mass and charge, exhibit a wave behavior as predicted by Louis de Broglie in 1923.

1.3.1 Diffraction of X Rays by a Crystal

Suppose that we irradiate a crystal with a quasi-monochromatic beam of X rays, i. e. an electromagnetic wave with a wavelength λ between 0.01 nm and 1 nm. We place a photographic plate beyond the crystal at a distance large compared with the crystal size. We then observe, in addition to the central spot corresponding to unaffected X rays, a set of spots which correspond to the intersection of the plate with *diffracted beams.*

The interpretation of this experiment is well known, and it follows the same lines as for the two-slit experiment. Consider a crystal whose unit cell has one atom. This cell is defined by the three vectors $\boldsymbol{a}_1$, $\boldsymbol{a}_2$ and $\boldsymbol{a}_3$. Suppose the crystal has N_i lattice sites in the directions x_i $(i = 1, 2, 3)$. The origin of space is chosen such that the position of an atom α of the crystal is given by

$$\boldsymbol{r}_\alpha = n_1\boldsymbol{a}_1 + n_2\boldsymbol{a}_2 + n_3\boldsymbol{a}_3 , \qquad \text{where} \qquad n_i = 0, 1, \dots , N_i - 1 . \tag{1.8}$$

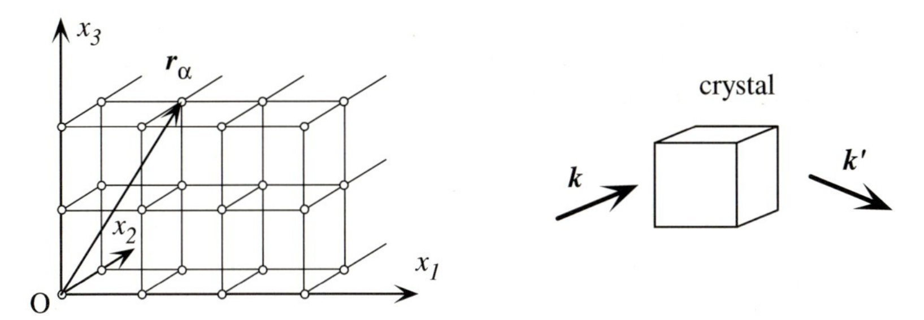

Fig. 1.5. Diffraction of X rays by a crystal

We assume that the incident wave is a plane wave, with a wave vector $\boldsymbol{k}$ and an amplitude $\psi(\boldsymbol{r}, t) = \psi_0\, \mathrm{e}^{\mathrm{i}(\boldsymbol{k}\cdot\boldsymbol{r}-\omega t)}$. We also assume that the atomic scattering is elastic (i.e. the modulus of $\boldsymbol{k}$ is not altered).

We now calculate the amplitude scattered by the atoms of the crystal in a direction defined by $\boldsymbol{k}'$ such that $|\boldsymbol{k}'| = |\boldsymbol{k}|$ (Fig. 1.5). The amplitude at $\boldsymbol{r}$ of the wave scattered by the atom α can be written

$$\psi_\alpha(\boldsymbol{r}, t) = F(\boldsymbol{k}, \boldsymbol{k}')\, \psi_0\, \mathrm{e}^{\mathrm{i}(\boldsymbol{k}'\cdot\boldsymbol{r}-\omega t+\varphi_\alpha)} .$$

The factor $F(\boldsymbol{k}, \boldsymbol{k}')$ is the scattering amplitude of the elementary process; it is the same for all atoms. The phase factor $\mathrm{e}^{\mathrm{i}\varphi_\alpha}$, where

$$\varphi_\alpha = \Delta\boldsymbol{k}\cdot\boldsymbol{r}_\alpha , \qquad \Delta\boldsymbol{k} = \boldsymbol{k} - \boldsymbol{k}' ,$$

accounts for the difference of lengths between the path involving scattering by the atom at the origin and the path where the wave is scattered by the atom α. If, for simplicity, we neglect multiple-scattering events, the total amplitude ψ of the wave scattered in the direction $\boldsymbol{k}'$ by the crystal is the sum of all these amplitudes, $\psi = \sum_\alpha \psi_\alpha$, which is proportional to

$$G(\Delta\boldsymbol{k}) = \sum_\alpha \mathrm{e}^{\mathrm{i}\,\Delta\boldsymbol{k}\cdot\boldsymbol{r}_\alpha} . \tag{1.9}$$

We now introduce a new system of axes, called the *reciprocal* lattice, defined by the vectors $\boldsymbol{a}_1^*, \boldsymbol{a}_2^*, \boldsymbol{a}_3^*$ such that $\boldsymbol{a}_i^*\cdot\boldsymbol{a}_j = \delta_{ij}$. Let Δk_i be the coordinates of $\Delta\boldsymbol{k}$ in this system. We have

$$\begin{aligned} G(\Delta\boldsymbol{k}) &= \sum_{n_1,n_2,n_3} \mathrm{e}^{\mathrm{i}(n_1\Delta k_1+n_2\Delta k_2+n_3\Delta k_3)} \\ &= G_1(\Delta k_1)\, G_2(\Delta k_2)\, G_3(\Delta k_3) , \end{aligned} \tag{1.10}$$

where

$$G_j(\Delta k_j) = \sum_{n_j=0}^{N_j-1} \mathrm{e}^{\mathrm{i}n_j\Delta k_j} = \frac{1-\mathrm{e}^{\mathrm{i}N_j\Delta k_j}}{1-\mathrm{e}^{\mathrm{i}\Delta k_j}} . \tag{1.11}$$

Any given point of the photographic plate corresponds to a direction $\boldsymbol{k}'$. The electromagnetic energy density detected at that point is proportional to the intensity diffracted in the direction $\boldsymbol{k}'$:

$$I(\boldsymbol{k}') = |F(\boldsymbol{k},\boldsymbol{k}')|^2\ |G_1|^2\ |G_2|^2\ |G_3|^2$$

$$\text{where}\quad |G_i|^2 = \frac{\sin^2(N_j\,\Delta k_i/2)}{\sin^2(\Delta k_i/2)}\ . \tag{1.12}$$

The appearance of diffracted spots on the photographic plate can be understood by noticing that the variation of $|G_i|^2$ with Δk_i is very sharp. For scattering directions such that $\Delta k_i = 2n\pi$, n integer, this function is equal to N_i^2, and it takes nonnegligible values only in intervals of width $2\pi/N_i$ centered on these particular values of Δk_i. These directions correspond to the diffracted beams, and the width of the corresponding spots on the screen is inversely proportional to the size of the crystal.

Remarks

1. The above calculation needs to be completed: the Δk_i are constrained by the relation $|\boldsymbol{k}| = |\boldsymbol{k}'|$, which leads to $\Delta\boldsymbol{k}\cdot(\Delta\boldsymbol{k} - 2\boldsymbol{k}) = 0$. The proper orientation conditions of the crystal with respect to $\boldsymbol{k}$ must be satisfied in order to ensure that this equation has solutions different from $\Delta\boldsymbol{k} = 0$.
2. This calculation shows that measurement of the diffraction peaks gives the reciprocal lattice $(\boldsymbol{a}_i^*)$, and therefore the crystal lattice $(\boldsymbol{a}_i)$.
3. More generally, if one studies the scattering of X rays by a sample of condensed matter, solid or liquid, of electron density $\rho(\boldsymbol{r})$, the discrete sum in (1.10) becomes $\int \mathrm{e}^{\mathrm{i}(\boldsymbol{k}-\boldsymbol{k}')\cdot\boldsymbol{r}}\,\rho(\boldsymbol{r})\,\mathrm{d}^3r$. The amplitude scattered in the direction $\boldsymbol{k}'$ is proportional to the Fourier transform of the density of scattering centers. This is the basis of X ray crystallography of solids, liquids, and organic and biological materials.

1.3.2 Electron Diffraction

The Davisson and Germer experiment consisted of irradiating a nickel crystal with a beam of electrons with a well-defined momentum $\boldsymbol{p}$. The outgoing electrons were collected in a detector (a Faraday box in the original experiment). In the same way as with X rays, outgoing diffracted beams were observed in well-defined directions, and the diffraction patterns obtained with X rays and with electrons were similar for a given crystal.

The diffraction pattern obtained with electrons is simply changed by a global scaling factor when one varies their momentum p. This pattern can be exactly superimposed on one obtained with X rays of wavelength λ_{X} if the de Broglie wavelength associated with the electrons $\lambda_{\mathrm{e}} = h/p$ coincides with λ_{X}. This identity originates from the interference term $\sum_\alpha \mathrm{e}^{\mathrm{i}\,\Delta\boldsymbol{k}\cdot\boldsymbol{r}_\alpha}$, which is the same in both cases, although the elementary scattering processes are different for X rays and for electrons.

Davisson, who was an engineer, had been Millikan's student. In 1919 he studied the emission of secondary electrons in electron tubes in order to improve their performance, in view of their use in the American transcontinental telephone system. Davisson noticed that a fraction of these secondary electrons were scattered elastically by the atoms of the electrodes. In 1921, he decided to probe the internal structure of atoms with these electrons. He constructed a very sophisticated apparatus, but he was unsuccessful in his attempts; the angular distribution of outgoing electrons did not show any significant structure. In 1926, he spent some vacation time in England, where fortuitously he heard about the developments of quantum theory and de Broglie's hypothesis. It took him only a few weeks to demonstrate the diffraction phenomenon. He worked with backward-scattered electrons from a nickel crystal at an energy of 60 to 80 eV. He published his results one month before G.P. Thomson, who used transmission through thin mica sheets at much higher energies (10^4 to 10^5 eV) (Debye–Scherrer diffraction).

Electron diffraction is now a common tool in industrial research. It is used, in particular, to study properties of materials and surfaces: corrosion, catalysis and chemical reactions, dislocations, etc. Figure 1.6 shows the principle of the measurement of the electron diffraction pattern of a thin slice of a solid material. The parallel electron beam crosses the slice M and an electron lens of focal length f_1. One measures the diffracted intensity in a given direction $\boldsymbol{k}$ by measuring the electron intensity in the focal plane S_1 of the lens. One can also obtain an image of the diffracting object by putting a second lens, of focal length f_2 (with $f_2 \gg f_1$) in the plane S_1. The electron detector is placed in the image plane of M, and this setup constitutes a microscope with magnification $f_2/(f_2 - f_1)$.

Some results obtained with this method are shown in Fig. 1.7. The material is an AlMnSi alloy obtained by rapid solidification. Fig. 1.7a shows the very inhomogeneous structure of the alloy, observed with an electron microscope at a small magnification. The dark "petals", whose sizes are of the order of a few microns, correspond to strongly diffracting regions, whose

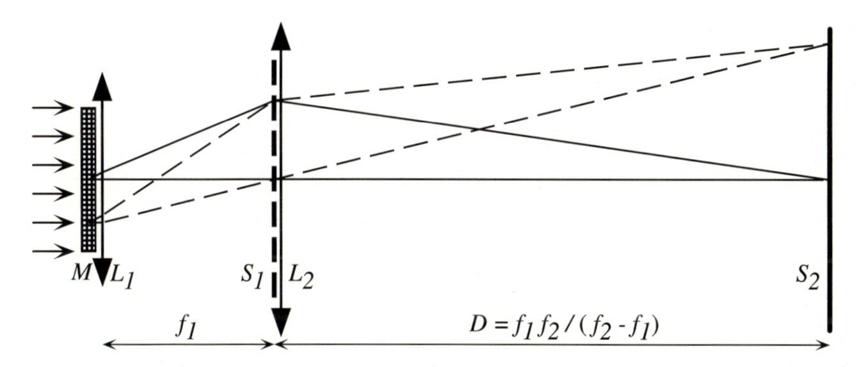

Fig. 1.6. Typical optical setup giving either the diffraction pattern of a slice of material M in the plane S_1 or a magnified image of the sample in the plane S_2. The converging lenses L_1 and L_2 have focal lengths f_1 and f_2, and the incident electron beam is collimated

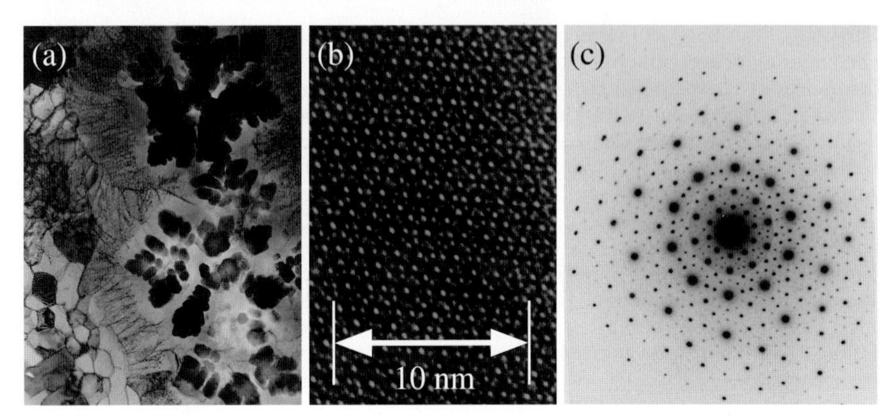

Fig. 1.7. Observation of the first quasicrystals. (**a**) Large-scale structure of an AlMnSi alloy, observed with an electron microscope. In the dark regions, the crystal planes have the proper orientation to diffract the electron beam. (**b**) Magnified image of a dark region, revealing the fivefold symmetry of the material (look for pentagons!). (**c**) Electron diffraction pattern obtained with 200 keV electrons. The fivefold symmetry is clearly visible (photographs courtesy of Denis Gratias)

crystal planes are well oriented with respect to the incident electron beam. Figure 1.7b is an image of a dark petal at a high magnification. One can observe that it is not a periodic, but a quasiperiodic lattice of atoms, with a pentagonal elementary cell. Figure 1.7c shows the diffraction pattern, obtained in the focal plane of the first lens of Fig. 1.6. The fivefold symmetry appears in a spectacular manner. This series of pictures[4] constitutes the first experimental observation of quasicrystals, and has created considerable interest in condensed-matter physics. It has revealed that a solid material can have order of a type different from the usual crystalline order, i. e. without the periodic repetition of an elementary cell. Electron diffraction was the only possible tool to study these very small samples since the X ray diffraction signal was too weak to be detected. It is now possible to prepare quasicrystals in which the quasiperiodic order extends over several centimeters, which renders an X ray study feasible.

Similar diffraction experiments are often performed with other material particles, such as neutrons. For electrons with a kinetic energy of 100 eV, the de Broglie wavelength is 0.124 nm, which is the order of magnitude of atom spacings in crystals. Since the neutron mass is roughly $2000\, m_e$, similar diffraction patterns are obtained with neutrons of energy 0.05 eV. Such thermal neutrons are produced abundantly (fluxes of the order of $10^{13}\,\mathrm{cm^{-2}\,s^{-1}}$) in nuclear reactors. A special high-flux reactor with a large spectrum of energies – and therefore of wavelengths – has been constructed by Germany and France in Grenoble (the Laue–Langevin Institute). Neutrons interact only

[4] D. Schechtman, I. Blech, D. Gratias and J.W. Cahn, Phys. Rev. Lett. **53**, 1951 (1984). We thank D. Gratias for enlightening discussions.

with nuclei, and penetrate matter, contrary to electrons. They constitute very clean probes of the structure of matter. Neutron diffraction gives information complementary to electron diffraction, which essentially probes the electron distribution of the scatterers.

1.4 Summary of a Few Important Ideas

In what follows we shall constantly refer to the quantum phenomena we have just discussed. Later on, we shall present other observations, such as the Stern–Gerlach experiment, which will guide us in building quantum theory. We shall keep in mind the following aspects, which are in contradiction to "common sense" classical concepts.

1. Quantum phenomena are probabilistic. One can only predict the results of a measurement in statistical form (for a large number of events) or in probabilistic form (for a single event).
2. The analysis of interference and diffraction phenomena shows that one cannot work directly with probability laws, as in random phenomena of the usual kind. One must introduce *probability amplitudes* whose moduli squared give the desired probabilities.
3. At the microscopic level, particles have a wave-like behavior. We must abandon the concept of a trajectory for a particle.
4. Some physical quantities, which classically can take a continuous set of values, take only discrete values in quantum mechanics. This is the case for the internal energy of an atom or a molecule.
5. In general, the act of measuring a physical system *affects* this system.

Further Reading

- For the historical development of quantum mechanics one can refer to M. Jammer, *The Conceptual Development of Quantum Mechanics*, McGraw-Hill, New York (1966); B.L. Van Der Waerden, *Sources of Quantum Mechanics*, North-Holland, Amsterdam (1967); J. Mehra and H. Rechenberg, *The Historical Development of Quantum Theory*, Springer New York (1982).
- For an fascinating introduction to quantum phenomena, see T. Leggett, "Quantum physics: weird and wonderful", Phys. World, December 1999, p. 73.
- Davisson's interesting story can be found in R.K. Gehrenbeck, Phys. Today **31**, January 1978, p. 34.
- A description of interference experiments with material particles can be found in:
  - for electrons: G. Matteuci et al., Am. J. Phys. **46**, 619 (1978);

– for neutrons: “Neutron scattering”, Phys. Today **38**, 25 (1985);
– for Na_2 molecules: M.S. Chapman et al., Phys. Rev. Lett. **74**, 4783 (1995);
– for fullerenes, C_{60} molecules: M. Arndt et al., Nature **401**, 680 (1999).

Exercises

1.1. Photoelectric effect in metals. If one irradiates a potassium photocathode with $\lambda = 253.7\,\text{nm}$ photons (the resonance line of mercury), the maximum energy of the ejected electrons is 3.14 eV. If visible radiation with $\lambda = 589\,\text{nm}$ is used (the resonance line of sodium), the maximum energy of the ejected electrons is then 0.36 eV.

a. Calculate the value of Planck's constant.
b. Calculate the energy for extraction of electrons in potassium.
c. What is the maximum wavelength of radiation which can produce a photoelectric effect on potassium?

1.2. Photon fluxes.

a. A radio antenna emits waves at a frequency of 1 MHz with a power of 1 kW. How many photons are emitted per second?
b. A first-magnitude star emits a light flux of $\sim 1.6 \times 10^{-10}\,\text{W}\,\text{m}^{-2}$ as measured on earth, at an average wavelength of 556 nm. How many photons pass through the pupil of an eye per second?

1.3. Orders of magnitude for de Broglie wavelengths. What is the de Broglie wavelength (a) of an electron of energy 100 eV, (b) of a thermal neutron? How do these compare with atomic sizes?

1.4. De Broglie relation in the relativistic domain. In high-energy physics, electron accelerator energies reach ~100 GeV. What is the de Broglie wavelength of such electrons? Why are such high energies necessary? Remember the relativistic relation between energy and momentum $E = \sqrt{p^2c^2 + m^2c^4}$.

2. The Wave Function and the Schrödinger Equation

Believe and you will understand;
faith precedes, intelligence follows.
Saint Augustine

Erwin Schrödinger learned about Louis de Broglie's work in 1925. He was attracted by the ideas, but he remained skeptical for fundamental reasons related to his prejudices about relativity. Nevertheless, several people, including Debye and Einstein, encouraged him to take advantage of his skill in partial differential equations in order to investigate further this hypothesis. This resulted in a series of eight outstanding articles published by Schrödinger in 1926, which founded what is now called *wave mechanics*. This version of quantum mechanics was developed a little later than the *matrix mechanics* of Heisenberg, Born, Jordan and Dirac, which we shall describe later on. The most decisive contribution of Schrödinger was to obtain a wave equation which governs the behavior of a particle placed in a potential. The calculation of energy levels then appears as a well-defined mathematical problem, of the same type as the determination of standing waves with given boundary conditions.

We choose to approach quantum mechanics by means of wave mechanics, where we shall study the simplest problem of classical physics, i. e. the motion of a point-like particle under the influence of a field of force derived from a potential. We shall proceed in a semideductive way by stating principles, which will be illustrated by their consequences.

Before entering the discussion, we define our terminology. Common language contains many implicit conclusions about the nature of things, and, although one cannot avoid using it, it may lead to wrong ideas if one does not define a few basic words unambiguously. Physics is based on experimental observations and measurements, which consist of characterizing aspects of reality by numbers. These aspects of reality are elaborated into concepts of *physical quantities* (energy, intensity of an electric current, etc.). We shall say that, under given physical conditions, a physical *system*, i. e. an object pertaining to reality, is at any time in a given *state*. The state of the system corresponds to the particular form of its physical reality. We possess information about the state of a system if we have performed some observations on it and have collected the corresponding set of measured numbers. In addition, we assume that, by acting on the system and by filtering out the values of some subset of physical quantities (yet to be defined), we can prepare the system to be in a given well-defined state.

There will be four steps in the theory.

1. We must *describe* the state of the system. This means giving the state a mathematical representation which defines it in an operational manner. For instance, in Newtonian mechanics, the state of a point-like particle of mass m is described at any time t by its position $\boldsymbol{r}(t)$ and its velocity $\boldsymbol{v}(t) = \mathrm{d}\boldsymbol{r}(t)/\mathrm{d}t$ or its linear momentum $\boldsymbol{p} = m\boldsymbol{v}$.
2. We need to know the *time evolution* of the state after it has been prepared under given initial conditions. This means we want to be able to predict the state at time t if we know it at time $t = 0$. In Newtonian mechanics, the evolution is given by the fundamental law $\mathrm{d}\boldsymbol{p}/\mathrm{d}t = \boldsymbol{f}$, where $\boldsymbol{f}$ is the force acting on the particle.
3. We want to predict the results of measurements of physical quantities on a given system. This means knowing the laws that lead from the mathematical representation of the state of the system to the numbers which appear on measuring devices (meters, counters, etc.). In classical mechanics, these are *functions* of the state variables $\boldsymbol{r}$ and $\boldsymbol{p}$.
4. Finally, as we anticipated in the first chapter, we must address questions which were absent in classical mechanics. What does a measurement process consist of? What do we know after performing a measurement?

In the present chapter, we shall study the first two questions in the case of a "point-like particle" evolving in space. By point-like particle we mean any physical system whose internal structure or internal degrees of freedom are not relevant in the experiments of interest. At the present stage, a particle will be characterized by its mass and its electric charge, and its classical motion in an electromagnetic or gravitational field is known. For instance a neon atom can be considered as a particle in the experiment shown in Fig. 1.3, although one may have to take into account its internal structure (10 electrons, 10 protons and 10 neutrons) in order to describe other experimental effects.

2.1 The Wave Function

The first concept we need is that of a wave function and its probabilistic interpretation.

2.1.1 Description of the State of a Particle

The basic principle by which the state of a particle is described is the following.

Principle 2.1

The description of the state of a particle of mass m at time t is performed with a complex *wave function* $\psi(\boldsymbol{r}, t)$. The probability $\mathrm{d}^3 P(\boldsymbol{r})$ to find the particle at time t in a volume $\mathrm{d}^3 r$ around the point $\boldsymbol{r}$ is given by

$$\mathrm{d}^3 P(\boldsymbol{r}) = |\psi(\boldsymbol{r}, t)|^2 \, \mathrm{d}^3 r \,. \tag{2.1}$$

Comments

1. The wave function $\psi(\boldsymbol{r}, t)$ is also called the *probability amplitude* of finding the particle at point $\boldsymbol{r}$. It is square integrable and normalized to unity. If we denote by $\mathcal{D}$ the domain in space accessible to the particle, the total probability of finding the particle at any point in $\mathcal{D}$ is equal to 1:

$$\int_{\mathcal{D}} |\psi(\boldsymbol{r}, t)|^2 \, \mathrm{d}^3 r = 1 \,. \tag{2.2}$$

2. A given wave function constitutes a *complete* description of the state of the particle at time t. Two different wave functions describe two different states, except when they differ only by a constant multiplicative phase factor. The wave functions $\psi_1(\boldsymbol{r}, t)$ and $\psi_2(\boldsymbol{r}, t) = \mathrm{e}^{\mathrm{i}\alpha} \psi_1(\boldsymbol{r}, t)$, where α is a constant, describe the same state. They obviously lead to the same spatial probability density and also, as we shall see later on, to the same predictions for any other physical quantity.
3. This is a nonclassical probabilistic description. We do not work directly with probabilities but with probability amplitudes, whose moduli squared are probabilities. We shall see below how this accounts for interference phenomena.

2.1.2 Position Measurement of the Particle

What do we find when measuring at some time t the position of the particle? This type of question is fundamental in quantum mechanics, and we shall come back to it constantly. The following important observations or assumptions are in order.

1. The measurement is made with a classical, or *macroscopic*, apparatus whose description does not require a quantum description.
2. The *accuracy* of the measuring apparatus is in principle arbitrarily good (although it is limited in practice by technical constraints).
3. Concerning a position measurement, the answer of the measuring apparatus consists in the statement "at time t, a particle has been detected in a vicinity $\delta^3 r$ of point $\boldsymbol{r}$ and nowhere else, $\delta^3 r$ being intrinsic to the measuring apparatus".

Therefore, in measuring the position of the particle at time t, one finds a *well-defined* value $\boldsymbol{r}$, up to the accuracy $\delta^3 r$ of the measuring apparatus. The probabilistic description by a wave function has the following meaning. If we prepare *independently* a large number N of particles *in the same state* – i.e. all of these N particles are described by the *same wave function* $\psi(\boldsymbol{r})$ when each position measurement is made – the N results $\boldsymbol{r}_i$, $i = 1, \ldots N$, will not all be the same, but will be distributed according to the probability law (2.1).

The "expectation value" of these results, which we denote by $\langle \boldsymbol{r} \rangle$, is

$$\langle \boldsymbol{r} \rangle = \int \boldsymbol{r}\, |\psi(\boldsymbol{r})|^2\, \mathrm{d}^3 r\,. \tag{2.3}$$

This is a set of three equations for the three coordinates $\{x, y, z\}$.

The dispersion of the results is characterized by a *mean square deviation.* Let Δx, Δy and Δz be the dispersions of the three coordinates of the position; we have, by definition,

$$(\Delta x)^2 = \langle x^2 \rangle - \langle x \rangle^2 = \int x^2\, |\psi(\boldsymbol{r})|^2\, \mathrm{d}^3 r - \langle x \rangle^2\,, \tag{2.4}$$

and similarly for Δy and Δz. The smaller these dispersions, the more accurately we know the position of the particle when it is prepared in the state $\psi(\boldsymbol{r})$.

2.2 Interference and the Superposition Principle

As is indicated above, the wave function provides us with a probabilistic description of quantum phenomena. The basic properties of wave functions can be inferred from the observation of interference experiments.

2.2.1 De Broglie Waves

We turn back to the interpretation of interference experiments. The simplest idea is to assume that particles of well-defined velocity $\boldsymbol{v}$ and momentum $\boldsymbol{p} = m\boldsymbol{v}$, moving freely in space, are described by wave functions close to monochromatic plane waves of the form

$$\psi(\boldsymbol{r}, t) = \psi_0\, \mathrm{e}^{\mathrm{i}(\boldsymbol{k}\cdot\boldsymbol{r} - \omega t)}\,, \tag{2.5}$$

where ψ_0 is a constant. In these plane waves, the wavelength $\lambda = h/p$ and, equivalently, the wave vector $\boldsymbol{k}$ satisfy the relations

$$\lambda = h/p\,, \qquad \boldsymbol{k} = \boldsymbol{p}/\hbar\,, \tag{2.6}$$

as predicted by de Broglie. If we apply the usual arguments which lead to acoustic or light interference, this should explain the experimental observations.

However, an interference experiment – be it a Young double-slit experiment or a Davisson–Germer-type experiment – does not tell us the frequency ω of these waves. Indeed the phase factor $\mathrm{e}^{-\mathrm{i}\omega t}$ factorizes out of the amplitude of the wave on a detector, and the measured signal does not depend on the particular choice of ω. The choice made by de Broglie in 1923 consists in relating this frequency to the energy of the particle in the same way as Einstein did for the photon:

$$\hbar\omega = E\,, \quad \text{where, for a free particle,} \quad E = \boldsymbol{p}^2/2m\ . \tag{2.7}$$

We then obtain the following form for what are called *de Broglie waves*:

$$\psi(\boldsymbol{r},t) = \psi_0\ \mathrm{e}^{\mathrm{i}(\boldsymbol{p}\cdot\boldsymbol{r}-Et)/\hbar}\,, \qquad \text{where} \qquad E = \boldsymbol{p}^2/2m\,. \tag{2.8}$$

2.2.2 The Superposition Principle

Consider the specific case of the Young slit experiment sketched in Fig. 1.2. By analogy with the usual interference phenomena, we can explain the observations provided the following condition is satisfied. We send a plane de Broglie wave from the left onto the screen pierced with two slits. Suppose we know the wave function $\psi_1(\boldsymbol{r}_C,t)$ for any point $\boldsymbol{r}_C$ of the detection screen if only S_1 is open. Similarly, suppose we know the wave function $\psi_2(\boldsymbol{r}_C,t)$ diffracted by the slit S_2. The interference phenomenon can be accounted for provided that, when both slits are open, the wave function on the screen is the sum of these two wave functions:

$$\psi(\boldsymbol{r}_C,t) \propto \psi_1(\boldsymbol{r}_C,t) + \psi_2(\boldsymbol{r}_C,t)\ . \tag{2.9}$$

Under this condition, de Broglie waves will account for the interference experiments on matter waves described in the previous chapter.

Equation (2.9) expresses the fundamental property of wave functions. We can state it as a fundamental principle of quantum mechanics:

Superposition Principle

Any linear superposition of wave functions is also a possible wave function.

Specifically, this means that if $\psi_1(\boldsymbol{r},t)$ and $\psi_2(\boldsymbol{r},t)$ describe possible states for the particle, any linear combination

$$\psi(\boldsymbol{r},t) \propto \alpha_1\,\psi_1(\boldsymbol{r},t) + \alpha_2\,\psi_2(\boldsymbol{r},t)\,, \tag{2.10}$$

where α_1 and α_2 are arbitrary complex coefficients, also represents a possible state; the proportionality coefficient in (2.10) is adjusted so that (2.2) is satisfied. The coefficients α_1 and α_2 need not be equal: one can attenuate or modify the phase of one of the beams.

This is a central principle in quantum theory. The additivity of probability amplitudes is the basis of interference phenomena. It goes beyond the particular wave equation obeyed by the wave functions $\psi(\boldsymbol{r}, t)$ that we shall give below. This wave equation must be linear in order to ensure that the superposition principle holds at all times. When we generalize quantum mechanics to systems more complicated than point-like particles, we shall see that this property is much more important than the notion of wave functions itself. In mathematical terms, this means that the family of wave functions of a system forms a vector space.

2.2.3 The Wave Equation in Vacuum

Consider the de Broglie waves in (2.8). These particular plane waves describe free particles of well-defined momentum $\boldsymbol{p}$, and energy $E = p^2/2m$. Taking the time derivative on one hand and the Laplacian on the other, we observe that de Broglie waves satisfy the partial differential equation

$$\mathrm{i}\hbar \frac{\partial}{\partial t}\psi(\boldsymbol{r}, t) = -\frac{\hbar^2}{2m}\Delta\psi(\boldsymbol{r}, t) \, .$$

In a similar way as for the principle of inertia in classical mechanics, we can take this as a principle for the propagation of particles in free space, i. e. in the absence of forces:

Principle 2.2.a: Motion of a Free Particle

If a particle is in vacuum and is not subject to any interaction, the wave function satisfies the partial differential equation

$$\mathrm{i}\hbar \frac{\partial}{\partial t}\psi(\boldsymbol{r}, t) = -\frac{\hbar^2}{2m}\Delta\psi(\boldsymbol{r}, t) \, . \tag{2.11}$$

This partial differential equation is nothing but the Schrödinger equation in the absence of forces, as we shall see in Sect. 2.5. It is a linear equation, in agreement with the superposition principle.

Energy–Frequency Relation. The result $E = \hbar\omega$ in (2.7) can also be obtained directly by requiring that the function (2.5) satisfies the wave equation (2.11). Indeed, if we substitute the form (2.5) into this wave equation we directly obtain the constraint

$$\hbar\omega = \frac{\hbar^2 k^2}{2m} = \frac{p^2}{2m} \, .$$

Therefore two equivalent points of view may be considered in order to find the dynamics of the wave function for a free particle. One may assume that the most general wave function is a linear combination of de Broglie waves (2.8). One can then show the result that the wave equation in vacuum is of the form (2.11). Alternatively, one can assume the Schrödinger equation for a free particle (2.11), and de Broglie waves appear as particular plane wave solutions of this equation.

Interference Phenomena. From a mathematical standpoint, the result of the Young double-slit experiment can be explained rigorously by solving the wave equation (2.11) with the following boundary conditions.

1. For any $\boldsymbol{r}$ on the surface of the pierced plate, $\psi(\boldsymbol{r}) = 0$. Indeed, since the probability of finding the particle at a point $\boldsymbol{r}_\mathrm{i}$ inside the material of the plate is zero, one has $\psi(\boldsymbol{r}_\mathrm{i}) = 0$. The value of ψ on the surface follows by continuity. Therefore, in the plane of the slits, the wave function takes nonzero values in the location of the slits only.
2. For $z \to -\infty$, where z is the propagation axis, $\psi(\boldsymbol{r})$ is the superposition of the incident plane wave $\mathrm{e}^{\mathrm{i}(\boldsymbol{p}\cdot\boldsymbol{r} - Et)/\hbar}$ and some reflected wave which is irrelevant in the present discussion.
3. For $z \to +\infty$, $\psi \to 0$.

One can show that this is a well-defined mathematical problem which has a unique solution. The exact solution is involved and requires computer calculations, but one can show analytically that, at large distances from the plate ($D \gg a$) and for small angular deviations x/D, the usual formula for interference calculations, in particular (2.9), applies.

Conservation of the Norm. Consider at time t_0 a function $\psi(\boldsymbol{r}, t_0)$ which is normalized to unity. This function describes a possible state of the particle at time t_0, and the wave equation (2.11) then allows one to determine $\psi(\boldsymbol{r}, t)$ at any other time. We can check that the quantity $\int |\psi(\boldsymbol{r}, t)|^2 \, \mathrm{d}^3 r$ is time-independent during the evolution (2.11). This guarantees that ψ will remain normalized at all times, which is of course essential in order for us to interpret $|\psi(\boldsymbol{r}, t)|^2$ as a probability density. To show this, we calculate the time derivative:

$$\begin{aligned} \frac{\mathrm{d}}{\mathrm{d}t} \int |\psi(\boldsymbol{r}, t)|^2 \, \mathrm{d}^3 r &= \int \psi^* \frac{\partial \psi}{\partial t} \, \mathrm{d}^3 r + \int \frac{\partial \psi^*}{\partial t} \psi \, \mathrm{d}^3 r \\ &= \frac{\mathrm{i}\hbar}{2m} \left(\int \psi^* \, \Delta\psi \, \mathrm{d}^3 r - \int \Delta\psi^* \, \psi \, \mathrm{d}^3 r \right) = 0 \, , \end{aligned}$$

where we have integrated by parts in the last step of the calculation, and we assume that ψ and ψ^* vanish at infinity.

2.3 Free Wave Packets

Realistic free wave functions are constructed as follows.

2.3.1 Definition of a Wave Packet

The monochromatic plane wave (2.8) cannot represent the state of a particle, since it is not normalizable. A physically acceptable state is a *wave packet*, consisting of a complex linear superposition of monochromatic plane waves of type (2.8), in agreement with the superposition principle:

$$\psi(\boldsymbol{r},t) = \int \varphi(\boldsymbol{p})\, \mathrm{e}^{\mathrm{i}(\boldsymbol{p}\cdot\boldsymbol{r}-Et)/\hbar}\, \frac{\mathrm{d}^3p}{(2\pi\hbar)^{3/2}} \,;$$

$$\text{for a free particle,} \quad E = \frac{\boldsymbol{p}^2}{2m}\,. \tag{2.12}$$

The constant $(2\pi\hbar)^{3/2}$ is introduced for normalization and dimensionality reasons, and $\varphi(\boldsymbol{p})$ is arbitrary except that we require that this expression exists and that it is suitably normalized.

Expression (2.12) is the general solution of the wave equation (2.11). In order to understand the physical properties of wave packets, we remark that, in (2.12), the functions $\psi(\boldsymbol{r},t)$ and $\varphi(\boldsymbol{p})\,\mathrm{e}^{-\mathrm{i}Et/\hbar}$ are *Fourier transforms* of one another.

2.3.2 Fourier Transformation

The properties of Fourier transformation are given in Appendix B. We outline here those which are relevant to the following discussion:

1. Two functions $f(\boldsymbol{r})$ and $g(\boldsymbol{p})$ are Fourier transforms of one another if

$$f(\boldsymbol{r}) = (2\pi\hbar)^{-3/2} \int \mathrm{e}^{\mathrm{i}\boldsymbol{p}\cdot\boldsymbol{r}/\hbar}\, g(\boldsymbol{p})\, \mathrm{d}^3p\,. \tag{2.13}$$

2. The inverse transformation is then

$$g(\boldsymbol{p}) = (2\pi\hbar)^{-3/2} \int \mathrm{e}^{-\mathrm{i}\boldsymbol{p}\cdot\boldsymbol{r}/\hbar}\, f(\boldsymbol{r})\, \mathrm{d}^3r\,. \tag{2.14}$$

   The product $\boldsymbol{p}\cdot\boldsymbol{r}$ has the dimension of an action, which explains the presence of the constant $\hbar$ in these expressions.

3. Differentiating (2.13) with respect to $x_j = x, y$ or z gives

$$\frac{\partial f}{\partial x_j} = (2\pi\hbar)^{-3/2} \int \mathrm{e}^{\mathrm{i}\boldsymbol{p}\cdot\boldsymbol{r}/\hbar}\, \frac{\mathrm{i}p_j}{\hbar}\, g(\boldsymbol{p})\, \mathrm{d}^3p\,. \tag{2.15}$$

   This means that the Fourier transform of $\partial f/\partial x_j$ is $\mathrm{i}p_j g(\boldsymbol{p})/\hbar$. A differentiation in $\boldsymbol{r}$ space corresponds to a multiplication by the relevant variable in $\boldsymbol{p}$ space (and vice versa, using (2.14)).

4. The Fourier transformation is "isometric". If $f_1(\boldsymbol{r})$ and $f_2(\boldsymbol{r})$ are the Fourier transforms of $g_1(\boldsymbol{p})$ and $g_2(\boldsymbol{p})$, respectively, the Parseval–Plancherel theorem implies

$$\int f_1^*(\boldsymbol{r})\, f_2(\boldsymbol{r})\, \mathrm{d}^3r = \int g_1^*(\boldsymbol{p})\, g_2(\boldsymbol{p})\, \mathrm{d}^3p\,. \tag{2.16}$$

5. The smaller the "width" of $|g(\boldsymbol{p})|^2$ (in the vicinity of a given value p_0), the larger the "width" of $|f(\boldsymbol{r})|^2$. More precisely, if we define

$$\langle p_x \rangle = \int p_x\, |g(\boldsymbol{p})|^2\, \mathrm{d}^3p\,, \qquad (\Delta p_x)^2 = \langle p_x^2 \rangle - \langle p_x \rangle^2 \tag{2.17}$$

and similarly for $\langle x \rangle$ and Δx starting from f, the product of the dispersions Δx and Δp_x is constrained by the inequality

$$\Delta x\, \Delta p_x \geq \hbar/2\,, \tag{2.18}$$

and similarly for the y and z axes. Here we implicitly assume that $|f|^2$ and therefore $|g|^2$ are probability distributions normalized to 1.

2.3.3 Structure of the Wave Packet

Owing to (2.16), the wave packet (2.12) satisfies

$$\int |\psi(\boldsymbol{r},t)|^2\, \mathrm{d}^3r = \int |\varphi(\boldsymbol{p})|^2\, \mathrm{d}^3p\,. \tag{2.19}$$

Therefore, $\psi(\boldsymbol{r},t)$ is square integrable and properly normalized if and only if $\varphi(\boldsymbol{p})$ is also square integrable and normalized:

$$\int |\varphi(\boldsymbol{p})|^2\, \mathrm{d}^3p = 1\,. \tag{2.20}$$

The construction of a wave packet consists in choosing a square-integrable, normalized function $\varphi(\boldsymbol{p})$. The resulting wave function $\psi(\boldsymbol{r},t)$ is also normalized at any time t. It is a linear superposition of plane waves which interfere destructively outside a localized region in space. It is useful to keep in mind the following two limiting cases, illustrated in Fig. 2.1:

1. *Approximation to a plane wave.* If $\varphi(\boldsymbol{p})$ is peaked in the vicinity of some value $\boldsymbol{p}_0$, $\psi(\boldsymbol{r},t)$ is close to a monochromatic plane wave in a large region of space, while it can be normalized (Fig. 2.1a).
2. *Approximation to a well-localized particle.* Conversely, we can construct wave packets for which $|\psi(\boldsymbol{r},t)|^2$ is very concentrated around some value $\boldsymbol{r}_0$ (Fig. 2.1c). Such wave packets involve a superposition of plane waves with a large dispersion in $|\boldsymbol{p}|$.

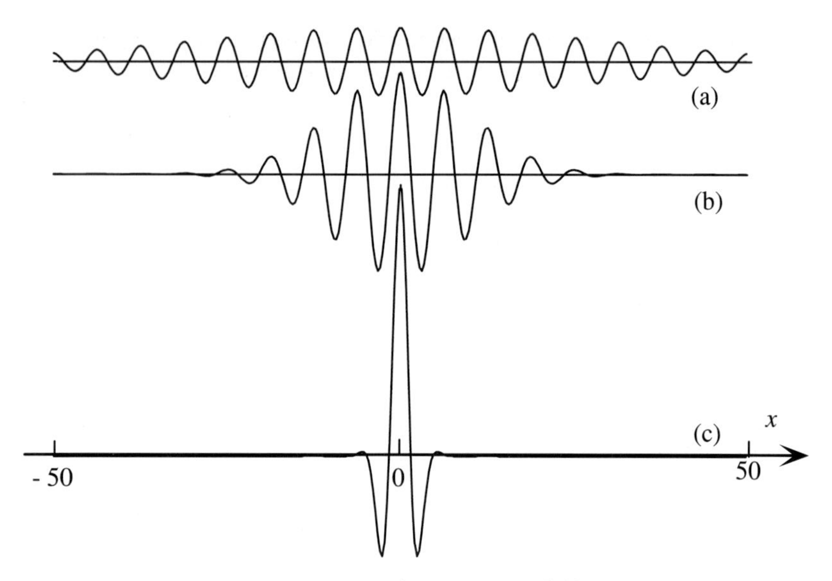

Fig. 2.1. Examples of wave packets (real part of $\psi(x)$) corresponding to a Gaussian function $\varphi(p)$: $\varphi(p) \propto \exp\{-(p-p_0)^2/[2(\hbar\sigma)^2]\}$. (**a**) Approximation to a monochromatic plane wave: large spread in x, obtained for $\varphi(p)$ peaked in the vicinity of p_0 ($\hbar\sigma = p_0/50$). (**b**) Intermediate case ($\hbar\sigma = p_0/10$). (**c**) Localized wave corresponding to a large dispersion of $\varphi(p)$ ($\hbar\sigma = p_0/2$). In order to keep the functions on the same graph, the vertical scales of (a), (b) and (c) differ

2.3.4 Propagation of a Wave Packet: the Group Velocity

Consider a linear superposition of plane waves of the form

$$f(x,t) = \int \mathrm{e}^{\mathrm{i}(kx-\omega t)}\, g(k)\,\mathrm{d}k\,, \tag{2.21}$$

where $\omega \equiv \omega(k)$ is a function of k and where $g(k)$ is chosen such that $\int |g(k)|^2\,\mathrm{d}k = \int |f(x,t)|^2\,\mathrm{d}x = 1$. For simplicity we restrict ourselves here to a one-dimensional situation, but the argument can easily be extended to three dimensions. In optics or acoustics, $|f|^2$ is proportional to the energy density. Here, $|f|^2$ is a probability density.

We assume that the spreading of the function $g(k)$ around its center $k_0 = \int k\,|g(k)|^2\,\mathrm{d}k$ is sufficiently small so that the linear expansion

$$\omega(k) \simeq \omega_0 + v_{\mathrm{g}}(k-k_0)\,, \quad \text{where} \quad \omega_0 = \omega(k_0)\;,\; v_{\mathrm{g}} = \left.\frac{\mathrm{d}\omega}{\mathrm{d}k}\right|_{k=k_0}\,, \tag{2.22}$$

is valid for all k's where $g(k)$ takes significant values. If we insert this expansion into (2.21) we obtain

$$\begin{aligned} &f(x,t) \simeq \mathrm{e}^{\mathrm{i}(k_0 v_{\mathrm{g}}-\omega_0)t}\, f(x-v_{\mathrm{g}}t,0) \\ &\Rightarrow\ |f(x,t)|^2 \simeq |f(x-v_g t,0)|^2\,. \end{aligned} \tag{2.23}$$

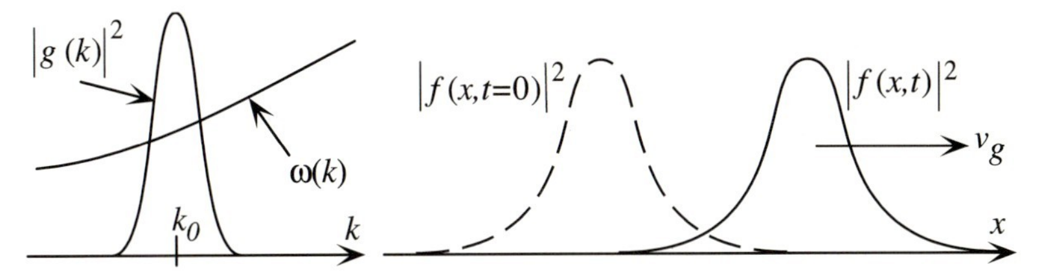

Fig. 2.2. For a distribution $g(k)$ concentrated around a value k_0, the evolution of the wave packet $f(x)$ during a relatively short time is a displacement at the group velocity $v_{\mathrm{g}} = \mathrm{d}\omega/\mathrm{d}k|_{k=k_0}$. For longer times the change in the width of the wave packet has to be taken into account (Sect. 2.3.5)

Equation (2.23) shows that the distribution $|f(x,t)|^2$ of the energy (for electromagnetic waves, acoustic waves, etc.) or of the probability density (in quantum mechanics) propagates along the x axis with a velocity v_{g} (Fig. 2.2). This quantity is called the group velocity and is in general different from the phase velocity of a monochromatic wave $v_\varphi = \omega/|\boldsymbol{k}|$.

For de Broglie waves (2.12), the angular frequency ω varies quadratically with $k = p/\hbar$:

$$\omega = \frac{\hbar k^2}{2m} = \frac{p^2}{2m\hbar}\,. \tag{2.24}$$

If the width Δp of $|\varphi(p)|^2$ is much smaller than the mean value p_0, the expansion (2.22) is valid and yields

$$v_{\mathrm{g}} = p_0/m\,. \tag{2.25}$$

The propagation velocity of the probability density is equal to the velocity of a classical particle with a momentum equal to the mean value p_0 of the wave packet. This is how the classical limit of the theory will emerge. If one uses the *relativistic* form of the energy $E = c(p^2 + m^2c^2)^{1/2}$, the group velocity is again equal to the velocity of the particle $\boldsymbol{v} = \boldsymbol{p}c^2/E$. This was actually the original formulation of de Broglie.

2.3.5 Propagation of a Wave Packet: Average Position and Spreading

Using the equation of motion (2.3), we can go one step further and determine the general time evolution of the center of a wave packet and of its width. We consider again a one-dimensional case for simplicity and examine the evolution of $\langle x\rangle_t$, which represents the expectation value of a position measurement at time t. The time derivative of (2.2) gives

$$\frac{\mathrm{d}\langle x\rangle_t}{\mathrm{d}t} = \int x\left(\psi^*\frac{\partial\psi}{\partial t}+\frac{\partial\psi^*}{\partial t}\psi\right)\mathrm{d}x = \frac{\mathrm{i}\hbar}{2m}\int x\left(\psi^*\frac{\partial^2\psi}{\partial x^2}-\frac{\partial^2\psi^*}{\partial x^2}\psi\right)\mathrm{d}x$$
$$= \frac{\mathrm{i}\hbar}{2m}\int\left(\psi\frac{\partial\psi^*}{\partial x}-\psi^*\frac{\partial\psi}{\partial x}\right)\mathrm{d}x\,, \tag{2.26}$$

where we have integrated by parts. The right-hand side is constant in time, as can be shown by taking its time derivative and again integrating by parts. If we set

$$v_0 = \frac{\mathrm{i}\hbar}{2m}\int\left(\psi\frac{\partial\psi^*}{\partial x}-\psi^*\frac{\partial\psi}{\partial x}\right)\mathrm{d}x = -\frac{\mathrm{i}\hbar}{m}\int\psi^*\frac{\partial\psi}{\partial x}\,\mathrm{d}x \tag{2.27}$$

where v_0 has the dimensions of a velocity, we obtain

$$\langle x\rangle_t = \langle x\rangle_0 + v_0 t\ . \tag{2.28}$$

The motion of the center of the wave packet is uniform, as is the motion of a free classical particle.

We can proceed along similar lines to obtain the evolution of the variance Δx_t^2 of the position probability distribution (see Exercise 2.2). We obtain

$$\Delta x_t^2 = \Delta x_0^2 + \xi_1 t + \Delta v^2 t^2\,, \tag{2.29}$$

where ξ_1 and Δv are constant coefficients. Notice that there exists a constraint on the relative values of Δx_0^2, ξ_1 and Δv^2, since $\Delta x_t^2 \geq 0$ at any time. In particular, Δv^2 is positive and is given by

$$\Delta v^2 = v_1^2 - v_0^2\,, \qquad \text{where} \qquad v_1^2 = \frac{\hbar^2}{m^2}\int\frac{\partial\psi}{\partial x}\frac{\partial\psi^*}{\partial x}\,\mathrm{d}x\,, \tag{2.30}$$

where v_1 also has the dimensions of a velocity.

The physical consequences of (2.29) are important: the variance Δx_t^2 of a wave packet varies quadratically with time. As a function of time, Δx_t^2 reaches a minimum value at some time t_1 before it expands indefinitely; for large $|t|$, we find $\Delta x_t \sim \Delta v\,|t|$. This is similar to the variation of the spatial extent of a focused acoustic or light wave, before and after its focal point. The spreading of the wave packet in time is a consequence of the quadratic dispersion law for de Broglie waves $\omega \propto k^2$. The result (2.23) found in the previous section is valid as long as this spreading is negligible. In the next section we shall obtain a physical interpretation for the coefficients v_0 and Δv which appear in the two equations (2.28) and (2.29).

2.4 Momentum Measurements and Uncertainty Relations

We now address the problem of finding the *momentum* probability amplitude of the particle.

2.4.1 The Momentum Probability Distribution

The wave function completely describes the state of the particle. Up to now, we have used it only as the probability distribution of the position. We want now to exploit the structure of the wave packet in order to infer the distribution of the results when we measure the *momentum* of the particle.

Consider the wave packet (2.12) at $t = 0$ for simplicity, and examine the following assumption:

In a measurement of the momentum, the probability of finding a result located in a volume d^3p *enclosing the value* $\boldsymbol{p}$ *is*

$$\mathrm{d}^3P(\boldsymbol{p}) = |\varphi(\boldsymbol{p})|^2\,\mathrm{d}^3p\,. \tag{2.31}$$

The statement (2.31) is consistent with what precedes it. Indeed, $|\varphi(\boldsymbol{p})|^2$ can be a probability density. It is a positive definite quantity and its integral is normalized to one (see (2.20)).

We have already seen that the more the support of $\varphi(\boldsymbol{p})$ is concentrated in the vicinity of $\boldsymbol{p}_0$, the closer the wave packet (2.12) is to a monochromatic plane wave with a wave vector $\boldsymbol{k}_0 = \boldsymbol{p}_0/\hbar$. In the limit of an infinitely narrow function $\varphi(p)$ we obtain a de Broglie plane wave associated with a particle of well-defined momentum $\boldsymbol{p}_0$. In Sect. 2.6 we shall give a proof of the statement (2.31), by investigating the results of a velocity measurement in a time-of-flight procedure.

Assuming that (2.31) is true, we can repeat all steps of Sect. 2.1.2 for momentum instead of position measurements. We define an expectation value:

$$\langle \boldsymbol{p}\rangle = \int \boldsymbol{p}\,|\varphi(\boldsymbol{p})|^2\,\mathrm{d}^3p\,. \tag{2.32}$$

This definition coincides with the constant $m\boldsymbol{v}_0$ introduced in (2.27), as a consequence of the Parseval–Plancherel theorem (2.16) applied to the pair of Fourier transforms (i) $f_1^*(\boldsymbol{r}) = \psi^*(\boldsymbol{r})$ and $g_1^*(\boldsymbol{p}) = \varphi^*(\boldsymbol{p})$ (ii) $f_2(\boldsymbol{r}) = -\mathrm{i}\hbar\,\partial\psi/\partial x$ and $g_2(\boldsymbol{p}) = p_x\varphi(\boldsymbol{p})$ (see 2.15):

$$-\mathrm{i}\hbar\int \psi^*(\boldsymbol{r})\,\frac{\partial\psi}{\partial x}\,\mathrm{d}^3r = \int \varphi^*(\boldsymbol{p})\,p_x\,\varphi(\boldsymbol{p})\,\mathrm{d}^3p\,. \tag{2.33}$$

We can also define the dispersions Δp_j $(j = x, y, z)$:

$$(\Delta p_j)^2 = \langle p_j^2\rangle - \langle p_j\rangle^2\,. \tag{2.34}$$

Using the Parseval–Plancherel theorem, we also find that the dispersion Δp_x coincides with the coefficient $m\,\Delta v$ appearing in the one-dimensional time evolution of the spatial width of a wave packet (2.29).

In this framework, the physical interpretation of the equations giving the time evolution of the center of a wave packet is simple. The wave packet is formed by a superposition of plane waves, each of which corresponds to a well defined momentum p. The center of the wave packet propagates with the average momentum deduced from the probability distribution $|\varphi(p)|^2$. In addition, the group velocity $\mathrm{d}\omega/\mathrm{d}k$ of each component is different, and the wave packet spreads out as time increases.

2.4.2 Heisenberg Uncertainty Relations

The previous observations, in particular the statement (2.31), lead to a cornerstone of quantum mechanics. As follows from Fourier analysis (cf. (2.18)), *whatever the wave function is*, the following inequalities hold:

$$\Delta x \, \Delta p_x \geq \hbar/2 \,, \quad \Delta y \, \Delta p_y \geq \hbar/2 \,, \quad \Delta z \, \Delta p_z \geq \hbar/2 \,. \tag{2.35}$$

These are called the *Heisenberg uncertainty relations*. The uncertainty relations are saturated, i. e. $\Delta x \Delta p_x = \hbar/2$, if and only if the wave function is a *Gaussian*. As we shall see in the next section, these relations remain valid when the particle is placed in a potential.

These uncertainty relations should be understood in the following way. Suppose we prepare $2N$ particles all in the same state $\psi(\boldsymbol{r})$, with $N \gg 1$ (see Fig. 2.3). For N of them, we measure the position. We obtain some distribution of results, with some mean value $\boldsymbol{r}_0$ and standard deviations Δx, Δy and Δz. For the remaining N particles, we measure the momentum. The distribution of the results has a mean value $\boldsymbol{p}_0$ and standard deviations Δp_x, Δp_y and Δp_z. The Heisenberg uncertainty relations state that one necessarily finds (2.35) if N is large enough that the statistics of the measured distributions are significant. This holds *whatever* $\psi(\boldsymbol{r})$ is, i. e. for any state of the system.

Theses inequalities are *intrinsic properties* of the quantum description of any system. They have nothing to do with any uncertainty of an individual

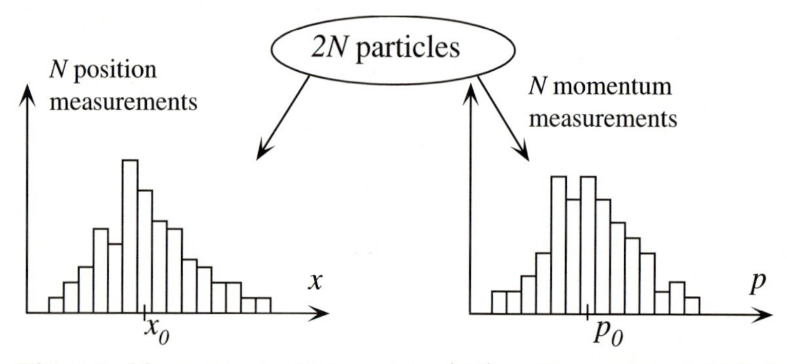

Fig. 2.3. Measurement of the position (*left*) and momentum (*right*) distributions for a collection of $2N \gg 1$ particles, all prepared in the same state. The Heisenberg uncertainty relation states that the product of the standard deviations of the two histograms is always larger than $\hbar/2$, whatever the accuracy of each individual measurement (i. e. the width of the channels of the histograms and the number of individual measurements)

measurement or with the accuracy of the measuring devices. They mean that a point particle cannot be conceived as being localized both in position and in momentum beyond the limit (2.35). The starting point of classical mechanics, where the state of a particle is described by a simultaneous knowledge of its position and momentum, is in contradiction with the uncertainty relations.

The classical limit corresponds to cases where Δx and Δp_x are both much smaller than the accuracy of the measurements. Given the value of $\hbar \sim 10^{-34}$ J s, this will be the case in most macroscopic observations. Consider for instance, an object of mass $m = 1$ gram, and a measuring apparatus that gives the position of the center of mass with a precision of 10^{-15} m (the diameter of a nucleus) and its velocity with a precision of $10^{-15}\,\mathrm{m\,s^{-1}}$ (a rather impressive apparatus!). This is still not accurate enough to detect the quantum uncertainty. Whenever we refer to the classical limit, it is this precise sense we mean, i. e. Δx and Δp_x are small but they still satisfy (2.35).

A plane wave is a limiting case of the uncertainty relations. It would correspond to a well-defined momentum, $\Delta p_x = 0$, but it is completely delocalized in space, $\Delta x = \infty$.

Finally, we note that the uncertainty relations, taken together with the relation (2.29) giving the time evolution of the spatial width of the wave packet, indicate that it is not possible to have an arbitrarily good spatial localization of a free quantum particle at two different times t_1 and t_2 if these two times are sufficiently far apart from one another. If the particle is localized at time t_1 within a short distance Δr_1, the momentum width $\Delta p_1 \geq \hbar/2\Delta r_1$ is quite large. Consequently, the spatial width at time t_2 will be dominated by the spreading term $\Delta p_1\,(t_2 - t_1)/m$, which is also quite large, except for very small values of $(t_2 - t_1)$.

It is instructive to examine quantitatively the phenomenon of wave packet spreading in two extreme cases:

1. Consider a free electron which is initially localized within an atomic size ($\Delta x_0 \sim 10^{-10}$ m) and suppose for simplicity that $\xi_1 = 0$ in (2.29). After one second, the result $\Delta x \sim 600$ km (!) shows that the wave function has literally exploded. This simply means that one second is a very long time on the atomic scale, and that a much shorter time is sufficient to ensure that an electron is delocalized over any relevant macroscopic length, such as the crystal size in electric-conduction phenomena, for instance.
2. Conversely, consider a mass of 1 mg of water initially localized with a precision $\Delta x_0 = 1$ mm. We find that the position uncertainty doubles in a time $t = 2 \times 10^{22}\,\mathrm{s} \sim 10^{15}$ years. This shows that, with great confidence one may usually neglect quantum effects on a macroscopic scale.

2.5 The Schrödinger Equation

The wave equation (2.11) for a free particle is a direct consequence of the structure of de Broglie waves. The first major contribution of Schrödinger was to find the wave equation which is satisfied when the particle is placed in a potential $V(\boldsymbol{r})$.

2.5.1 Equation of Motion

This wave equation is the *Schrödinger equation*:

Principle 2.2.b: The Schrödinger Equation

When a particle is placed in a potential $V(\boldsymbol{r},t)$, the time evolution of the wave function is governed by the Schrödinger equation:

$$i\hbar\frac{\partial}{\partial t}\psi(\boldsymbol{r},t) = -\frac{\hbar^2}{2m}\Delta\psi(\boldsymbol{r},t) + V(\boldsymbol{r},t)\,\psi(\boldsymbol{r},t)\,. \tag{2.36}$$

This equation is also linear, in agreement with the superposition principle. It is a *first order* partial differential equation in time. Therefore, it completely determines the wave function $\psi(\boldsymbol{r},t)$ at any time t, if it is known at some initial time t_0. Naturally, it reduces to (2.11) if the potential vanishes or is constant (one passes from a constant to a zero potential under a simple change of phase of the wave function).

The justification for the Schrödinger equation lies in its consequences. As we shall see throughout the following chapters, it gives results in agreement with experiments as long the average velocities of the particles are small compared with the velocity of light (i. e. in the nonrelativistic limit).

Schrödinger first tried other equations which he found more sensible, because they incorporated relativity.[1] Unfortunately, they did not give the correct relativistic corrections to the Bohr–Balmer formula, and he was somewhat discouraged. He noticed one day that a "nonrelativistic approximation" (which he said he didn't really understand) gave the correct result.[2] Schrödinger was not aware of spin, which played a major role in the relativistic corrections he was looking for, which were called the fine structure of the hydrogen atom. Schrödinger gave the symbol ψ to the wave function. He mistook its physical interpretation. It was Max Born who, at the end of 1926, found the correct interpretation of ψ as a probability amplitude, by analyzing experiments on the scattering of electrons on nuclei. The use of a new experimental technology, i. e. Geiger–Müller counters, which allowed one to count numbers of electrons, as opposed to Faraday boxes, which measured electrical intensities or charges, played an important role in this conceptual progress: in the *same* experiment, electrons could behave as waves (when they interacted with the nuclei) and as particles (when they interacted with the detector).

2.5.2 Particle in a Potential: Uncertainty Relations

The wave function $\psi(\boldsymbol{r},t)$ of a particle moving in a potential can be written as

[1] Schrödinger was the first to write down what is now known as the Klein–Gordon equation.

[2] See M. Jammer, *The Conceptual Development of Quantum Mechanics*, Chap. 5, McGraw-Hill, New York (1966).

$$\psi(\boldsymbol{r},t) = \int \varphi(\boldsymbol{p},t)\, \mathrm{e}^{\mathrm{i}\boldsymbol{p}\cdot\boldsymbol{r}/\hbar}\, \frac{\mathrm{d}^3 p}{(2\pi\hbar)^{3/2}} \,, \tag{2.37}$$

where $\varphi(\boldsymbol{p},t)$ is the Fourier transform of $\psi(\boldsymbol{r},t)$.

It is a fact that the modulus squared of this Fourier transform $|\varphi(\boldsymbol{p},t)|^2$ gives the probability density for the momentum distribution at time t (see Sect. 2.6) as it does for free particles (we come back to this point in Sect. 2.6). Therefore, as a consequence of the results of Fourier analysis, the uncertainty relations remain valid for a particle in a potential.

Equation (2.35) is the rigorous form of the uncertainty relations, which are saturated if and only if the wave function is a Gaussian. However, in addition to this exact result, it is useful to remember that for an arbitrary potential microscopic well $V(\boldsymbol{r})$ and for a quantum particle bound by this potential in one of the lowest energy states, one always has the relation

$$\Delta x\, \Delta p_x \sim \gamma\, \hbar \,, \tag{2.38}$$

where the geometrical constant γ is of order one. This provides a very simple means to calculate the orders of magnitude of the velocities and energies of many physical systems, once their sizes are known.

Consider for example a nucleus with A nucleons (protons and neutrons). Measurements show (i) that the nucleus can be treated as a sphere of radius $r_0\, A^{1/3}$, with $r_0 \sim 1.2 \times 10^{-15}$ m, and (ii) that the proton and neutron densities are (roughly) constant inside this sphere. Therefore each nucleon is confined in a sphere of radius $\sim r_0$ and we must have $\Delta p \sim \hbar/r_0$, which corresponds to a mean momentum of $|p| \sim 140$ MeV/c. This agrees with experimental values (typically $|p| \sim 200$ MeV/c). The mean kinetic energy of a nucleon is $E_{\mathrm{k}} = \Delta p^2/2m_{\mathrm{p}}$, where m_{p} is the proton (or neutron) mass, which gives $E_{\mathrm{k}} \sim 10$ MeV. Since each nucleon is bound in the nucleus, we conclude that its (negative) potential energy is larger in absolute value than this kinetic energy $|\langle V\rangle| \geq 10$ MeV, and that the *binding energy* $(V + E_{\mathrm{k}})$ is likely to be also of the order of a few MeV. This is indeed the correct order of magnitude: for large nuclei, i. e. $A \geq 20$, the binding energy per nucleon is roughly a constant, approximately 8 MeV.

2.5.3 Stability of Matter

The uncertainty relations provide the solution to a fundamental inconsistency of classical physics concerning the stability of matter made up of point particles. Consider the simple case of the hydrogen atom, with an electron moving in the Coulomb field of the proton $V(r) = -q^2/4\pi\epsilon_0 r$. We consider for simplicity a classical circular orbit of radius r. The mechanical equilibrium condition is $m_{\mathrm{e}} v^2/r = q^2/4\pi\epsilon_0 r^2$, and the electron energy is therefore

$$E = \frac{p^2}{2m_{\mathrm{e}}} + V(r) = -\frac{1}{2}\,\frac{q^2}{4\pi\epsilon_0 r} \,.$$

This energy is not bounded from below, since it tends to $-\infty$ as the radius shrinks to 0. In its circular motion, the electron is accelerating. It is an inevitable consequence of Maxwell's equations that an accelerating electric charge radiates. Therefore, from a classical point of view, the electron should lose energy continuously and an atom should be unstable. It should radiate an infinite amount of energy and the electron should collapse onto the nucleus.

The uncertainty relations save us from this catastrophic fate. Let $\langle r \rangle$ be the mean distance between the proton (which we consider as fixed) and the electron. This distance measures the uncertainty in the electron position. The Coulomb potential energy is of the order of $q^2/4\pi\epsilon_0\langle r \rangle$. Using the order of magnitude (2.38), the kinetic energy is $E_\mathrm{k} \geq \hbar^2/2m_\mathrm{e}\langle r \rangle^2$. The total average energy is therefore of the order of or larger than

$$E \geq \frac{\hbar^2}{2m_\mathrm{e}\langle r \rangle^2} - \frac{q^2}{4\pi\epsilon_0\langle r \rangle}\,. \tag{2.39}$$

The right-hand side of this expression is bounded from below. Its minimum, obtained for $\langle r \rangle = 4\pi\epsilon_0\hbar^2/(m_\mathrm{e}q^2) \sim 0.53 \times 10^{-10}\,\mathrm{m}$, is

$$E_\mathrm{min} = -\frac{m_\mathrm{e}}{2\hbar^2}\left(\frac{q^2}{4\pi\epsilon_0}\right)^2 = -13.6\,\mathrm{eV}\,.$$

This is a fundamental result. The uncertainty relations put a lower bound on the mean distance and the binding energy between the electron and proton. This explains the stability of matter.

The above argument is not mathematically rigorous. Later on (Chap. 9), we shall prove other uncertainty relations such as $\langle p^2 \rangle \geq \hbar^2\langle 1/r \rangle^2$ for any system. Applying this result to $\langle E \rangle = \langle p^2 \rangle/2m_\mathrm{e} - (q^2/4\pi\epsilon_0)\langle 1/r \rangle$, one can make the argument exact, i. e. demonstrate that $\langle 1/r \rangle$ is bounded from above.

2.6 Momentum Measurement in a Time-of-Flight Experiment

In order to show that the Fourier transform $\varphi(\boldsymbol{p}, t)$ of the wave function is the probability amplitude for the momentum distribution, we examine an experimental procedure for performing such a measurement. This is based on the "time-of-flight" method, which relies on the determination of the macroscopic distance over which the particle travels freely during a given macroscopic time interval.

We start at time $t = 0$ with a particle whose wave function is $\psi(\boldsymbol{r}, t = 0)$. Either the particle is free, or it moves in a potential $V(\boldsymbol{r})$. In order to determine the momentum distribution $\mathcal{P}(\boldsymbol{p})$ at $t = 0$, we assume we can switch off suddenly the potential and let the wave packet of the particle evolve freely for some macroscopic time t. This can easily be achieved if the potential is created by external sources, e. g. electric fields, magnetic fields, light waves. It is not possible to do so if the potential is due to microscopic interactions, such as the Coulomb field created by a nucleus, but our ambition here is to give a proof of principle and not a universal practical method. In recent years, gaseous Bose–Einstein condensates have provided remarkable physical systems on which this method has been tested directly (see, e. g., Fig. 16.4).

At time $t = 0$, we assume that the state of the particle is such that $\langle \boldsymbol{r}_0 \rangle = 0$ (this is a convention) with some uncertainty $\delta \boldsymbol{r}_0$ due to the extension of its wave function. This means that we know the initial position of the particle within $\delta \boldsymbol{r}_0$ (in full rigor, within a small multiple of δx_0 in the x variable, of δy_0 in the y variable and of δz_0 in the z variable). At some later time t we perform a measurement of the position of the particle. When we detect it at a point $\boldsymbol{r}$ with a precision $\delta \boldsymbol{r} \equiv (\delta x, \delta y, \delta z)$, we obtain a measurement of its velocity $\boldsymbol{v}$ or its momentum $\boldsymbol{p} = m\boldsymbol{v}$, since these quantities are constant between 0 and t (free motion). The result of such a measurement is $\boldsymbol{p} = m\boldsymbol{r}/t$, with an uncertainty $\delta \boldsymbol{p}$ involving both the initial ($\delta \boldsymbol{r}_0$) and final ($\delta \boldsymbol{r}$) uncertainties in position. We assume the experimental conditions are such that $\delta r \gg \delta r_0$ so that $\delta \boldsymbol{p} \sim m\, \delta \boldsymbol{r}/t$. We are free to choose the time t as large as necessary so that δp is as small as we wish.

Therefore, the probability of obtaining the result $\boldsymbol{p}$ (with uncertainty $\delta \boldsymbol{p}$) in this time-of-flight momentum measurement is

$$\delta^3 \mathcal{P}(\boldsymbol{p}) = |\psi(\boldsymbol{r}, t)|^2 \, \delta x \, \delta y \, \delta z \,, \quad \text{where} \quad \boldsymbol{r} = \boldsymbol{p}t/m \quad (\text{for large } t) \,. \tag{2.40}$$

In order to show that this definition is related to the Fourier transform $\varphi(\boldsymbol{p})$ of $\psi(\boldsymbol{r}, 0)$, we first establish the following result:

After a free propagation during the time interval t, the position probability density at a point $\boldsymbol{r}$ is given by

$$|\psi(\boldsymbol{r}, t)|^2 = (m/t)^3 \, |\tilde{\varphi}(m\boldsymbol{r}/t)|^2 \,, \tag{2.41}$$

where the function $\tilde{\varphi}$ is defined by

$$\tilde{\varphi}(\boldsymbol{p}) = \frac{1}{(2\pi\hbar)^{3/2}} \int \mathrm{e}^{-\mathrm{i}\boldsymbol{r}' \cdot \boldsymbol{p}/\hbar} \, \mathrm{e}^{\mathrm{i}mr'^2/(2t\hbar)} \, \psi(\boldsymbol{r}', 0) \, \mathrm{d}^3 r' \,. \tag{2.42}$$

In order to prove this result, we rewrite the probability density as

$$\begin{aligned} |\psi(\boldsymbol{r}, t)|^2 &= \frac{1}{(2\pi\hbar)^3} \iint \mathrm{d}^3 p_1 \, \mathrm{d}^3 p_2 \, \mathrm{e}^{\mathrm{i}(\boldsymbol{p}_1 - \boldsymbol{p}_2) \cdot \boldsymbol{r}/\hbar} \mathrm{e}^{\mathrm{i}(p_2^2 - p_1^2)t/2m\hbar} \varphi(\boldsymbol{p}_1) \varphi^*(\boldsymbol{p}_2) \\ &= \frac{1}{(2\pi\hbar)^3} \iint \mathrm{d}^3 p \, \mathrm{d}^3 p' \, \mathrm{e}^{\mathrm{i}\boldsymbol{p}' \cdot (\boldsymbol{r} - \boldsymbol{p}t/m)/\hbar} \varphi(\boldsymbol{p} + \boldsymbol{p}'/2) \varphi^*(\boldsymbol{p} - \boldsymbol{p}'/2) \,, \end{aligned} \tag{2.43}$$

where we have set $\boldsymbol{p} = (\boldsymbol{p}_1 + \boldsymbol{p}_2)/2$ and $\boldsymbol{p}' = \boldsymbol{p}_1 - \boldsymbol{p}_2$. We can express $\varphi(\boldsymbol{p} + \boldsymbol{p}'/2)\varphi^*(\boldsymbol{p} - \boldsymbol{p}'/2)$ in terms of the wave function at $t = 0$:

$$\varphi(\boldsymbol{p}+\boldsymbol{p}'/2)\,\varphi^*(\boldsymbol{p}-\boldsymbol{p}'/2)$$
$$= \frac{1}{(2\pi\hbar)^3}\iint \mathrm{d}^3r_1\,\mathrm{d}^3r_2\,\mathrm{e}^{-\mathrm{i}\boldsymbol{r}_1\cdot(\boldsymbol{p}+\boldsymbol{p}'/2)/\hbar}\,\mathrm{e}^{\mathrm{i}\boldsymbol{r}_2\cdot(\boldsymbol{p}-\boldsymbol{p}'/2)/\hbar}\,\psi(\boldsymbol{r}_1,0)\,\psi^*(\boldsymbol{r}_2,0)\,.$$

Inserting this expression into (2.43), we evaluate the integral over $\boldsymbol{p}'$:

$$\int \exp\left\{\mathrm{i}\boldsymbol{p}'\cdot\left[\boldsymbol{r}-\boldsymbol{p}t/m-(\boldsymbol{r}_1+\boldsymbol{r}_2)/2\right]/\hbar\right\}\,\mathrm{d}^3p'$$
$$= (2\pi\hbar)^3\delta(\boldsymbol{r}-\boldsymbol{p}t/m-(\boldsymbol{r}_1+\boldsymbol{r}_2)/2)\,.$$

The integral over $\boldsymbol{p}$ in (2.43) is then straightforward and leads to

$$|\psi(\boldsymbol{r},t)|^2 = \frac{m^3}{(2\pi\hbar t)^3}\iint \mathrm{d}^3r_1\,\mathrm{d}^3r_2\,\mathrm{e}^{\mathrm{i}m\boldsymbol{r}\cdot(\boldsymbol{r}_2-\boldsymbol{r}_1)/(\hbar t)}\,\mathrm{e}^{\mathrm{i}m(r_1^2-r_2^2)/(2\hbar t)}$$
$$\times\quad \psi(\boldsymbol{r}_1,0)\,\psi^*(\boldsymbol{r}_2,0)\,,$$

which proves the result stated in (2.41).

As the time t increases, the function $\tilde{\varphi}$ tends to the Fourier transform φ. In fact, these functions differ only by the extra factor $\mathrm{e}^{\mathrm{i}mr'^2/(2t\hbar)}$ in the integrand which defines $\tilde{\varphi}$. This factor provides an effective cutoff in the r' integration of (2.42) above $r' \geq \sqrt{2t\hbar/m}$. If δr_0 is the size of the region where $\psi(\boldsymbol{r},0)$ takes a nonnegligible value, this cutoff factor therefore plays a negligible role if $t \gg m\,\delta r_0^2/\hbar$. Physically, this also amounts to considering a time-of-flight duration t large enough that the initial position distribution plays a negligible role in the determination of the momentum. We have, consequently,

$$t \gg m\,\delta r_0^2/\hbar \qquad \Rightarrow \qquad |\psi(\boldsymbol{r},t)|^2 \simeq (m/t)^3\,|\varphi(m\boldsymbol{r}/t)|^2\,.$$

If we take this equation into account in (2.40) we obtain, using $\delta\boldsymbol{p} = m\,\delta\boldsymbol{r}/t$,

$$\delta^3\mathcal{P}(\boldsymbol{p}) = |\varphi(\boldsymbol{p})|^2\,\delta p_x\,\delta p_y\,\delta p_z\,. \tag{2.44}$$

This is the result we anticipated. The probability density for the momentum distribution corresponding to a wave function $\psi(\boldsymbol{r},0)$ is given by the modulus squared of the Fourier transform $\varphi(\boldsymbol{p})$ of this wave function.

Notice, finally, that this method is by no means in contradiction with the uncertainty relations. The momentum distribution of the measured particles at time t is obviously such that the uncertainty relations for the initial state hold.

Further Reading

- In this chapter we have deliberately refrained from defining physical reality itself, and from getting involved in the corresponding debates. On this subject, see for instance H. Margenau, *The Nature of Physical Reality*, McGraw-Hill, New York (1950); *Quantum Theory and Measurement*, edited by J.A. Wheeler and W.H. Zurek, Princeton University Press, Princeton (1983).

• D. Cassidy, "W. Heisenberg and the uncertainty principle," Sci. Am., June 1992.

Exercises

2.1. Phase velocity and group velocity. The Klein–Gordon equation

$$\left(\frac{1}{c^2}\frac{\partial^2}{\partial t^2} - \Delta + \frac{m^2c^2}{\hbar^2}\right)\psi(\boldsymbol{r},t) = 0$$

is a relativistic wave equation for free particles.

a. What relation between ω and $\boldsymbol{k}$ must hold in order for a plane wave $\mathrm{e}^{\mathrm{i}(\boldsymbol{k}\cdot\boldsymbol{r}-\omega t)}$ to satisfy this equation? Can all frequencies propagate freely?
b. If $\boldsymbol{p} = \hbar\boldsymbol{k}$ is interpreted as the momentum of a free particle of mass m, what is the relation between the energy E and the frequency ω?
c. What is the group velocity v_g of the corresponding wave packets and how is v_g related to the phase velocity of such waves?

2.2. Spreading of a wave packet of a free particle.

a. Consider a free particle moving along the x axis. Show that the time derivative of $\langle x^2\rangle_t$ can be written as

$$\frac{\mathrm{d}\langle x^2\rangle_t}{\mathrm{d}t} = A(t)\,, \qquad \text{where} \qquad A(t) = \frac{\mathrm{i}\hbar}{m}\int x\left(\psi\frac{\partial\psi^*}{\partial x} - \psi^*\frac{\partial\psi}{\partial x}\right)\mathrm{d}x\,.$$

b. Calculate the time derivative of $A(t)$ and show that:

$$\frac{\mathrm{d}A}{\mathrm{d}t} = B(t)\,, \qquad \text{where} \qquad B(t) = \frac{2\hbar^2}{m^2}\int\frac{\partial\psi}{\partial x}\frac{\partial\psi^*}{\partial x}\,\mathrm{d}x\,.$$

c. Show that $B(t)$ is constant.
d. By setting

$$v_1^2 = \frac{\hbar^2}{m^2}\int\frac{\partial\psi}{\partial x}\frac{\partial\psi^*}{\partial x}\,\mathrm{d}x$$

and $\xi_0 = A(0)$, show that $\langle x^2\rangle_t = \langle x^2\rangle_0 + \xi_0 t + v_1^2 t^2$.
e. Show that (2.29) holds, where

$$\xi_1 = \frac{\mathrm{i}\hbar}{m}\int x\left(\psi_0\frac{\partial\psi_0^*}{\partial x} - \psi_0^*\frac{\partial\psi_0}{\partial x}\right)\mathrm{d}x - 2x_0v_0\,.$$

Here $\psi_0 \equiv \psi(\boldsymbol{r},0)$. The coefficient ξ_1 can be interpreted physically, using the results of the next chapter, as the correlation at time 0 between position and velocity: $\xi_1/2 = \langle xv\rangle_0 - x_0v_0$. One can verify that the constraint on ξ_1 resulting from the fact that $(\Delta x_t)^2 > 0$ is equivalent to the condition that $\Delta x_t\,\Delta p_t \geq \hbar/2$ at all times t.

2.3. The Gaussian wave packet. Consider the wave packet given by

$$\varphi(p) = (\pi\sigma^2\hbar^2)^{-1/4} \exp\left(-\frac{(p-p_0)^2}{2\sigma^2\hbar^2}\right) . \tag{2.45}$$

a. For $t=0$, show that $\Delta x \, \Delta p = \hbar/2$.
b. Show that the spatial width of the wave packet at time t is given by

$$\Delta x^2(t) = \frac{1}{2}\left(\frac{1}{\sigma^2} + \frac{t^2\sigma^2\hbar^2}{m^2}\right) . \tag{2.46}$$

2.4. Characteristic size and energy in a linear or quadratic potential Using an argument similar to that of Sect. 2.5.3 evaluate the characteristic size of the wave function, and energy of a particle with mass m placed in (i) a one-dimensional harmonic potential $V(x) = m\omega^2 x^2/2$; (ii) a one-dimensional linear potential $V(x) = \alpha|x|$, when the particle is in its ground state.

3. Physical Quantities and Measurements

If you want knowledge, you must take part
in the practice of changing reality.
Mao Zedong

It seems obvious that one acquires information about a system when one measures physical quantities. The question we address in this chapter is: what information do we acquire when we perform a measurement in quantum mechanics?

There are several different aspects to this seemingly simple question.

- First, suppose we know the state of the system, i. e. its wave function $\psi(\boldsymbol{r}, t)$. How can we *predict* the result of the measurement of a given physical quantity A, i. e. the set of possible outcomes and the corresponding probabilities? With the probabilistic interpretation of the wave function, we already know the answer as far as the position and momentum variables are concerned, but this is not yet the case in general, for other physical quantities.
- Secondly, suppose we perform an experiment. For instance, we may want to demonstrate an experimental fact or verify some theoretical prediction. In the outcome of the experimental procedure, the system under consideration will be in some state, and we wish to obtain all the possible *information* about this state.
- Thirdly, suppose we want to perform an experiment on a *given* state of a system. The experiment will consist (schematically) in a process where the system interacts with a series of devices and we observe (i. e. measure) its final state. In this case, we want to be confident that the initial state has been *prepared* experimentally so that it has well-defined properties.

In classical physics, all three aspects boil down to the same point. All we need to do is to determine or define the position and momentum of the particle under consideration. The values of physical quantities are all completely determined as functions of these state variables.

In quantum physics, the evolution of the state of a system is deterministic in the sense that it is completely defined by the Schrödinger equation, which allows one to calculate the state $\psi(\boldsymbol{r}, t)$ at any time t in terms of the initial state $\psi(\boldsymbol{r}, t_0)$. However, we emphasized in the previous chapter that even though the state of the system can be well known, there exists indeterminacy in the possible results of position or momentum measurements. This property is very general in quantum mechanics: the measured values of a given physical quantity are usually distributed at random and only the probability law can be determined.

Naturally, we shall concentrate here on the particular case of a point-like particle, for which the wave mechanics presented in Chap. 2 applies. However, all the concepts introduced in this chapter will be reexamined in Chap. 5 when we work in the more general Hilbert space formalism. Therefore we shall not give proofs of all of the results presented in this chapter. Such proofs, as well as the correct statement of the measurement postulate, will be much easier to perform when the algebraic structure of Chap. 5 is available.

3.1 Measurements in Quantum Mechanics

Before plunging into the mathematical description of physical quantities and of their measurement, a few general comments are in order.

3.1.1 The Measurement Procedure

In Chap. 2 we have presented the possible outcomes of a position or momentum measurement, and we have discussed the procedure one must follow in order to acquire the maximum available knowledge about the system from the measurement of these two particular quantities $\boldsymbol{r}$ and $\boldsymbol{p}$. This procedure remains essentially valid for any other physical quantity, so that we outline it here as a starting point for our discussion.

We prepare $N \gg 1$ systems independently in the same state, i.e. they have the same wave function $\psi(\boldsymbol{r}, t)$ at the time t at which the measurement is performed. The result of the measurement of a physical quantity A is, most of the time, not unique: there is a set $\{a_i\}$ of possible results, or outcomes. This set $\{a_i\}$ may be continuous, as when we measure the position or the momentum of a free particle. It can also be discrete, for instance when we measure the energy of a bound atomic electron. In the first case it is characterized by a probability density $\mathcal{P}(a)$, and in the second case by the set of probabilities $\{p_i\}$. The N measurements lead to an expectation value $\langle a \rangle$, a dispersion Δa, etc., where for instance, $\langle a \rangle = \int a\, \mathcal{P}(a)\, \mathrm{d}a$ or $\langle a \rangle = \sum_i a_i\, p_i$.

The complete result of an experimental measurement of the quantity A on the system consists of a determination of the possible outcomes $\{a_i\}$ and the corresponding probabilities $\mathcal{P}(a)$ or p_i. By assumption, the wave function $\psi(\boldsymbol{r}, t)$ contains all the physical information about the system. Therefore the theory consists in a prescription which can extract the numbers $\{a_i, p_i\}$ from the function $\psi(\boldsymbol{r}, t)$.

3.1.2 Experimental Facts

When we construct the general framework of the measurement procedure, the following two very general facts will serve as guidelines.

Outcomes and Probabilities. A fundamental experimental observation is that the set of possible outcomes a_i does not depend on the state $\psi(\boldsymbol{r},t)$ of the system, but only on its nature. If one measures the binding energy of the electron of a hydrogen atom, one always finds a result in the set $\{-E_\mathrm{I}/n^2, n = 1,2,\dots\}$ with $E_\mathrm{I} \simeq 13.6\,\mathrm{eV}$, whatever the state of the bound electron. There exist particular states for which some possible outcomes (i. e. some values of n) are missing, but no state $\psi(\boldsymbol{r})$ of the electron can lead to any outcome not belonging to the set. Consequently, the nature of the system (the mass of the particle and the potential $V(\boldsymbol{r})$) determines the possible outcomes $\{a_i\}$, while the state $\psi(\boldsymbol{r},t)$ of the system determines the probability law p_i.

Repeatability of a Measurement. Consider a single system in the state $\psi(\boldsymbol{r},t)$. Suppose that we measure at time t_1 the quantity A and that we find the result a_i. If we repeat the same measurement at a second time t_2 arbitrarily close to t_1, then we always find the *same* result a_i. Consequently, immediately after we have performed a measurement yielding the result a_i, the state of the system has *changed*; it has been transformed into a new state on which a measurement of A gives the result a_i with certainty (i. e. probability one).

This sudden change of the state of the system is a consequence of the simplifying assumption (presented in the Sect. 2.1.2) that the measuring apparatus is described by classical mechanics and only the system is described by quantum mechanics. If we were aiming at giving a quantum description of the measuring apparatus, we would have to deal with the global quantum state of the system plus the measuring device, and the measurement process would result from the evolution of this global state.

In particular, we can conclude that:

- There exist particular wave functions $\psi(\boldsymbol{r})$ for which a measurement of A gives a unique answer, with no uncertainty (probability 1).
- For each possible outcome a_i of the first measurement, there must correspond at least one state $\psi_i(\boldsymbol{r})$ which has this property of giving with probability 1 the same result a_i in the second measurement.

3.1.3 Reinterpretation of Position and Momentum Measurements

To conclude this introductory section, we come back to the position and momentum measurements that we have investigated in the previous chapter. Our goal here is to reinterpret the results that we have already derived in order to fulfill the requirement outlined above, that the theory should provide us with a law to extract numbers (here $\langle \boldsymbol{r} \rangle$ or $\langle \boldsymbol{p} \rangle$) from the state of the system ψ.

According to the previous chapter, the wave function $\psi(\boldsymbol{r},t)$ and its Fourier transform $\varphi(\boldsymbol{p},t)$ provide us with the probability laws for the position $\boldsymbol{r}$ and the momentum $\boldsymbol{p}$. In particular, the expectation values of $\boldsymbol{r}$ and $\boldsymbol{p}$ read

$$\langle \boldsymbol{r} \rangle_t = \int \boldsymbol{r}\, |\psi(\boldsymbol{r},t)|^2 \, \mathrm{d}^3 r\,, \tag{3.1}$$

$$\langle \boldsymbol{p} \rangle_t = \int \boldsymbol{p}\, |\varphi(\boldsymbol{p},t)|^2 \, \mathrm{d}^3 p\,. \tag{3.2}$$

However, we do not yet know the probability laws for physical quantities which are functions of *both* $\boldsymbol{r}$ *and* $\boldsymbol{p}$, such as the angular momentum $\boldsymbol{L} = \boldsymbol{r} \times \boldsymbol{p}$.

It would not be convenient to resort to a new operation similar to the Fourier transform for each new physical quantity we can think of. Consequently, we want to express all expectation values directly in terms of the wave function $\psi(\boldsymbol{r},t)$ in the simplest possible way. Concerning $\langle \boldsymbol{r} \rangle_t$ the expression (3.1) is fine. However, the expression for $\langle \boldsymbol{p} \rangle_t$, needs to be transformed. Actually, we have already done this in the previous chapter when we obtained the fundamental result (see (2.33)):

$$\langle p_x \rangle_t = -\mathrm{i}\hbar \int \psi^*(\boldsymbol{r},t)\, \frac{\partial}{\partial x} \psi(\boldsymbol{r},t) \, \mathrm{d}^3 r\,. \tag{3.3}$$

This formula has the desired form, since the expectation value of the momentum is expressed *directly* in terms of the wave function $\psi(\boldsymbol{r},t)$. By applying the linear operator $-\mathrm{i}\hbar\, \partial/\partial x$, we transform the wave function $\psi(\boldsymbol{r},t)$ into another function. Multiplying the result by the complex conjugate of the wave function and integrating over all space, we obtain the desired expectation value.

If we collect the two relations analogous to (3.3) for the other components p_x, p_y and p_z of the momentum $\boldsymbol{p}$, we obtain

$$\langle \boldsymbol{p} \rangle_t = \int \psi^*(\boldsymbol{r},t)\, \frac{\hbar}{\mathrm{i}}\, \boldsymbol{\nabla} \psi(\boldsymbol{r},t) \, \mathrm{d}^3 r\,. \tag{3.4}$$

The purpose of the next section is to generalize the structure of this expression to any measurable physical quantity.

3.2 Physical Quantities and Observables

We now consider the question of how to extract the predictions of measurements from a given wave function.

3.2.1 Expectation Value of a Physical Quantity

As a generalization of the results (3.1) and (3.4), we introduce the following principle:

Principle 3.1

With any physical quantity A we associate an *observable* $\hat{A}$, which is a linear Hermitian operator acting in the space of wave functions. For a particle whose state is described by the wave function $\psi(\boldsymbol{r}, t)$, the expectation value $\langle a\rangle_t$ of a measurement at time t of the quantity A is given by

$$\langle a\rangle_t = \int \psi^*(\boldsymbol{r}, t) \left[\hat{A}\, \psi(\boldsymbol{r}, t)\right] \mathrm{d}^3 r\,. \tag{3.5}$$

Comments

- A linear operator is a linear mapping of the space onto itself, $\psi(\boldsymbol{r}, t) \to \chi(\boldsymbol{r}, t) = \hat{A}\psi(\boldsymbol{r}, t)$. In this context $\hat{A}$ is said to be Hermitian if

$$\int \left[\hat{A}\psi_2\right]^* \psi_1 \, \mathrm{d}^3 r = \int \psi_2^* \left[\hat{A}\psi_1\right] \mathrm{d}^3 r \tag{3.6}$$

  for any pair of functions ψ_1, ψ_2. This guarantees that the expectation value (3.5) is a real number.
- If the observable $\hat{A}$ is associated with the physical quantity A, the operator $\hat{A}^2$ is associated with the square of the quantity A. Consequently, we can also use (3.5) to calculate the expectation value $\langle a^2\rangle_t$ and the dispersion Δa_t of the results:

$$\Delta a_t^2 = \int \psi^*(\boldsymbol{r}, t) \left[\hat{A}^2\, \psi(\boldsymbol{r}, t)\right] \mathrm{d}^3 r - \langle a\rangle_t^2\,. \tag{3.7}$$

3.2.2 Position and Momentum Observables

As far as the position $\boldsymbol{r}$ is concerned, the above principle simply consists of rewriting (3.1) in the form

$$\langle \boldsymbol{r}\rangle_t = \int \psi^*(\boldsymbol{r}, t) \left[\boldsymbol{r}\, \psi(\boldsymbol{r}, t)\right] \mathrm{d}^3 r\,. \tag{3.8}$$

The operator $\hat{\boldsymbol{r}}$ associated with the position is simply the multiplication of the wave function by $\boldsymbol{r}$ (this means of course the set of three quantities x, y and z). The momentum observable $\hat{\boldsymbol{p}}$ can be read off from the result (3.4). It has the fundamental form

$$\hat{\boldsymbol{p}} = \frac{\hbar}{\mathrm{i}} \boldsymbol{\nabla}\,. \tag{3.9}$$

3.2.3 Other Observables: the Correspondence Principle

In this chapter, we shall only deal with observables which have classical analogs that are functions of the state variables $\boldsymbol{r}$ and $\boldsymbol{p}$. For these observables, the *correspondence principle* consists of choosing operators which are the same functions of the position and momentum operators $\hat{\boldsymbol{r}}$ and $\hat{\boldsymbol{p}}$ as in classical mechanics. This leads to the results listed in Table 3.1.

Table 3.1. Observables corresponding to common physical quantities

Physical quantity A	Observable $\hat{\boldsymbol{A}}$
Position $x, y, z, \boldsymbol{r}$	Multiplication by $x, y, z, \boldsymbol{r}$
Momentum $p_x, p_y, p_z, \boldsymbol{p}$	$\hat{p}_x = \frac{\hbar}{\mathrm{i}}\frac{\partial}{\partial x}, \quad \hat{p}_y = \frac{\hbar}{\mathrm{i}}\frac{\partial}{\partial y}, \quad p_z = \frac{\hbar}{\mathrm{i}}\frac{\partial}{\partial z},$ $\boldsymbol{p} = \frac{\hbar}{i}\boldsymbol{\nabla}$
Kinetic energy $E_\mathrm{k} = \frac{\boldsymbol{p}^2}{2m}$	$\hat{E}_k = -\frac{\hbar^2}{2m}\boldsymbol{\nabla}^2 = -\frac{\hbar^2}{2m}\Delta$
Potential energy $V(\boldsymbol{r})$	Multiplication by $V(\boldsymbol{r})$
Total energy $E = E_\mathrm{k} + V(\boldsymbol{r})$	$\hat{H} = -\frac{\hbar^2}{2m}\Delta + V(\boldsymbol{r})$
Angular momentum $\boldsymbol{L} = \boldsymbol{r} \times \boldsymbol{p}$	$\hat{\boldsymbol{L}} = \hat{\boldsymbol{r}} \times \hat{\boldsymbol{p}} = \frac{\hbar}{i}\boldsymbol{r} \times \boldsymbol{\nabla}$

The operator $\hat{H}$ associated with the total energy of the system is called the *Hamiltonian* of the system. In nonrelativistic quantum mechanics, *time* is not an observable, but a parameter on which the state of the system depends. In other words, we assume there is a clock, external to the system. At some instant, read on this clock, we perform measurements or observations with a macroscopic measuring apparatus.

3.2.4 Commutation of Observables

We notice a fundamental property of position and momentum observables: *these operators do not commute.* For instance, $\hat{x}\hat{p}_x$ and $\hat{p}_x\hat{x}$ are not equal. Indeed,

$$\hat{x}\hat{p}_x\,\psi = \hat{x}\,[\hat{p}_x\psi] = -\mathrm{i}\hbar x\frac{\partial\psi}{\partial x},$$

$$\hat{p}_x\hat{x}\,\psi = \hat{p}_x\,[\hat{x}\psi] = -\mathrm{i}\hbar\frac{\partial}{\partial x}(x\psi) = \hat{x}\hat{p}_x\psi - \mathrm{i}\hbar\psi\,.$$

This result can be written by introducing the *commutator* $[\hat{A}, \hat{B}] = \hat{A}\hat{B} - \hat{B}\hat{A}$ of any two observables:

$$[\hat{x}, \hat{p}_x] \equiv \hat{x}\hat{p}_x - \hat{p}_x\hat{x} = \mathrm{i}\hbar\hat{I}\,,$$

where $\hat{I}$ is the identity operator. In contrast, it is clear that $\hat{x}$ and $\hat{p}_y$ commute. Denoting the components of $\hat{\boldsymbol{r}}$ and $\hat{\boldsymbol{p}}$ by $\hat{x}_i$ and $\hat{p}_i$, $i = 1, 2, 3$, we have in general

$$[\hat{x}_i, \hat{x}_j] = [\hat{p}_i, \hat{p}_j] = 0\,, \quad [\hat{x}_j, \hat{p}_k] = \mathrm{i}\hbar\delta_{jk}\,, \tag{3.10}$$

where, for simplicity, we do not write explicitly the identity operator $\hat{I}$.

The correspondence between physical quantities and operators is simple when the quantity is a function of either the position or the momentum alone. When the quantity is a function of both $\hat{\boldsymbol{r}}$ and $\hat{\boldsymbol{p}}$, some care has to be taken, since products of operators depend in general on the order in which the terms are written. This does not happen for the angular momentum $\hat{\boldsymbol{L}}$, since $\hat{x}\hat{p}_y$ is equal to $\hat{p}_y\hat{x}$, for instance. In contrast, if we consider the classical quantity xp_x, the two operators $\hat{x}\hat{p}_x$ and $\hat{p}_x\hat{x}$ are not the same. In simple cases, one usually obtains the appropriate result by symmetrizing the expression, i. e. by taking the observable $(\hat{x}\hat{p}_x + \hat{p}_x\hat{x})/2$ to describe the physical quantity xp_x. We shall come back to the general form of the correspondence principle in Sect. 15.3.

3.3 Possible Results of a Measurement

We now examine how one finds the possible outcomes of a measurement and the corresponding probabilities.

3.3.1 Eigenfunctions and Eigenvalues of an Observable

Consider an observable $\hat{A}$. A function $\psi_\alpha(\boldsymbol{r})$ is called an eigenfunction of this operator, and a_α the corresponding eigenvalue, if ψ_α is not identically zero and if the following relation is satisfied:

$$\hat{A}\psi_\alpha(\boldsymbol{r}) = a_\alpha \psi_\alpha(\boldsymbol{r})\,. \tag{3.11}$$

We assume that the eigenfunctions $\psi_\alpha(\boldsymbol{r})$ are normalized, i. e. $\int |\psi_\alpha(\boldsymbol{r})|^2 \, \mathrm{d}^3 r = 1$. We note that the eigenvalues of a Hermitian operator are real. Indeed, if we multiply (3.11) by $\psi_\alpha^*(\boldsymbol{r})$ and integrate over $\boldsymbol{r}$ we obtain

$$a_\alpha = \frac{\int \psi_\alpha^*(\boldsymbol{r}) \left[\hat{A}\,\psi_\alpha(\boldsymbol{r})\right] \mathrm{d}^3 r}{\int |\psi_\alpha(\boldsymbol{r})|^2 \, \mathrm{d}^3 r}\,, \tag{3.12}$$

which is real.

The following theorem plays an important role in our discussion.

A measurement at time t of a physical quantity A yields a result a with certainty (i. e. with a probability equal to 1) if and only if the wave function of the particle at time t is an eigenfunction $\psi_\alpha(\boldsymbol{r})$ of the observable $\hat{A}$. The result a is then the eigenvalue a_α associated with $\psi_\alpha(\boldsymbol{r})$.

One form of the equivalence is easy to prove. If ψ is an eigenfunction of $\hat{A}$ with eigenvalue a_α, clearly the expectation value $\langle a \rangle$ given in (3.5) is equal to a_α (we recall that ψ is normalized so that $\int |\psi|^2 \, \mathrm{d}^3r = 1$). We also see from (3.11) that ψ is an eigenfunction of $\hat{A}^2$ with eigenvalue a_α^2. Therefore the variance Δa^2 given in (3.7) vanishes. This means that the dispersion of the results is zero, or in other words that we are certain to find the result a_α when we perform a measurement of the physical quantity A on a system prepared in the state ψ_α. The proof of the converse form of the equivalence is slightly more complicated to write with wave functions. We shall give it in Chap. 5, within the general formalism of Hilbert space.

The concept of eigenfunctions can be extended to the notion of (non-normalizable) eigendistributions. For instance, the eigendistributions of the momentum operator $-\mathrm{i}\hbar\boldsymbol{\nabla}$ are the plane waves $\psi_{\boldsymbol{p}_0}(\boldsymbol{r}) = \mathrm{e}^{\mathrm{i}\boldsymbol{p}_0\cdot\boldsymbol{r}/\hbar}$. The eigenvalue corresponding to $\psi_{\boldsymbol{p}_0}$ is $\boldsymbol{p}_0$ (see Appendix Appendix C for more details).

3.3.2 Results of a Measurement and Reduction of the Wave Packet

We have already mentioned in Sect. 3.1.2 that after a measurement of A yielding the result a_i, a subsequent immediate measurement must yield the same result with probability 1. This consistency condition of the theory, associated with the theorem given in the previous subsection, implies the following essential result:

The outcome a_i of the measurement of the physical quantity A must be an eigenvalue of $\hat{A}$, and the state of the system after the first measurement must be an eigenfunction of $\hat{A}$ corresponding to the eigenvalue a_i.

We can therefore make the following statement, which is deliberately vague at this stage:

Principle 3.2

When a measurement is performed on a system, the state of the system just after the measurement is in general different from the state $\psi(\boldsymbol{r})$ just before the measurement.

Consequently, a measurement is destructive in the sense that the state of the system is modified in an irreversible way by the measurement process. It is changed "instantaneously" from $\psi(\boldsymbol{r},t)$ to $\psi_\alpha(\boldsymbol{r})$. This aspect of the measurement process is referred to as *the reduction of the wave packet.*

Probabilities. At this stage, we have only discussed the expectation value $\langle a \rangle$ of the measurement (3.5). This quantity can also be written

$$\langle a \rangle = \sum_\alpha p_\alpha a_\alpha \,, \tag{3.13}$$

where the a_α's are the possible outcomes of the measurement (i. e. the eigenvalues of $\hat{A}$) and the p_α's are the probabilities of obtaining the result a_α ($\sum_\alpha p_\alpha = 1$). Using the spectral theorem of Riesz, we shall show in Chap. 5 that the probability p_α is given by

$$p_\alpha = \left| \int \psi_\alpha^*(\boldsymbol{r})\, \psi(\boldsymbol{r},t)\, \mathrm{d}^3r \right|^2 \Big/ \int |\psi_\alpha(\boldsymbol{r})|^2\, \mathrm{d}^3r \tag{3.14}$$

where $\psi_\alpha(\boldsymbol{r})$ is the eigenfunction of $\hat{A}$ with eigenvalue α, if the eigenvalue α is not degenerate, i. e. if all the eigenfunctions associated with a_α are proportional to each other. In the case of a degenerate eigenvalue, the expression for p_α is slightly more involved and will be given in Chap. 5.

3.3.3 Individual Versus Multiple Measurements

Since the state of the system is generally changed by a measurement, an individual measurement on a single system cannot provide any detailed information about the state of the system $\psi(\boldsymbol{r},t)$ *before* the measurement process. We get from the measurement a single number, i. e. the indication a_α of a meter or of a counter. Such an individual measurement only provides information about the state of the system *after* the measurement, since we know that the state is then an eigenfunction ψ_α. The measurement can be viewed as a means to *prepare* the system in a known state with some known physical characteristics, or, equivalently, as a filtering operation on the possible values of A.

In order to obtain precise information about the wave function $\psi(\boldsymbol{r},t)$ itself *before* the measurement, it is necessary to perform multiple measurements, i. e. perform the same measurement on a large number N of systems all prepared independently in the state $\psi(\boldsymbol{r},t)$. With this procedure, one can determine the possible outcomes a_α and their probability distribution.

3.3.4 Relation to Heisenberg Uncertainty Relations

Consider again a position measurement performed on a free particle. Suppose that this measurement gives the result x with some accuracy δx arising from the measuring apparatus. We then can make the statement that "the position of the particle is x within an accuracy δx". This means that the wave function just after the measurement is localized in a neighborhood δx of the point x. This new wave function can be very different from the wave function before the measurement. If, for instance, δx is very small compared with the spatial extension Δx of the initial wave function, the measurement will transform the wave function into another one for which the spread in momentum Δp will be much larger than it was initially.

Originally, some confusion occurred in the interpretation of the uncertainty relations. Heisenberg first presented his principle by stating that "If we measure the position of a particle with some accuracy Δx, we must modify its momentum in an unknown random way by an amount $\Delta p \sim h/\Delta x$ and, consequently, we cannot know this momentum with an accuracy better than $h/\Delta x$". It is true that by measuring the position of the particle, one modifies its momentum distribution. Nevertheless, the relation $\Delta x\, \Delta p_x \geq \hbar/2$ is an intrinsic property of the quantum description of physical quantities and should not be confused with the reduction of the wave packet, which is a consequence of the act of measurement.

3.3.5 Measurement and Coherence of Quantum Mechanics

To conclude this section we note that the measurement problem is still a controversial question concerning the foundations of quantum mechanics. We shall give some details of the relevant problems in Chap. 5 since our formalism will then be applicable to systems more general that a point-like particle moving in space. However, we can already point out one of the main difficulties associated with the approach followed here. We have referred to a classical measuring apparatus, for which a quantum description is not necessary. This allowed us to make some claims such as the repeatability of a given measurement, which leads us to the "reduction of the wave packet".

Therefore the situation is ill founded in the sense that, on one hand, quantum theory should contain classical physics as a limiting case, while on the other hand, the theory needs its own limiting case, i. e. classical physics, in order to establish its own foundations.

3.4 Energy Eigenfunctions and Stationary States

In all quantum systems the energy observable, or Hamiltonian, plays a central role. Notice that using the form of the Hamiltonian $\hat{H}$

$$\hat{H} = \frac{\hat{\boldsymbol{p}}^2}{2m} + \hat{V} = -\frac{\hbar^2}{2m}\Delta + V(\boldsymbol{r}) , \tag{3.15}$$

one can rewrite the Schrödinger equation in the form

$$\mathrm{i}\hbar \frac{\partial}{\partial t}\psi(\boldsymbol{r},t) = \hat{H}\psi(\boldsymbol{r},t) . \tag{3.16}$$

This form will be extended later on to all quantum systems. It is quite remarkable that the energy observable "governs" the time evolution of a system. Actually, this is also a result of analytical classical mechanics, as we shall recall in Chap. 15.

3.4.1 Isolated Systems: Stationary States

Assume that the system is isolated, in other words that the potential V does not depend explicitly on time (this would not be the case for a charged particle placed in an oscillating electric field). Consider the eigenfunctions of the Hamiltonian, which are defined by the eigenvalue equation

$$\hat{H}\,\psi_\alpha(\boldsymbol{r}) = E_\alpha\,\psi_\alpha(\boldsymbol{r})\,, \tag{3.17}$$

where the E_α are the energy eigenvalues, which are real as a consequence of (3.12).

These particular wave functions correspond to states of the system which have well-defined values E_α of the energy. Using these eigenfunctions, we obtain solutions of the Schrödinger equation (3.16), by choosing

$$\psi(\boldsymbol{r},t) = \psi_\alpha(\boldsymbol{r})\,\mathrm{e}^{-\mathrm{i}E_\alpha t/\hbar}\,. \tag{3.18}$$

The time dependence of such states is periodic, with angular frequency $\omega = E_\alpha/\hbar$. These are called *stationary states*. Indeed, the following are true.

- The corresponding probability distribution for the position is time independent:

$$|\psi(\boldsymbol{r},t)|^2 = |\psi_\alpha(\boldsymbol{r})|^2\,.$$

- The expectation value of any observable $\hat{A}$ which does not depend explicitly on time is also time independent:

$$\langle a\rangle = \int \psi^*(\boldsymbol{r},t)\,\left[\hat{A}\,\psi(\boldsymbol{r},t)\right]\,\mathrm{d}^3r = \int \psi_\alpha^*(\boldsymbol{r})\,\left[\hat{A}\,\psi_\alpha(\boldsymbol{r})\right]\,\mathrm{d}^3r\,. \tag{3.19}$$

We therefore discover a very remarkable property. A state whose energy is well defined “does not evolve”. Neither the probability law nor the expectation value of its position evolves with time. In order for a system to evolve with time, it must be a *superposition* of at least two stationary states with different energies. For instance, one can readily check that in the superposition $\psi(\boldsymbol{r},t) = \lambda\,\psi_1(\boldsymbol{r})\,\mathrm{e}^{-\mathrm{i}E_1 t/\hbar} + \mu\,\psi_2(\boldsymbol{r})\,\mathrm{e}^{-\mathrm{i}E_2 t/\hbar}$, the cross term proportional to $\psi_2^*\psi_1$ in $|\psi|^2$ depends on time. For such a state, one finds that $\Delta E \neq 0$, i. e. the energy is not well defined. We shall prove later on that if τ is a characteristic time of evolution of a system and if ΔE is the dispersion of the energy of the system, then

$$\Delta E\;\tau \geq \hbar/2\,,$$

which is called the time–energy uncertainty relation and will be discussed in Chap. 17. In the limit of stationary states, we have $\Delta E = 0$ and $\tau = \infty$.

3.4.2 Energy Eigenstates and Time Evolution

If we use the explicit form of the Hamiltonian, the eigenvalue equation (3.17) becomes

$$-\frac{\hbar^2}{2m}\Delta\psi_\alpha(\boldsymbol{r}) + V(\boldsymbol{r})\,\psi_\alpha(\boldsymbol{r}) = E_\alpha\,\psi_\alpha(\boldsymbol{r})\,, \tag{3.20}$$

which is a partial differential equation involving only the space variables $\boldsymbol{r}$. This equation is called the *time-independent Schrödinger equation.* Its solutions define the set $\{E_\alpha, \psi_\alpha(\boldsymbol{r})\}$ of eigenvalues and eigenfunctions of the energy. The numbers E_α are the *energy levels* of the system.

Consider a wave function $\psi(\boldsymbol{r}, t)$ which, at time $t = 0$, is a linear superposition of stationary states:

$$\psi(\boldsymbol{r}, t = 0) = \sum_\alpha C_\alpha\,\psi_\alpha(\boldsymbol{r})\,. \tag{3.21}$$

Using the linearity of the Schrödinger equation (2.36), the evolution in time of this wave function can be written readily without solving any further equation:

$$\psi(\boldsymbol{r}, t) = \sum_\alpha C_\alpha\,\mathrm{e}^{-\mathrm{i}E_\alpha t/\hbar}\,\psi_\alpha(\boldsymbol{r})\,. \tag{3.22}$$

As we shall see in Chap. 5, owing to the spectral theorem of Riesz, any wave function can be expanded in the form (3.21) with

$$C_\alpha = \int \psi_\alpha^*(\boldsymbol{r})\,\psi(\boldsymbol{r}, 0)\,\mathrm{d}^3r\,, \tag{3.23}$$

where the ψ_α's are normalized. Consequently, the time evolution of an isolated system is immediately known if one knows the stationary solutions of the Schrödinger equation. Therefore, for an isolated system, the solution of a quantum mechanical problem consists first of finding the eigenvalues and eigenfunctions of the Hamiltonian. The time evolution of any given state follows immediately.

3.5 The Probability Current

Consider the probability density $\rho(\boldsymbol{r}, t)$ of the particle at point $\boldsymbol{r}$:

$$\rho(\boldsymbol{r}, t) = \psi^*(\boldsymbol{r}, t)\,\psi(\boldsymbol{r}, t)\,, \qquad \text{for which} \qquad \int \rho(\boldsymbol{r}, t)\,\mathrm{d}^3r = 1\,. \tag{3.24}$$

We can calculate its time evolution using the Schrödinger equation (2.36) and its complex conjugate:

$$\begin{aligned}\frac{\partial}{\partial t}\rho(\boldsymbol{r},t) &= \psi^*\frac{\partial\psi}{\partial t} + \frac{\partial\psi^*}{\partial t}\psi \\ &= \frac{1}{\mathrm{i}\hbar}\psi^*\left(-\frac{\hbar^2}{2m}\Delta\psi + V(\boldsymbol{r})\psi\right) - \frac{1}{\mathrm{i}\hbar}\left(-\frac{\hbar^2}{2m}\Delta\psi^* + V(\boldsymbol{r})\psi^*\right)\psi\,. \\ &= \frac{\mathrm{i}\hbar}{2m}(\psi^*\Delta\psi - \psi\Delta\psi^*)\,. \end{aligned} \tag{3.25}$$

We can introduce the *probability current* $\boldsymbol{J}(\boldsymbol{r},t)$:

$$\boldsymbol{J}(\boldsymbol{r},t) = \frac{\hbar}{2im}(\psi^*\,\boldsymbol{\nabla}\psi - \psi\,\boldsymbol{\nabla}\psi^*)\,. \tag{3.26}$$

This can also be written

$$\boldsymbol{J} = \frac{1}{m}\mathrm{Re}\left(\psi^*\frac{\hbar}{i}\boldsymbol{\nabla}\psi\right) = \mathrm{Re}\left(\psi^*\frac{\hat{\boldsymbol{p}}}{m}\psi\right), \tag{3.27}$$

where $\hat{\boldsymbol{p}}/m$ is the velocity observable $\hat{\boldsymbol{v}}$. A simple calculation yields

$$\frac{\partial}{\partial t}\rho(\boldsymbol{r},t) + \boldsymbol{\nabla}\cdot\boldsymbol{J}(\boldsymbol{r},t) = 0\,. \tag{3.28}$$

This equation, whose structure is identical to the mass conservation equation for the flow of a fluid, is the *local* expression of the conservation of probability. Consider for instance a closed surface S and the corresponding enclosed volume V. We have

$$\frac{\mathrm{d}}{\mathrm{d}t}\int_V \rho(\boldsymbol{r},t)\,\mathrm{d}^3r = -\int_S \boldsymbol{J}\cdot\mathrm{d}\boldsymbol{S}\,. \tag{3.29}$$

The left-hand side gives the time variation of the probability of finding the particle inside the volume V, or in other words the difference between the probabilities per unit time that the particle crosses the surface S outwards and inwards. As a consequence, the equality (3.29) indicates that this balance is given directly by the flux of $\boldsymbol{J}$ through the surface S.

For a stationary state, which describes a steady-state probability distribution, ρ is independent of time, and we obtain

$$\text{Stationary state:}\begin{cases} \text{2 or 3 dimensions: } \boldsymbol{\nabla}\cdot\boldsymbol{J} = 0 \\ \text{1 dimension:} \quad \dfrac{\mathrm{d}J}{\mathrm{d}x} = 0 \quad \Rightarrow J = \text{constant.}\end{cases} \tag{3.30}$$

In particular, in the case of de Broglie waves $\psi_0\,\mathrm{e}^{\mathrm{i}(\boldsymbol{k}\cdot\boldsymbol{r}-\omega t)}$, the probability current is

$$\boldsymbol{J} = \frac{\hbar\boldsymbol{k}}{m}|\psi_0|^2\,. \tag{3.31}$$

3.6 Crossing Potential Barriers

In this last section we consider examples of one-dimensional motions of particles in simple potentials $V(x)$ which are made up of step functions. The solution of the Schrödinger equation is then the same as in the usual problems of wave physics.

We shall consider the propagation of plane waves in such potentials. In regions where the potential is a constant, and where the kinetic energy is positive, the eigenstates of the Hamiltonian can be written as a single plane wave e^{ikx} or as a superposition of two plane waves $\mathrm{e}^{\pm ikx}$. Once these eigenstates are known, the physical interpretation of the results can be performed in two equivalent ways, which lead to identical conclusions:

1. One can consider a plane wave as representing a continuous flux of particles and use the notion of the probability current given in the previous section. One can then calculate the relevant reflection and transmission coefficients from the ratios between the various fluxes.
2. One can superimpose these eigenstates in order to construct a wave packet whose time evolution can be studied subsequently.

In what follows we shall restrict ourselves to the first interpretation, but it is useful to keep in mind the second in order to gain insight into the physics of wave packets.

3.6.1 The Eigenstates of the Hamiltonian

In a region of space where the potential is a constant ($V(x) = V$), the time-independent Schrödinger equation has a simple form:

$$-\frac{\hbar^2}{2m}\,\psi''(x) + (V - E)\,\psi(x) = 0\,. \tag{3.32}$$

For simplicity, we omit in this section the index α in the function $\psi_\alpha(x)$ and the energy E_α (see, e.g., (3.20)). This equation is readily integrated:

$$\psi(x) = \xi_+\,\mathrm{e}^{\mathrm{i}px/\hbar} + \xi_-\,\mathrm{e}^{-\mathrm{i}px/\hbar}\,, \tag{3.33}$$

where $\xi_\pm$ are constant and where p is related to E by

$$p^2/2m = E - V\,. \tag{3.34}$$

We remark that:

1. If $E - V > 0$, p is real and the wave function is a superposition of monochromatic plane waves propagating to the right and to the left:

   $$\psi(x,t) = \xi_+\,\mathrm{e}^{\mathrm{i}(px-Et)/\hbar} + \xi_-\,\mathrm{e}^{-\mathrm{i}(px+Et)/\hbar}\,. \tag{3.35}$$

   In this region, the particle propagates freely; p is the modulus of its momentum, and $p^2/2m$ is its kinetic energy.

2. If $E-V < 0$, the quantity $p^2/2m$ is negative and the wave function (3.33) is a sum of real exponentials. Classically, the particle cannot penetrate a region where its total energy is lower than its potential energy. In quantum physics, as we shall see, although a particle cannot propagate, there is nevertheless a nonvanishing probability of finding the particle in that region.

3.6.2 Boundary Conditions at the Discontinuities of the Potential

In order to simplify the calculations, in this chapter as well as in the following ones, we substitute for the actual potentials (which are usually continuous functions) simplified forms which can have discontinuities or infinite values. One must impose continuity relations on the solutions, which are direct consequences of the differential equation (3.32).

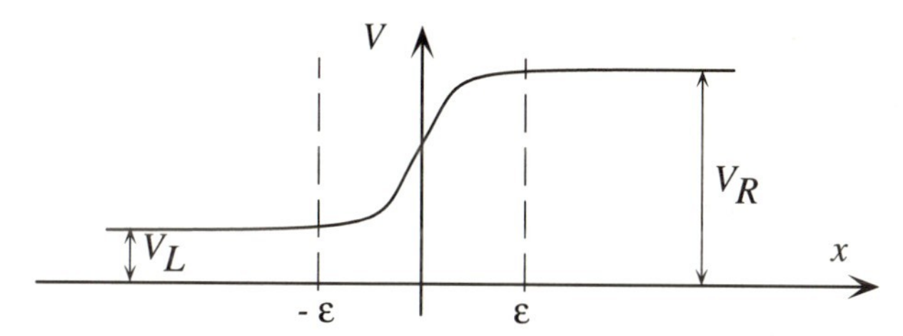

Fig. 3.1. Rapid variation of the potential in the vicinity of $x = 0$

Consider a potential $V(x)$ which varies rapidly around $x = 0$ (Fig. 3.1). Integrating the differential equation (3.32) between $x = -\epsilon$ and $x = +\epsilon$, one finds

$$\psi'(+\epsilon) - \psi'(-\epsilon) = \frac{2m}{\hbar^2} \int_{-\epsilon}^{+\epsilon} [V(x) - E]\, \psi(x)\, \mathrm{d}x\,.$$

When the width 2ϵ of the interval in which the potential varies from V_{L} to V_{R} tends to 0, the integral tends to zero. Therefore the derivative ψ' of the wave function is continuous, as well as the wave function itself.

This property can be derived rigorously in the context of distribution theory: if ψ were discontinuous at $x = 0$, we could write it as $\psi(x) = \tilde{\psi}(x) + b\,\theta(x)$, where $\tilde{\psi}(x)$ is continuous, $\theta(x)$ is the Heaviside function, and b is the discontinuity of ψ. Therefore ψ' would contain a term $b\,\delta(x)$ and ψ'' a term $b\,\delta'(x)$. However, the equation $\psi'' = 2m\ (V - E)\ \psi/\hbar^2$ would then impose $b = 0$ (no term on the right-hand side can compensate $b\,\delta'(x)$). Therefore ψ is continuous and so is ψ'.

This continuity of ψ and ψ' at $x = 0$ is valid if the potential $V(x)$ has a finite discontinuity at $x = 0$. Later on, we shall study cases such as an infinite well or a δ function potential, where only the wave function ψ is continuous.

3.6.3 Reflection and Transmission on a Potential Step

Consider the potential of Fig. 3.2, $V = 0$ for $x < 0$, $V = V_0 > 0$ for $x \geq 0$. Classically, there are two possibilities: (a) if the particle, coming from the left, has an energy larger than V_0, it continues on its way to the right; (b) if its energy is smaller than V_0, the particle bounces back to the left.

In the quantum problem, consider a solution of the form (3.18), i.e. $\psi(x,t) = \psi(x)\,\mathrm{e}^{-\mathrm{i}Et/\hbar}$, where E is the energy, and set

$$k = \sqrt{2mE}/\hbar\,. \tag{3.36}$$

As in classical mechanics, we must consider two cases.

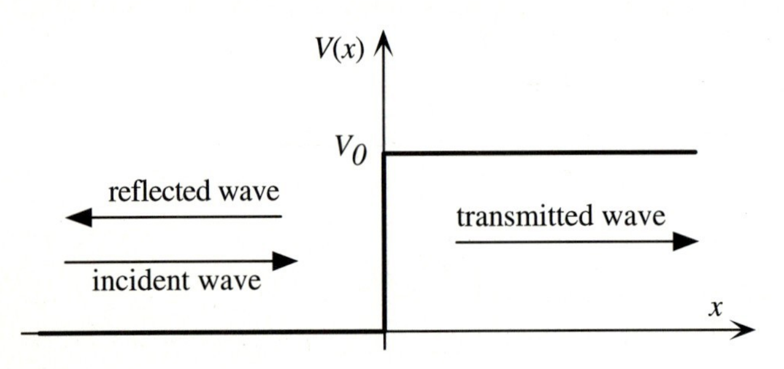

Fig. 3.2. Potential step

The case $E < V_0$. We set

$$\kappa' = \sqrt{2m(V_0 - E)}/\hbar\,. \tag{3.37}$$

The general form of the solution of the differential equation (3.32) is, up to a multiplicative factor,

$$x < 0 \;:\; \psi(x) = \mathrm{e}^{\mathrm{i}kx} + \xi\,\mathrm{e}^{-\mathrm{i}kx}\,; \quad x > 0 \;:\; \psi(x) = \beta\,\mathrm{e}^{-\kappa' x}\,. \tag{3.38}$$

For $x > 0$ we keep only the decreasing exponential for obvious normalization reasons (it is impossible to construct normalized wave packets by superposing waves which diverge as $\mathrm{e}^{\kappa' x}$ when $x \to \infty$). For $x < 0$ we normalize to 1 the term e^{ikx}, which represents the incoming wave.

The continuity conditions for ψ and ψ' give

$$1 + \xi = \beta\,, \qquad \mathrm{i}k(1 - \xi) = -\kappa'\beta\,, \tag{3.39}$$

from which we deduce

$$\xi = \frac{k - \mathrm{i}\kappa'}{k + \mathrm{i}\kappa'}\,. \tag{3.40}$$

We notice in particular that $|\xi|^2 = 1$, which means that the reflected wave has the same flux as the incoming wave. There is a *total* reflection of the wave, as in the classical case. However, the wave function does not vanish inside the potential step. It decreases exponentially with a penetration length proportional to $\hbar/\sqrt{2m(V_0 - E)}$, which is reminiscent of the skin effect in electromagnetism. The penetration length vanishes in the three limits $\hbar \to 0$, $m \to \infty$, $V_0 - E \to \infty$, in agreement with our expectations for the classical limit.

The Case $E > V_0$. We set

$$k' = \sqrt{2m(E - V_0)}/\hbar\,. \tag{3.41}$$

The general solution of (3.32) can be written

$$x < 0: \quad \psi(x) = \xi_+\,\mathrm{e}^{ikx} + \xi_-\,\mathrm{e}^{-ikx}\,; \tag{3.42}$$

$$x > 0: \quad \psi(x) = \beta_+\,\mathrm{e}^{ik'x} + \beta_-\,\mathrm{e}^{-ik'x}\,. \tag{3.43}$$

The continuity of ψ and ψ' at $x = 0$ gives two relations between the four coefficients $\xi_\pm$, $\beta_\pm$:

$$\xi_+ + \xi_- = \beta_+ + \beta_-\,, \qquad k(\xi_+ - \xi_-) = k'(\beta_+ - \beta_-)\,, \tag{3.44}$$

which allows us to express two coefficients, e. g. ξ_- and β_+, in terms of the two other ones, ξ_+ and β_-. We remark that the physical meaning of the wave function is not changed if we multiply it by a constant (it cannot be normalized anyway), therefore we can always set one coefficient to unity, as we did in (3.38). We are still left with one undetermined coefficient, which is simply an illustration of the superposition principle. Indeed, two physical situations can be considered for $E > V_0$: (i) one can send the particle in from the left (e^{ikx}) and obtain a reflected wave (e^{-ikx}) and a transmitted wave ($\mathrm{e}^{ik'x}$), in which case $\beta_- = 0$; (ii) one can also send the particle in from the right ($\mathrm{e}^{-ik'x}$) and obtain a reflected wave ($\mathrm{e}^{ik'x}$) and a transmitted wave (e^{-ikx}), in which case $\xi_+ = 0$. The general solution of (3.32) is an arbitrary linear superposition of these two particular solutions. We concentrate here on the situation (i) and set $\xi_+ = 1$, $\beta_- = 0$ so that the wave coming from the left is normalized to unity, and there is only an outgoing wave on the right.

We obtain

$$\xi_- = \frac{k - k'}{k + k'}\,, \qquad \beta_+ = \frac{2k}{k + k'}\,. \tag{3.45}$$

We see that the reflection coefficient $|\xi_-|^2$ is never zero. This is contrary to the classical observation that the particle passes across the potential step and goes to the right if its energy is larger than V_0. In quantum mechanics there is always a nonvanishing probability $|\xi_-|^2$ that the particle is reflected.

To interpret (3.45) physically we calculate the probability currents J_i, J_r and J_t associated with the incident wave e^{ikx}, the reflected wave $\xi_-\mathrm{e}^{-ikx}$ and the transmitted wave $\beta_+\mathrm{e}^{ik'x}$, respectively. We obtain

$$J_{\mathrm{i}} = \frac{\hbar k}{m}, \qquad J_{\mathrm{r}} = -\frac{\hbar k}{m}|\xi_-|^2, \qquad J_{\mathrm{t}} = \frac{\hbar k'}{m}|\beta_+|^2. \tag{3.46}$$

We can check using (3.45) that $J_{\mathrm{i}} = |J_{\mathrm{r}}| + J_{\mathrm{t}}$ which means that the current is conserved, as expected for a stationary state in one dimension (cf. (3.30)). One can therefore define a reflection coefficient R and a transmission coefficient T for this barrier:

$$R = \frac{|J_{\mathrm{r}}|}{J_{\mathrm{i}}} = |\xi_-|^2, \qquad T = \frac{J_{\mathrm{t}}}{J_{\mathrm{i}}} = |\beta_+|^2 \frac{k'}{k}, \tag{3.47}$$

with $R + T = 1$.

One may wonder why the reflection coefficient R does not vanish in the "classical" limit $\hbar \to 0$. This is an artifact due to our choice of a discontinuous potential, which would correspond to an infinite force $-V_0\,\delta(x)$ at $x = 0$. For a smoother and more realistic potential one does recover full transmission for $E > V_0$ and $\hbar \to 0$. For instance, for the potential $V(x) = V_0/(1+\mathrm{e}^{-\alpha x})$, which becomes square in the limit $\alpha \to +\infty$, the reflection coefficient is proportional to $R \sim \exp[-4\pi\sqrt{2m\,E}/(\alpha\hbar)]$, which does vanish for any finite value of α as $\hbar \to 0$, but which keeps a constant value if one takes first the limit $\alpha \to +\infty$ (see L. Landau and E. Lifshitz, *Quantum Mechanics*, Chap. 3, Sect. 25, Pergamon, Oxford (1965)).

3.6.4 Potential Barrier and Tunnel Effect

We turn now to a bona fide quantum effect of great importance. Consider the potential barrier shown in Fig. 3.3, and suppose that the energy of the incident particle is *smaller* than V_0. If $E < V_0$, we know that, classically, the particle cannot cross the potential barrier. Let us examine the quantum case. Again we set $\kappa' = [2m(V_0 - E)]^{1/2}/\hbar$, and we look for a solution of (3.32) of the form

$$\psi(x) = \begin{cases} \mathrm{e}^{\mathrm{i}kx} + \xi\,\mathrm{e}^{-\mathrm{i}kx} & x < 0 \\ \gamma\,\mathrm{e}^{-\kappa' x} + \delta\,\mathrm{e}^{\kappa' x} & 0 \le x \le a\,. \\ \beta\,\mathrm{e}^{\mathrm{i}kx} & a < x \end{cases} \tag{3.48}$$

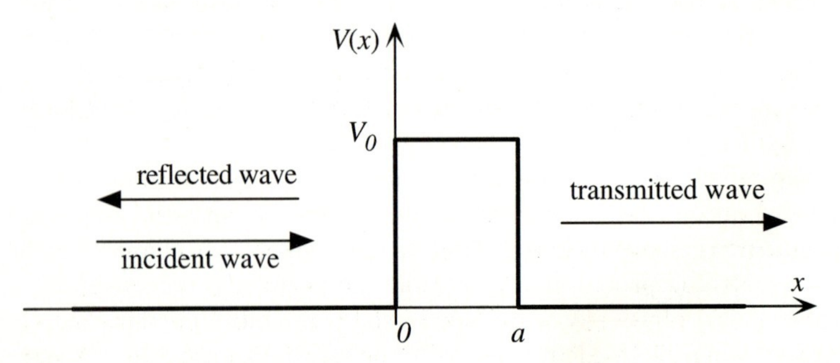

Fig. 3.3. Potential barrier

As above, we consider here the particular case of a wave incident from the left. The continuity conditions for ψ and ψ' give

$$\begin{cases} 1+\xi=\gamma+\delta \\ \mathrm{i}k(1-\xi)=\kappa'(\delta-\gamma) \end{cases} \qquad \begin{cases} \gamma\,\mathrm{e}^{-\kappa' a}+\delta\,\mathrm{e}^{\kappa' a}=\beta\,\mathrm{e}^{\mathrm{i}ka} \\ \kappa'(\delta\,\mathrm{e}^{\kappa' a}-\gamma\,\mathrm{e}^{-\kappa' a})=\mathrm{i}k\beta\,\mathrm{e}^{\mathrm{i}ka} \end{cases} \tag{3.49}$$

From these, we obtain

$$\beta=\frac{4\mathrm{i}k\kappa'\,\mathrm{e}^{-\mathrm{i}ka}}{(k+\mathrm{i}\kappa')^2\,\mathrm{e}^{\kappa' a}-(k-\mathrm{i}\kappa')^2\,\mathrm{e}^{-\kappa' a}}\,. \tag{3.50}$$

In the case $\kappa' a \gg 1$, which is of particular interest, we have simply

$$|\beta|^2 \simeq \frac{16\,k^2k'^2}{(k^2+k'^2)^2}\,\mathrm{e}^{-2\kappa' a}\,. \tag{3.51}$$

The probability $|\beta|^2$ that the particle crosses the barrier is nonzero although the incident kinetic energy is smaller than the height of the barrier! This phenomenon is called the *tunnel effect* and is unknown in classical mechanics. We note that this probability tends to zero *exponentially* in the limits (a) $m \gg \hbar^2/(V_0-E)a^2$ (classical limit), (b) $V_0 - E \gg \hbar^2/ma^2$ (very high barrier), (c) $a^2 \gg \hbar^2/m(V_0-E)$ (very broad barrier).

The tunnel effect plays a fundamental role in physics. It is, for instance, responsible for α decay of nuclei, which classically should be stable; it plays a major role in nuclear fission and fusion reactions and it is responsible for chemical binding, etc. Two spectacular applications of quantum tunneling have received the Nobel prize in the last few decades. In 1973 Josephson was awarded the prize, together with Esaki and Giaever, for a discovery of a junction which couples coherently two macroscopic wave functions in superconducting materials, separated by a thin insulating junction. In 1985 the prize was awarded to Binnig and Rohrer, inventors of the tunneling-effect microscope. In this device one displaces a very sharp conducting tip close to the surface of a conducting sample. Electrons can tunnel between the tip and the sample, and the corresponding *macroscopic* current allows a very accurate mapping of the surface of the sample. The extremely fast variation of the exponential function $\mathrm{e}^{-2\kappa' a}$ entering into the expression for $|\beta|^2$ (proportional to the current) allows a resolution of the order of 0.01 nm, as shown in Fig. 3.4. By extending this technique, one can also manipulate atoms or molecules deposited on the surface of a crystal. This technology is the basis of considerable developments in electronics.

3.7 Summary of Chapters 2 and 3

- The state of a particle in space is described by a *wave function* $\psi(\boldsymbol{r},t)$ whose modulus squared gives the probability density for finding the particle around point $\boldsymbol{r}$ at time t.

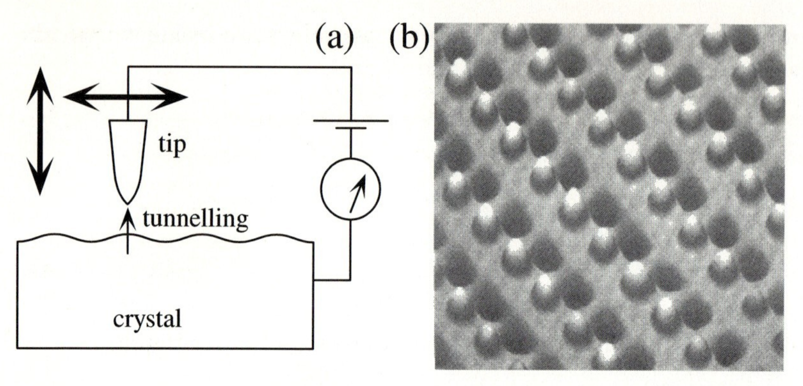

Fig. 3.4. (**a**) Principle of a tunneling microscope: a sharp tip is displaced in the vicinity of a crystal using piezoelectric transducers. A feedback loop adjusts the distance from the tip to the crystal such that the current associated with the tunneling of the electrons is constant. The error signal of the feedback loop gives a direct mapping of the distribution of the electron density (more precisely, the electrostatic potential) at the surface of the crystal. An example is shown in (**b**), which displays the surface of an InSb crystal. The antimony atoms appear raised. The size of the image is $\sim$ 3 nm (picture by Y. Liang et al., J. Vac. Sci. Technol. **B9**, 730 (1991))

- The time evolution of the wave function of a particle placed in a potential $V(\boldsymbol{r})$ is given by the *Schrödinger equation*,

$$\mathrm{i}\hbar\frac{\partial}{\partial t}\psi(\boldsymbol{r},t) = \hat{H}\psi(\boldsymbol{r},t)\,,$$

where the energy observable $\hat{H}$, or the Hamiltonian of the system, is

$$\hat{H} = -\frac{\hbar^2}{2m}\Delta + V(\boldsymbol{r})\,.$$

- The probability amplitude for the momentum of the particle is given by the Fourier transform of the wave function,

$$\varphi(\boldsymbol{p},t) = \int \mathrm{e}^{-\mathrm{i}\boldsymbol{p}\cdot\boldsymbol{r}/\hbar}\,\psi(\boldsymbol{r},t)\,\frac{\mathrm{d}^3r}{(2\pi\hbar)^{3/2}}\,.$$

- This results in the *Heisenberg uncertainty relations*, which relate the mean square deviations of position and momentum measurements:

$$\Delta x\;\Delta p_x \geq \hbar/2\,.$$

- To each physical quantity A is associated an *observable* $\hat{A}$, which is a linear Hermitian operator acting on the wave functions. The expectation value $\langle a\rangle_t$ at the time t of the measurement of the quantity A is

$$\langle a\rangle_t = \int \psi^*(\boldsymbol{r},t)\,\left[\hat{A}\,\psi(\boldsymbol{r},t)\right]\,\mathrm{d}^3r\,.$$

The position observable $\hat{\boldsymbol{r}}$ corresponds to multiplying the wave function by $\boldsymbol{r}$; the momentum observable is:

$$\hat{\boldsymbol{p}} = \frac{\hbar}{\mathrm{i}} \boldsymbol{\nabla} .$$

These observables do not commute; we have, for instance,

$$[\hat{x}, \hat{p}_x] = \mathrm{i}\hbar .$$

- If the wave function is an eigenfunction of the observable $\hat{A}$ corresponding to the eigenvalue a_α, the result of a measurement of A is equal to a_α with probability one.
- For an isolated system placed in a time-independent potential, the *stationary states* are the eigenstates of the energy, whose wave functions are of the form

$$\psi(\boldsymbol{r}, t) = \psi_\alpha(\boldsymbol{r}) \mathrm{e}^{-\mathrm{i}E_\alpha t/\hbar} ,$$

where ψ_α is a solution of the time-independent Schrödinger equation

$$\hat{H} \psi_\alpha(\boldsymbol{r}) = E_\alpha \, \psi_\alpha(\boldsymbol{r}) .$$

The time evolution of any wave function $\psi(\boldsymbol{r}, t)$ can be written directly, provided one knows the stationary solutions, as

$$\psi(\boldsymbol{r}, t) = \sum_\alpha C_\alpha \mathrm{e}^{-\mathrm{i}E_\alpha t/\hbar} \psi_\alpha(\boldsymbol{r}) ,$$

$$\text{where} \quad C_\alpha = \int \psi_\alpha^*(\boldsymbol{r}) \, \psi(\boldsymbol{r}, t=0) \, \mathrm{d}^3\boldsymbol{r} .$$

Further Reading

- S. Hawking and R. Penrose, “The nature of space and time”, Sci. Am., August 1996.
- C.F. Quate, “Vacuum tunneling: a new technique for microscopy”, Phys. Today **39**, August 1986; K. Likarov and T. Claeson, “The ultimate electronics”, Sci. Am., July 1992; J. Gimzewski, “Molecules, nanophysics and nanoelectronics”, Phys. World, June 1998, p. 29; J. Bernholc, “Computational material science: the era of applied quantum mechanics”, Phys. Today, September 1999, p. 30.
- The question of the “tunneling” time raises interesting problems addressed for instance by R.Y. Chiao, P.G. Kwiat, and A.M. Steinberg, “Faster than light?”, Sci. Am., August 1993, p. 38.

Exercises

3.1. Expectation values and variances. Consider the one-dimensional wave function $\psi(x) = \sqrt{2/a}\sin(\pi x/a)$ if $0 \le x \le a$, and $\psi(x) = 0$ otherwise. Calculate $\langle x\rangle, \Delta x, \langle p\rangle, \Delta p$ and the product $\Delta x\,\Delta p$.

3.2. The mean kinetic energy is positive. Verify that for any wave function $\psi(x)$, the expectation value $\langle p^2\rangle$ is positive.

3.3. Real wave functions. Consider a *real* one-dimensional wave function $\psi(x)$. Show that $\langle p\rangle = 0$.

3.4. Translation in momentum space. Consider a one-dimensional wave function $\psi(x)$ such that $\langle p\rangle = q$ and $\Delta p = \sigma$. What are the values of $\langle p\rangle$ and Δp for the wave function $\psi(x)\mathrm{e}^{\mathrm{i}p_0x/\hbar}$?

3.5. The first Hermite function. Show that the wave function $\psi(x) = \mathrm{e}^{-x^2/2}$ is an eigenfunction of the operator $(x^2 - \partial^2/\partial x^2)$ with eigenvalue 1.

3.6. Ramsauer effect. In 1921, Ramsauer noticed that for some particular values of the incident energy, rare gases such as helium, argon or neon were transparent to low-energy electron beams. This can be explained in the following one-dimensional model. Consider a stationary solution of the Schrödinger equation of positive energy E, for a particle of mass m in the following one-dimensional potential ($V_0 > 0$):

$$V(x) = 0 \quad \text{for } |x| > a\,, \qquad V(x) = -V_0 \quad \text{for } |x| \le a\,.$$

We set $q^2 = 2m(V_0 + E)/\hbar^2$, $k^2 = 2mE/\hbar^2$ and we are interested in a solution of the form

$$\begin{aligned}
\psi(x) &= \mathrm{e}^{\mathrm{i}kx} + A\,\mathrm{e}^{-\mathrm{i}kx} && x \le -a\,,\\
\psi(x) &= B\,\mathrm{e}^{\mathrm{i}qx} + C\,\mathrm{e}^{-\mathrm{i}qx} && -a < x \le a\,,\\
\psi(x) &= D\,\mathrm{e}^{\mathrm{i}kx} && x > a\,.
\end{aligned}$$

a. Write down the continuity relations at $x = -a$ and $x = a$.
b. Setting $\Delta = (q+k)^2 - \mathrm{e}^{4\mathrm{i}qa}(q-k)^2$, calculate the transmission probability $T = |D|^2$. Calculate the reflection probability $R = |A|^2$. Check that $R + T = 1$.
c. Show that $T = 1$ for some values of the energy. Interpret this result and the Ramsauer effect.

d. In helium, the lowest energy at which the phenomenon occurs is $E = 0.7\,\mathrm{eV}$. Assuming that the radius of the atom is $a = 0.1\,\mathrm{nm}$, calculate the depth V_0 of the potential well inside the atom in this model.
e. How does the reflection coefficient behave as the ratio E/V_0 tends to zero? When one directs very slow hydrogen atoms onto a liquid-helium surface, these atoms bounce back elastically instead of being adsorbed. Explain this phenomenon qualitatively.

4. Quantization of Energy in Simple Systems

Why do simple when one can do complicated?
Second Shadok principle
(Jacques Rouxel, *Les Shadoks*, Circumflexe, Paris (1994))

As beautiful as it may be, a theory can only be accepted if its predictions stand the test of experiment. The hydrogen atom, which we shall study later on, was the first testing ground for quantum mechanics. The spectrum of the hydrogen atom was calculated nearly simultaneously by Pauli (at the end of 1925 with matrix mechanics), by Schrödinger (in the beginning of 1926 with wave mechanics) and by Dirac (in 1926 with his mechanics of noncommuting observables). Other simple problems, such as the Stark effect in this atom, the harmonic oscillator and its nonharmonic corrections, convinced the scientific community of the validity of the theory.

In this chapter we consider some simple applications of wave mechanics. We aim to illustrate its principles by treating a few examples of motions of particles in time-independent potentials. These potentials are sufficiently simple that the Schrödinger equation can be solved analytically. Starting from the solution of such simple models one can understand qualitatively, and often quantitatively, the structure of atomic, molecular or nuclear systems.

4.1 Bound States and Scattering States

In classical mechanics, one can make a distinction between two regimes in the motion of a particle in a potential which has a finite limit V_1 at infinity. When the energy E of the particle, which is a constant of the motion, is smaller than V_1 the trajectory is confined in a finite region of space at all times. In contrast, if $E > V_1$ the trajectory goes to infinity as $t \to \pm\infty$. The first situation is called a bound state, and the second one a scattering state (Fig. 4.1). The same distinction exists in quantum mechanics. According to whether the value of the energy is smaller or larger than V_1, we shall deal either with bound states or with scattering states.

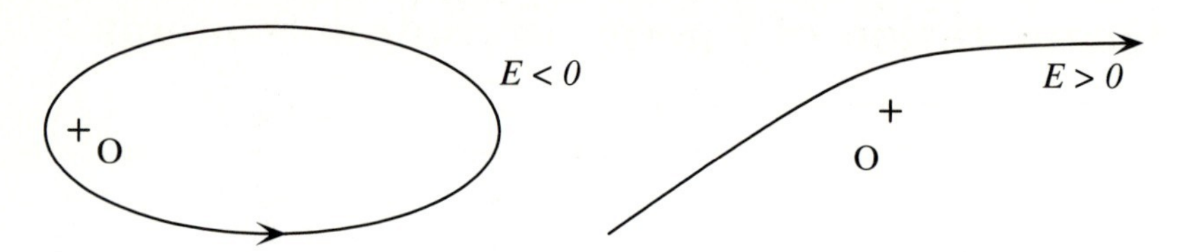

Fig. 4.1. Classical motion for the attractive Kepler problem ($1/r$ potential, with $V_1 = 0$). A particle with $E < 0$ remains confined within a finite region of space, while a particle with $E > 0$ escapes to infinity

4.1.1 Stationary States of the Schrödinger Equation

In this chapter, we shall study the quantum motion of a particle of mass m, placed in a potential $V(\boldsymbol{r})$, so that the Hamiltonian $\hat{H}$ has the form

$$\hat{H} = -\frac{\hbar^2}{2m}\Delta + V(\boldsymbol{r})\,. \tag{4.1}$$

We have seen in Chap. 3 that the first step in this study consists of finding the eigenfunctions (or "stationary states") of the Hamiltonian $\hat{H}$, given by

$$\hat{H}\psi_\alpha(\boldsymbol{r}) = E_\alpha \psi_\alpha(\boldsymbol{r})\ , \tag{4.2}$$

and the corresponding eigenvalues. In fact, an arbitrary state $\psi(\boldsymbol{r}, 0)$ can always be written as a linear combination of the $\psi_\alpha(\boldsymbol{r})$:

$$\psi(\boldsymbol{r}, 0) = \sum C_\alpha\, \psi_\alpha(\boldsymbol{r})\ , \tag{4.3}$$

where the symbol $\sum$ can stand for either a discrete sum or an integral. The time evolution of this state can be written immediately as

$$\psi(\boldsymbol{r}, t) = \sum C_\alpha\, \psi_\alpha(\boldsymbol{r})\, \mathrm{e}^{-\mathrm{i}E_\alpha t/\hbar}\,. \tag{4.4}$$

The boundary conditions which the functions $\psi_\alpha(\boldsymbol{r})$ satisfy at infinity are different according to whether we consider bound states or scattering states.

4.1.2 Bound States

A bound state is defined as a solution of (4.2) for which the function $\psi_\alpha(\boldsymbol{r})$ is *square integrable*, and therefore normalizable:

$$\int |\psi_\alpha(\boldsymbol{r})|^2\, \mathrm{d}^3r = 1\,. \tag{4.5}$$

One can prove that this happens only for a *discrete* set $\{E_n\}$ of values of the energy, which are called the *energy levels*. This is the origin of the quantization of energy.

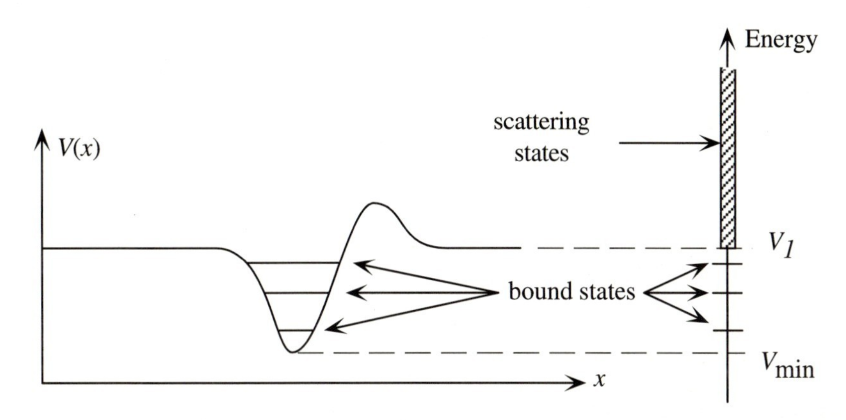

Fig. 4.2. For a potential which tends to a finite value V_1 at infinity and which has a minimum value $V_{\min}$, the eigenvalues E of the Hamiltonian form (i) a discrete set E_n between $V_{\min}$ and V_1 corresponding to bound states and (ii) a continuous set of values above V_1 corresponding to scattering states. Any physical state of the system can be written as a linear superposition of these eigenfunctions

The fact that the functions $\psi_n \equiv \psi_\alpha$ can be normalized is associated with the classical property that the trajectory of the particle remains confined in a finite region of space at all times. Each E_n is smaller than the value of the potential at infinity V_1, and larger than the minimal value $V_{\min}$ over all space (see Fig. 4.2):

$$V_{\min} < E_n < V_1 \,. \tag{4.6}$$

It may happen that one of these two bounds becomes formally infinite. For instance the left-hand side is $-\infty$ for the Kepler problem, and the right-hand side is $+\infty$ for a harmonic oscillator.

Since the ψ_n's can be normalized, each of them represents a possible state of the particle. If we choose $\psi(\boldsymbol{r},0) = \psi_n(\boldsymbol{r})$, the particle has a well-defined energy. The time evolution of the state is simply $\psi(\boldsymbol{r},t) = \psi_n(\boldsymbol{r}) \exp(-\mathrm{i}E_n t/\hbar)$ and the expectation value of any physical quantity is time independent: there is no motion in the usual sense. Motion appears only when we construct linear combinations of stationary states with different energies.

4.1.3 Scattering States

The time-independent Schrödinger equation (4.2) also has solutions for a continuous set of energies E larger than V_1. Asymptotically, such solutions are plane waves, since the potential is constant as $\boldsymbol{r} \to \infty$. If one performs a proper analysis of such solutions, one finds that these correspond to scattering states of an asymptotically free particle, scattered by the potential $V(\boldsymbol{r})$.

We have already encountered such cases in Chap. 3 when we studied transmission and reflection at a one-dimensional potential barrier. The solutions of (4.2) can no longer be labeled by an integer n, since they form a continuous set. In addition, as we noticed in Chap. 3, they are not square integrable and they cannot represent physical states of the particle. One must form wave packets in order to obtain physical states (the sum in (4.3) is then replaced by an integral).

As we noted in Chap. 3, a great deal of physical information can nevertheless be extracted from the expression for the stationary solutions ψ_α themselves, even though they are not eligible to be physical states for the particle. In one dimension, one can deduce, for instance, the reflection and transmission coefficients of a potential barrier as a function of the incident energy. In three dimensions, the solution of scattering problem proceeds along the same general lines although it is technically more complicated. From a knowledge of the ψ_α's, one can derives scattering amplitudes and cross sections, which are the relevant physical quantities in scattering processes. This type of problem will be analyzed in Chap. 18.

4.2 The One Dimensional Harmonic Oscillator

Our first example is the harmonic oscillator which has a large variety of applications.

4.2.1 Definition and Classical Motion

A harmonic oscillator is a system consisting of a particle of mass m elastically bound to a center x_0, with a restoring force $F = -K(x - x_0)$ proportional to the distance from the center. The coefficient K is the spring constant of the oscillator, and the potential energy reads $V(x) = V_0 + K(x - x_0)^2/2$.

Many systems can be approximated by harmonic oscillators. For instance when a classical particle is in stable equilibrium at $x = x_0$, its potential energy is a minimum. Therefore $\partial V(x)/\partial x|_{x=x_0} = 0$ and a Taylor expansion in the vicinity of x_0 yields

$$V(x) = V_0 + \frac{K}{2}(x - x_0)^2 + C(x - x_0)^3 + \dots . \tag{4.7}$$

For small displacements around x_0 ($|x - x_0| \ll K/C$), the cubic term is negligible and this system can be approximated by a harmonic oscillator.

The classical equation of motion is $m\ddot{x} = -K(x - x_0)$. The motion is sinusoidal with an angular frequency $\omega = \sqrt{K/m}$ independent of the amplitude. For simplicity we shall assume in the following that the origins of energy and position have been chosen such that $V_0 = 0$ and $x_0 = 0$. The total energy (kinetic + potential) of the classical particle is then

$$E = \frac{1}{2}m\dot{x}^2 + \frac{1}{2}m\omega^2 x^2 . \tag{4.8}$$

This energy is always positive and there are only bound states, since the potential tends to infinity as $|x| \to \infty$.

4.2.2 The Quantum Harmonic Oscillator

In the quantum problem the Hamiltonian has the form

$$\hat{H} = \frac{\hat{p}_x^2}{2m} + \frac{1}{2}m\omega^2\hat{x}^2 , \tag{4.9}$$

and we want to solve the eigenvalue equation

$$\left(-\frac{\hbar^2}{2m}\frac{\mathrm{d}^2}{\mathrm{d}x^2} + \frac{1}{2}m\omega^2 x^2\right)\psi(x) = E\,\psi(x) . \tag{4.10}$$

Since the potential tends to infinity when $x \to \pm\infty$, there are only bound states in this problem. The only relevant values of the energy E are those for which the solutions are square integrable.

In the quantum problem, the combination of the physical parameters of the problem m and ω together with Planck's constant $\hbar$ yields the natural scales for energies and lengths. These scales are $\hbar\omega$ and $a = \sqrt{\hbar/(m\omega)}$, respectively. As a consequence, it is natural to work with dimensionless quantities ε and y, defined as

$$\varepsilon = \frac{E}{\hbar\omega} , \qquad y = \frac{x}{a} , \tag{4.11}$$

so that the time-independent Schrödinger equation reads

$$\frac{1}{2}\left(y^2 - \frac{\mathrm{d}^2}{\mathrm{d}y^2}\right)\phi(y) = \varepsilon\,\phi(y) , \tag{4.12}$$

where we have put $\phi(y) = \psi(x)\sqrt{a}$.

This differential equation is well known in mathematics. Its square integrable solutions are proportional to the Hermite functions

$$\phi_n(y) = c_n\,\mathrm{e}^{-y^2/2}\,H_n(y) , \tag{4.13}$$

where $c_n = \left(\sqrt{\pi}\,2^n\,n!\right)^{-1/2}$ and where $H_n(y)$ is a polynomial of degree n, containing only even powers of y if n is even, and odd powers of y if n is odd, defined by

$$H_n(y) = (-1)^n\,\mathrm{e}^{y^2}\,\frac{\mathrm{d}^n}{\mathrm{d}y^n}\left(\mathrm{e}^{-y^2}\right) .$$

For instance

$$H_0(y) = 1, \quad H_1(y) = 2y, \quad H_2(y) = 4y^2 - 2, \quad H_3(y) = 8y^3 - 12y .$$

The corresponding eigenvalues are

$$\varepsilon_n = n + \frac{1}{2} , \qquad n \text{ nonnegative integer} . \tag{4.14}$$

Consequently, the quantized energy levels of the one-dimensional harmonic oscillator are

$$E_n = \left(n + \frac{1}{2}\right) \hbar\omega \, . \tag{4.15}$$

The additive constant $\hbar\omega/2$, called the *zero-point energy*, is crucial in order to satisfy the Heisenberg uncertainty relations, as shown in Exercise 4.1.

The normalized eigenfunctions $\psi_n(x)$ are

$$\psi_n(x) = \frac{\pi^{-1/4}}{\sqrt{2^n \, n! \, a}} \mathrm{e}^{-x^2/2a^2} \, H_n(x/a) \, . \tag{4.16}$$

These functions, the first four of which are plotted in Fig. 4.3 for the first four values of n, are real and orthogonal, i.e.

$$\int \psi_n^*(x) \, \psi_{n'}(x) \, \mathrm{d}x = \delta_{n,n'} \, . \tag{4.17}$$

Using the definition (4.13), one can check that these functions satisfy the following recursion relations:

$$x\sqrt{2} \, \psi_n(x) = a\sqrt{n+1} \, \psi_{n+1}(x) + a\sqrt{n} \, \psi_{n-1}(x) \, , \tag{4.18}$$

$$a\sqrt{2} \, \frac{\mathrm{d}}{\mathrm{d}x}\psi_n(x) = \sqrt{n} \, \psi_{n-1}(x) - \sqrt{n+1} \, \psi_{n+1}(x) \, . \tag{4.19}$$

These relations are very useful in practice, as we shall see. They give the action of the operators $\hat{x}$ and $\hat{p}_x = -\mathrm{i}\hbar\partial/\partial x$ on the set $\psi_n(x)$.

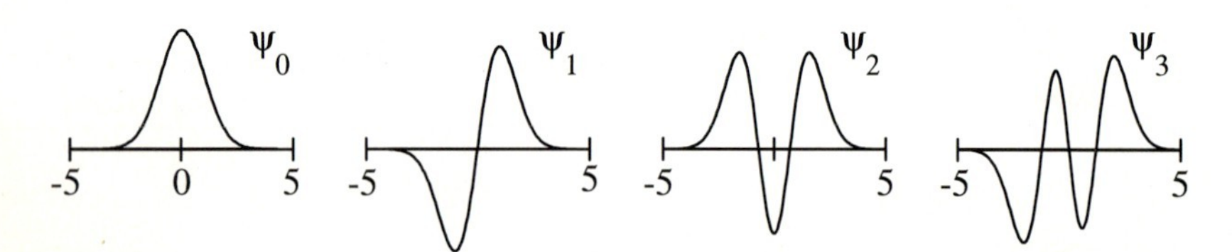

Fig. 4.3. The first four Hermite functions (abscissa x/a): $\psi_0(x)$ is a Gaussian, $\psi_1(x)$ is this Gaussian multiplied by $\sqrt{2}x/a$, etc.

Remark. We can separate the eigenfunctions $\psi_n(x)$ into two classes: the symmetric solutions $\psi_n(x) = \psi_n(-x)$ obtained for n even (see, e.g., (4.13)), and the antisymmetric solutions $\psi_n(x) = -\psi_n(-x)$ obtained for n odd. This originates from the invariance of $\hat{H}$ under the transformation $x \to -x$ (see (4.9)). Therefore, if $\psi(x)$ is a solution of (4.10), $\psi(-x)$ is also a solution with the same eigenvalue. Consequently, the symmetric and antisymmetric functions $\psi(x) \pm \psi(-x)$ are either solutions corresponding to the same eigenvalue or identically zero. This is the first example of a very important property in quantum mechanics, which we shall meet again. If the Hamiltonian possesses invariance properties, these are reflected in the symmetry properties of the eigenfunctions.

4.2.3 Examples

Molecular Physics. Consider a diatomic molecule such as CO, which we encountered in Sect. 1.1. Besides the rotation of the molecule, the two atoms can vibrate with respect to one another in their center-of-mass frame. These atoms are bound by a chemical-binding force. Let x be the distance between the two nuclei. The potential $V(x)$ from which the force is derived is difficult to calculate exactly. However, we can certainly approximate it by the shape represented in Fig. 4.4. The potential must go to infinity if x tends to zero (in which case the two atoms are on top of each other). Also, $V(x)$ goes to a constant if the atoms are far apart. Since the system is bound, the potential has some minimum at the classical equilibrium position x_0. We therefore replace the "true" potential $V(x)$ by a parabola (dotted line in Fig. 4.4). Intuitively, we expect that this will give a good approximation for the energy levels whose wave functions are concentrated in the region where $V(x)$ and its parabolic approximation are close.

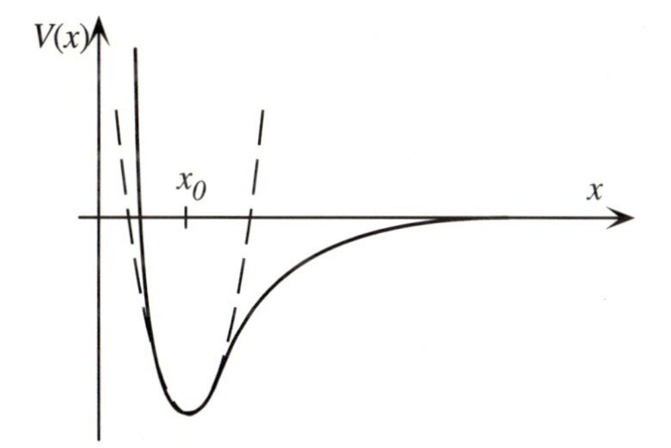

Fig. 4.4. Molecular potential (*full line*) and its harmonic approximation (*dashed line*)

Black-Body Radiation and Planck Oscillators. The classical model of an oscillator formed by an elastically bound charge, due to H.A. Lorentz, was well known to physicists at the end of the 19th century. As early as 1895, Planck had tried to tackle the problem of the thermodynamic equilibrium between electromagnetic radiation and a set of such oscillators forming the inner surface of a closed container (an oscillator of frequency ν absorbs and emits light at this frequency ν). In order to interpolate between the two regimes of low and high frequencies of the spectral distribution of the radiation inside the container, Planck had introduced an empirical two-parameter formula relating the entropy and the internal energy of the radiation field. From this formula, he was able to obtain a relation between the energy and the entropy of the oscillators. Planck then realized that he could fix these two parameters using Boltzmann's statistical theory. Indeed, in the first chapter of his 1877 memoir, Boltzmann had considered the "academic" case of the equilibrium distribution of a system of N molecules with a total energy E split into n

discrete and *equal* energy amounts E/n. Assuming that for an oscillator of frequency ν, the discrete energy amounts were equal to $h\nu$, Planck calculated the value of the fundamental constant h, and obtained his celebrated formula for black-body radiation. To some extent, Max Planck had guessed that the quantization of the energy of an oscillator occurs in integer multiples of $h\nu$.

Trapping Charged Particles. A *Penning trap* consists in a superposition of a uniform magnetic field $\boldsymbol{B}$ and a quadrupole electric field. In such a device, a charged particle can be confined around the center of the quadrupole by harmonic forces. In this way one can construct an artificial atom, called "geonium". This allows one to measure very precisely constants such as the electron magnetic moment, the fine structure constant (see Chap. 11) and the ratio of the proton and electron masses $m_{\mathrm{p}}/m_{\mathrm{e}}$.

Quantization of a Field. A crystal containing N atoms may be considered as a set of $3N$ harmonic oscillators. Also, the classical stationary electromagnetic waves in a container with reflecting walls can be shown to be equivalent to a set of harmonic oscillators. These are the starting points of quantum field theory, which gives rise to the concepts of *phonons* for the vibrations of the crystal and of *photons* in the case of the electromagnetic field. The harmonic oscillator is one of the foundation stones of relativistic quantum field theory.

4.3 Square-Well Potentials

In the rest of this chapter, we shall consider piecewise constant potentials, for which a simple analytical determination of the bound states can be performed.

4.3.1 Relevance of Square Potentials

Although the use of such potentials might seem at first sight quite academic it is worth emphasizing that they constitute accurate models of reality in many physical situations. Let us mention two of them.

- Nuclear forces are very strong but they have a *short range*, i. e. they act only at small distances. The corresponding potentials are quite different from the Coulomb potential $qq'/4\pi\epsilon_0 r$. A first approximation can be obtained by using potentials of the form

$$V = V_0 \quad \text{for } 0 < r \leq r_0\,, \qquad V = 0 \quad \text{for } r > r_0 \quad (V_0 < 0)\,,$$

  where r_0 is of the order of 10^{-15} m, i. e. the dimensions of a nucleus. In many cases, with an appropriate adjustment of the parameters V_0 and r_0, such potentials give a good account of nuclear phenomena at low energies.

- In modern microelectronics, such simple piecewise constant potentials also have numerous applications (see Fig. 4.5). When an electron moves in a semiconductor such as GaAs or GaAlAs, it is subject to a constant potential whose value depends on the nature of the semiconductor. Consequently, by constructing "sandwiches" of alternating thin layers of different semiconductors, one can form square wells and barriers with typical widths of 2 to 5 nm. The confinement of electrons in these domains, named *quantum wells*, has led to unprecedented technological developments in electronics and in computer technologies. Such devices are also very promising in the development of optoelectronics; the transitions between the corresponding bound states ($\Delta E \sim 50$ to 200 meV) are located in the infrared part of the spectrum.

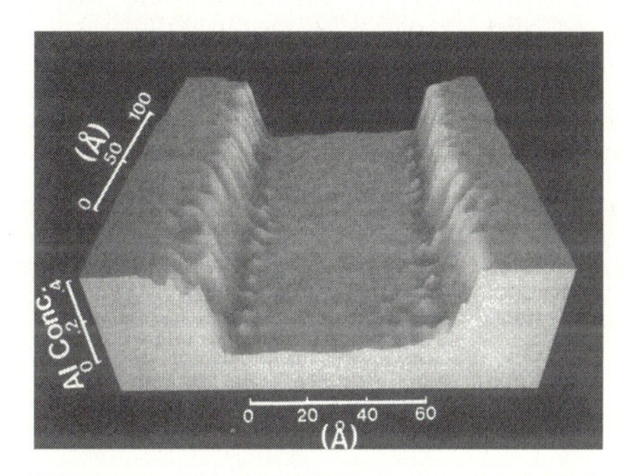

Fig. 4.5. Sandwich composed of AlGaAs – GaAs – AlGaAs. The central part, consisting of GaAs, is 6 nm wide. The vertical scale indicates the concentration of aluminum (from 0 to 40%), which controls the potential (averaged over a distance of the order of the lattice spacing) seen by a conduction electron. Photograph courtesy of Abbas Ourmazd, ATT Bell Labs

4.3.2 Bound States in a One-Dimensional Square-Well Potential

Consider a symmetric square-well potential of depth V_0 and width $2a$ (Fig. 4.6a). In regions I′ ($x < -a$) and I ($x > a$), the potential is constant, equal to V_0. Inside the well ($-a \leq x \leq a$), the potential is zero (this is a mere choice of the origin of the energy). The eigenvalue E of the Hamiltonian which we want to calculate is therefore the kinetic energy inside the potential well. We are only interested here in bound states, such that $0 < E < V_0$ (the classical analog is a particle confined inside the potential well).

In a region where the potential is a constant, the time-independent Schrödinger equation (4.2) boils down to

$$\psi'' + \frac{2m}{\hbar^2}(E - V)\psi = 0 . \tag{4.20}$$

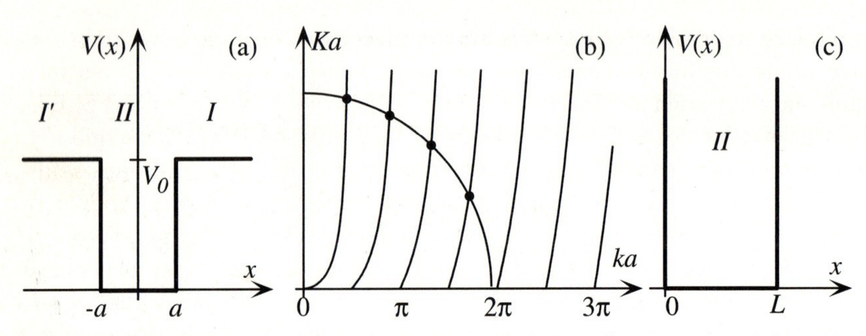

Fig. 4.6. Square well potential: (**a**) form of the potential well; (**b**) graphical determination of the energy levels; (**c**) the limit of an infinite square well

This equation has solutions whatever the value of E. The solutions have different forms according to the sign of $E - V$:

1. If $E - V > 0$, the solutions are sinusoids.
2. If $E - V < 0$, the solutions are exponentials.

In the present case $V = 0$ for $|x| \leq a$ and $V = V_0$ for $|x| > a$. Consequently, for $0 < E < V_0$ the solutions of (4.2) are sinusoids in the middle region II, increasing exponentials in I′ and decreasing exponentials in I, as follows

$$
\begin{aligned}
\text{I}' \qquad & \psi(x) = D\,\mathrm{e}^{Kx}\,, \\
\text{II} \qquad & \psi(x) = A\,\sin kx + B\,\cos kx\,, \\
\text{I} \qquad & \psi(x) = C\,\mathrm{e}^{-Kx}\,,
\end{aligned}
\tag{4.21}
$$

where $K = \sqrt{2m(V_0 - E)}/\hbar$ and $k = \sqrt{2mE}/\hbar$. The constants A, B, C and D are to be determined from the continuity of ψ and ψ' at $\pm a$. Note that we discard solutions with an increasing exponential in I or a decreasing exponential in I′, since a contribution of such a function to the expansion (4.3) would not lead to solutions which can be normalized.

The continuity of the function $\psi(x)$ at $x = \pm a$ yields

$$A \sin ka + B \cos ka = C\mathrm{e}^{-Ka}\,, \tag{4.22}$$

$$-A \sin ka + B \cos ka = D\mathrm{e}^{-Ka}\,. \tag{4.23}$$

Similarly the continuity of the derivative of $\psi(x)$ at $\pm a$ gives

$$A\,k \cos ka - Bk \sin ka = -K\;C\mathrm{e}^{-Ka}\,, \tag{4.24}$$

$$A\,k \cos ka + Bk \sin ka = K\;D\mathrm{e}^{-Ka}\,. \tag{4.25}$$

This set of four equations has no solutions for which the two coefficients A and B are both different from 0. If such a solution existed the two relations $k \cot ka = -K$ and $k \tan ka = K$ would be satisfied simultaneously and the elimination of K between the two would yield $\tan^2 2ka = -1$, which is absurd.

Therefore we classify the solutions into two categories:

$$
\begin{aligned}
&A = 0 \quad \text{and} \quad C = D\,, \qquad k \tan ka = K \quad \text{(even solutions)}\,;\\
&B = 0 \quad \text{and} \quad C = -D \quad k \cot ka = -K \text{ (odd solutions)}\,.
\end{aligned}
\tag{4.26}
$$

In exactly the same manner as for the harmonic oscillator, the potential satisfies $V(x) = V(-x)$, and the Hamiltonian is invariant under the transformation $x \to -x$. We can therefore classify its eigenfunctions according to their parity.

The conditions (4.26) express a quantization of k, and therefore of the energy. These transcendental equations have a simple graphical solution (Fig. 4.6b). We must have

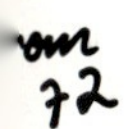

$$K^2a^2 + k^2a^2 = \frac{2ma^2V_0}{\hbar^2} = \text{constant}\,. \tag{4.27}$$

In the plane $\{ka,\, Ka\}$, this is the equation of a circle. We must find the intersections of this circle with the curves $Ka = ka \tan ka$ and $Ka = -ka \cot ka$. These intersections form a finite set and they relate alternately to even and odd solutions. Suppose a is fixed; the number of intersections increases with V_0. If V_0 is smaller than a minimum value given by

$$\frac{a\sqrt{2mV_0}}{\hbar} < \frac{\pi}{2} \qquad \Rightarrow \qquad V_0 < \frac{\pi^2\hbar^2}{8ma^2}\,, \tag{4.28}$$

there is only one bound state.

A theorem due to Sturm and Liouville states that one can classify the levels in order of increasing energy using the number of nodes of their wave function. The wave function of the ground state has no node, the wave function of the first excited state has a single node, and so on. This theorem holds if the potential is sufficiently regular (which will be the case in all of this book).

4.3.3 Infinite Square Well

The limit where V_0 becomes infinite is particularly simple and interesting. In this limit, the particle is confined in region II, where the potential is zero; $V(x)$ is infinite in regions I and I$'$.

For convenience, we place this infinite square well between $x = 0$ and $x = L$ as shown in Fig. 4.6c. The calculation is straightforward: the coefficient K introduced in Sect. 4.3.2 is infinite and we must have $\psi(x) = 0$ in regions I and I$'$. By continuity, we deduce

$$\psi(0) = 0 \qquad \psi(L) = 0\,. \tag{4.29}$$

The eigenfunctions of the Hamiltonian which satisfy these boundary conditions are

$$\psi_n(x) = A\,\sin(n\pi x/L)\,, \qquad n \text{ integer } > 0\,, \tag{4.30}$$

and the normalization condition gives $A = \sqrt{2/L}$.

The energy levels are given by

$$E_n = n^2 \frac{\pi^2\hbar^2}{2mL^2}, \qquad n \text{ integer } > 0 . \tag{4.31}$$

In this case, only the wave function $\psi(x)$ is continuous at $x = 0$ and $x = L$. Contrary to what happens for a finite potential well V_0, $\psi'(x)$ is discontinuous at these points.[1] Notice that the quantization condition can be written as $kL = n\pi$, where k is the wave number defined by $E = \hbar^2 k^2/2m$: the quantization of energy appears as an elementary stationary-wave phenomenon.

4.3.4 Particle in a Three-Dimensional Box

We now extend the previous case to three dimensions, and consider the problem of a particle of mass m confined inside a parallelepiped of sides L_1, L_2, L_3. The confining potential can be written

$$V^{(3)}(x, y, z) = V(x) + V(y) + V(z) , \tag{4.32}$$

where

$$V(x_i) = 0 \quad \text{if} \quad 0 \leq x_i \leq L_i ,$$
$$V(x_i) = \infty \quad \text{if} \quad x_i < 0 \text{ or } x_i > L_i ,$$

and $x_i = x, y, z$ for $i = 1, 2, 3$.

Separation of the Variables. We want to solve the eigenvalue equation

$$\hat{H}\,\Psi(\boldsymbol{r}) = E\,\Psi(\boldsymbol{r}) , \tag{4.33}$$

where

$$\hat{H} = -\frac{\hbar^2}{2m}\left(\frac{\partial^2}{\partial x^2} + \frac{\partial^2}{\partial y^2} + \frac{\partial^2}{\partial z^2}\right) + V^{(3)}(x, y, z) . \tag{4.34}$$

We look for particular solutions which are factorized as follows:

$$\Psi(\boldsymbol{r}) = \psi_1(x)\,\psi_2(y)\,\psi_3(z) . \tag{4.35}$$

One can verify that such a factorized function is a solution of (4.33) with the eigenvalue E if the following conditions are fulfilled:

$$-\frac{\hbar^2}{2m}\psi_i''(x_i) = [E_i - V(x_i)]\,\psi_i(x_i) \tag{4.36}$$

for $i = 1, 2, 3$, and

$$E = E_1 + E_2 + E_3 . \tag{4.37}$$

We then recover a one-dimensional problem and we can use the solutions that we found in the previous subsection (Sect. 4.3.3). One can prove that this procedure determines *all* the eigenvalues and that it provides a *basis of eigenfunctions.*

[1] The continuity of ψ and the discontinuity of ψ' at the edges of the potential well can be obtained directly by taking the limit $V_0 \to +\infty$ in the solutions (4.26).

Solution of the Eigenvalue Equation. In order to obtain a solution of (4.33), we can combine three solutions of the type (4.30). Such a solution depends on three quantum numbers n_1, n_2 and n_3 (strictly positive integers) and on the dimensions L_1, L_2, L_3 of the volume under consideration:

$$\Psi_{n_1,n_2,n_3}(\boldsymbol{r}) = \frac{\sqrt{8}}{\sqrt{L_1 L_2 L_3}} \sin\left(\frac{n_1 \pi x}{L_1}\right) \sin\left(\frac{n_2 \pi y}{L_2}\right) \sin\left(\frac{n_3 \pi z}{L_3}\right), \quad (4.38)$$

$$E = E_{n_1,n_2,n_3} = \frac{\hbar^2 \pi^2}{2m} \left(\frac{n_1^2}{L_1^2} + \frac{n_2^2}{L_2^2} + \frac{n_3^2}{L_3^2}\right). \quad (4.39)$$

Degeneracies. It may happen that for a given eigenvalue E, there are *several* different eigenfunctions. In the present case, for instance, if $L_1 = L_2$, the two solutions obtained by interchanging n_1 and n_2 in (4.38) correspond to the same eigenvalue E. When such a situation occurs, we say that the eigenvalue is *degenerate*, and the *degree of degeneracy* is equal to the dimension of the eigensubspace.

Remark. Despite their great simplicity, the results of this calculation can be used in many cases. For instance, molecules in a container cannot penetrate through the walls of the container. Free electrons in a conductor are confined by the attractive potential of the crystal lattice. The neutrons of a neutron star are confined by the gravitational field inside a sphere of radius $\sim 10\,\text{km}$, for a total mass of the order of the solar mass (see Sect. 19.3.2).

4.4 Periodic Boundary Conditions

In many circumstances, it is convenient to use plane waves in order to describe the states of free particles, or of particles confined in a large box. This happens, for instance, in statistical mechanics when dealing with gases, in which particles move freely within some volume. It is also convenient to use plane waves when one is considering collision processes between particles (Chap. 18). Before and after the collision, the particles have well-defined energies and momenta. The question arises as to how to deal with such plane waves, which cannot be normalized. A convenient method consists in considering a problem similar to the infinite potential well, but where *periodic* boundary conditions are imposed on the wave functions.

4.4.1 A One-Dimensional Example

Consider again the one-dimensional problem studied in Sect. 4.3.3, and suppose that we replace the boundary conditions $\psi(0) = 0$ and $\psi(L) = 0$ by the following ones:

$$\psi(L) = \psi(0) \qquad \text{and} \qquad \psi'(L) = \psi'(0). \quad (4.40)$$

These *periodic boundary conditions* correspond to a particle confined on a circle of circumference L, rather than a segment of length L. In practice, they are used in situations where the predictions for physical quantities (equations of state in statistical physics, cross sections in scattering theory) do not depend on L in the limit $L \to \infty$. The mathematical distinction between confinement on a circle and on a segment is then unimportant.

It is possible to find a set of normalized eigenfunctions of the momentum operator $\hat{p} = -\mathrm{i}\hbar\, \partial/\partial x$ satisfying these boundary conditions. These eigenfunctions are

$$\psi_n(x) = \frac{1}{\sqrt{L}}\, \mathrm{e}^{\mathrm{i}p_n\, x/\hbar}\,, \tag{4.41}$$

where the eigenvalue p_n of $\hat{p}$ associated with the state ψ_n is

$$p_n = \frac{2\pi\hbar}{L} n\,, \qquad n \text{ negative or positive integer}\,. \tag{4.42}$$

Each wave function ψ_n is also an eigenfunction of the kinetic-energy operator $\hat{p}^2/2m$, with the eigenvalue

$$E_n = \frac{p_n^2}{2m} = \frac{4\pi^2\hbar^2}{2mL^2} n^2\,. \tag{4.43}$$

The wave functions (4.41) are therefore normalizable eigenfunctions of both the momentum and the energy (whereas in a closed box, for instance, the energy eigenfunctions are not eigenfunctions of the momentum).

Counting the Number of Quantum States. In statistical physics as well as in scattering theory, one often expresses the prediction $\mathcal{P}$ for a physical quantity as a sum of a given function $f(p)$ over possible momenta, e. g.

$$\mathcal{P} = \sum_n f(p_n)\,. \tag{4.44}$$

Suppose that L is very large in the sense that the splitting between two successive possible momenta $\Delta p = 2\pi\hbar/L$ is very small compared with the typical momentum scale in the problem under consideration (the thermal momentum, for instance). It is then possible to transform a discrete summation such as (4.44) over microscopic states into an integral, by introducing the number $\mathrm{d}N$ of energy-momentum quantum eigenstates whose momenta are located in a neighborhood $\mathrm{d}p$ of some given value p. This number can be obtained readily from (4.42) as $\mathrm{d}N/\mathrm{d}p = L/2\pi\hbar$. One obtains

$$\mathcal{P} \simeq \frac{L}{2\pi\hbar} \int f(p)\, \mathrm{d}p\,. \tag{4.45}$$

4.4.2 Extension to Three Dimensions

The extension to three dimensions is straightforward. In a cube of side L with periodic boundary conditions for the three variables (x, y, z), the normalized eigenstates of the momentum are

$$\psi_{\boldsymbol{n}}(\boldsymbol{r}) = \frac{1}{\sqrt{L^3}}\, \mathrm{e}^{\mathrm{i}\boldsymbol{p}_{\boldsymbol{n}}\cdot\boldsymbol{r}/\hbar}\,, \qquad \boldsymbol{p}_{\boldsymbol{n}} = \frac{2\pi\hbar}{L}\,\boldsymbol{n}\,, \tag{4.46}$$

where $\boldsymbol{n}$ stands for a triplet (n_1, n_2, n_3) of positive or negative integers. These momentum eigenstates are orthogonal to one another:

$$\int_{L^3} \psi^*_{\boldsymbol{n}}(\boldsymbol{r})\, \psi_{\boldsymbol{n}'}(\boldsymbol{r})\, \mathrm{d}^3r = \delta_{n_1,n_1'}\, \delta_{n_2,n_2'}\, \delta_{n_3,n_3'}\,. \tag{4.47}$$

Density of States. As done previously in the one-dimensional case, we can replace a discrete sum over the momentum eigenstates

$$\mathcal{P} = \sum_{\boldsymbol{n}} f(\boldsymbol{p}_{\boldsymbol{n}}) \tag{4.48}$$

by an integral over $\boldsymbol{p}$ if the function $f(\boldsymbol{p})$ varies slowly on the scale $2\pi\hbar/L$. The number of independent quantum states (i. e. eigenstates of $\hat{\boldsymbol{p}}$) in a volume d^3p of momentum space is

$$\mathrm{d}^3N = \frac{L^3}{(2\pi\hbar)^3}\, \mathrm{d}^3p\,, \tag{4.49}$$

so that

$$\mathcal{P} \simeq \frac{L^3}{(2\pi\hbar)^3} \int f(\boldsymbol{p})\, \mathrm{d}^3p\,. \tag{4.50}$$

Quite often one meets situations where the function $f(\boldsymbol{p})$ is a function only of the energy $E = p^2/2m$: $f(\boldsymbol{p}) \equiv g(E)$. In this case the integral (4.50) can be written in spherical coordinates and integrated over the polar angles defining the direction of $\boldsymbol{p}$. The result can be cast into the form

$$\mathcal{P} \simeq \int_0^{+\infty} g(E)\, \rho(E)\, \mathrm{d}E\,. \tag{4.51}$$

The *density of states* $\rho(E)$ is defined as the ratio $\mathrm{d}N/\mathrm{d}E$, where $\mathrm{d}N$ is the number of independent quantum states in the (narrow) energy slice $\mathrm{d}E$:

$$\rho(E) = \frac{\mathrm{d}N}{\mathrm{d}E} = \frac{mL^3\sqrt{2mE}}{2\pi^2\hbar^3}\,. \tag{4.52}$$

For particles of spin s (see Chap. 12), these formulas generalize to $\mathrm{d}^3N = (2s+1)\,(L/2\pi\hbar)^3\, \mathrm{d}^3p$ and $\rho(E) = (2s+1)mL^3\sqrt{2mE}/(2\pi^2\hbar^3)$.

4.4.3 Introduction of Phase Space

In classical mechanics the state of a particle is defined at time t by a point in a six-dimensional space called phase space. The coordinates of this point are the components of the position and the momentum x, y, z, p_x, p_y, p_z. We wish to transpose to phase space the result (4.49). This formula indicates that the number of independent *states* is equal to the accessible volume of phase space $(L^3)\,(\Delta p_x \Delta p_y \Delta p_z)$ divided by the cube of Planck's constant $h = 2\pi\hbar$.

Assuming all necessary care is taken about the orders of magnitude which must be satisfied, we can generalize this result. In an arbitrary volume of phase space

$$\Delta^6 \mathcal{V} = (\Delta x \Delta y \Delta z) \times (\Delta p_x \Delta p_y \Delta p_z) \ , \tag{4.53}$$

the number of independent quantum states is given by the relation

$$\Delta^6 \mathcal{N} = \frac{\Delta^6 \mathcal{V}}{(2\pi\hbar)^3} \, . \tag{4.54}$$

This formula is essential in statistical mechanics. We have derived it in the context of periodic quantization conditions, but its validity can be established on much more general grounds.

The accuracy of this formula can be tested on the simple case of the one-dimensional harmonic oscillator of angular frequency ω. Let us estimate, for instance, the number $\mathcal{N}(E_0)$ of independent states whose energy is lower than some given value E_0, much larger than $\hbar\omega$. The accessible phase space domain is the area within an ellipse in the x–p plane:

$$\mathcal{N}(E_0) = \int_{E(x,p)\leq E_0} \frac{\mathrm{d}x\,\mathrm{d}p}{2\pi\hbar} \, , \quad \text{where} \quad E(x,p) = \frac{p^2}{2m} + \frac{1}{2} m\omega^2 x^2 \, . \tag{4.55}$$

The half-axes of the ellipse in the x and p directions are $(2E_0/m\omega^2)^{1/2}$ and $(2mE_0)^{1/2}$, respectively, so that the one-dimensional version of (4.54) then yields

$$\mathcal{N}(E_0) \simeq \frac{E_0}{\hbar\omega} \, . \tag{4.56}$$

This is in excellent agreement with the exact result deduced from (4.15), which says that $\mathcal{N}(E_0)$ is the integer nearest to $E_0/\hbar\omega$.

4.5 The Double Well Problem and the Ammonia Molecule

We shall now consider a problem which at first sight is similar to the square well studied above, but will turn out to be much more subtle and will bring new physics.

4.5.1 Model of the NH_3 Molecule

The ammonia molecule NH_3 has the shape of a pyramid (Fig. 4.7a), where the nitrogen atom is at the apex and the three hydrogen atoms form the base in the shape of an equilateral triangle. The plane of the three hydrogen atoms is denoted P and the perpendicular to this plane passing through the nitrogen atom is denoted $\mathcal{D}$. The distance x represents the position of the intersection of P with $\mathcal{D}$. The position of the nitrogen atom is chosen as the origin of the x axis. For low excitation energies, the molecule preserves its pyramidal shape and the nitrogen atom remains fixed.

Qualitatively, the variations of the potential energy $\mathcal{V}(x)$ with x are as follows. At the equilibrium position $x = b$, $\mathcal{V}(x)$ has a minimum (Fig. 4.7b). If we force x to become smaller, the energy increases; it goes through a maximum for $x = 0$, which corresponds to an unstable state where the four atoms are in the same plane. If x becomes negative, the molecule is turned over like an umbrella in the wind. For symmetry reasons there exists another minimum for $x = -b$ and the potential energy satisfies $\mathcal{V}(x) = \mathcal{V}(-x)$.

In the following we replace the actual potential $\mathcal{V}(x)$ by the simplified square-well potential $V(x)$ represented by dotted lines in Fig. 4.7b. For this potential, which reproduces the main interesting features of $\mathcal{V}(x)$, we study the quantum motion of a "particle" representing the collective motion of the three hydrogen atoms, assuming that they stay in the same plane. The mass m of the particle is equal to $3\,m_{\mathrm{H}}$, where m_{H} is the mass of a hydrogen atom.

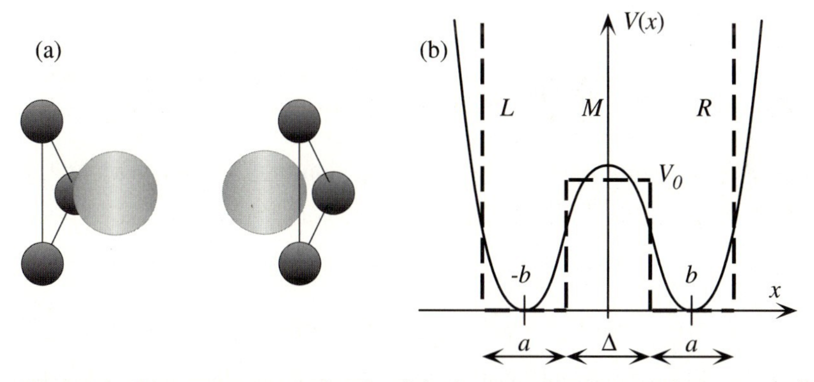

Fig. 4.7. The ammonia molecule: (**a**) the two classical configurations; (**b**) the actual potential (*full line*) and the simplified potential (*dotted line*) which describes the inversion of the molecule

4.5.2 Wave Functions

Following the same procedure as in Sect. 4.3, it is straightforward to find the stationary states in this problem. We concentrate on the case $E < V_0$, for which the classical motion of the "particle" is confined in one of the potential wells (left or right), i.e. the molecule cannot turn over classically. The solutions to the quantum problem are sinusoids in the regions L and R and exponentials in the middle region M. Since the wave functions have to vanish for $x = \pm(b + a/2)$, the eigenstates of the Hamiltonian can be written

$$\begin{aligned} \psi(x) &= \pm\lambda \sin k(b + a/2 + x) && \text{region L,} \\ \psi(x) &= \left\{ \begin{array}{ll} \mu \cosh Kx & \text{symmetric solution} \\ \mu \sinh Kx & \text{antisymmetric solution} \end{array} \right\} && \text{region M,} \\ \psi(x) &= \lambda \sin k(b + a/2 - x) && \text{region R,} \end{aligned} \tag{4.57}$$

where we set, as previously, $k = \sqrt{2mE}/\hbar$ and $K = \sqrt{2m(V_0 - E)}/\hbar$. These two types of solutions are represented in Fig. 4.8.

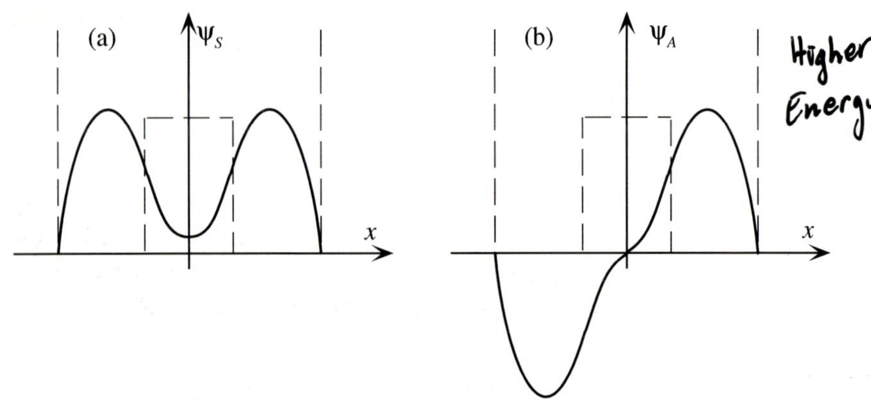

Fig. 4.8. Symmetric solution (**a**) and antisymmetric solution (**b**), in the symmetric double-well model of the ammonia molecule

The continuity equations for the wave function and its derivative at the points $x = \pm(b - a/2)$ lead to the conditions

$$\tan ka = -\frac{k}{K} \coth K(b - a/2) \qquad \text{for a symmetric solution } \psi_{\mathrm{S}}\,,$$

$$\tan ka = -\frac{k}{K} \tanh K(b - a/2) \qquad \text{for an antisymmetric solution } \psi_{\mathrm{A}}\,.$$

In order to obtain some physical insight with simple algebra, we consider the case where the ground-state energy E is very small compared with the height V_0 of the potential barrier. This leads to $K \sim \sqrt{2mV_0}/\hbar \gg k$. In addition, we assume that the central potential barrier is wide enough that $K\Delta \gg 1$, where $\Delta = 2b - a$ is the width of this barrier. These assumptions hold in the case of the ammonia molecule, as we shall see below. We then have

$$\tan ka \simeq -\frac{k}{K} \left(1 \pm 2\mathrm{e}^{-K\Delta}\right) \,, \tag{4.58}$$

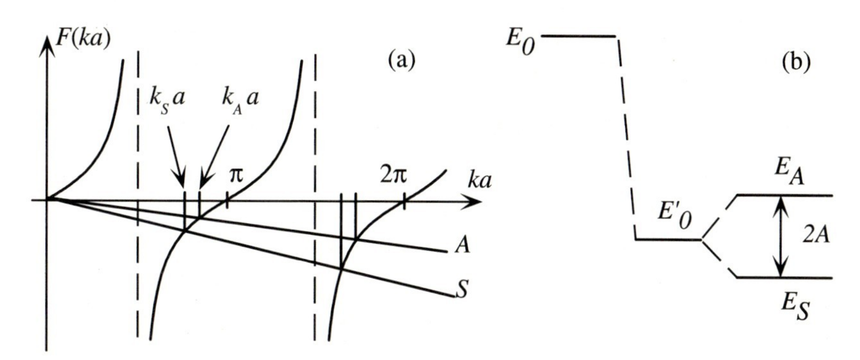

Fig. 4.9. (**a**) Graphical determination of the energy levels in the double well; (**b**) positions of the two first energy levels. They are both lower than the ground-state energy of a single potential well similar to L or R ($E_0 \to E_0'$), and we observe a splitting of these levels ($E_0' \to E_\mathrm{A}$ and E_S) owing to the coupling between the two wells due to quantum tunneling

where the + sign corresponds to ψ_S and the − sign to ψ_A. This equation allows us to calculate the quantized values of ka. These values appear on the graph in Fig. 4.9 as the abscissae of the intersections of the successive branches of $y = \tan ka$ with the two straight lines $y = -\varepsilon_\mathrm{A} ka$ and $y = -\varepsilon_\mathrm{S} ka$. These intersections are located in the vicinity of $ka \sim \pi$. The two constants ε_A and ε_S are

$$\varepsilon_\mathrm{A} = \frac{1}{Ka}\left(1 - 2\mathrm{e}^{-K\Delta}\right) , \qquad \varepsilon_\mathrm{S} = \frac{1}{Ka}\left(1 + 2\mathrm{e}^{-K\Delta}\right) . \tag{4.59}$$

They are close to each other and such that $\varepsilon_\mathrm{A} < \varepsilon_\mathrm{S} \ll 1$, since $Ka \gg ka \sim \pi$.

4.5.3 Energy Levels

We denote by k_S and k_A the two (close) values of k corresponding to the eigenstates ψ_S and ψ_A of lowest energy. The graph in Fig. 4.9 shows the following:

1. The two quantities k_S and k_A are slightly smaller than π/a, which is the lowest value of the wave number in an individual well, similar to L or R, of width a with infinitely high and thick walls.
2. The quantity k_S is slightly smaller than k_A; consequently the respective energies of the two lowest-lying levels

$$E_\mathrm{S} = \hbar^2 k_\mathrm{S}^2/2m , \qquad E_\mathrm{A} = \hbar^2 k_\mathrm{A}^2/2m \tag{4.60}$$

are such that $E_\mathrm{S} < E_\mathrm{A}$.

In the range of parameters considered here ($K \gg k$, $Ka \gg 1$), we find

$$k_{\mathrm{S}} \sim \frac{\pi}{a(1+\varepsilon_{\mathrm{S}})}, \qquad k_{\mathrm{A}} \sim \frac{\pi}{a(1+\varepsilon_{\mathrm{A}})}, \tag{4.61}$$

with ε_{S} and $\varepsilon_{\mathrm{A}} \ll 1$. Putting together (4.58), (4.60) and (4.61), we obtain the mean energy $E_0' = (E_{\mathrm{A}} + E_{\mathrm{S}})/2$:

$$E_0' \simeq \frac{\hbar^2\pi^2}{2ma^2}\left(1 - \frac{2}{Ka}\right). \tag{4.62}$$

The splitting $E_{\mathrm{A}} - E_{\mathrm{S}}$ between these two energy levels will be of particular interest. It is given by

$$E_{\mathrm{A}} - E_{\mathrm{S}} \equiv 2A \simeq \frac{\hbar^2\pi^2}{2ma^2}\left[\frac{1}{(1+\varepsilon_{\mathrm{A}})^2} - \frac{1}{(1+\varepsilon_{\mathrm{S}})^2}\right], \tag{4.63}$$

where

$$A \simeq \frac{\hbar^2\pi^2}{2ma^2}\,\frac{4\mathrm{e}^{-K\Delta}}{Ka}. \tag{4.64}$$

Since K is approximately equal to $\sqrt{2mV_0}/\hbar$, we see that A decreases *exponentially* when the width Δ or the height V_0 of the intermediate potential barrier increases. We also note that A vanishes as $\exp(-\text{const.}/\hbar)$ in the limit $\hbar \to 0$.

4.5.4 The Tunnel Effect and the Inversion Phenomenon

Classically, for $E < V_0$, the plane of the three hydrogen atoms in the molecule is either on the right or on the left. No transition L $\leftrightarrow$ R is possible. There are two ground states of equal energy, one in the L configuration, the other in the R configuration. In contrast, the two lowest-lying energy states of the quantum molecule have different energies. The two corresponding wave functions have well-defined parities: one is symmetric (ψ_{S}), the other is antisymmetric (ψ_{A}). In both cases the probabilities (modulus squared of ψ) that the particle (or the triangle of hydrogen atoms) is on the right and on the left are *equal.*

For both eigenstates ψ_{S} and ψ_{A} the probability density is nonzero in the M region, which is classically forbidden. Again we are facing the possibility that a quantum particle can be located in regions where its total energy is less than the local potential energy. This results in a lowering of the energies of the two lowest eigenstates of the Hamiltonian with respect to the case $V_0 = \infty$. Indeed, in that case there would be two possible ground states for the molecule, corresponding to the L and R configurations (or to any linear combination of these states), with the same energy $E_0 = \hbar^2\pi^2/(2ma^2)$. Because V_0 is finite the molecule sees an effective size of each well (L or R) which is slightly larger than a ($a_{\mathrm{eff}} \sim a + K^{-1}$); this explains the lowering of the mean energy $E_0 \to E_0'$.

This global lowering is followed by a splitting $E_0' \to E_0' \pm A$ into two sublevels. The physical origin of this splitting is the tunneling effect, i. e. the possibility for the particle to cross the potential barrier and pass from one well to the other. We now investigate this very important phenomenon in more detail.

The wave functions $\psi_{\rm S}$ and $\psi_{\rm A}$ are eigenstates of the Hamiltonian. We can combine them to form other physically acceptable states of the system. Two linear combinations are particularly interesting:

$$\psi_{\rm L} = (\psi_{\rm S} - \psi_{\rm A})/\sqrt{2} \qquad \text{and} \qquad \psi_{\rm R} = (\psi_{\rm S} + \psi_{\rm A})/\sqrt{2}\,. \tag{4.65}$$

These wave functions describe states for which the probability density is concentrated nearly entirely on the left for $\psi_{\rm L}$ and on the right for $\psi_{\rm R}$. These correspond to the "classical" configurations, for which the molecule is oriented towards either the left- or the right-hand side (Fig. 4.10).

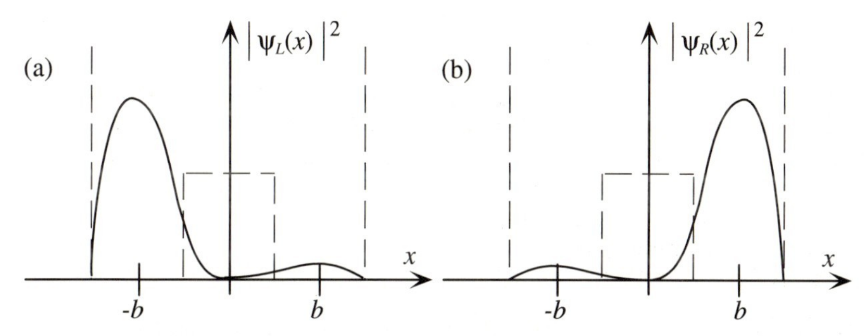

Fig. 4.10. Classical configurations of the ammonia molecule

Consider a wave function $\psi(x,t)$ equal to $\psi_{\rm R}$ at time $t = 0$. It describes a molecule localized in the "right" configuration. Its time evolution is

$$\begin{aligned}\psi(x,t) &= \frac{1}{\sqrt{2}}\left(\psi_{\rm S}(x)\,{\rm e}^{-{\rm i}E_{\rm S}t/\hbar} + \psi_{\rm A}(x)\,{\rm e}^{-{\rm i}E_{\rm A}t/\hbar}\right)\\ &= \frac{{\rm e}^{-{\rm i}E_{\rm S}t/\hbar}}{\sqrt{2}}\left(\psi_{\rm S}(x) + \psi_{\rm A}(x){\rm e}^{-{\rm i}\omega t}\right)\,,\end{aligned} \tag{4.66}$$

where we have introduced the *Bohr frequency* $\hbar\omega = E_{\rm A} - E_{\rm S} = 2A$.

We notice that after a time $t = \pi/\omega = \pi\hbar/(2A)$, the wave function $\psi(x,t)$ is, up to a phase factor, proportional to $\psi_{\rm L}$, and the molecule is in the *left* configuration! At time $t = 2\pi/\omega$ the wave function $\psi(x,t)$ is again proportional to $\psi_{\rm R}$: the molecule is back to the right configuration. In other words the superposition (4.66) represents a state of the molecule which oscillates from right to left at the Bohr frequency $\nu = \omega/2\pi$. The ammonia molecule prepared in a classical configuration at $t = 0$, *turns over* periodically because of quantum tunneling. This phenomenon, called the *inversion* of the

NH_3 molecule, plays a fundamental role in the principle of the ammonia *maser*, which we shall discuss in Chap. 6.

The quantity A controls the frequency at which the transition from one minimum of the potential to the other occurs. Comparing the expression (4.64) for A with the tunneling probability found in Sect. 3.6.4, we notice that the two expressions are very similar, the essential point being the presence of the exponential term. For ammonia, the energy difference $2A$ is small compared with typical binding energies in atomic and molecular physics: $2A \sim 10^{-4}$ eV. The frequency ν and period T of the oscillation are

$$\nu = \frac{\omega}{2\pi} = \frac{2A}{h} \simeq 24\,\mathrm{GHz}\,, \qquad T = \frac{1}{\nu} = \frac{h}{2A} \simeq 4.2 \times 10^{-11}\,\mathrm{s}\,.$$

As we shall see in Chap. 6, the oscillation is associated with the emission or absorption of electromagnetic radiation. The corresponding wavelength is $\lambda = c/\nu = 1.25\,\mathrm{cm}$. This wavelength can be measured with great accuracy, and it constitutes a "fingerprint" of ammonia, which, for instance, allows us to detect the presence of this molecule in the interstellar medium.

4.6 Other Applications of the Double Well

The general formalism we have just developed for the ammonia molecule can be extended to many other symmetric double-well situations. Consider for instance two identical atoms A_1 and A_2 at a distance Δ from one another. An electron sees a double well, sketched in Fig. 4.11, each minimum being centered on one of the atoms. We choose the origin of energy such that $V \to 0$ for $x \to \infty$. If Δ is sufficiently large, one may, to a good approximation, consider that $V \sim 0$ in the middle region between the two atoms.

An electron bound to one atom in a given energy level $E_0 < 0$ must cross a potential barrier of height $|E_0|$ and width Δ in order to jump to the other atom. We want to estimate, in terms of Δ and E_0, the order of magnitude of the typical time T needed for this transition.

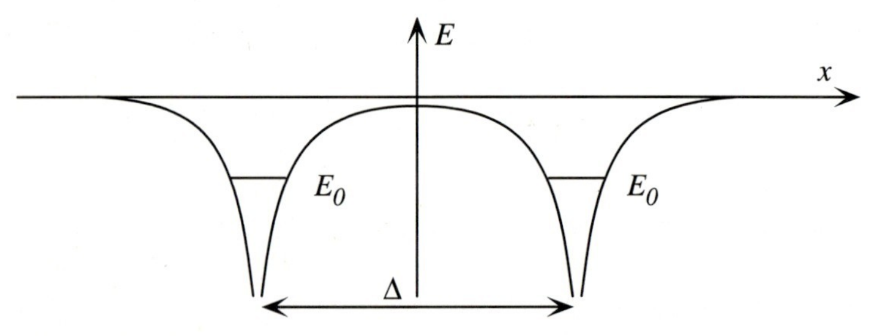

Fig. 4.11. Double well seen by an electron when two atoms are separated by a distance Δ

We shall assume here that the kinetic energy E_k of the bound electron is of the order of $|E_0|$ (for hydrogen atoms, this is exact because of the virial theorem applied to the Coulomb potential). In the exponential related to the tunnel effect, we have $K = \sqrt{2m|E_0|}/\hbar$. Since the electron is bound in an atom, we have $Ka \sim \pi$ (see Chap. 11). The essential result of the previous section, i. e. the exponential dependence of the oscillation frequency on the parameter $K\Delta$, remains valid. Therefore we rewrite (4.64) in the form $A \sim E_k e^{-K\Delta} \sim |E_0| e^{-K\Delta}$, where we neglect a numerical factor of order unity.

In a molecule or in a solid, the interatomic distance is of the order of one or a few angstroms. In a gas at room temperature and atmospheric pressure, the interatomic distance is ~ 30 Å. The least bound electrons in an atom (*valence electrons*) have binding energies of a few eV; we then find

$$\begin{aligned} \text{Solid:}\quad & \Delta = 2\ \text{Å}\,,\ |E_0| = 4\,\text{eV}\,, \qquad && A = 1\ \text{eV}\,;\ T = 10^{-15}\,\text{s}\,,\\ \text{Gas:}\quad & \Delta = 30\,\text{Å}\,;\ |E_0| = 4\,\text{eV} && A = 10^{-12}\,\text{eV}\,;\ T = 10^{-3}\,\text{s}\,. \end{aligned}$$

We see that tunneling is important for valence electrons inside a molecule or in a solid. These electrons jump rapidly from one atom to another, and they are delocalized in the global molecular structure. In contrast, the corresponding phenomenon is completely negligible in gases. Indeed, because of the thermal motion, two given atoms or molecules in a gas remain at a relative distance of the order of 30 Å only for a time shorter than 10^{-10} s. The oscillation associated with the tunneling effect has a period of 10^{-3} s and it cannot have an appreciable effect on a timescale as short as 10^{-10} s. In a gas, it is justified to consider that even the least bound electrons "belong" to a given atom.

The essential ingredient in the above reasoning is the exponential variation of A with Δ and $K = \sqrt{2m|E_0|}/\hbar$. This very large variation explains why, in going from a system to another one which seems similar, the characteristic times may be extremely different. For a system where $K\Delta$ is slightly too large, the oscillation period T can become so incredibly large that tunneling may safely be neglected.

The particularly interesting case of NH_3 and the similar molecules ND_3, PH_3, AsH_3, etc. is treated in detail by Townes and Schawlow,[2] who give more realistic forms of the potentials. Consider, for instance, the passage from NH_3 to AsH_3:

$$\begin{aligned} NH_3:\ & V_0 = 0.25\ \text{eV}\ ,\ b = 0.4\ \text{Å}\ : && \nu_0 = 2.4 \times 10^{10}\,\text{Hz}\,;\\ AsH_3:\ & V_0 = 1.5\ \text{eV}\ ,\ b = 2\ \text{Å}\ : && \nu_0 = 1.6 \times 10^{-8}\,\text{Hz}\,. \end{aligned}$$

A change by a factor 6 in V_0 and a factor 5 in b induces a dramatic decrease of the inversion frequency, by 18 orders of magnitude! The frequency found for AsH_3 corresponds to one oscillation every two years, and its detection

[2] C.H. Townes and A.L. Schawlow, *Microwave Spectroscopy*, Chap. 12, McGraw-Hill, New York (1955).

is completely beyond the reach of current experimental techniques. In other words AsH_3, which seems to differ only moderately from NH_3, behaves as a classical object from the point of view of the tunneling phenomenon that we have considered here, simply because the As atom is ~ 5 times larger than the nitrogen atom.

The stability of systems which do not have definite symmetry properties is frequently encountered on the microscopic scale. Among many examples, there is the case of optical isomers in organic chemistry. The simplest example is the molecule CHClFBr. The tetrahedral structure of the bonding of carbon results in the fact that two nonequivalent configurations exist. They are represented in Fig. 4.12 and are called optical isomers. Such isomers have different optical, chemical and biological properties. The situation for these isomers is similar to the situation we have just described. Such molecules should, in principle, oscillate from one configuration to the other. However, both types of such molecules are perfectly stable in practice. This is due to the fact that the inversion period T is so large that one cannot detect the oscillation.

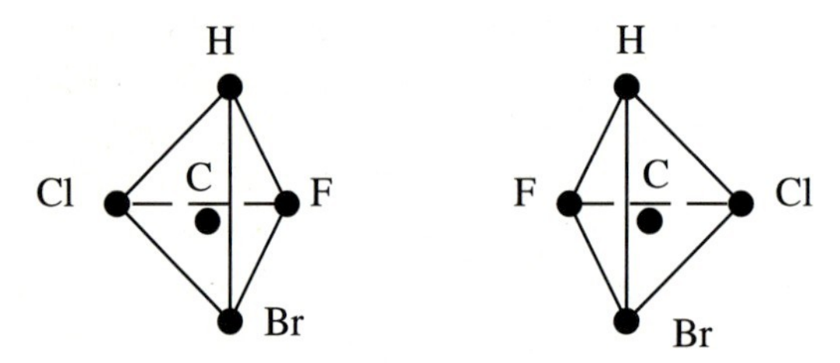

Fig. 4.12. Two optical isomers: can one detect the tunneling oscillation between these two configurations?

Further Reading

- The history of Planck's procedure for explaining the black-body radiation is presented in J. Mehra and H. Rechenberg, *The Historical Development of Quantum Theory* Vol. 1, Chap. 1, Springer, New York (1982).
- The principle and applications of the Penning trap are presented by L.S. Brown and G. Gabrielse, "Geonium theory: physics of a single electron or ion in a Penning trap", Rev. Mod. Phys. **58**, 223 (1986). See also G. Gabrielse, "Ultracold antiprotons", Sci. Am., January 1993; J.-L. Basdevant and J. Dalibard, "The Quantum Mechanics Solver: How to Apply Quantum Theory to Modern Physics", Springer, Berlin, Heidelberg (2000), Chap. 13.
- Square-well potentials in semiconductor physics: M.A. Reed, "Quantum dots", Sci. Am., February 1993, p. 98; L.L. Chang and L. Esaki, "Semicon-

ductor quantum heterostructures", Phys. Today **45**, 36 (1992); L. Kouwenhoven and C. Marcus, "Quantum dots", Phys. World, June 1998, p. 35; C. Dekker, "Carbon nanotubes as molecular quantum wires", Phys. Today, May 1999, p. 22.

- Application of the square-well model to explain the properties of colored centers in crystals: J.-L. Basdevant and J. Dalibard, "The Quantum Mechanics Solver: How to Apply Quantum Theory to Modern Physics", Springer, Berlin, Heidelberg (2000), Chap. 1.

Exercises

4.1. Uncertainty relation for the harmonic oscillator. Using the recursion relations satisfied by the Hermite functions (4.18), show that, in a state of energy E_n given by (4.16), $\langle x\rangle = \langle p\rangle = 0$. Calculate $\langle x^2\rangle$ and $\langle p^2\rangle$ and show that the zero-point energy is essential in order to preserve the uncertainty relations.

4.2. Time evolution of a one-dimensional harmonic oscillator. Consider a harmonic oscillator with the Hamiltonian $\hat{H} = \hat{p}^2/2m + m\omega^2\hat{x}^2/2$ and its first two normalized eigenfunctions $\phi_0(x)$ and $\phi_1(x)$.
Consider a system which at time $t = 0$ has the wave function

$$\psi(x, t = 0) = \cos\theta\ \phi_0(x) + \sin\theta\ \phi_1(x)\,, \quad \text{where} \quad 0 \le \theta < \pi\,.$$

a. What is the wave function $\psi(x, t)$ at time t?
b. Calculate the expectation values $\langle E\rangle$, $\langle E^2\rangle$ and $\Delta E^2 = \langle E^2\rangle - \langle E\rangle^2$. Explain their time dependence.
c. Calculate the time evolution of $\langle x\rangle$, $\langle x^2\rangle$ and Δx.

4.3. Three-dimensional harmonic oscillator. Consider in three dimensions a particle of mass m and the Hamiltonian $\hat{H} = \hat{\boldsymbol{p}}^2/2m + m\omega^2\hat{\boldsymbol{r}}^2/2$, where $\hat{\boldsymbol{r}}^2 = \hat{x}^2 + \hat{y}^2 + \hat{z}^2$.

a. What are the energy levels and their degeneracies?
b. How do these results change in the case of an anisotropic potential

$$V = m\left(\omega_1^2 x^2 + \omega_2^2 y^2 + \omega_3^2 z^2\right)/2\,?$$

4.4. One-dimensional infinite potential well. Consider an infinite potential well of width a: $V(x) = 0$ for $0 < x < a$ and $V = \infty$ otherwise.

a. Show that in the energy eigenstate $\psi_n(x)$, one has $\langle x\rangle = a/2$ and $\Delta x^2 = a^2(1 - 6/n^2\pi^2)/12$.
b. Consider the wave function $\psi(x) = Ax(a - x)$.
   (i) What is the probability p_n of finding the particle in the nth excited state using (3.15)?

(ii) From this set of probabilities, calculate the expectation values $\langle E \rangle$ and $\langle E^2 \rangle$ for that wave function.

Recall that $\sum_{k=0}^{\infty}(2k+1)^{-2n} = \pi^2/8$ for $n=1$, $\pi^4/96$ for $n=2$ and $\pi^6/960$ for $n=3$.

c. Check that if one applies blindly the correspondence principle, i. e. if one uses $\hat{H}^2 = (\hbar^2/2m)^2 \mathrm{d}^4/\mathrm{d}x^4$ in the definition of $\langle E^2 \rangle$, one obtains the absurd result $\Delta E^2 < 0$. What is the reason for this?

4.5. Isotropic states of the hydrogen atom. The energy levels of spherically symmetric states of the hydrogen atom can be obtained by means of the following one-dimensional calculation. Consider an electron of mass m in a potential $V(x)$ such that $V = \infty$ if $x \leq 0$ and $V = -A/x$ if $x > 0$, where $A = q^2/4\pi\varepsilon_0$ and q is the elementary charge. We set $\alpha = q^2/(4\pi\varepsilon_0 \hbar c) \simeq 1/137$ (a dimensionless constant), where c is the velocity of light.

a. Show that the wave function $\psi(x) = Cx\mathrm{e}^{-x/a}$ for $x \geq 0$ and $\psi(x) = 0$ for $x < 0$ is an eigenfunction of the Hamiltonian, with energy E for a given value of a. Express E and a in terms of m, α, $\hbar$ and c.
b. Calculate the numerical values of E and a. You can use $mc^2 = 5.11 \times 10^5\,\mathrm{eV}$ and $\hbar c = 197\,\mathrm{eV\,nm}$.
c. Determine the normalization constant C in terms of a.
d. Calculate the expectation value of $1/x$ in the state $|\psi\rangle$ and deduce from that the expectation value of the kinetic energy. What is the relation, valid also in classical mechanics, between these two quantities?

4.6. δ-function potentials.

a. Consider a particle of mass m in the one-dimensional potential $V(x) = \alpha\delta(x)$, $\alpha < 0$. We are interested in bound states ($E < 0$).
   (i) Assuming that the wave function $\psi(x)$ is continuous at $x = 0$ (which can be proven), find the relation between the discontinuity of its derivative and $\psi(0)$ by integrating the Schrödinger equation between $x = -\varepsilon$ and $x = +\varepsilon$.
   (ii) How many bound states are there? With what energies? Set $K = \sqrt{-2mE}/\hbar$ and $\lambda_0 = -\hbar^2/m\alpha$.
b. Consider the double-δ-function potential

$$V(x) = \alpha\left[\delta(x + d/2) + \delta(x - d/2)\right] .$$

   (i) Write down the general form of the bound-state wave functions. What is the quantization condition?
   (ii) Discuss the number of bound states as a function of the distance d between the two wells.

4.7. Localization of internal atomic electrons. Estimate the tunneling time between adjacent atoms for highly bound internal electrons in a molecule or a solid ($E_0 = -1\,\mathrm{keV}$).

5. Principles of Quantum Mechanics

The everlasting silence of these infinite spaces frightens me.
Blaise Pascal

Wave mechanics started in 1923 with de Broglie's pioneering idea, but it became a respectable theory only in 1926 with the work of Schrödinger. However, as early as 1924, Heisenberg, in a brilliant inspiration largely based on philosophical ideas,[1] had developed a somewhat mysterious theoretical scheme whose results were amazingly efficient. Quite rapidly, Max Born recognized, in the *symbolic multiplication* introduced by Heisenberg, the rules of matrix multiplication. At the beginning of 1925, the Göttingen school (Heisenberg, Born and Jordan, who were joined by Pauli) had set out the basis of *matrix mechanics*, called *Quanten Mechanik* for the first time by Born. After hearing a lecture that Heisenberg gave in Cambridge in July 1925, Dirac developed independently his own formulation of the theory, which was based on the property that quantum variables are noncommutative. This formulation was equivalent to the Göttingen formulation, but it was more general and more elegant.

The controversy between the two types of quantum theory – *matrix theory* and *wave theory* – disappeared when Schrödinger, at the end of 1926, and Dirac, at the beginning of 1927, showed that the two approaches were equivalent. The unifying concept was Hilbert space analysis, and the mathematical foundations of quantum mechanics, as we use them now, were set down by Hilbert and Von Neumann in 1927. In this chapter, we are interested in this unifying process, which leads to a clearer formulation of the basic principles.

As far as the motion of a particle in space is concerned, this is a mere rewriting of the theory in a different language. However, this recasting of the theory will prove to be a considerable conceptual improvement. It will allow us to extend the theory to systems or quantities which do not possess classical analogs. It will also simplify many problems by eliciting the important structures and parameters.

---

[1] One can talk only about what is observable and *measurable*, i. e. the positions and intensities of spectral lines, and not the position and velocity of an electron in an atom.

After introducing the basic notion of a *state vector* and the notation due to Dirac, we shall see how the matrices of Heisenberg emerge in a natural way. Then we shall come back to observables and physical quantities. This will enable us to expound the general principles of quantum mechanics.

5.1 Hilbert Space

We now recall the general framework of Hilbert spaces, in which we shall formulate the theory.

5.1.1 The State Vector

The fundamental property of wave functions $\psi(\boldsymbol{r}, t)$ in wave mechanics is that they belong to a *Hilbert space* $\mathcal{E}_{\mathrm{H}}$. For a particle moving in three-dimensional space, the Hilbert space $\mathcal{E}_{\mathrm{H}}$ is the space of square-integrable functions in three real variables (x, y, z) called $\mathcal{L}^2(R^3)$ (where R^3 denotes the set of real triplets x, y, z).

The description of the state of a particle in terms of $\psi(\boldsymbol{r}, t)$ is not the only one possible. The Fourier transform $\varphi(\boldsymbol{p}, t)$ provides an equivalent description of this state since $\psi(\boldsymbol{r}, t)$ and $\varphi(\boldsymbol{p}, t)$ have a one-to-one relationship. Actually, there exist an infinite number of other equivalent descriptions of this state. The situation is similar to what happens in the usual Euclidean geometry, where a point or a vector can be represented by an infinite set of different coordinates, according to which basis one chooses. These are different *representations* of a single mathematical object, in our case a vector of the Hilbert space $\mathcal{E}_{\mathrm{H}}$ of interest. From now on, we shall say that a physical system is described at any time t by a *state vector*, which we write, according to the notation introduced by Dirac, as

$$|\psi(t)\rangle\,, \qquad \text{which is an element of the Hilbert space } \mathcal{E}_{\mathrm{H}}\,. \tag{5.1}$$

Dirac called these vectors *kets*.

It was a remarkable idea of some mathematicians of the early 20th century, such as Banach, Fréchet and Hilbert that one could use a geometric language in order to solve problems of analysis by considering functions as *vectors* of appropriate spaces. The starting point for this idea was actually contained in Fourier's work, one century earlier.

5.1.2 Scalar Products and the Dirac Notations

A Hilbert space $\mathcal{E}_{\mathrm{H}}$ is a complex vector space which possesses a positive definite Hermitian scalar product. In this book, we assume that the Hilbert space $\mathcal{E}_{\mathrm{H}}$ is complete (all "Cauchy sequences" converge) and that it is separable (there exists a "sequence" which is everywhere "dense" in $\mathcal{E}_{\mathrm{H}}$).

The Hermitian scalar product of $|\psi_1\rangle$ and $|\psi_2\rangle$ is denoted by $\langle\psi_2|\psi_1\rangle$. It possesses the Hermitian symmetry:[2]

$$\langle\psi_2|\psi_1\rangle = (\langle\psi_1|\psi_2\rangle)^* \, .$$

It is linear in $|\psi_1\rangle$ and antilinear in $|\psi_2\rangle$. The norm of a vector, denoted by $||\,|\psi\rangle||$ or $||\psi||$, is defined as

$$||\psi|| = \sqrt{\langle\psi|\psi\rangle} \, .$$

By definition, owing to the probabilistic interpretation as we shall see, the norm of a state vector is equal to one:

$$\langle\psi(t)|\psi(t)\rangle = 1 \, . \tag{5.2}$$

5.1.3 Examples

In a finite-dimensional Hilbert space of dimension n (which is also called a Hermitian space), the vectors can be represented as column matrices:

$$|u\rangle = \begin{pmatrix} u_1 \\ u_2 \\ \vdots \\ u_n \end{pmatrix} , \qquad |v\rangle = \begin{pmatrix} v_1 \\ v_2 \\ \vdots \\ v_n \end{pmatrix} , \tag{5.3}$$

where the u_i and v_i are complex numbers. The scalar product $\langle v|u\rangle$ is then simply

$$\langle v|u\rangle = \sum_{i=1}^{n} v_i^* u_i \, , \tag{5.4}$$

i. e. it is the matrix product of the row matrix

$$(v_1^*, v_2^*, \dots, v_n^*) \tag{5.5}$$

with the column matrix $|u\rangle$.

In the space $\mathcal{L}^2(R^3)$, which is infinite dimensional, we define the scalar product of $|\psi_1\rangle$ and $|\psi_2\rangle$ as

$$\langle\psi_2|\psi_1\rangle = \int \psi_2^*(\boldsymbol{r})\, \psi_1(\boldsymbol{r})\, \mathrm{d}^3r \, . \tag{5.6}$$

---

[2] Unfortunately, the notations of physicists and of mathematicians, often differ. The complex conjugate of a number z is written $\overline{z}$ by many mathematicians while we write it as z^* here. Similarly, many mathematicians, including Hilbert himself, would write scalar products the other way around, i. e. $(\psi_1|\psi_2)$, which is linear on the left and antilinear on the right.

5.1.4 Bras and Kets, Brackets

One can show that there is a bijection between the space $\mathcal{E}_\mathrm{H}$ and its dual space $\mathcal{E}_\mathrm{H}^*$, i. e. the space of continuous "linear forms" defined on $\mathcal{E}_\mathrm{H}$. Dirac's notations profit from this isomorphism in the following way:

- With any ket $|\phi\rangle$ of $\mathcal{E}_\mathrm{H}$, we associate an element of $\mathcal{E}_\mathrm{H}^*$ denoted by $\langle\phi|$, called a *bra*.
- The action of the bra $\langle\phi|$ on any ket $|\psi\rangle$ is equal to the scalar product of $|\psi\rangle$ and $|\phi\rangle$:

$$\langle\phi|\,(|\psi\rangle) = \langle\phi|\psi\rangle\,.$$

Dirac's notation therefore appears as a very simple grammatical rule, which is identical to the usual rules of matrix multiplication in finite-dimensional spaces that we recalled in (5.3) and (5.4), The bra $\langle v|$ is nothing but (5.5) in this case. In Dirac's notation, the bra $\langle\phi|$ acts on the ket $|\psi\rangle$ to give a *bracket*, i. e. a complex number equal to the scalar product $\langle\phi|\psi\rangle$.

5.2 Operators in Hilbert Space

Let us give a few definitions together with their analogs in finite-dimensional spaces.

5.2.1 Matrix Elements of an Operator

Consider a linear operator $\hat{A}$ acting in the Hilbert space $\mathcal{E}_\mathrm{H}$. It transforms any given ket $|\psi\rangle$ into another ket $\hat{A}|\psi\rangle$ of $\mathcal{E}_\mathrm{H}$. We shall frequently consider the scalar product of $\hat{A}|\psi\rangle$ with another ket $|\phi\rangle$:

$$\langle\phi|(\hat{A}|\psi\rangle)\,. \tag{5.7}$$

(a) Operators in Finite-Dimensional Spaces. Consider first a space of finite dimension n. The operator $\hat{A}$ is an $n \times n$ square matrix, and the above expression can be interpreted in the following way. We first multiply the column vector $|\psi\rangle$ by the $n \times n$ matrix in order to obtain another column vector, and we then take the scalar product of this second vector with the vector $\langle\phi|$. However, we know that this product is associative; we can just as well do the operation

$$(\langle\phi|\hat{A})|\psi\rangle$$

first, i. e. calculate first the product of the row matrix $\langle\phi|$ and the square matrix which represents $\hat{A}$. This results in a new row matrix, and we then take the scalar product with the column matrix $|\psi\rangle$. In other words, in finite-dimensional spaces, owing to the associativity of matrix products, there is no reason to use any parentheses in (5.7), and we can set

$$\langle \phi | \hat{A} | \psi \rangle = \langle \phi | \big(\hat{A} | \psi \rangle \big) = \big(\langle \phi | \hat{A} \big) | \psi \rangle \,. \tag{5.8}$$

The quantity $\langle \phi | \hat{A} | \psi \rangle$ is called the matrix element of $\hat{A}$ between $\langle \phi |$ and $| \psi \rangle$.

(b) Operators in Infinite-Dimensional Spaces. In infinite-dimensional spaces, we shall use the same simplification, although there exist "pathological" operators for which, in full rigor, some care must be taken. In Appendix Appendix C, we justify our assertion that one can use a notation similar to (5.8) for the "nonpathological" operators we shall deal with.

In what follows, we shall mention in a few places some mathematical difficulties that one may meet in manipulating operators in Hilbert space, but most of the time we shall just forget these difficulties. In other words, we work as if we were dealing with finite-dimensional spaces. It is possible to use a rigorous formalism, but this is quite cumbersome and does not bring anything really new to the physics, at least at this level.

(c) Expectation Value of an Operator. Consider a state vector $|\psi\rangle$. The expectation value of an operator $\hat{A}$ in the state $|\psi\rangle$ (which is supposed to be normalized, cf. (5.2)) is defined as

$$\langle a \rangle = \langle \psi | \hat{A} | \psi \rangle \,. \tag{5.9}$$

5.2.2 Adjoint Operators and Hermitian Operators

Consider an operator $\hat{A}$ acting in $\mathcal{E}_{\mathrm{H}}$. The adjoint of $\hat{A}$, noted $\hat{A}^\dagger$, is defined by the relations:

$$\langle \psi_2 | \hat{A}^\dagger | \psi_1 \rangle = \big(\langle \psi_1 | \hat{A} | \psi_2 \rangle \big)^* \quad \text{for any } |\psi_1\rangle, |\psi_2\rangle \text{ in } \mathcal{E}_{\mathrm{H}} \,. \tag{5.10}$$

An operator $\hat{A}$ is *Hermitian*, or *self-adjoint*, if

$$\hat{A} = \hat{A}^\dagger \,, \quad \text{i.e.} \quad \langle \psi_2 | \hat{A} | \psi_1 \rangle = \langle \psi_1 | \hat{A} | \psi_2 \rangle^* \,. \tag{5.11}$$

If $\hat{A}$ is self-adjoint the *expectation value* of the observable $\hat{A}$, defined in (5.9), is *real*:

$$\hat{A} = \hat{A}^\dagger \Rightarrow \langle a \rangle = \langle a \rangle^* \,. \tag{5.12}$$

Conversely, one can show that if an operator $\hat{A}$ is such that $\langle \psi | \hat{A} | \psi \rangle$ is real for any ψ, this operator is Hermitian. This is the reason why the observables are self-adjoint operators, since the measured quantities (and therefore their mean value) are real numbers.[3]

---

[3] One can of course also consider complex combinations of observables, for instance when one deals with the amplitude and phase of a wave. On the other hand, the Hamiltonian is always self-adjoint in order to ensure the conservation of the norm and of probabilities (see Sect. 5.5).

Examples.

1. In the finite-dimensional case, an operator is Hermitian if and only if its matrix representation $[A_{ij}]$ in an orthonormal basis satisfies $A_{ij} = A^*_{ji}$.
2. Consider the case of an infinite-dimensional space, such as $\mathcal{L}^2(R)$, and consider the *position* operator $\hat{x}$ introduced in Chap. 3: if $\psi(x)$ is the wave function associated with the ket $|\psi\rangle$, $x\psi(x)$ is the wave function associated with $\hat{x}|\psi\rangle$. We obtain:
$$\langle\psi_2|\hat{x}^\dagger|\psi_1\rangle = (\langle\psi_1|\hat{x}|\psi_2\rangle)^* = \left(\int \psi_1^*(x)\, x\, \psi_2(x)\, \mathrm{d}x\right)^*$$
$$= \int \psi_2^*(x)\, x\, \psi_1(x)\, \mathrm{d}x = \langle\psi_2|\hat{x}|\psi_1\rangle\,.$$
The position operator is therefore Hermitian. Consider now the momentum operator $\hat{p}_x$. We know from the results of Chap. 3 that this operator corresponds to $\hbar/\mathrm{i}$ times differentiation with respect to x. In other words, if the function $\psi(\boldsymbol{r})$ is associated with the ket $|\psi\rangle$, then $(\hbar/\mathrm{i})(\partial\psi/\partial x)$ is associated with $\hat{p}_x|\psi\rangle$. We obtain:
$$\langle\psi_2|\hat{p}_x^\dagger|\psi_1\rangle = (\langle\psi_1|\hat{p}_x|\psi_2\rangle)^* = \left(\int \psi_1^*\, \frac{\hbar}{i}\frac{\partial\psi_2}{\partial x}\, \mathrm{d}^3r\right)^*$$
$$= -\frac{\hbar}{\mathrm{i}}\int \frac{\partial\psi_2^*}{\partial x}\, \psi_1\, \mathrm{d}^3r\,.$$
After integrating by parts, we obtain
$$\langle\psi_2|\hat{p}_x^\dagger|\psi_1\rangle = \frac{\hbar}{\mathrm{i}}\int \psi_2^*\, \frac{\partial\psi_1}{\partial x}\, \mathrm{d}^3r = \langle\psi_2|\hat{p}_x|\psi_1\rangle\,. \tag{5.13}$$
The momentum operator is therefore also Hermitian (notice that the factor i is crucial).

5.2.3 Eigenvectors and Eigenvalues

A nonzero vector $|\psi_a\rangle$ is said to be an *eigenvector* of the operator $\hat{A}$ if
$$\hat{A}|\psi_a\rangle = a_\alpha|\psi_a\rangle\,. \tag{5.14}$$
The number a_α is the *eigenvalue* associated with this eigenvector.

Writing (5.12) with $|\psi\rangle = |\psi_\alpha\rangle$ we deduce that the eigenvalues a_α of Hermitian operators are *real*. In addition, if $|\psi_\alpha\rangle$ and $|\psi_\beta\rangle$ are two eigenvectors of a Hermitian operator associated with different eigenvalues a_α and a_β, they are orthogonal. We have, indeed,
$$\langle\psi_\alpha|\hat{A}|\psi_\beta\rangle = \langle\psi_\alpha|\left(\hat{A}|\psi_\beta\rangle\right) = a_\beta\, \langle\psi_\alpha|\psi_\beta\rangle$$
$$= \left(\langle\psi_\alpha|\hat{A}\right)|\psi_\beta\rangle = a_\alpha\, \langle\psi_\alpha|\psi_\beta\rangle\,,$$
which implies $\langle\psi_\alpha|\psi_\beta\rangle = 0$ if $a_\alpha \neq a_\beta$.

5.2.4 Summary: Syntax Rules in Dirac's Formalism

1. When a bra is on the left of a ket, they contract to give a number:

$$(\langle\psi_2|)(|\psi_1\rangle) = \langle\psi_2|\psi_1\rangle\,.$$

2. The Hermitian conjugate of an expression is obtained by

– reversing the order of the terms
– transforming

(a) the operators into their adjoints,
(b) the kets into bras and vice versa,
(c) the numbers into their complex conjugates.

For example, the Hermitian conjugate of $\lambda|\phi\rangle\langle\psi|\hat{A}^\dagger\hat{B}$ is $\lambda^*\hat{B}^\dagger\hat{A}|\psi\rangle\langle\phi|\,.$

5.3 The Spectral Theorem

The notion of Hilbertian bases and the spectral theorem of F. Riesz play a central role in the physical interpretation of quantum mechanics.

5.3.1 Hilbertian Bases

In nonrelativistic quantum mechanics, the Hilbert spaces which are used are always separable. This implies the existence of at least one denumerable orthonormal basis, also called a Hilbertian basis. Consider the example of $\mathcal{L}^2(R)$. The family $\{\phi_n\}$ of Hermite functions defined in (4.13),

$$\phi_n(x) = c_n\,\mathrm{e}^{x^2/2}\left(\frac{\mathrm{d}}{\mathrm{d}x}\right)^n \mathrm{e}^{-x^2}\,,\ n = 0, 1, 2,\ \ldots\,, \tag{5.15}$$

is a particular Hilbertian basis of this space. We have

$$\int_{-\infty}^{+\infty} \phi_n^*(x)\,\phi_m(x)\,\mathrm{d}x = \delta_{n,m}\,, \tag{5.16}$$

and any square-integrable function $\psi(x)$ can be expanded in this basis as

$$\psi(x) = \sum_{n=0}^{\infty} C_n\phi_n(x)\,, \quad \text{where} \quad C_n = \int \phi_n^*(x)\,\psi(x)\,\mathrm{d}x\,. \tag{5.17}$$

We notice that there are usually an infinite number of nonvanishing coefficients C_n. More formally and more generally, let $\{|n\rangle, n = 1, 2, \ldots\}$ be a Hilbertian basis with

$$\langle m|n\rangle = \delta_{n,m}\,. \tag{5.18}$$

Any ket $|\psi\rangle$ or bra $\langle\psi|$ can be decomposed in this basis as

$$|\psi\rangle = \sum_n C_n |n\rangle \,, \qquad \langle\psi| = \sum_n C_n^* \langle n| \,, \tag{5.19}$$

where $C_n = \langle n|\psi\rangle$ and $C_n^* = \langle\psi|n\rangle$. We have, therefore, $\langle\psi|\psi\rangle = \sum_n |C_n|^2$. We remark that the set of coordinates $\{C_n\}$ in the basis $\{|n\rangle\}$ defines $|\psi\rangle$ (or $\psi(x)$) completely. Therefore it is a new *representation* of the state $|\psi\rangle$, as we anticipated above in (5.3). Moreover, if $|\psi\rangle = \sum_n C_n|n\rangle$ and $|\chi\rangle = \sum_n B_n|n\rangle$, then

$$\langle\chi|\psi\rangle = \sum_n B_n^* C_n \,. \tag{5.20}$$

5.3.2 Projectors and Closure Relation

The expression $|u\rangle\langle v|$ can be considered as an operator if we use Dirac's multiplication rules:

$$(|u\rangle\langle v|)|\psi\rangle = |u\rangle\langle v|\psi\rangle = \lambda|u\rangle \,, \tag{5.21}$$

where λ is the complex number $\langle v|\psi\rangle$. Consider a Hilbertian basis $\{|n\rangle \,, n = 1, 2, ...\}$. The operator

$$\hat{P}_n = |n\rangle\langle n| \tag{5.22}$$

is a projection operator, or *projector*, onto the state $|n\rangle$, i.e.

$$\hat{P}_n|\psi\rangle = (|n\rangle\langle n|)|\psi\rangle = \langle n|\psi\rangle|n\rangle = C_n|n\rangle \tag{5.23}$$

and $\hat{P}_n^2 = \hat{P}_n$.

This can be extended to any subspace $\mathcal{E}_\nu$ of $\mathcal{E}_{\mathrm{H}}$ generated by a subset $\{|n\rangle, n \in \{\nu\}\}$ of basis vectors. We define the projector $\hat{P}_\nu$ onto this subspace as

$$\hat{P}_\nu = \sum_{n\in\{\nu\}} \hat{P}_n \,. \tag{5.24}$$

The projection of $|\psi\rangle$ onto the entire Hilbert space is naturally $|\psi\rangle$ itself. Therefore we have the important *closure relation*, or *decomposition of the identity*,

$$\sum_{\text{all } n} \hat{P}_n = \sum_{\text{all } n} |n\rangle\langle n| = \hat{I} \,, \tag{5.25}$$

where $\hat{I}$ is the identity operator.

5.3.3 The Spectral Decomposition of an Operator

Consider a Hermitian operator $\hat{A} = \hat{A}^\dagger$. The set of its eigenvalues is denoted by $\{a_\alpha, \alpha = 1, 2, \dots\}$, and the corresponding normalized eigenvectors are $|\alpha, r\rangle$:

$$\hat{A}|\alpha, r\rangle = a_\alpha |\alpha, r\rangle . \tag{5.26}$$

We recall that the eigenvalues a_α are real since $\hat{A} = \hat{A}^\dagger$, and that two eigenvectors corresponding to two different eigenvalues are orthogonal. Here, by convention, two different indices α and β correspond to two different eigenvalues $a_\alpha \neq a_\beta$. In order to write the eigenvectors, we introduce the extra label r, because the eigensubspace corresponding to the eigenvalue a_α may be of dimension $n_\alpha > 1$; r then takes the values $r = 1, 2, \ldots, n_\alpha$. When $n_\alpha > 1$, we say that the eigenvalue a_α is *degenerate* with a degree of degeneracy n_α. We choose these vectors orthonormal:

$$\langle \beta, r'|\alpha, r\rangle = \delta_{\alpha,\beta}\delta_{r,r'} . \tag{5.27}$$

We shall use frequently the following fundamental theorem of Hilbertian analysis, called the *spectral theorem* and due to Frederic Riesz:

The set $\{|\alpha, r\rangle\}$ *of orthonormal eigenvectors of a Hermitian operator is a Hilbertian basis of* $\mathcal{E}_\mathrm{H}$.

Consequences.
(*i*) Any vector $|\psi\rangle$ can be expanded in the set of the eigenvectors of $\hat{A}$, $\{|\alpha, r\rangle\}$. Equivalently, we have a spectral decomposition of the identity:

$$\hat{I} = \sum_\alpha \sum_{r=1}^{n_\alpha} |\alpha, r\rangle\langle\alpha, r| . \tag{5.28}$$

(*ii*) The operator $\hat{A}$ possesses a *spectral decomposition*, i. e. we can write it as

$$\hat{A} = \sum_\alpha \sum_{r=1}^{n_\alpha} a_\alpha\, |\alpha, r\rangle\langle\alpha, r| . \tag{5.29}$$

Strictly speaking, our formulation of the spectral theorem is valid in finite-dimensional spaces only. For infinite-dimensional spaces, the formulation needs in principle to be refined, and based rather on the properties (5.28) and (5.29). It is nevertheless convenient to work in the above terminology. Using a mathematically rigorous language would not change the conclusions that we shall reach, but would certainly obscure the discussion.

5.3.4 Matrix Representations

Consider a Hermitian basis $\{|n\rangle,\ n = 1, 2, ...\}$ and the formulas (5.19) and (5.20). In this basis, $|\psi\rangle$ can be represented by a (possibly infinite) column vector, and $\langle\psi|$ can be represented by a row vector, as follows:

$$|\psi\rangle \Leftrightarrow \begin{pmatrix} C_1 \\ C_2 \\ \vdots \\ C_n \\ \vdots \end{pmatrix}, \qquad \langle\psi| \Leftrightarrow (C_1^*, C_2^*, \ldots, C_n^*, \ldots) . \tag{5.30}$$

Similarly, any operator $\hat{A}$ is represented, in this basis, by the *matrix* whose elements $A_{n,m}$ are

$$A_{n,m} = \langle n|\hat{A}|m\rangle . \tag{5.31}$$

Indeed, if $|\chi\rangle = \hat{A}|\psi\rangle$, and if $\{B_n\}$ and $\{C_n\}$ are the coefficients of the expansions of $|\chi\rangle$ and $|\psi\rangle$, we have

$$B_n = \langle n|\chi\rangle = \langle n| \left(\hat{A}|\psi\rangle\right) \tag{5.32}$$

and, by inserting this in the expansion of $|\psi\rangle$, we obtain

$$B_n = \sum_m \langle n|\hat{A}|m\rangle \, C_m \tag{5.33}$$

Example. Consider a harmonic oscillator of mass m and angular frequency ω. The matrix representation of the operators $\hat{x}$ and $\hat{p}$ in the basis of the energy eigenfunctions can be obtained easily by using the recursion relations of the Hermite functions given in Chap. 4, in (4.18). We obtain

$$\hat{x} \;\Rightarrow\; \sqrt{\frac{\hbar}{2m\omega}} \begin{pmatrix} 0 & \sqrt{1} & 0 & 0 & \ldots \\ \sqrt{1} & 0 & \sqrt{2} & 0 & \ldots \\ 0 & \sqrt{2} & 0 & \sqrt{3} & \ldots \\ 0 & 0 & \sqrt{3} & 0 & \ldots \\ \vdots & \vdots & \vdots & \vdots & \end{pmatrix}, \tag{5.34}$$

$$\hat{p} \;\Rightarrow\; -\mathrm{i}\sqrt{\frac{m\omega\hbar}{2}} \begin{pmatrix} 0 & \sqrt{1} & 0 & 0 & \ldots \\ -\sqrt{1} & 0 & \sqrt{2} & 0 & \ldots \\ 0 & -\sqrt{2} & 0 & \sqrt{3} & \ldots \\ 0 & 0 & -\sqrt{3} & 0 & \ldots \\ \vdots & \vdots & \vdots & \vdots & \end{pmatrix}. \tag{5.35}$$

In this basis, the matrix which represents the Hamiltonian is *diagonal.*

The matrices given above are nothing but the (infinite) matrices that, in his remarkable inspiration, Heisenberg introduced as early as 1924. The previous pages reflect the work done by Schrödinger and Dirac when they unified wave mechanics and matrix mechanics.

5.4 Measurement of Physical Quantities

In Sect. 5.5, when stating the general principles of quantum mechanics, we shall generalize a statement made in Chap. 3 with respect to wave mechanics: with each physical quantity A we associate a Hermitian operator $\hat{A}$. The expectation value $\langle a \rangle$ and the root mean square (r.m.s.) dispersion Δa of the possible results are then given by

$$\langle a \rangle = \langle \psi | \hat{A} | \psi \rangle \,, \qquad \Delta a^2 = \langle \psi | \hat{A}^2 | \psi \rangle - \langle a \rangle^2 \,. \tag{5.36}$$

Here $|\psi\rangle$ designates the state of the system when the measurement is performed. The purpose of this section is to analyze the consequence of this statement in terms of the possible results of an individual measurement, and also to give a constraint on the state of the system after the measurement has been performed. The property we wish to establish is the following:

In a measurement of a quantity A, the only possible results of the measurement are the eigenvalues a_α of the observable $\hat{A}$.

We denote by $\{|\alpha\rangle\}$ a normalized set of eigenvectors of $\hat{A}$, associated with the eigenvalues a_α. For simplicity we restrict ourselves here to the case where the eigenvalues a_α are not degenerate. Owing to the spectral theorem, the normalized state vector $|\psi\rangle$ of the system before the measurement is performed can be expanded as follows:

$$|\psi\rangle = \sum_\alpha C_\alpha \, |\alpha\rangle \,, \qquad C_\alpha = \langle \alpha | \psi \rangle \,, \qquad \sum_\alpha |C_\alpha|^2 = 1 \,. \tag{5.37}$$

The expectation value $\langle a \rangle$ of the physical quantity A is

$$\langle a \rangle = \langle \psi | \hat{A} | \psi \rangle = \sum_\alpha a_\alpha \, |C_\alpha|^2 \,. \tag{5.38}$$

This formula can be read as the expectation value of a random variable with possible outcomes a_α and probabilities $|C_\alpha|^2$. This is in agreement with the property that we want to establish and, in addition, it yields the probability $|C_\alpha|^2$ of finding the value a_α as a result of the measurement. It is plausible that $|C_\alpha|^2$ is the probability of finding a_α. This probability law is properly normalized. Furthermore, if the particle is in the state $|\alpha_0\rangle$, then $|C_\alpha|^2 = \delta_{\alpha,\alpha_0}$, and we are sure to find the result a_{α_0} and no other result.

In order to be convinced that the only possible results of the measurement of the quantity A are the eigenvalues a_α, we first prove the following theorem:

The mean square deviation Δa of an observable $\hat{A}$ vanishes if and only if the state $|\psi\rangle$ is an eigenstate of $\hat{A}$.

The proof is simple. If $|\psi\rangle$ is an eigenstate of $\hat{A}$ with eigenvalue a_α, then $\hat{A}|\psi\rangle = \alpha|\psi\rangle$ and $\hat{A}^2|\psi\rangle = \alpha^2|\psi\rangle$. Therefore the expectation values of A and A^2 are $\langle a \rangle = \alpha$ and $\langle a^2 \rangle = \alpha^2$, respectively: the variance $\Delta a^2 = \langle a^2 \rangle - \langle a \rangle^2$ cancels.

Conversely, consider the norm of the vector $(\hat{A} - \langle a \rangle \hat{I})|\psi\rangle$, where $\hat{I}$ is the identity operator. We obtain

$$\begin{aligned} \|(\hat{A} - \langle a \rangle \hat{I})|\psi\rangle\|^2 &= \langle\psi|(\hat{A} - \langle a \rangle \hat{I})^2|\psi\rangle \\ &= \langle\psi|\hat{A}^2|\psi\rangle - \langle a \rangle^2 = \Delta a^2 \,. \end{aligned}$$

If the mean square deviation vanishes, this means that the vector $(\hat{A} - \langle a \rangle \hat{I})|\psi\rangle$ is the null vector, or, in other words, $\hat{A}|\psi\rangle = \langle a \rangle \, |\psi\rangle$. Therefore if $\Delta a = 0$, $|\psi\rangle$ is necessarily an eigenstate of $\hat{A}$ with eigenvalue $\langle a \rangle$.

Consider a measurement of the physical quantity A and suppose that we find the result a with as good an accuracy as we wish. If the act of measuring gives us information, which is the basis of physics, we are sure that *immediately* after this measurement, i. e. before the system has evolved appreciably, a new measurement of the same quantity performed on the state of the system will give us the same answer a with probability one. In other words, after a measurement, the system is in a state for which the quantity A is *well defined.* Owing to the above theorem, this can only happen if a belongs to the set of eigenvalues $\{a_\alpha\}$.

We can deduce the probability $\mathcal{P}(a_\alpha)$ of finding the result a_α in a measurement of A. We simply use the result (5.38), which can be generalized to any power of A, assuming that the operator $\hat{A}^n$ is associated with the physical quantity A^n:

$$\langle a^n \rangle = \langle\psi|\hat{A}^n|\psi\rangle = \sum_\alpha a_\alpha^n \, |C_\alpha|^2 \,. \tag{5.39}$$

A probability law is completely determined if one knows its possible outcomes and its moments $\langle a^n \rangle$. In the present case the set of possible outcomes is the set of eigenvalues $\{a_\alpha\}$. Consequently, the only possible set of probabilities which can lead to (5.39) is $\mathcal{P}(a_\alpha) = |C_\alpha|^2 = |\langle\alpha|\psi\rangle|^2$.

Remarks

1. The last formula must be modified when the eigenvalue a_α is degenerate. We shall come back to this technical point when stating precisely the principles of quantum mechanics.

2. In the cases of the position and momentum observables $\hat{x}$ and $\hat{p}$, the spectrum, i. e. the set of eigenvalues, is a *continuous* set, and the previous statement must be modified appropriately, in order to recover the probability laws of Chaps. 2 and 3. This point will also be considered below.

5.5 Statement of the Principles of Quantum Mechanics

We now state the principles of quantum mechanics. These are valid for any system and generalize the principles that we gave for wave mechanics, which was concerned with the special case of a point particle. The only restriction is that we are dealing here with physical systems which are in "pure states". The notion of "statistical mixtures" is considered in Appendix Appendix D. This distinction is well known in optics: completely polarized light is in a pure polarization state (linear, circular, etc.), whereas nonpolarized or partially polarized light is a statistical mixture of polarization states.

First Principle: the Superposition Principle

With each physical system one can associate an appropriate Hilbert space $\mathcal{E}_{\mathrm{H}}$. At each time t, the state of the system is completely determined by a normalized vector $|\psi(t)\rangle$ of $\mathcal{E}_{\mathrm{H}}$.

This principle entails that any normalized linear superposition $|\psi\rangle = \sum C_i|\psi_i\rangle$ of state vectors $|\psi_i\rangle$, where the C_i are complex and such that $\langle\psi|\psi\rangle = \sum_{i,j} C_j^* C_i \langle\psi_j|\psi_i\rangle = 1$, is an accessible state vector. Notice that the convention $\|\psi\| = 1$ leaves an indeterminacy: a state vector is defined up to a phase factor $\mathrm{e}^{\mathrm{i}\delta}$ (δ real). This phase factor is arbitrary: it is not possible to make a distinction between $|\psi\rangle$ and $\mathrm{e}^{\mathrm{i}\delta}|\psi\rangle$ in a measurement or in the evolution of the state.

However, the *relative* phases of the various states of the system are not arbitrary. If $|\psi_1'\rangle = \mathrm{e}^{\mathrm{i}\delta_1}|\psi_1\rangle$ and $|\psi_2'\rangle = \mathrm{e}^{\mathrm{i}\delta_2}|\psi_2\rangle$, the superposition of states $C_1|\psi_1'\rangle + C_2|\psi_2'\rangle$ represents a state different from $C_1|\psi_1\rangle + C_2|\psi_2\rangle$.

Second Principle: Measurements of Physical Quantities

(a) With each physical quantity A one can associate a linear Hermitian operator $\hat{A}$ acting in $\mathcal{E}_{\mathrm{H}}$: $\hat{A}$ is the *observable* which represents the quantity A.

(b) We denote by $|\psi\rangle$ the state of the system before the measurement of A is performed. Whatever $|\psi\rangle$ may be, the only possible results of the measurement are the eigenvalues a_α of $\hat{A}$.

(c) We denote by $\hat{P}_\alpha$ the projector onto the subspace associated with the eigenvalue a_α. The probability of finding the value a_α in a measurement of A is

$$\mathcal{P}(a_\alpha) = \|\psi_\alpha\|^2\,, \qquad \text{where} \quad |\psi_\alpha\rangle = \hat{P}_\alpha|\psi\rangle\,. \tag{5.40}$$

(d) Immediately after the measurement of A has been performed and has given the result a_α, the new state $|\psi'\rangle$ of the system is

$$|\psi'\rangle = |\psi_\alpha\rangle/\|\psi_\alpha\|\,. \tag{5.41}$$

In the case of a nondegenerate eigenvalue a_α, the projector P_α is simply $|\alpha\rangle\langle\alpha|$. If the eigenvalue is n_α times degenerate, we introduce as in Sect. 5.3.3 the n_α orthonormal eigenstates $|\alpha, r\rangle$, with $r = 1, 2, ..., n_\alpha$, which span the eigensubspace $\mathcal{E}_\alpha$. The projector $\hat{P}_\alpha$ onto $\mathcal{E}_\alpha$ is then

$$\hat{P}_\alpha = \sum_{r=1}^{n_\alpha} |\alpha, r\rangle\langle\alpha, r| \,. \tag{5.42}$$

Equation (5.40) can be written in the equivalent forms

$$\mathcal{P}(a_\alpha) = \langle\psi|\hat{P}_\alpha|\psi\rangle = |\langle\psi|\psi_\alpha\rangle|^2 \,. \tag{5.43}$$

Terminology. Statement (b) of the Second Principle is called the *quantization principle.* Statement (c) is called the *principle of spectral decomposition.* Statement (d) is the principle of the *reduction of the wave packet.* It provides the quantitative formulation of the fact that a measurement perturbs the system.

The Case of Variables with a Continuous Spectrum. In this case, the only meaningful prediction is that the result falls in some range of values $[a, a+\mathrm{d}a[$. The discrete probability law (5.40) is then replaced by a continuous law. In the case of the position observable x of a point particle, for instance, this law is

$$\mathcal{P}(x)\,\mathrm{d}x = |\psi(x)|^2\,\mathrm{d}x \,, \tag{5.44}$$

where the function $\psi(x)$ is the wave function introduced in Chap. 2.

This law can be cast into a form analogous to (5.43) if we introduce the eigenstates $|x\rangle$ of the position operator (see Appendix Appendix C):

$$\mathcal{P}(x)\,\mathrm{d}x = |\langle x|\psi\rangle|^2\,\mathrm{d}x \,.$$

The "states" $|x\rangle$ are not normalizable and do not belong to the Hilbert space.

Expectation Value of a Measurement. Knowing the probability $\mathcal{P}(a_\alpha)$ of finding a_α in a measurement of A, we can calculate the expectation value $\langle a\rangle$ for a system in a state $|\psi\rangle$. By definition, we have

$$\langle a\rangle = \sum_\alpha a_\alpha \mathcal{P}(a_\alpha) \,. \tag{5.45}$$

Among the various equivalent forms of $\mathcal{P}(a_\alpha)$ (see (5.43)), we choose $\mathcal{P}(a_\alpha) = \langle\psi|\hat{P}_\alpha|\psi\rangle$. We therefore obtain

$$\langle a\rangle = \sum_\alpha a_\alpha \langle\psi|\hat{P}_\alpha|\psi\rangle \,. \tag{5.46}$$

After applying the spectral theorem (5.29), we know that $\sum_\alpha a_\alpha \hat{P}_\alpha = \hat{A}$, and therefore

$$\langle a \rangle = \langle \psi | \hat{A} | \psi \rangle \,. \tag{5.47}$$

The statement (5.36), which we made at the beginning of the Sect. 5.4 as a direct transcription of the principles of wave mechanics, can therefore be viewed as a consequence of the more general formulation of the Second Principle.

Third Principle: Time Evolution

We denote by $|\psi(t)\rangle$ the state of the system at time t. As long as the system does not undergo any observation, its time evolution is given by the Schrödinger equation:

$$\mathrm{i}\hbar \frac{\mathrm{d}}{\mathrm{d}t} |\psi(t)\rangle = \hat{H} |\psi(t)\rangle \,, \tag{5.48}$$

where $\hat{H}$ is the energy observable, or Hamiltonian, of the system.

Conservation of the Norm. The norm $\|\psi\| = \sqrt{\langle\psi|\psi\rangle}$ of a state vector is time independent. This is a consistency condition for the theory, which follows from the fact that the Hamiltonian is Hermitian: $\hat{H} = \hat{H}^\dagger$. Let us write down the relation (5.48) and its Hermitian conjugate:

$$\mathrm{i}\hbar \frac{\mathrm{d}}{\mathrm{d}t} |\psi\rangle = \hat{H} |\psi\rangle \,, \quad -\mathrm{i}\hbar \frac{\mathrm{d}}{\mathrm{d}t} \langle\psi| = \langle\psi| \hat{H}^\dagger = \langle\psi|\hat{H} \,.$$

Multiplying the first relation on the left by $\langle\psi|$ and the second on the right by $|\psi\rangle$ and subtracting, we obtain

$$\mathrm{i}\hbar \left[\langle\psi| \left(\frac{\mathrm{d}}{\mathrm{d}t} |\psi\rangle \right) + \left(\frac{\mathrm{d}}{\mathrm{d}t} \langle\psi| \right) |\psi\rangle \right] = 0 \,, \qquad \text{i.e.} \quad \frac{\mathrm{d}}{\mathrm{d}t} \langle\psi|\psi\rangle = 0 \,.$$

Time Dependence of a State Vector. We assume the system is isolated, i. e. that $\hat{H}$ does not depend on time. The energy eigenstates are the eigenstates of the operator $\hat{H}$:

$$\hat{H} |\psi_\alpha\rangle = E_\alpha |\psi_\alpha\rangle \,.$$

We assume, for simplicity, that the eigenvalues E_α are not degenerate. The set of eigenvectors $\{|\psi_\alpha\rangle\}$ forms a basis of the space $\mathcal{E}_{\mathrm{H}}$, in which we can expand any state vector $|\psi\rangle$. At $t = 0$ we have

$$|\psi(t=0)\rangle = \sum_\alpha C_\alpha |\psi_\alpha\rangle \,, \qquad \text{where} \qquad C_\alpha = \langle \psi_\alpha | \psi(t=0) \rangle \,.$$

At any later time t we can write $|\psi(t)\rangle = \sum_\alpha \lambda_\alpha(t)|\psi_\alpha\rangle$, where $\lambda_\alpha(0) = C_\alpha$. The Schrödinger equation is

$$\mathrm{i}\hbar \sum \frac{\mathrm{d}}{\mathrm{d}t}\lambda_\alpha(t)|\psi_\alpha\rangle = \sum \lambda_\alpha(t)E_\alpha|\psi_\alpha\rangle \,.$$

Since the vectors $|\psi_\alpha\rangle$ are orthogonal, this reduces, for any α, to

$$\mathrm{i}\hbar \frac{\mathrm{d}}{\mathrm{d}t}\lambda_\alpha(t) = E_\alpha \lambda_\alpha(t) \,.$$

We therefore obtain the fundamental relation

$$|\psi(t)\rangle = \sum C_\alpha \mathrm{e}^{-\mathrm{i}E_\alpha t/\hbar}|\psi_\alpha\rangle \,. \tag{5.49}$$

5.6 Structure of Hilbert Space

Let us return to the First Principle and to the terminology that we have used of an "appropriate" Hilbert space. The question we address is: what is the structure of the Hilbert space in which one describes a given system?

5.6.1 Tensor Products of Spaces

For the one-dimensional motion of a particle along an axis x, the Hilbert space is $\mathcal{L}^2(R)$, for which the Hermite functions $\{\phi_n(x),\ n \text{ integer } \geq 0\}$ form a basis. Consider a particle moving in the xy plane. The appropriate Hilbert space is $\mathcal{L}^2(R^2)$, i.e. the square-integrable functions $\psi(x,y)$ of two variables. A particular basis of this space $\mathcal{L}^2(R^2)$ is constituted by the set $\{\phi_m(x)\phi_n(y),\ m,n \text{ integer } \geq 0\}$. In other words, any function $\Psi(x,y)$ of $\mathcal{L}^2(R^2)$ can be expanded as

$$\Psi(x,y) = \sum_{m,n} C_{m,n}\, \phi_m(x)\, \phi_n(y) \,. \tag{5.50}$$

Mathematically, this operation reflects the fact that the space $\mathcal{L}^2(R^2)$ can be considered as the tensor product of the space $\mathcal{L}^2(R)$ in which we describe the motion along x and of the space $\mathcal{L}^2(R)$ in which we describe the motion along y. Using Dirac's notation, (5.50) can be written as

$$|\Psi\rangle = \sum_{m,n} C_{m,n}\, |\phi_m\rangle \otimes |\phi_n\rangle \,, \tag{5.51}$$

where $|\phi_m\rangle \otimes |\phi_n\rangle$ is by definition the ket of $\mathcal{L}^2(R^2)$ corresponding to $\phi_m(x)\, \phi_n(y)$.

In order to define the notion of a tensor product in general, we consider two Hilbert spaces $\mathcal{E}$ and $\mathcal{F}$. We associate with $\mathcal{E}$ and $\mathcal{F}$ a third Hilbert space $\mathcal{G}$ and a bilinear mapping T of the direct product $\mathcal{E} \times \mathcal{F}$ on $\mathcal{G}$ such that the following apply.

1. $T(\mathcal{E} \times \mathcal{F})$ generates $\mathcal{G}$, i. e. any element of $\mathcal{G}$ is a (possibly infinite) sum of elements of the form

$$T(|u\rangle, |v\rangle); \qquad |u\rangle \in \mathcal{E}, \qquad |v\rangle \in \mathcal{F}.$$

2. Consider a basis $\{|e_m\rangle\}$ of $\mathcal{E}$ and a basis $\{|f_n\rangle\}$ of $\mathcal{F}$. Then the family $\{T(|e_m\rangle, |f_n\rangle)\}$ is a basis of $\mathcal{G}$.

The space $\mathcal{G}$ is called the *tensor product* of $\mathcal{E}$ and $\mathcal{F}$; this relation is denoted $\mathcal{G} = \mathcal{E} \otimes \mathcal{F}$. The mapping T is written as $T(|u\rangle, |v\rangle) = |u\rangle \otimes |v\rangle$. The elements of $\mathcal{E} \otimes \mathcal{F}$ are called tensors; they have the general form

$$|\Psi\rangle = \sum_{m,n} C_{m,n} \, |e_m\rangle \otimes |f_n\rangle \, . \tag{5.52}$$

Elements of the form $|u\rangle \otimes |v\rangle$ are said to be *factorized.* Any tensor can be written in a nonunique way as a (possibly infinite) sum of factorized tensors.

5.6.2 The Appropriate Hilbert Space

In order to define the Hilbert space in which we describe the state of a quantum system, we introduce the notion of *degrees of freedom.* A particle moving in space has three translational degrees of freedom, corresponding to motions along x, y and z. A system of two particles in space has six degrees of freedom. We shall see later that a particle may possess an intrinsic angular momentum (its spin), which results in a new degree of freedom.

Each degree of freedom is described in a particular Hilbert space. For instance the motion along x is described in the space of square-integrable functions of the variable x, $\mathcal{L}^2(R)$, as we have just said. We postulate that any given system involving N degrees of freedom can be described in the Hilbert space $\mathcal{E}$ which is the *tensor product* of the Hilbert spaces $\mathcal{E}_i$, $i = 1, 2, ..., N$, in which each of these N degrees of freedom is described:

$$\mathcal{E} = \mathcal{E}_1 \otimes \mathcal{E}_2 \otimes \ldots \otimes \mathcal{E}_N \, .$$

5.6.3 Properties of Tensor Products

1. If the dimensions $N_\mathcal{E}$ and $N_\mathcal{F}$ of $\mathcal{E}$ and $\mathcal{F}$ are finite, the dimension of $\mathcal{G} = \mathcal{E} \otimes \mathcal{F}$ is $N_\mathcal{G} = N_\mathcal{E} N_\mathcal{F}$.
2. When there are no ambiguities, it is useful to use the compact notations $|u\rangle \otimes |v\rangle \equiv |u\rangle|v\rangle \equiv |u, v\rangle$.
3. The Hermitian scalar product of two factorized kets $|\psi\rangle = |u\rangle \otimes |v\rangle$ and $|\chi\rangle = |u'\rangle \otimes |v'\rangle$ factorizes as follows:

$$\langle\chi|\psi\rangle = \langle u'|u\rangle \, \langle v'|v\rangle \, . \tag{5.53}$$

5.6.4 Operators in a Tensor Product Space

Consider two operators $\hat{A}_E$ and $\hat{B}_F$ acting in $\mathcal{E}$ and $\mathcal{F}$, respectively. We can define the tensor product of the operators $\hat{A}_E$ and $\hat{B}_F$,

$$\hat{C}_G = \hat{A}_E \otimes \hat{B}_F ,$$

by the rule

$$(\hat{A}_E \otimes \hat{B}_F)(|u\rangle \otimes |v\rangle) = (\hat{A}_E|u\rangle) \otimes (\hat{B}_F|v\rangle) . \tag{5.54}$$

This allows us to define the action of $\hat{C}_G$ on the elements of the factorized basis $\{|m\rangle \otimes |n\rangle\}$ and, consequently, on any element of $\mathcal{G}$. In particular, we can define the *continuation, or extension*, of the operator $\hat{A}_E$ in $\mathcal{G}$ by $\hat{A}_G = \hat{A}_E \otimes \hat{I}_F$, where $\hat{I}_F$ is the identity operator in $\mathcal{F}$.

5.6.5 Simple Examples

A Two-Particle System. Consider two particles, labeled 1 and 2, of masses m_1 and m_2, performing a one-dimensional harmonic motion. The appropriate Hilbert space is the tensor product of individual Hilbert spaces $\mathcal{E}_\mathrm{H} = \mathcal{E}^{(1)} \otimes \mathcal{E}^{(2)} = \mathcal{L}^2(R) \otimes \mathcal{L}^2(R)$. The Hamiltonian of the system can be written as

$$\hat{H} = \left(\frac{\hat{p}_1^2}{2m_1} + \frac{1}{2}m_1\omega_1^2\hat{x}_1^2\right) \otimes \hat{I}_2 \;+\; \hat{I}_1 \otimes \left(\frac{\hat{p}_2^2}{2m_2} + \frac{1}{2}m_2\omega_2^2\hat{x}_2^2\right) .$$

When there is no ambiguity, this can be written in the more compact form

$$\hat{H} = \frac{\hat{p}_1^2}{2m_1} + \frac{1}{2}m_1\omega_1^2\hat{x}_1^2 \;+\; \frac{\hat{p}_2^2}{2m_2} + \frac{1}{2}m_2\omega_2^2\hat{x}_2^2 .$$

From the wave-function point of view, the state of the system is described by a function $\Psi(x_1, x_2)$ which is square integrable in both of the variables x_1 and x_2. The operators $\hat{p}_1$ and $\hat{p}_2$ are $-\mathrm{i}\hbar\partial/\partial x_1$ and $-\mathrm{i}\hbar\partial/\partial x_2$, respectively. We know the solution of the eigenvalue problem for a one-dimensional harmonic oscillator (Sect. 4.2). The eigenfunctions are the Hermite functions $\phi_n(x/a)$ (with $a = \sqrt{\hbar/m\omega}$) and the eigenvalues are $(n + 1/2)\hbar\omega$. In the problem considered here, an eigenbasis of $\hat{H}$ is therefore given by the set

$$\Phi_{n_1,n_2}(x_1, x_2) = \phi_{n_1}(x_1/a_1)\,\phi_{n_2}(x_2/a_2) , \qquad n_1, n_2 \text{ integers} ,$$

where $a_i = (\hbar/m_i\omega_i)^{1/2}$. The associated eigenvalues are

$$E_{n_1,n_2} = \left(n_1 + \frac{1}{2}\right)\hbar\omega_1 + \left(n_2 + \frac{1}{2}\right)\hbar\omega_2 .$$

These eigenvalues are nondegenerate, except if ω_1/ω_2 is rational. Any function $\Psi(x_1, x_2)$ of $\mathcal{L}^2(R) \otimes \mathcal{L}^2(R)$ can be written as

$$\Psi(x_1, x_2) = \sum_{n_1,n_2} C_{n_1,n_2}\,\phi_{n_1}(x_1/a_1)\,\phi_{n_2}(x_2/a_2) .$$

Particle in a Box. When we studied in Chap. 4 the motion of a particle in a three-dimensional cubic box of side L, it was convenient to separate the motions of the particle along x, y, z and to seek particular solutions of the form

$$\Psi_{n_1,n_2,n_3}(x,y,z) = \psi_{n_1}(x)\,\psi_{n_2}(y)\,\psi_{n_3}(z) \\ \propto \sin(n_1\pi x/L)\,\sin(n_2\pi y/L)\,\sin(n_3\pi z/L)\,.$$

In the present language these are nothing but factorizable tensors. A general wave function is of the form

$$\Psi(x,y,z) = \sum C_{n_1,n_2,n_3}\,\Psi_{n_1,n_2,n_3}(x,y,z)\,.$$

More subtle examples will occur when we study particles with internal degrees of freedom, such as the magnetic moment of an atom in the Stern–Gerlach experiment or the spin 1/2 of the electron.

5.7 Reversible Evolution and the Measurement Process

The principles we have just stated imply that there are two different types of evolution for a quantum system. When the system is not subject to a measurement, its evolution can be considered as reversible since a knowledge of the state vector $|\psi(t)\rangle$ at time t and of the Hamiltonian between an initial time t_i and the time t allows one to determine[4] the state vector at the initial time $|\psi(t_\mathrm{i})\rangle$. The reduction of the wave packet which occurs in a measurement (Second Principle part (d) in Sect. 5.5) is, in contrast, fundamentally irreversible: after a single measurement has been made on a given system, one *cannot* reconstruct the state vector $|\psi\rangle$ of the system *before* the measurement. We only know the projection $\hat{P}_\alpha|\psi\rangle$.

The coexistence of two different types of evolution is very paradoxical. Indeed, it should in principle be possible to describe the set of atoms which constitute the detector by quantum mechanics, and to determine the Hamiltonian which couples this detector to the system $\mathcal{S}$ on which the measurement is made. The evolution of the large system $\{\mathcal{S} + \text{detector}\}$ should then be governed by the Schrödinger equation during the measurement process, in contradiction with the principle of wave packet reduction.

Actually, the principle of such a procedure was outlined by Von Neumann in the early days of quantum mechanics. Consider a system $\mathcal{S}$ on which we wish to measure a quantity A corresponding to the operator $\hat{A}$. We denote by $\{|\alpha\rangle\}$ the eigenstates of $\hat{A}$. To perform this measurement we couple this system to a quantum detector $\mathcal{D}$. The Hilbert space in which we describe the state of the ensemble "$\mathcal{S}$ and $\mathcal{D}$" is the tensor product of the space associated

[4] This does not depend on whether the Hamiltonian is time dependent.

with the system $\mathcal{S}$ and the space describing the state of the detector, corresponding to the various outcomes of the measurement. Initially, when no measurement has been made, the detector is in the state $|D_0\rangle$. For instance, if the detector is a photomultiplier, $|D_0\rangle$ is the state for which no photon has been counted. When we couple the system and the detector, we arrange $\hat{H}$ so that the interaction between $\mathcal{S}$ and $\mathcal{D}$ leads to the following evolution:

$$\begin{aligned} |\alpha\rangle \otimes |D_0\rangle &\longrightarrow |\alpha\rangle \otimes |D_\alpha\rangle\,, \\ |\alpha'\rangle \otimes |D_0\rangle &\longrightarrow |\alpha'\rangle \otimes |D_{\alpha'}\rangle\,, \quad \ldots\,, \end{aligned} \tag{5.55}$$

where the two states $|D_\alpha\rangle$ and $|D_{\alpha'}\rangle$ are "macroscopically" different from each other. Each state $|D_\alpha\rangle$ corresponds, for instance, to a given position of the needle on a meter, or to a given number of counts written in the memory of a computer. The evolution given in (5.55) expresses the intuitive fact that the detector should give an indication D_α when the system $\mathcal{S}$ is with certainty in the state $|\alpha\rangle$.

If the system $\mathcal{S}$ is in a state $\sum_\alpha c_\alpha|\alpha\rangle$ before the measurement, the linearity of the Schrödinger equation implies that the evolution of the ensemble $\mathcal{S}+\mathcal{D}$ reads

$$\left(\sum_\alpha c_\alpha|\alpha\rangle\right) \otimes |D_0\rangle \longrightarrow \sum_\alpha c_\alpha|\alpha\rangle \otimes |D_\alpha\rangle\,. \tag{5.56}$$

This result is quite different from one would expect from part (d) of the Second Principle, which would be, rather,

$$\left(\sum_\alpha c_\alpha|\alpha\rangle\right) \otimes |D_0\rangle \longrightarrow \begin{cases} |\alpha\rangle \otimes |D_\alpha\rangle \;\; \text{with a probability } |c_\alpha|^2 \\ |\alpha'\rangle \otimes |D_{\alpha'}\rangle \text{ with a probability } |c_{\alpha'}|^2 \\ \ldots \end{cases}. \tag{5.57}$$

This difference is not surprising, since the Schrödinger equation is perfectly deterministic, and cannot generate the nondeterministic evolution (5.57). The relevant question at this stage is the following: what are the consequences of the difference between (5.56) and (5.57), both in terms of the interpretation of quantum mechanics and in terms of the prediction of experimental results.

We first note that (5.56) raises a new and fundamental question. The state of the ensemble $\mathcal{S}+\mathcal{D}$ is a linear superposition of macroscopically different states, and it is quite difficult to obtain intuitive insight into its meaning: what is the significance of the sentence "the oscilloscope is in a linear superposition of the two results 1 volt and 2 volts"? The most celebrated example of a quantum superposition of macroscopic states is the Schrödinger cat paradox. By considering a cat shut up in a chamber with a radioactive substance, a small flask of hydrocyanic acid and a "diabolical device", one arrives at the conclusion that the state of the whole system after some time has "the living and the dead cat (pardon the expression) mixed or smeared out in equal parts".

The possible solutions to these paradoxical situations involving superpositions of macroscopic states are at present the subject of many public debates and discussions. One can, of course, assume that quantum mechanics does not apply to large objects such as a detector or a cat, which must be sufficiently complex that it can be read or seen by a human eye. Notice, however, that there does not exist at present any experiment that can tell us the critical size above which quantum mechanics would not apply.

A possible interpretation of (5.56) is to say that the human observer should be included as a part of the detector. In this case (5.56) expresses the fact that the global wave function is a superposition of several possible "worlds", each of which corresponds to a possible result of the measurement: as a result of observation, the world splits into several new worlds. As long as it is not possible to travel from one world to another, this interpretation does not lead to any contradiction with the postulates presented above. It is also an essential ingredient of several science-fiction novels.

However this "many worlds" interpretation is rejected by several authors, who consider that it is meaningless to consider a very large or infinite number of worlds with which we have no contact, and which have no influence on us. An opposite view considers that the discontinuity in the measurement operation (part (d) of the Second Principle) indeed exists, and that it takes place in the mind of the observer. Before the observer becomes aware of the results, the system is in the complicated superposition of states (5.56), which includes all possible results of the measurement. It is when the observer becomes aware of the result that the state vector is projected onto the state $\hat{P}_\alpha|\psi\rangle$ (or that the poor cat is killed); one is then left with (5.57).

Between these two extreme points of view, one finds a large class of physicists who do not ask quantum mechanics to give a proper description of "reality", but simply to give an operational method for calculating the possible outcomes of an experiment. In this respect, an appropriate analysis of the measuring apparatus along the lines initiated by Von Neumann indeed permits one to "circumvent" part (d) of the Second Principle. To summarize these ideas, we first note that, from a purely operational point of view, part (d) of the Second Principle becomes of interest only if one performs at least two consecutive measurements on a system $\mathcal{S}$. In a single measurement, it is sufficient to know the various possible outcomes and the corresponding probabilities, which are given by parts (a), (b) and (c) of the Second Principle. However, in the case of two measurements, it is essential to know the state of $\mathcal{S}$ after the first measurement in order to evaluate the probabilities obtained in the second measurement. Suppose we perform on $\mathcal{S}$ a sequence of measurements corresponding to the observables $\hat{A}, \hat{B}, \ldots$; we want to know the probability of finding the sequence of results $a_\alpha, b_\beta, \ldots$. According to part (d) of the Second Principle, when the first measurement is made, we must project the state vector onto $|\alpha\rangle$, then let it evolve under the Schrödinger equation until we perform the second measurement, then project it again onto $|\beta\rangle$, etc.

However, without invoking the reduction of the wave packet, and by performing an appropriate quantum analysis of the detectors along lines similar to (5.56), including their coupling to the outside environment, one can calculate the probability that detector A displays the value a_α, detector B the value b_β, etc., after the sequence of measurements has been performed. One finds that this probability coincides with the probability one obtains by assuming the reduction of the wave packet (see Exercise 5.5).

In this context, the reduction of the wave packet appears to be a *convenient tool* rather than a new principle. It allows one to evaluate results of multiple measurements without getting involved in a complicated description of all the detectors which participate in the series of measurements. This approach, which has undergone much recent development, is called *decoherence theory*. The reason for this name is that the theory is based on the fact that quantum correlations (coherences), which appear between the various macroscopically different states $|\alpha\rangle \otimes |D_\alpha\rangle$ as a consequence of the measurement process, are very rapidly washed out owing to the coupling of the detectors to their environment.

Further Reading

- The problem of measurement in quantum mechanics has caused debates which started immediately after the theory was founded and which are still going on. The book *Quantum Theory and Measurement*, edited by J.A. Wheeler and W.H. Zurek, Princeton University Press, Princeton (1983), contains the essential papers on this subject published before 1982. In particular, an English translation of the paper in which Schrödinger discussed the fate of a cat in front of a diabolical device is reproduced in this book. See also S. Goldstein, "Quantum theory without observers", Phys. Today, March 1998, p. 42.
- The book entitled *The Many-World Interpretation of Quantum Mechanics*, by B.S. DeWitt and N. Graham, eds., Princeton University Press, Princeton (1973) presents several papers centered on this particular theory.
- The possible role of human consciousness in wave packet reduction was advocated in particular by E.P. Wigner, *Symmetries and Reflection*, Indiana University Press, Bloomington (1967).
- For recent discussions of the decoherence approach, or rather of the decoherence approaches since different conceptions of the notion of reality emerge among different authors, see for instance W.H. Zurek, Phys. Today **44**, 36 (1991); M. Gell-Mann, *The Quark and the Jaguar*, Little, Brown, London (1994); S. Haroche, "Entanglement, decoherence, and the quantum/classical boundary", Phys. Today, July 1998, p. 36; R. Omnès, *Understanding Quantum Mechanics*, Princeton University Press, Princeton (1999); F. Laloë, "Do we really understand quantum mechanics?", Am. J. Phys. **69**, 655 (2001).

- Recent experimental results on quantum superposition of mesoscopic states ("Schrödinger kittens") are presented in M. Brune et al., Phys. Rev. Lett. **77**, 4887 (1996). See also P. Yam, "Bringing Schrödinger cat to life", Sci. Am., June 1997, p. 104; P. Kwiat, H. Weinfurter and A. Zeilinger, "Quantum seeing in the dark", Sci. Am., November 1996, p. 52.

Exercises

5.1. Translation and rotation operators.

a. Consider a one-dimensional problem and a wave function $\psi(x)$ which can be expanded in a Taylor series. Show that the operator $\hat{T}(x_0) = \mathrm{e}^{-\mathrm{i}x_0\hat{p}/\hbar}$, where x_0 is a length and $\hat{p}$ is the momentum operator, is such that

$$\hat{T}(x_0)\psi(x) = \psi(x - x_0)\,.$$

N.B. The expansion $\mathrm{e}^{\mathrm{i}\hat{u}} = \sum_{n=0}^{\infty} (i\hat{u})^n/n!$ is mathematically legitimate.

b. We now consider a two-dimensional problem in the xy plane and define the z component of the angular momentum operator by (cf. Table 3.1)

$$\hat{L}_z = \hat{x}\hat{p}_y - \hat{y}\hat{p}_x = -\mathrm{i}\hbar\left(x\frac{\partial}{\partial y} - y\frac{\partial}{\partial x}\right) = -\mathrm{i}\hbar\frac{\partial}{\partial\theta}\,,$$

where the polar coordinates r, θ are defined by $r = \left(x^2 + y^2\right)^{1/2}$ and $\theta = \arctan(y/x)$. Show that the operator $\hat{R}(\varphi) = \mathrm{e}^{-\mathrm{i}\varphi\hat{L}_z/\hbar}$, where φ is dimensionless, is such that

$$\hat{R}(\varphi)\,\psi(r,\theta) = \psi(r,\theta-\varphi)\,.$$

5.2. The evolution operator. Consider a system whose Hamiltonian does not depend on time (an isolated system). Show that the state vector at time t, denoted $|\psi(t)\rangle$, can be deduced from the state vector $|\psi(t_0)\rangle$ at the initial time using

$$|\psi(t)\rangle = \hat{U}(t-t_0)\,|\psi(t_0)\rangle\,, \qquad \text{where} \qquad U(\tau) = \mathrm{e}^{-\mathrm{i}\hat{H}\tau/\hbar}\,. \tag{5.58}$$

Show that $\hat{U}(\tau)$ is unitary, i. e. $\hat{U}^\dagger = \hat{U}^{-1}$.

5.3. Heisenberg representation. Consider an isolated system whose Hamiltonian is $\hat{H}$. We denote by $|\psi(0)\rangle$ the state vector of the system at time $t = 0$. We want to calculate the expectation value $a(t)$ of the results of the measurement of an observable $\hat{A}$ at time t.

a. Express $a(t)$ in terms of $|\psi(0)\rangle$, $\hat{A}$ and the evolution operator $\hat{U}(t)$ defined in the previous exercise.

b. Show that $a(t)$ can be written as the expectation value of an operator $\hat{A}(t)$ for the state $|\psi(0)\rangle$. Show that $\hat{A}(t)$ can be determined from

$$\mathrm{i}\hbar\frac{\mathrm{d}\hat{A}(t)}{\mathrm{d}t} = [\hat{A}(t), \hat{H}] \qquad \text{and} \qquad \hat{A}(0) = \hat{A}\,. \tag{5.59}$$

This approach is called the Heisenberg representation (or Heisenberg picture): the state vector is time independent, and the operators obey the Heisenberg equation (5.59).

5.4. Dirac formalism with a two-state problem. Consider two normalized eigenstates $|\psi_1\rangle$ and $|\psi_2\rangle$ of a Hamiltonian $\hat{H}$ corresponding to different eigenvalues E_1 and E_2 (one can set $E_1 - E_2 = \hbar\omega$).

a. Show that $|\psi_1\rangle$ and $|\psi_2\rangle$ are orthogonal.
b. Consider the state $|\psi_-\rangle = \{|\psi_1\rangle - |\psi_2\rangle\}/\sqrt{2}$, calculate the expectation value $\langle E\rangle$ of the energy and the dispersion ΔE in this state.
c. Assume that at $t = 0$ the system is in the state $|\psi(t=0)\rangle = |\psi_-\rangle$. What is the state of the system $|\psi(t)\rangle$ at time t?
d. Consider an observable $\hat{A}$ defined by $\hat{A}|\psi_1\rangle = |\psi_2\rangle$, and $\hat{A}|\psi_2\rangle = |\psi_1\rangle$. What are the eigenvalues a of $\hat{A}$ in the subspace generated by $|\psi_1\rangle$ and $|\psi_2\rangle$?
e. Construct the corresponding combinations of $|\psi_1\rangle$ and $|\psi_2\rangle$, which are eigenvectors of $\hat{A}$.
f. Assume that at $t = 0$ the system is in the state $|\psi_-\rangle$ corresponding to the eigenvalue $a = -1$. What is the probability of finding $a = -1$ in a measurement of A at a later time t?

5.5. Successive measurements and the principle of wave packet reduction. Consider a quantum system $\mathcal{S}$ which is prepared in a state $|\psi_0\rangle$ at time $t = 0$, and two observables $\hat{A}$ and $\hat{B}$ associated with this system. We are interested in the (joint) probability $P(\alpha_i, 0;\ \beta_j, t)$ that a measurement of A at time $t = 0$ gives the result α_i *and* that a measurement of B at time t gives the result β_j.

We note using $\{|a_i\rangle\}$ (resp. $\{|b_j\rangle\}$) a basis of eigenvectors of $\hat{A}$ (resp. $\hat{B}$) associated with the eigenvalues $\{\alpha_i\}$ (resp. $\{\beta_u\}$). We assume for simplicity that the spectra of $\hat{A}$ and $\hat{B}$ are nondegenerate.

a. In this question, we assume we can apply part (d) of Second Principle (reduction of the wave packet).
   i) In terms of $\langle a_i|\psi_0\rangle$, express the probability that a measurement of A gives the result α_i at time $t = 0$.
   ii) We assume that the measurement of A has given the result α_i.
      i. What is the state $|\psi(0_+)\rangle$ of the system just after this measurement?

ii. Write the state of the system at time t using the evolution operator $e^{-i\hat{H}_S t/\hbar}$ (see Exercise 2), where $\hat{H}_S$ is the Hamiltonian of the system $\mathcal{S}$.

iii. Express in terms of $\langle b_j|e^{-i\hat{H}_S t/\hbar}|a_i\rangle$ the probability that a measurement of B at time t gives the result β_j.

iii) Calculate the desired probability $P(\alpha_i, 0; \beta_j, t)$.

b. In this section, we do *not* apply the reduction part (d) of the Second Principle. We consider two quantum detectors $\mathcal{A}$ and $\mathcal{B}$ which measure, respectively, the physical quantities A and B. The initial state of the ensemble $\mathcal{E}$, formed by $\mathcal{S}$, $\mathcal{A}$ and $\mathcal{B}$, can be written as:

$$|\Psi_0\rangle = |\psi_0\rangle \otimes |\mathcal{A}_0\rangle \otimes |\mathcal{B}_0\rangle ,$$

where the states $|\mathcal{A}_0\rangle$ and $|\mathcal{B}_0\rangle$ correspond to detectors which have not yet measured anything. We assume that the measurement operation of A amounts to coupling $\mathcal{S}$ and $\mathcal{A}$ in order to generate the following evolution:

$$|a_i\rangle \otimes |\mathcal{A}_0\rangle \otimes |\mathcal{B}\rangle \rightarrow |a_i\rangle \otimes |\mathcal{A}_i\rangle \otimes |\mathcal{B}\rangle \quad (|\mathcal{B}\rangle : \text{ arbitrary state of } \mathcal{B}) .$$

We make the similar assumption concerning the measurement of B:

$$|b_j\rangle \otimes |\mathcal{A}\rangle \otimes |\mathcal{B}_0\rangle \rightarrow |b_j\rangle \otimes |\mathcal{A}\rangle \otimes |\mathcal{B}_j\rangle \quad (|\mathcal{A}\rangle : \text{ arbitrary state of } \mathcal{A}) .$$

Here the states $|\mathcal{A}_i\rangle$ (resp. $|\mathcal{B}_j\rangle$) are macroscopically different from each other and they correspond to the different possible result α_i (resp. β_j) displayed on the detector $\mathcal{A}$ (resp. $\mathcal{B}$). Apart from time intervals when measurements of A and B are performed, and where couplings are present, the evolution of $\mathcal{E}$ occurs via the Hamiltonian

$$\hat{H} = \hat{H}_\mathcal{S} + \hat{H}_\mathcal{A} + \hat{H}_\mathcal{B} .$$

We assume that the states $|\mathcal{A}_0\rangle$ and $|\mathcal{A}_i\rangle$ (resp. $|\mathcal{B}_0\rangle$ and $|\mathcal{B}_j\rangle$) are eigenstates of $\hat{H}_\mathcal{A}$ (resp. $\hat{H}_\mathcal{B}$). We note A_0 and A_i (resp. B_0 and B_j) the corresponding eigenvalues. The time interval of the coupling corresponding to the measurement of A or B is assumed to be sufficiently small so that we can neglect the action of $\hat{H}$ during that interval.

(a) Write the state of $\mathcal{E}$ at time 0_+, i.e. just after the coupling of $\mathcal{S}$ and $\mathcal{A}$ which results in a measurement of A.

(b) Write the state of $\mathcal{E}$ at time t, just before the measurement of B. In order to do that, introduce the evolution operator $e^{-i\hat{H}t/\hbar}$ and use the fact that $\hat{H}_\mathcal{S}$, $\hat{H}_\mathcal{A}$ and $\hat{H}_\mathcal{B}$ all commute.

(c) Write the state of $\mathcal{E}$ at time t_+, just after the measurement of B.

(d) What is the probability of finding the detector $\mathcal{A}$ in the state $|\mathcal{A}_i\rangle$ *and* the detector $\mathcal{B}$ in the state $|\mathcal{B}_j\rangle$? Compare the result with what one obtains by applying the principle of wave packet reduction.

c. Discuss the differences between the two approaches, both in their principles and in their practical aspects.

6. Two-State Systems, Principle of the Maser

Act as a primitive and predict as a strategist.
René Char

In Chaps. 2 to 4, we became familiar with quantum theory in the form of wave mechanics. This approach is well suited for studying the quantum motion of a point particle in space. The physics of this problem is simple and intuitive. However, the mathematical tools turn out to be somewhat cumbersome: the Hilbert space has an infinite number of dimensions, we make use of the Fourier transform, etc. In the present chapter we want to exploit the matrix formulation presented in the previous chapter, in order to consider a physical problem which, in contrast, can be described in the simplest possible mathematical structure: a *two-dimensional* Hilbert space.

We choose to study a particular problem, the ammonia maser, for which we shall make use of the physical results obtained for the NH_3 molecule in Chap. 4. However, the results we shall obtain actually have a much larger class of applications than this particular example, which we have simplified intentionally. Numerous physical systems can be described in two- or three-dimensional Hilbert spaces, either exactly (the electron spin or magnetic moment that we shall meet later on; light polarization, which we briefly consider in Sect. 6.2; the physics of neutral K and B mesons in particle physics, the problem of neutrino masses; etc.) or approximately (masers, laser physics and many atomic-physics problems).

The vectors of a two-dimensional space can all be constructed as linear superpositions of two independent basis vectors, and one frequently calls such systems *two-state systems.* Of course, these systems can be in an infinite number of states, but these states are all linear superpositions of two of them. This is the reason for this terminology.

6.1 Two-Dimensional Hilbert Space

Consider two orthonormal basis vectors $|\psi_1\rangle$ and $|\psi_2\rangle$. We can use a matrix representation of the form

$$|\psi_1\rangle \Leftrightarrow \begin{pmatrix} 1 \\ 0 \end{pmatrix}, \qquad |\psi_2\rangle \Leftrightarrow \begin{pmatrix} 0 \\ 1 \end{pmatrix}, \tag{6.1}$$

$$\langle\psi_1| \Leftrightarrow (1 \quad 0), \qquad \langle\psi_2| \Leftrightarrow (0 \quad 1), \tag{6.2}$$

and the general form of any vector $|\psi\rangle$ is

$$|\psi\rangle = \alpha|\psi_1\rangle + \beta|\psi_2\rangle \Leftrightarrow \begin{pmatrix} \alpha \\ \beta \end{pmatrix}, \tag{6.3}$$

$$\langle\psi| = \alpha^*\langle\psi_1| + \beta^*\langle\psi_2| \Leftrightarrow (\alpha^* \quad \beta^*), \tag{6.4}$$

with the normalization condition $\langle\psi|\psi\rangle = |\alpha|^2 + |\beta|^2 = 1$.

A linear operator in this space can be represented by a 2×2 complex matrix. The *most general* 2×2 Hermitian matrix $\hat{M}$ can be written as

$$\hat{M} = \begin{pmatrix} a+d & b-\mathrm{i}c \\ b+\mathrm{i}c & a-d \end{pmatrix} = a\hat{I} + b\hat{\sigma}_1 + c\hat{\sigma}_2 + d\hat{\sigma}_3\,, \qquad a, b, c, d \text{ real numbers,}$$

where $\hat{I}$ is the unit matrix and where the Hermitian matrices $\hat{\sigma}_k$, which are called the *Pauli matrices*, are defined as

$$\hat{\sigma}_1 = \begin{pmatrix} 0 & 1 \\ 1 & 0 \end{pmatrix}, \qquad \hat{\sigma}_2 = \begin{pmatrix} 0 & -\mathrm{i} \\ \mathrm{i} & 0 \end{pmatrix}, \qquad \hat{\sigma}_3 = \begin{pmatrix} 1 & 0 \\ 0 & -1 \end{pmatrix}. \tag{6.5}$$

6.2 A Familiar Example: the Polarization of Light

There exists a familiar phenomenon for which a description in a two-dimensional Hilbert space is natural: the polarization of light.

6.2.1 Polarization States of a Photon

Light waves are transverse. Classically, the polarization of light describes the behavior of the electric-field vector, transverse to the direction of propagation. There exist various types of polarization; in general it is elliptical, with the limiting cases of circular or linear. Natural light is not polarized; more precisely, it is in a statistical mixture of polarization states. Light with an arbitrary polarization can be prepared by means of a suitable combination of polarizers and quarter- or half-wave plates. In practice, two types of polarizers are used. One-way polarizers transmit light which has a polarization parallel to a given axis, and absorb light polarized orthogonally to that axis. Two-way polarizers transmit light that is incident with a given linear polarization, and deflect in another direction the light with the orthogonal polarization component (Fig. 6.1).

Rather than working with the classical description of the electric field vector, we shall consider here the *polarization states* of the individual photons which constitute the light beam. These photons are elementary particles and

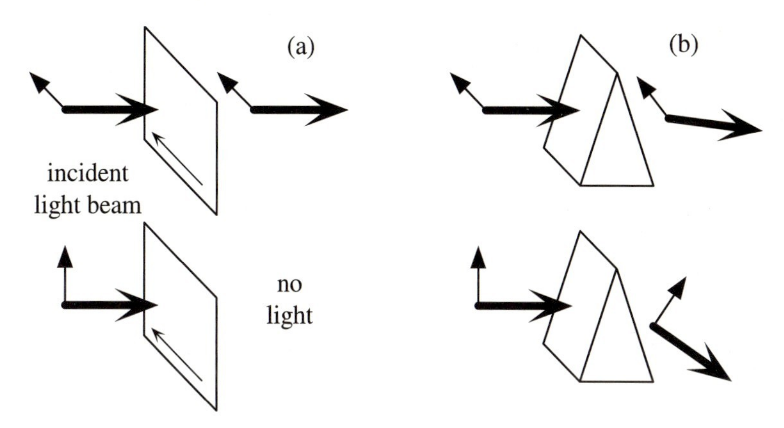

Fig. 6.1. (**a**) A one-way polarizer: incident light with a polarization parallel to the axis of the polarizer is transmitted, and light with a polarization orthogonal to the axis is absorbed; (**b**) a two-way polarizer (e. g. a calcite prism): the two polarizations emerge in different directions. This latter polarizer, when considered for a single photon, is very similar to the Stern–Gerlach apparatus for a spin of 1/2 (see Chaps. 8 and 12)

cannot be split into pieces. A beam of red light with a power of 1 watt carries 3×10^{18} photons per second. Such photons do not interact with one another because their mutual separations are much too large. Therefore, a light beam is a collection of independent photons.

One describes the polarization states of a photon in a two-dimensional Hilbert space. In this space, we can choose the basis states corresponding to horizontal and vertical polarization states, which we write as

$$|\rightarrow\rangle \quad \text{and} \quad |\uparrow\rangle\,, \tag{6.6}$$

respectively. These states are defined physically by the fact that if the photon is in the polarization state $|\rightarrow\rangle$, it is transmitted with probability 1 by a one-way polarizer whose axis is horizontal. In the opposite case, if the polarization state of the photon is $|\uparrow\rangle$, it is absorbed by this same polarizer, i. e. it is transmitted with probability zero. By definition, these states are such that $\langle\uparrow|\rightarrow\rangle = 0$.

We denote by $|\theta\rangle$ the state of the photon corresponding to a linear polarization in the direction making an angle θ with respect to the horizontal axis $(0 \le \theta < \pi)$. This state is a linear combination of the basis states (6.6)

$$|\theta\rangle = \cos\theta\,|\rightarrow\rangle + \sin\theta\,|\uparrow\rangle\,, \tag{6.7}$$

with real components.

If there existed only states with a linear polarization, there would be no need to introduce a Hilbert space formalism. Geometry in a two-dimensional Euclidean space, consisting of linear combinations of $|\rightarrow\rangle$ and $|\uparrow\rangle$ with real coefficients, would suffice.[1] However it is known in optics that one must also consider states with *complex* components, such as

$$|\Psi_{\mathrm{L,R}}\rangle = \frac{1}{\sqrt{2}}|{\rightarrow}\rangle \pm \frac{\mathrm{i}}{\sqrt{2}}|{\uparrow}\rangle\,. \tag{6.8}$$

One can easily check that these states keep the same mathematical form if they are expressed in any other basis of linear polarization states. These states correspond to left or right circular polarization. More generally, states of the type

$$|\psi\rangle = \alpha|{\rightarrow}\rangle + \beta|{\uparrow}\rangle\,,$$

where α and β are arbitrary complex numbers (satisfying $|\alpha|^2 + |\beta|^2 = 1$), correspond to elliptical polarization states and are physically relevant. Therefore it is essential to use Hilbert space (and not simply the transverse two-dimensional Euclidean space) in order to describe all polarization states of the photon.

6.2.2 Measurement of Photon Polarizations

After having introduced the Hilbert space in which the polarization of the photon is described, we turn to measurements. Consider first a linearly polarized light beam, obtained by filtering light with a one-way polarizer. The direction of the polarization of the beam is that of the axis of the polarizer, say horizontal. If the beam, of intensity I_0, meets a second polarizer (also called an analyzer) whose axis is at an angle θ to the first one, one finds that the transmitted intensity is $I_0 \cos^2\theta$ (Malus's law).

Let us describe this result in terms of photons, which cannot be divided, as mentioned above. When a photon encounters a polarizer, whatever its polarization state is, there are only two possibilities; either it crosses the polarizer or it doesn't (i. e. it is ejected in another direction or it is absorbed). Therefore Malus's law can be interpreted by saying that a linearly polarized photon has a *probability* $\cos^2\theta$ of passing through a polarizer at an angle θ, so that a fraction $\cos^2\theta$ of the incident photons are transmitted.

This result can be recovered from the general principles presented in the previous chapter, applied to the Hilbert space that we have just introduced. If we measure whether a photon is transmitted or not by a polarizer with its axis at an angle θ, we perform a measurement which can yield two results: 1 (transmission) or 0 (no transmission). The operator $\hat{A}_\theta$ associated with this

[1] Polarization comes from the fact that the photon is a spin-one particle. As we shall see in Chap. 10, the description of a spin-one degree of freedom should be done in a three-dimensional Hilbert space. The fact that here this space is two-dimensional instead of three-dimensional (i. e. that the polarization state parallel to the momentum of the photon does not exist) is related to gauge invariance and to the fact that the photon is a zero-mass particle.

measurement is the projector onto the state $|\theta\rangle$, i. e. the 2×2 Hermitian matrix

$$\hat{A}_\theta = |\theta\rangle\langle\theta| = \begin{pmatrix} \cos^2\theta & \cos\theta\sin\theta \\ \cos\theta\sin\theta & \sin^2\theta \end{pmatrix} .$$

This operator has two eigenstates: $|\theta\rangle$, associated with the eigenvalue 1, and $|\bar{\theta}\rangle$, associated with the eigenvalue 0 (with $\bar{\theta} = \theta \pm \pi/2$, so that $\langle\theta|\bar{\theta}\rangle = 0$). In particular, the probability for a photon initially in state $|\rightarrow\rangle$ to be transmitted is given by

$$P(\theta) = |\langle\theta|\rightarrow\rangle|^2 = \cos^2\theta ,$$

as expected from Malus's law.

We can also study the case of a circularly polarized photon, with an initial state given by (6.8). We obtain in this case

$$P_{\mathrm{L,R}}(\theta) = |\langle\theta|\Psi_{\mathrm{L,R}}\rangle|^2 = \left|\frac{\mathrm{e}^{\pm\mathrm{i}\theta}}{\sqrt{2}}\right|^2 = \frac{1}{2} .$$

The result is independent of the angle of the analyzer θ, as expected for circularly polarized light.

6.2.3 Successive Measurements and "Quantum Logic"

We set two polarizers at right angles (Fig. 6.2a). No light can get across, since two states polarized in perpendicular directions are orthogonal. Between these two crossed polarizers, we introduce another polarizer whose axis is at an angle θ (Fig. 6.2b). Light appears after the last polarizer, although we have introduced an object which absorbs or, at most, is transparent to light! It is easy to interpret this striking result in terms of polarization states. Consider for simplicity a photon initially polarized horizontally, so that the transmission probability at the first polarizer is $P_1 = 1$. The photon is transmitted by the second polarizer with a probability $P_2 = \cos^2\theta$. After crossing this polarizer, it is in the *new polarization state* $|\theta\rangle$ (reduction of the wave packet). Therefore, the probability that it crosses the third vertical polarizer is $P_3 = |\langle\uparrow|\theta\rangle|^2 = \sin^2\theta$. Altogether, the probability that a photon initially polarized horizontally crosses the set of three successive polarizers is $P = P_1P_2P_3 = \sin^2(2\theta)/4$, which vanishes only for $\theta = 0$ and $\theta = \pi/2$, i. e. if the axis of the intermediate polarizer is parallel to one of the two others.

It is clear in this example that classical probabilistic logic does not apply. Suppose we place two "contradictory", or mutually exclusive, logical gates one after the other, say "heads" and "tails". The probability that a coin crosses both gates is zero. Suppose that, between these two gates, we insert a logical gate that selects coins according to another criterion such as "North American" or "European". In a quantum financial world, one would observe that half of the coins which fell onto "heads" and which we know are Eu-

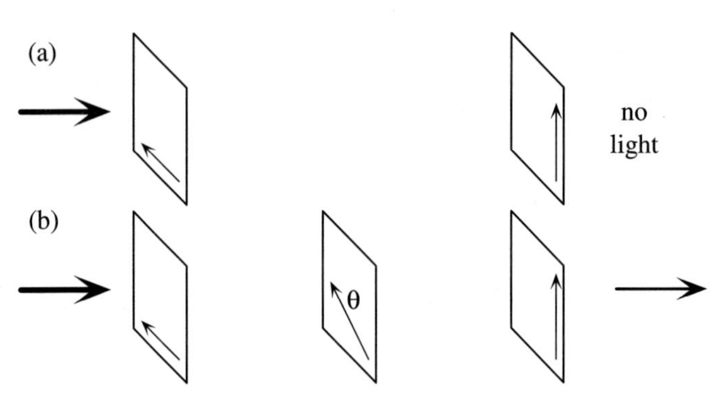

Fig. 6.2. (**a**) No light is transmitted by a set of two polarizers whose axes are orthogonal to each other. (**b**) If one inserts another polarizer between the two polarizers of (a), light emerges at the output of the setup. The relative intensity of the transmitted light is $\sin^2(2\theta)/4$

ropean had actually flipped onto "tails"! Business would be rather involved. This is a mere reflection of the superposition principle superimposed on the noncommutation of the relevant observables $\hat{A}_0$, $\hat{A}_\theta$, $\hat{A}_{\pi/2}$ of the problem, as we shall see in Chap. 7. For instance, the result is different according to whether the polarizer with axis θ is placed before the vertical polarizer (as in Fig. 6.2b) or after it.

6.3 The Model of the Ammonia Molecule

As an example we examine here how the ammonia molecule studied in Chap. 4 is described in the matrix formalism and, at low temperature, as a two-state system.

6.3.1 Restriction to a Two-Dimensional Hilbert Space

We recall first the results of the calculation done in Sect. 4.5 concerning the double-potential-well model of the NH_3 molecule. The two lowest energy levels correspond to symmetric and antisymmetric wave functions, denoted by $\psi_S(x)$ and $\psi_A(x)$, respectively. The corresponding energies are $E_S = E_0 - A$ and $E_A = E_0 + A$, with $A > 0$. All other energy levels E_α of the molecule are such that $E_\alpha - E_0 \gg A$. Therefore the splitting $2A$ of the two levels E_S and E_A is very small compared with the spacing between these levels and the other levels of the molecule. More precisely, for NH_3 one finds $2A \sim 10^{-4}$ eV; the first excited state is also split into two sublevels $E'_S = E_1 - A_1$, and $E'_A = E_1 + A_1$ with $2A_1 \sim 5 \times 10^{-3}$ eV and $E_1 - E_0 \sim 0.12$ eV.

In this chapter, we shall be interested *only* in states which are linear combinations of the two lowest-lying energy states $|\psi_S\rangle$ and $|\psi_A\rangle$, i.e.

$$|\psi\rangle = \lambda|\psi_S\rangle + \mu|\psi_A\rangle \,. \tag{6.9}$$

The physical meaning of this restriction is the following. The NH_3 molecule is a very complicated system and, in full generality, a state of this molecule is of the form

$$|\psi\rangle = \sum a_n |n\rangle \,, \tag{6.10}$$

where we have denoted by $\{\varepsilon_n\}$ the ordered sequence of its energy levels and $|n\rangle$ the corresponding eigenstates. For a molecule in the state $|\psi\rangle$, the probability of finding ε_n in an energy measurement is $P(\varepsilon_n) = |a_n|^2$. Physically, it is rather easy to impose constraints on the energy of NH_3 molecules in a gas, for instance if this gas is in thermal equilibrium with its environment. At a temperature T, the ratio of the populations of the energy levels E_i and E_j is given by Boltzmann's law:

$$N(E_j)/N(E_i) = \mathrm{e}^{-(E_j - E_i)/k_\mathrm{B}T} \,,$$

where k_B is the Boltzmann constant. Inserting the numerical values given above, one can readily check that, at a temperature of 100 K, $N(E_\mathrm{S})/N(E_\mathrm{A}) \sim 1$, whereas $N(E'_\mathrm{S})/N(E_\mathrm{S}) \sim N(E'_\mathrm{A})/N(E_\mathrm{S}) < 10^{-6}$. Therefore the probability of finding the molecule in a higher-energy state is very small, and we can safely approximate the state (6.10) of each molecule in a gas at 100 K by a combination of only the two states $|\psi_\mathrm{S}\rangle$ and $|\psi_\mathrm{A}\rangle$ such as (6.9).

6.3.2 The Basis $\{|\psi_\mathrm{S}\rangle, |\psi_\mathrm{A}\rangle\}$

Consider the basis of energy eigenstates $\{|\psi_\mathrm{S}\rangle, |\psi_\mathrm{A}\rangle\}$. A state vector $|\psi\rangle = \lambda|\psi_\mathrm{S}\rangle + \mu|\psi_\mathrm{A}\rangle$, where λ and μ are complex numbers such that $|\lambda|^2 + |\mu|^2 = 1$, is written in this basis as

$$|\psi\rangle = \begin{pmatrix} \lambda \\ \mu \end{pmatrix} . \tag{6.11}$$

The quantities $|\lambda|^2$ and $|\mu|^2$ are the probabilities of finding E_S and E_A, respectively, in an energy measurement. The Hamiltonian operator $\hat{H}$ is a diagonal matrix in this basis:

$$\hat{H} = \begin{pmatrix} E_0 - A & 0 \\ 0 & E_0 + A \end{pmatrix} . \tag{6.12}$$

Of course, this is not the total Hamiltonian of the NH_3 molecule, which in full generality is an infinite matrix acting on the states (6.10). Instead, the above operator is the *restriction* of the general Hamiltonian to the two-dimensional subspace which we are interested in.

The time evolution of a state vector which is defined at $t = 0$ by (6.11) can be calculated by solving the Schrödinger equation

$$\mathrm{i}\hbar \frac{\mathrm{d}}{\mathrm{d}t} |\psi(t)\rangle = \hat{H} |\psi(t)\rangle \tag{6.13}$$

and can be written directly as

$$|\psi(t)\rangle = \exp(-\mathrm{i}E_0 t/\hbar) \begin{pmatrix} \lambda \exp(\mathrm{i}\omega_0 t/2) \\ \mu \exp(-\mathrm{i}\omega_0 t/2) \end{pmatrix} , \tag{6.14}$$

where we have introduced the Bohr frequency ω_0 of the system, given by

$$2A = \hbar\omega_0 . \tag{6.15}$$

We can define the states $|\psi_\mathrm{R}\rangle$ and $|\psi_\mathrm{L}\rangle$ which correspond to the *classical* right and left configurations, respectively, as

$$|\psi_\mathrm{R}\rangle = \frac{1}{\sqrt{2}} (|\psi_\mathrm{S}\rangle + |\psi_\mathrm{A}\rangle) , \qquad |\psi_\mathrm{L}\rangle = \frac{1}{\sqrt{2}} (|\psi_\mathrm{S}\rangle - |\psi_\mathrm{A}\rangle) , \tag{6.16}$$

or, in matrix form,

$$|\psi_\mathrm{R}\rangle = \frac{1}{\sqrt{2}} \begin{pmatrix} 1 \\ 1 \end{pmatrix} , \qquad |\psi_\mathrm{L}\rangle = \frac{1}{\sqrt{2}} \begin{pmatrix} 1 \\ -1 \end{pmatrix} . \tag{6.17}$$

These states $|\psi_\mathrm{R}\rangle$ and $|\psi_\mathrm{L}\rangle$ are the eigenstates of the matrix $\hat{\sigma}_1$ given in (6.5), with eigenvalues ± 1. We thus define an observable $\hat{X}$, which we shall call the "position" operator of the particle representing the plane of hydrogen atoms, or, more precisely, the *location of the particle with respect to the center*:

$$\hat{X} = \begin{pmatrix} 0 & 1 \\ 1 & 0 \end{pmatrix} ; \qquad \hat{X}|\psi_\mathrm{R}\rangle = |\psi_\mathrm{R}\rangle ; \qquad \hat{X}|\psi_\mathrm{L}\rangle = -|\psi_\mathrm{L}\rangle . \tag{6.18}$$

This observable has eigenvalues ± 1. If the result of a measurement of X is $+1$, the particle is "in the right-hand well"; if the result is -1, it is "in the left-hand well". The observable $\hat{X}$ is not the position operator in the sense we used when dealing with wave functions. It only measures the side of the double well in which the particle is located. One can verify that, owing to the Heisenberg inequalities, if we require that the position of the particle be defined more precisely than the half-width of one of the wells, it is inconsistent to restrict the problem to two states. Indeed, a precise knowledge of the position is accompanied by a large spread in momentum and in kinetic energy, and more energy levels must be taken into account. Note that (6.18) is similar to the first 2×2 block of (5.34), which gives the matrix representation of the position operator $\hat{x}$ in the energy eigenbasis of a harmonic oscillator.

A straightforward calculation gives the expectation value $\langle x \rangle$ of the observable $\hat{X}$ in the state $|\psi(t)\rangle$ of (6.14) as

$$\langle x \rangle = \lambda^* \mu \mathrm{e}^{-\mathrm{i}\omega_0 t} + \lambda \mu^* \mathrm{e}^{\mathrm{i}\omega_0 t} . \tag{6.19}$$

In particular, if the particle is "on the right" at $t = 0$ ($\lambda = \mu = 1/\sqrt{2}$), we obtain

$$\langle x \rangle = \cos \omega_0 t . \tag{6.20}$$

As already pointed out in Chap. 4, the particle oscillates between right and left with an angular frequency ω_0. This *inversion phenomenon* of the NH_3 molecule occurs with a period $2\pi/\omega_0 = \pi\hbar/A$.

6.3.3 The Basis $\{|\psi_R\rangle, |\psi_L\rangle\}$

The two vectors $|\psi_L\rangle$ and $|\psi_R\rangle$ also form a basis of our two-dimensional Hilbert space. In this new basis, which corresponds to the classical configurations, it is interesting to see how the inversion phenomenon occurs. We can write the basis vectors as

$$|\psi_R\rangle = \begin{pmatrix} 1 \\ 0 \end{pmatrix}, \qquad |\psi_L\rangle = \begin{pmatrix} 0 \\ 1 \end{pmatrix}. \tag{6.21}$$

In this basis, the states $|\psi_S\rangle$ and $|\psi_A\rangle$ read

$$|\psi_S\rangle = \frac{1}{\sqrt{2}}\begin{pmatrix} 1 \\ 1 \end{pmatrix}, \qquad |\psi_A\rangle = \frac{1}{\sqrt{2}}\begin{pmatrix} 1 \\ -1 \end{pmatrix}. \tag{6.22}$$

These expressions reveal a very simple interference phenomenon: $|\psi_R\rangle$ is a linear superposition of the states $|\psi_S\rangle$ and $|\psi_A\rangle$, which interferes destructively on the left, and $|\psi_L\rangle$ is a superposition that interferes destructively on the right. Using (6.18) we find that in this basis, the observable $\hat{X}$ is diagonal:

$$\hat{X} = \begin{pmatrix} 1 & 0 \\ 0 & -1 \end{pmatrix}. \tag{6.23}$$

On the other hand, the Hamiltonian $\hat{H}$ is no longer diagonal. It has the form:

$$\hat{H} = \begin{pmatrix} E_0 & -A \\ -A & E_0 \end{pmatrix}, \tag{6.24}$$

whose eigenvectors are $|\psi_S\rangle$ and $|\psi_A\rangle$, with eigenvalues $E_0 \mp A$ as expected.

The nondiagonal terms in the above Hamiltonian are called *transition terms*. These terms are responsible for the inversion phenomenon, i. e. the periodic oscillation of the molecule between the right and left configurations $|\psi_R\rangle$ and $|\psi_L\rangle$. The larger A is, the faster this oscillation (i. e. $T = 2\pi/A$).

The matrix formalism simplifies things considerably, compared with the calculations of Chap. 4. Of course, when comparing the results with experiment, a fundamental ingredient of this matrix approach is the numerical value of the parameter A. In the double-square-well model we could calculate this parameter A. However, such a calculation is, in itself, an approximation, and the only accurate access we have to this number is through an experimental measurement. Once this value is known, the result of the matrix calculation is straightforward.

6.4 The Ammonia Molecule in an Electric Field

We are going to use the two-state formalism to present the principle of *masers*. These devices have brought about decisive progress in microwave physics, in telecommunications and in astrophysics. We shall concentrate here on the ammonia maser.

There exist a large variety of masers and lasers, whose principles are similar. In spite of their diversity, they are all governed essentially by the same mathematics as here. Some cases require more basis states (three- or four-state systems), but qualitatively the physical results can be understood starting from the prototype of the two-state ammonia molecule.

6.4.1 The Coupling of NH_3 to an Electric Field

In the classical configurations of the molecule represented in Fig. 6.3, the molecule possesses an electric dipole moment $\boldsymbol{D}$. This is a consequence of the large electron affinity of the nitrogen atom, which displaces the center of mass of the negative charges (electrons) with respect to that of the positive charges (nuclei). For symmetry reasons this electric dipole moment lies along the axis of the molecule, and it changes sign when the molecule flips from the right to the left configuration.

Suppose we apply a static electric field $\boldsymbol{\mathcal{E}}$ to the molecule. Classically, the potential energy W of the molecule in this field $\boldsymbol{\mathcal{E}}$ is

$$W = -\boldsymbol{D} \cdot \boldsymbol{\mathcal{E}} . \tag{6.25}$$

For simplicity, we assume that the field $\boldsymbol{\mathcal{E}}$ is parallel to the x axis. We want to find the form of the corresponding potential-energy observable $\hat{W}$.

From now on, we work with the basis $\{|\psi_S\rangle, |\psi_A\rangle\}$. In the problem under consideration (i.e. the two-state system), the natural choice for the *electric-dipole moment observable* of the molecule is to assume it is proportional to the position observable $\hat{X}$ that we have defined previously:

$$\hat{D} = d_0 \hat{X} = \begin{pmatrix} 0 & d_0 \\ d_0 & 0 \end{pmatrix} , \tag{6.26}$$

where d_0 is a characteristic measurable parameter of the molecule (experimentally, $d_0 \sim 3 \times 10^{-11}\,\mathrm{eV/(V/m)}$). In other words, we assume that if, in a measurement of the "position" (right or left) of the molecule, we find the values ± 1 with some probabilities, then when we measure its electric dipole moment, we shall find the two values $\pm d_0$ with the *same* probabilities. This

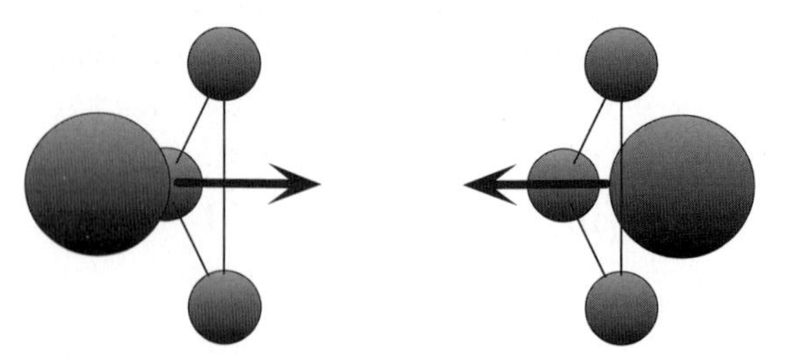

Fig. 6.3. The two classical configurations of the NH_3 molecule and the corresponding electric dipole moment

fulfills the condition that the electric dipole moment D flips if the molecule flips.

The natural choice for the potential-energy observable of the molecule in an electric field then consists in copying the classical form (6.25), i. e. taking the product of the observable $\hat{D}$ and the value of the field $\mathcal{E}$:

$$\hat{W} = -\mathcal{E}\hat{D} = \begin{pmatrix} 0 & -\eta \\ -\eta & 0 \end{pmatrix}, \qquad \text{where} \quad \eta = \mathcal{E} d_0 . \tag{6.27}$$

Of course, the real justification for this choice, a stronger justification than these plausibility arguments, is that it leads to a convincing explanation of experimental results.

The NH_3 molecule is a complex system containing four nuclei and ten electrons. For any configuration of these charged particles, one can define the centers $\boldsymbol{r}_+$ and $\boldsymbol{r}_-$ of the positive and negative charge distributions. The corresponding electric dipole moment is then $\boldsymbol{D} = Q(\boldsymbol{r}_+ - \boldsymbol{r}_-)$, where $Q = 10q$ and q is the unit charge. Using the correspondence principle, the observable $\hat{W}$ is $\hat{W} = -\boldsymbol{\mathcal{E}} \cdot \hat{\boldsymbol{D}} = -Q\,\boldsymbol{\mathcal{E}} \cdot (\hat{\boldsymbol{r}}_+ - \hat{\boldsymbol{r}}_-)$. The form (6.27) is the restriction of this operator to the two-dimensional subspace under consideration.

6.4.2 Energy Levels in a Fixed Electric Field

In an electric field $\mathcal{E}$, the Hamiltonian of the molecule is therefore

$$\hat{H} = \begin{pmatrix} E_0 - A & -\eta \\ -\eta & E_0 + A \end{pmatrix} = E_0 \hat{I} - \sqrt{A^2 + \eta^2} \begin{pmatrix} \cos 2\theta & \sin 2\theta \\ \sin 2\theta & -\cos 2\theta \end{pmatrix}, \tag{6.28}$$

where $\tan 2\theta = \eta / A$, $-\pi/4 < \theta < \pi/4$, and $\hat{I}$ is the unit matrix. The eigenvalues and eigenvectors of $\hat{H}$ can be calculated without difficulty:

$$E_- = E_0 - \sqrt{A^2 + \eta^2}, \qquad |\psi_-\rangle = \begin{pmatrix} \cos\theta \\ \sin\theta \end{pmatrix}, \tag{6.29}$$

$$E_+ = E_0 + \sqrt{A^2 + \eta^2}, \qquad |\psi_+\rangle = \begin{pmatrix} -\sin\theta \\ \cos\theta \end{pmatrix}. \tag{6.30}$$

Validity of the Calculation. The rule of the game is to remain in the two-dimensional subspace of the lowest-energy states. Therefore, there is a limit on the value of the field $\mathcal{E}$ that we can apply. This field must be such that $E_+ \ll E_1$, the first excited state. Otherwise it would be necessary to take into account higher excited states of the molecule and perform four-state, six-state, etc. calculations. Given the values of A, d_0 and $E_1 - E_0$ for NH_3, this approximation is valid for all fields achievable in the laboratory. Two limits are particularly interesting.

Weak Field. In the weak-field limit, the mixing angle θ is small and the levels and corresponding eigenstates are, to the lowest nontrivial order in $\mathcal{E}$,

$$E_{\mp} \simeq E_0 \mp \left(A + \frac{d_0^2\mathcal{E}^2}{2A}\right), \tag{6.31}$$

$$|\psi_-\rangle \simeq |\psi_{\mathrm{S}}\rangle + \frac{d_0\mathcal{E}}{2A}|\psi_{\mathrm{A}}\rangle, \qquad |\psi_+\rangle \simeq |\psi_{\mathrm{A}}\rangle - \frac{d_0\mathcal{E}}{2A}|\psi_{\mathrm{S}}\rangle. \tag{6.32}$$

In the absence of an electric field, each eigenstate of $\hat{H}$ is symmetric or antisymmetric, and the expectation value of the electric dipole moment D vanishes if the molecule is prepared in one of these two states. In the presence of an electric field, the eigenstates of $\hat{H}$ correspond to a partially polarized molecule. The expectation value of the dipole moment in the states $|\psi_-\rangle$ and $|\psi_+\rangle$ is proportional to $\mathcal{E}$:

$$\langle D\rangle_{\mp} \simeq \pm\frac{d_0^2}{A}\mathcal{E}.$$

The quantity d_0^2/A is called the *polarizability* of the molecule. It is large for NH_3 because the value of A is comparatively small. The shift of the energy levels varies quadratically with the field $\mathcal{E}$. This is understandable, since the interaction energy is proportional to the product of the field $\mathcal{E}$ and the induced dipole moment, which is itself proportional to the field.

Strong Field. If the field is strong, i.e. if $\eta \gg A$, the mixing angle θ approaches $\pi/4$ and the eigenvalues and eigenstates are:

$$E_{\pm} \simeq E_0 \pm d_0\mathcal{E} \qquad |\psi_-\rangle \simeq |\psi_{\mathrm{R}}\rangle \qquad |\psi_+\rangle \simeq -|\psi_{\mathrm{L}}\rangle. \tag{6.33}$$

In the strong field regime, the eigenstates of the Hamiltonian correspond to a molecule which is completely polarized:

$$\langle D\rangle_{\mp} \simeq \pm d_0.$$

The energies are *linear* in the field, as is the case for a classical electric dipole.

Between these two limits, two effects compete to determine the eigenstates of the Hamiltonian. The transition term A, which is due to the tunneling between right and left, tends to symmetrize the molecule and to favor the states of well defined symmetry $|\psi_{\mathrm{S}}\rangle$ and $|\psi_{\mathrm{A}}\rangle$. The presence of the field polarizes the molecule and tends to pull it towards the classical configurations $|\psi_{\mathrm{R}}\rangle$ and $|\psi_{\mathrm{L}}\rangle$. Figure 6.4 shows the evolution of the energy levels as a function of the intensity of the applied field (in units of A/d_0). The value of the electric dipole moment of ammonia is $d_0 \sim 3\times 10^{-11}\,\mathrm{eV/(V\,m^{-1})}$. The borderline between the two regimes is $\mathcal{E}_{\mathrm{c}} = A/d_0 \sim 1.7\times 10^6\ \mathrm{V\,m^{-1}}$ (an extremely high lab. field). This can be compared with the case of PH_3 molecule, where the splitting A is much smaller and where $A/d_0 \sim 30\,\mathrm{V\,m^{-1}}$.

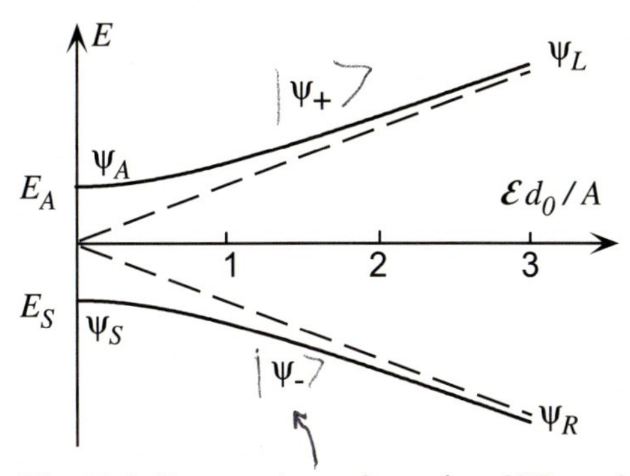

Fig. 6.4. Energy eigenvalues of an NH_3 molecule in an electric field

6.4.3 Force Exerted on the Molecule by an Inhomogeneous Field

Let us assume that we are in the weak-field regime, where the energy levels are given by (6.31). The term $d_0^2\mathcal{E}^2/2A$ can be interpreted as a potential-energy term of the molecule in the field. This potential energy has *opposite signs* according to whether the internal state of the molecule is $|\psi_-\rangle$ or $|\psi_+\rangle$.

Suppose that we prepare a molecular beam traveling along some direction x, and that the beam crosses a region where we apply an inhomogeneous field. The molecules are "large" objects, and, to a good approximation, their motion in space can be treated as classical.[2] When the molecules cross this inhomogeneous field, they are subjected to a force

$$\boldsymbol{F}_{\mp} = \pm\boldsymbol{\nabla}\left(\frac{d_0^2\mathcal{E}^2}{2A}\right) . \tag{6.34}$$

The sign of this force depends on the *internal state* ($|\psi_-\rangle$ or $|\psi_+\rangle$) of each molecule. Consequently, when the molecules leave the field zone, the initial beam is split into two outgoing beams: one beam contains molecules in the $|\psi_-\rangle$ state, and the other contains molecules in the $|\psi_+\rangle$ state.

In the specific case of an ammonia maser, the inhomogeneous field is such that $\mathcal{E}^2 \propto y^2 + z^2$. The molecules in the state $|\psi_+\rangle$ are therefore in a potential $d_0^2\mathcal{E}^2/2A$ which is harmonic in the yz plane transverse to the beam. Their trajectories are superpositions of a linear uniform motion along x and small oscillations in the transverse plane (Fig. 6.5). In contrast, molecules in the $|\psi_-\rangle$ state are in a reversed harmonic potential $-d_0^2\mathcal{E}^2/2A$ and are expelled from the vicinity of the x axis. The field inhomogeneity is a means of selecting molecules whose (internal) state is $|\psi_+\rangle \sim |\psi_\mathrm{A}\rangle$, by separating them from molecules in the state $|\psi_-\rangle \sim |\psi_\mathrm{S}\rangle$.

[2] The ensuing argument will be justified rigorously by means of the Ehrenfest theorem in the analogous case of the motion of silver atoms in the Stern–Gerlach experiment (Chap. 8).

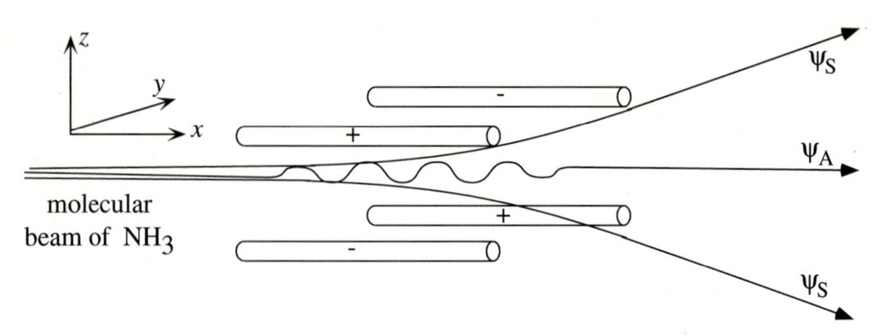

Fig. 6.5. Stabilization of the beam $|\psi_+\rangle$ and divergence of the beam $|\psi_-\rangle$ in an electrostatic quadrupole field ($\mathcal{E}^2 \propto y^2 + z^2$)

The resulting beam is no longer in thermal equilibrium, since the most populated state is not the lowest-energy state $|\psi_S\rangle$. This type of operation is called *population inversion.* The simple device described above is only one of the many techniques to achieve population inversion. Population inversion destroys the thermal equilibrium between $|\psi_S\rangle$ and $|\psi_A\rangle$, which may have existed in the initial molecular beam.

Remarks

a. A beam in the pure state $|\psi_-\rangle$ will be defocused and a beam in the pure state $|\psi_+\rangle$ will be channeled. It is by no means obvious what happens to a beam in the quantum superposition $\alpha|\psi_-\rangle + \beta|\psi_+\rangle$. We shall study this problem in the similar case of the Stern–Gerlach experiment in Chap. 8. The net result is that two outgoing beams of relative intensities $|\alpha|^2$ and $|\beta|^2$ are defocused and channeled, respectively. The molecules in the defocused beam are in the internal state $|\psi_-\rangle$ and the molecules in the channeled beam are in the state $|\psi_+\rangle$.
b. The hypothesis that the electric field is parallel to the axis of the molecule might seem questionable. To address this point correctly, one must consider the real structure of the symmetric and antisymmetric states, which requires some knowledge about the properties of angular momentum in quantum mechanics (Chap. 10). One finds that each energy level E_S and E_A is actually degenerate. One can choose a basis of the corresponding subspaces $\mathcal{E}_S$ and $\mathcal{E}_A$ such that each basis state corresponds to a given projection of the angular momentum of the molecule along a given axis, say the direction of the electric field at the location of the molecule. The exact treatment is more involved[3] than what we have presented here, but the main conclusion remains valid. The internal state of the molecule can be expanded in these basis states, of which some are focused (as $|\psi_+\rangle$) and the others are expelled from the center of the beam (as $|\psi_-\rangle$).

[3] See J.P. Gordon et al., Phys. Rev. **99**, 1264 (1955), which also takes into account the hyperfine splitting of the energy levels E_S and E_A.

6.5 Oscillating Fields and Stimulated Emission

In order to achieve the maser effect, we are going to force those molecules which we have selected to be in the state $|\psi_A\rangle$ to release their energy $2A$ by falling down to the ground state $|\psi_S\rangle$. The molecules can decay spontaneously and emit a photon, but they do so with a lifetime of the order of a month, which is much too long for our purpose. However, one can *stimulate* this emission by acting on them with an oscillating field of angular frequency ω, $\mathcal{E} = \mathcal{E}_0 \cos \omega t$, provided ω is tuned to the Bohr frequency ω_0 of the system.

Let us again set $\eta = d_0 \mathcal{E}_0$; the Hamiltonian is

$$\hat{H} = \begin{pmatrix} E_0 - A & -\eta \cos \omega t \\ -\eta \cos \omega t & E_0 + A \end{pmatrix} . \tag{6.35}$$

Because of the time dependence of this Hamiltonian, the notion of stationary states is inadequate. We must solve the Schrödinger equation $\mathrm{i}\hbar (\mathrm{d}/\mathrm{d}t)|\psi(t)\rangle = \hat{H}|\psi(t)\rangle$ in order to determine the time evolution of the system. We write the state vector of a given molecule as

$$|\psi(t)\rangle = \begin{pmatrix} a(t) \\ b(t) \end{pmatrix} .$$

The Schrödinger equation reduces to the first-order coupled linear differential system

$$\mathrm{i}\hbar \dot{a} = (E_0 - A)a - \eta b \cos \omega t \,, \tag{6.36}$$

$$\mathrm{i}\hbar \dot{b} = (E_0 + A)b - \eta a \cos \omega t \,. \tag{6.37}$$

If we set $a(t) = \mathrm{e}^{-\mathrm{i}(E_0 - A)t/\hbar} \alpha(t)$ and $b(t) = \mathrm{e}^{-\mathrm{i}(E_0 + A)t/\hbar} \beta(t)$, we obtain:

$$2\mathrm{i}\dot{\alpha} = -\,\omega_1\, \beta \left(\mathrm{e}^{\mathrm{i}(\omega - \omega_0)t} + \mathrm{e}^{-\mathrm{i}(\omega + \omega_0)t} \right) , \tag{6.38}$$

$$2\mathrm{i}\dot{\beta} = -\,\omega_1\, \alpha \left(\mathrm{e}^{-\mathrm{i}(\omega - \omega_0)t} + \mathrm{e}^{\mathrm{i}(\omega + \omega_0)t} \right) . \tag{6.39}$$

This system involves three angular frequencies:

$$\omega, \quad \omega_0 = 2A/\hbar \quad \text{and} \quad \omega_1 = \eta/\hbar = d_0 \mathcal{E}_0/\hbar \,. \tag{6.40}$$

Physically, this system corresponds to forced oscillations with a resonance[4] at $\omega = \omega_0$. It is not possible to solve this system analytically. However, we can obtain a good approximation in the vicinity of the resonance $\omega \sim \omega_0$ if we neglect the rapidly oscillating terms $\mathrm{e}^{\pm \mathrm{i}(\omega + \omega_0)t}$ whose effect averages to zero after a time $\sim 2\pi/\omega$. The system then reduces to an analytically soluble problem, which we shall study in greater detail in Chap. 12, in the context of

[4] There are actually two resonances, at $\omega = \pm\omega_0$, but these two values are equivalent for our purposes.

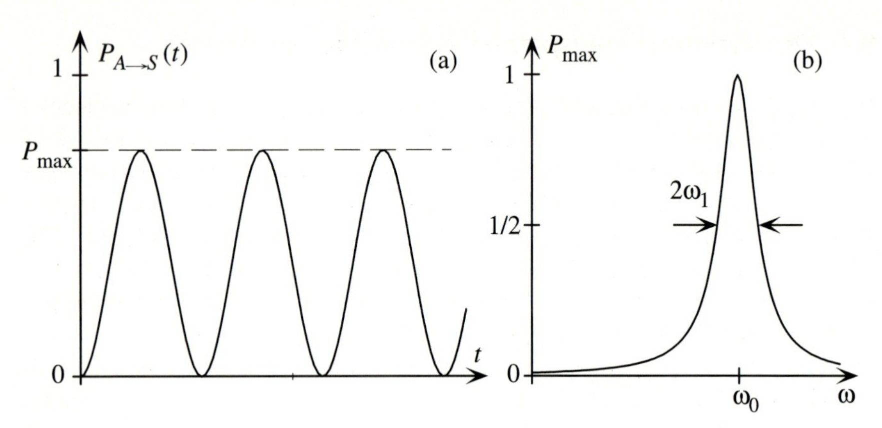

Fig. 6.6. Rabi oscillation. (**a**) Probability of finding the molecule in state $|\psi_S\rangle$ as a function of time. (**b**) Resonance curve, showing the maximum transition probability as a function of the angular frequency ω of the external field

magnetic resonance. Here, we only give the solution of this simplified problem. The transition probability $P_{A\to S}(t)$ that the molecule is found at time t in the state $|\psi_S\rangle$, having thus released its energy $2A = E_A - E_S$, is given by

$$P_{A\to S}(t) \simeq \frac{\omega_1^2}{(\omega-\omega_0)^2+\omega_1^2} \sin^2\left(\sqrt{(\omega-\omega_0)^2+\omega_1^2}\,\frac{t}{2}\right) . \tag{6.41}$$

This formula is due to Rabi. As shown in Fig. 6.6a, the probability $P_{A\to S}(t)$ oscillates in time between 0 and a maximum value P_{max} given by

$$P_{max} = \frac{\omega_1^2}{(\omega-\omega_0)^2+\omega_1^2} .$$

When we vary the frequency ω of the applied field (Fig. 6.6b), the maximum probability P_{max} has a characteristic resonant behavior, with a maximum equal to 1 for $\omega = \omega_0$. The half width at half maximum of the resonance curve is ω_1.

If the frequency of the field is tuned close to resonance, i. e. $|\omega - \omega_0| \ll \omega_1$, practically all the molecules have released their energy $2A$ at a time $T = \pi/\omega_1$. This energy release occurs through the emission of an electromagnetic radio wave of frequency $\nu = \omega_0/(2\pi) \simeq 24\,\text{GHz}$, and is called *stimulated emission*. The smaller ω_1, the narrower the resonance curve of Fig. 6.6b will be, and the longer the release time.

In the absence of an external field, the molecules can undergo spontaneous transitions from the state $|\psi_A\rangle$ to $|\psi_S\rangle$. The lifetime for spontaneous emission is very long (1 month). The mechanism of stimulated emission allows this transition to occur very rapidly ($T \sim 7 \times 10^{-8}$ s for a field $\mathcal{E}_0 \sim 10^3$ V m^{-1}). It was Einstein who, in 1917, in his celebrated analysis of the equilibrium of radiation and matter in the black-body problem, first understood the stimulated emission effect (see Chap. 17 for more details).

6.6 Principle and Applications of Masers

A schematic description of a maser (microwave amplification by stimulated emission of radiation) is given in Fig. 6.7. Starting from a molecular beam of mean velocity v ($v \simeq \sqrt{kT/m}$), one first separates the molecules in the state $|\psi_A\rangle$ by means of an electrostatic quadrupole field (see Sect. 6.4.3). The resulting beam then enters a high-frequency cavity, where the oscillating field $\mathcal{E}_0 \cos\omega_0 t$ is applied. The length L of the cavity is adjusted[5] so that $L/v = T = (2n+1)\pi/\omega_1$. When they leave the cavity, the molecules are in the state $|\psi_S\rangle$ and have released their energy $2A$ in the form of electromagnetic radiation of angular frequency ω_0. There are basically three ways to use such a device.

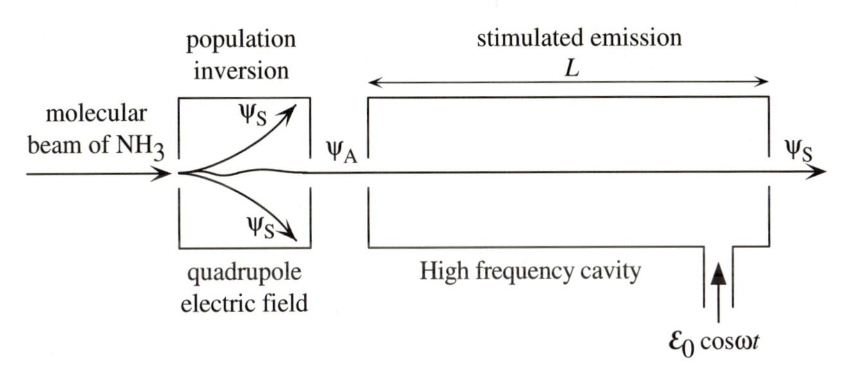

Fig. 6.7. Sketch of an NH_3 maser device

6.6.1 Amplifier

One can *amplify*, in a selective way and with little background noise, a very weak signal (for NH_3, a bandwidth of 3 kHz is possible, i.e. $\delta\omega/\omega \sim 10^{-7}$). This technique resulted in a major revolution in radio astronomy. Masers have been the starting point for studying the interstellar medium.

In his first experiments, Townes used a beam of 10^{14} molecules per second, giving a power of 10^{-9} W at resonance. Nowadays maser amplifiers containing solid-state materials such as ruby (an Al_2O_3 crystal doped with Cr^{+3} ions at about 0.05% concentration) have gains of about 36 dB. Such masers were used in 1965 by Penzias and Wilson when they discovered the 2.73 K cosmic background radiation, which is one of the clearest observational signatures of the Big Bang.

---

[5] It is not necessary for L to have *exactly* the correct value; the transition probability has a significant value provided one does not have the bad luck to fall on the unfavorable values $T = 2n\pi/\omega_1$. In practice, a servo mechanism constantly tunes the cavity so that the signal is a maximum.

Figure 6.7 does not in fact show an amplifier, but rather an emitter since the output energy does not depend on the intensity of the input signal. In order to see how this device can work as an amplifier, one must calculate its response to an incoherent signal with a spread of frequencies. Our calculation was done explicitly for a coherent, monochromatic external field.

6.6.2 Oscillator

A field of angular frequency ω_0 will maintain itself in the cavity. If we extract the electromagnetic wave produced, we have a very stable oscillator.

6.6.3 Atomic Clocks

Atomic clocks, which are our official timekeepers at present, function according to a very similar scheme. Such devices are used to define the time standard. They use cesium atomic beams (the isotope ^{133}Cs). The ground state of this atom is split by a hyperfine interaction (see Chap. 13) originating from the interaction between the magnetic moment of the valence electron and the magnetic moment of the nucleus. This interaction results in a splitting of the atomic ground-state level into a two-level structure similar to what we have studied here, except that we are now dealing with a magnetic interaction instead of an electric one. The two sublevels $|g_1\rangle$ and $|g_2\rangle$ have energies E_1 and E_2. The splitting $\nu_{12} = (E_2 - E_1)/h$ is equal to 9 192 631 770 Hz by definition.

In order to build an atomic clock, one first prepares a beam of cesium atoms in the state $|g_1\rangle$. These atoms cross a cavity where one applies a radio wave of frequency ν, and one adjusts ν in order to maximize the flux of outgoing atoms in the state $|g_2\rangle$. The frequency ν is therefore locked in the vicinity of ν_{12}. One measures any length of time by counting the number of oscillations of the wave of frequency ν during this period of time.

Present-day cesium clocks have a relative accuracy of 10^{-15}, which makes the time standard the most accurate of all. Such an accuracy is mandatory in positioning and navigation in the Global Positioning System. It is also crucial in fundamental physics, in astrophysics and in testing the theory of relativity.

Further Reading

- For the ammonia maser, see C.H. Townes and A.L. Schawlow, *Microwave Spectroscopy*, McGraw-Hill, New York (1954), Chap. 15; M. Brotherton, *Masers and Lasers*, McGraw-Hill, New York (1964).
- M. Elitzur, "Masers in the sky", Sci. Am., February 1995, p. 52; R. Vessot et al., "Test of relativistic gravitation with a space-borne hydrogen maser", Phys. Rev. Lett. **45**, 2081 (1980).

- For atomic clocks: J. Vanier and C. Audouin, *The Quantum Physics of Atomic Frequency Standards*, Adam Hilger, Bristol (1989); W. Itano and N. Ramsey, "Accurate measurement of time", Sci. Am., July 1993, p. 46.

Exercises

6.1. Linear three-atom molecule. We consider the states of an electron in a linear three-atom molecule (such as N_3 or C_3) with equally spaced atoms L, C, R at distances d from one another.

Let $|\psi_{\rm L}\rangle$, $|\psi_{\rm C}\rangle$ and $|\psi_{\rm R}\rangle$ be the eigenstates of an observable $\hat{B}$, corresponding to an electron localized in the vicinity of the atoms L, C and R, respectively:

$$\hat{B}|\psi_{\rm L}\rangle = -d|\psi_{\rm L}\rangle\,, \qquad \hat{B}|\psi_{\rm C}\rangle = 0\,, \qquad \hat{B}|\psi_{\rm R}\rangle = +d|\psi_{\rm R}\rangle\,.$$

In the basis $\{|\psi_{\rm L}\rangle, |\psi_{\rm C}\rangle, |\psi_{\rm R}\rangle\}$, the Hamiltonian of the system is represented by the matrix

$$\hat{H} = \begin{pmatrix} E_0 & -a & 0 \\ -a & E_0 & -a \\ 0 & -a & E_0 \end{pmatrix}\,, \qquad a > 0\,.$$

a. Calculate the energy levels and eigenstates of $\hat{H}$.
b. Consider the ground state. What are the probabilities of finding the electron in the vicinity of L, C and R?
c. Suppose the electron is in the state $|\psi_{\rm L}\rangle$, and we measure its energy. What values can we find, with what probabilities? Calculate $\langle E\rangle$ and ΔE in this state.

6.2. Crystallized violet and malachite green. The active ingredient of the dye Crystal Violet (C.I. number 42555) is the organic monovalent cation $C[C_6H_4N(CH_3)_2]_3^+$. The skeleton of this ion is made of three identical branches (Fig. 6.8). The electron deficit responsible for the positive charge can be taken from either of these three branches. One can treat the electronic state of this ion as a three-state system. The Hamiltonian $\hat{H}$ is not diagonal in the basis $\{|1\rangle, |2\rangle, |3\rangle\}$ (which we assume orthonormal), because of tunneling between these classical configurations.

a. We work in the basis $\{|1\rangle, |2\rangle, |3\rangle\}$ corresponding to the "classical configurations". We choose the origin of the energy such that $\langle 1|\hat{H}|1\rangle = \langle 2|\hat{H}|2\rangle = \langle 3|\hat{H}|3\rangle = 0$. We set $\langle 1|\hat{H}|2\rangle = \langle 2|\hat{H}|3\rangle = \langle 3|\hat{H}|1\rangle = -A$, where A is real and positive ($A > 0$).

   Write the matrix $\hat{H}$ in this basis. Comparing the present case with the case of the ammonia molecule NH_3, justify briefly the choice of this matrix.

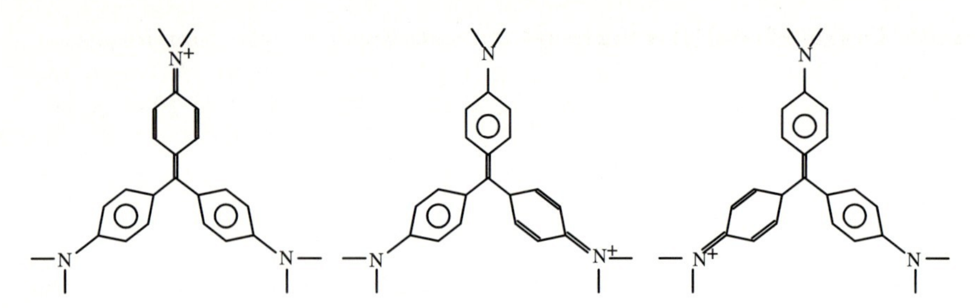

Fig. 6.8. The three possible configurations of the molecule of crystallized violet

b. Consider the states $|\phi_1\rangle = (|1\rangle + |2\rangle + |3\rangle)/\sqrt{3}$ and $|\phi_2\rangle = (|2\rangle - |3\rangle)/\sqrt{2}$. Calculate the expectation value $\langle E\rangle$ and the dispersion ΔE in each of these states. Interpret the result.

c. Determine the energy levels of the system. Give a corresponding orthonormal eigenbasis. Is this basis unique?

d. The value of A is $A \approx 0.75$ eV. Why is this ion violet?

   We recall that the colors of the spectrum of natural light are, for increasing energy ($E = hc/\lambda$), red (from ≈ 1.65 to 2.0 eV), orange (from ≈ 2.0 to 2.1 eV), yellow (from ≈ 2.1 to 2.3 eV), green (from ≈ 2.3 to 2.55 eV), blue (from ≈ 2.55 to 2.65 eV) and violet (from ≈ 2.65 to 3.1 eV). The main pairs of "complementary colors", which produce white light when the members of the pair are combined, are yellow–violet, red–green and blue–orange.

e. One of the $N(CH_3)_2$ groups is replaced by a hydrogen atom. We assume that the sole effect of this substitution is to increase $\langle 1|\hat{H}|1\rangle$ by an amount $\Delta > 0$, and that it leaves the other matrix elements of $\hat{H}$ unchanged.
   (i) Show that A is still an eigenvalue of the Hamiltonian. What are the other energy levels of this new system?
   (ii) How do they behave in the limits $\Delta \ll A$ and $\Delta \gg A$?

f. This modified ion (dye Malachite green (C.I. number 42000)) absorbs light of two wavelengths: 620 and 450 nm. Calculate Δ and comment on the agreement between theory and experiment.
   You can use $hc \approx 1240$ eV nm.

7. Commutation of Observables

God was wise when he placed birth before death;
otherwise, what would we know about life?
Alphonse Allais

When Dirac studied Heisenberg's papers in the summer of 1925, he realized that all ingredients of the newborn theory boiled down to the noncommutativity of physical quantities. An amazing property appeared. In quantum physics, multiplying A by B is not the same as multiplying B by A, the order is of great importance. Such a property of physical quantities had never occurred before in physics, and Heisenberg was worried by what appeared to be a very unpleasant feature of his theoretical framework. He did his best to conceal this aspect when he presented his theory. Dirac, who knew of the existence of noncommutative algebras, thought that, after all, no fundamental principle imposes a requirement that physical quantities should be commutative. He therefore attempted to construct a new approach to quantum theory. This consisted in modifying the classical equations in order to incorporate this noncommutativity. Independently of the Göttingen group, he built his own formalism, proving the relation $\hat{x}\hat{p} - \hat{p}\hat{x} = \mathrm{i}\hbar$ at the same time as Born and Jordan did. He published his own results, which were actually quite close to those of Born, Heisenberg and Jordan, who had founded *Quanten Mechanik*, in the fall of 1925.

The principles of quantum mechanics which we have laid down in Chap. 5 are only a general framework. In order to treat a specific problem, it is necessary to specify the form of the Hamiltonian and of the various observables of the system under consideration. In wave mechanics these are provided by the correspondence principle. In the abstract Hilbert space formalism, what plays a crucial role is not so much the specific form of the observables, but the *algebraic* relations between them, and in particular their *commutation relations.* For systems which have a simple classical analog, we shall of course make use of the correspondence principle in order to obtain the commutation relations. In many problems, however, once the symmetries and invariance laws (translation, rotation) have been taken into account in the first step, there is no substitute for confronting a given prescription with experiment.

In this chapter, after presenting some general considerations on commutation relations, we shall study four points which play a crucial role. We shall first derive the general form of the uncertainty relations. Then we shall demonstrate the important Ehrenfest theorem, which, among its many applications, shows how quantum mechanics transforms into Newtonian mechanics in the classical limit. We shall then introduce the notion of a complete set of commuting observables (CSCO). We shall finish with a concrete illustration of the algebraic method, due to Dirac, which will enable us to find the solution of the harmonic-oscillator problem in a simple and elegant way. This method, which relies on the notion of creation and annihilation operators, is a foundation stone of quantum field theory. Note the historical fact that the first correct calculation of the hydrogen atom was performed in 1925 by Pauli using the algebraic methods of *Quanten Mechanik*, before Schrödinger's 1926 calculation based on the solution of a partial differential equation with boundary conditions.

7.1 Commutation Relations

In the wave function formalism of Chaps. 2, 3 and 4, we have investigated the practical consequences of principles such as the uncertainty relations and the quantization of energy. Our analysis was based on the explicit form (differential or multiplicative) of the operators associated with physical quantities. In the general formalism of Hilbert space, this role is played by the *commutation relations* of the various observables.

The principles do not give us the form of the commutation relations. We must postulate their form. For quantities with a classical analog, we shall make use of the correspondence principle. Thus, the definition of a pair of conjugate position and momentum observables lies in the fundamental relations

$$[\hat{x}, \hat{p}_x] = \mathrm{i}\hbar\,, \qquad [\hat{y}, \hat{p}_y] = \mathrm{i}\hbar\,, \qquad [\hat{z}, \hat{p}_z] = \mathrm{i}\hbar\,. \tag{7.1}$$

From these relations, we can derive all the commutation relations for quantities which have a simple classical analog, since the classical observables are functions of positions and momenta. Consider for instance the angular momentum observable $\hat{\boldsymbol{L}} = \hat{\boldsymbol{r}} \times \hat{\boldsymbol{p}}$, which, in the wave function formalism, is given by

$$\hat{L}_z = \frac{\hbar}{\mathrm{i}}\left(x\frac{\partial}{\partial y} - y\frac{\partial}{\partial x}\right),$$

with two analogous relations for $\hat{L}_x$ and $\hat{L}_y$. From the definition of the three observables $\hat{L}_x$, $\hat{L}_y$ and $\hat{L}_z$, one can easily prove, using the fundamental relation (7.1), the commutation relation

$$[\hat{L}_x, \hat{L}_y] = \mathrm{i}\hbar\hat{L}_z \tag{7.2}$$

and two others deduced by cyclic permutation. We can group these relations together in the form

$$\hat{\boldsymbol{L}} \times \hat{\boldsymbol{L}} = \mathrm{i}\hbar \hat{\boldsymbol{L}} . \tag{7.3}$$

When we study angular momentum in Chap. 10, we shall use the commutation relations (7.3) as our starting point. Such relations will define the angular-momentum observables.

7.2 Uncertainty Relations

Our first use of the commutation relations is the general derivation of the uncertainty relations. Consider two quantities A and B, and the corresponding observables $\hat{A}$ and $\hat{B}$. Let $|\psi\rangle$ be the state of the system. The measurement of A and B yields the expectation values $\langle a\rangle$ and $\langle b\rangle$, and the mean-square deviations Δa and Δb. More precisely, we prepare N systems ($N \gg 1$) in the state $|\psi\rangle$. For half of them we perform a measurement of A and derive from the distribution of the results the two quantities $\langle a\rangle$ and Δa. For the other half, we measure B and find the values of $\langle b\rangle$ and Δb. We now want to relate Δa and Δb, knowing the state $|\psi\rangle$ and the two observables $\hat{A}$ and $\hat{B}$.

We first define centered variables, i. e. we set $\hat{A}' = \hat{A} - \langle a\rangle$, so that $\langle \hat{A}'\rangle = 0$ and we have

$$(\Delta a)^2 = \langle \psi | \hat{A}'^2 | \psi \rangle .$$

Similarly, for $\hat{B}$, $(\Delta b)^2 = \langle \psi | B'^2 | \psi\rangle$ with $\hat{B}' = \hat{B} - \langle b\rangle$.

Consider an arbitrary state $|\psi\rangle$ and the vector $(\hat{A}' + i\lambda\hat{B}')|\psi\rangle$, where λ is real. The square of the norm of this vector is

$$\begin{aligned} \|(\hat{A}' + \mathrm{i}\lambda\hat{B}')|\psi\rangle\|^2 &= \langle\psi|(\hat{A}' - \mathrm{i}\lambda\hat{B}')(\hat{A}' + \mathrm{i}\lambda\hat{B}')|\psi\rangle \\ &= \langle\psi|\hat{A}'^2|\psi\rangle + \lambda^2\langle\psi|\hat{B}'^2|\psi\rangle + \mathrm{i}\lambda\langle\psi|[\hat{A}', \hat{B}']|\psi\rangle \\ &= \Delta a^2 + \lambda^2 \Delta b^2 + \mathrm{i}\lambda\langle\psi|[\hat{A}', \hat{B}']|\psi\rangle . \end{aligned}$$

Since $\hat{A}'$ and $\hat{B}'$ are Hermitian, the operator $i[\hat{A}', \hat{B}']$ is also Hermitian, and the last term is real. Since the above expression is the square of a norm, it must be positive or zero for any λ. The discriminant of the trinomial in λ must be negative (Schwarz inequality), and therefore

$$\Delta a \; \Delta b \geq \frac{1}{2} |\langle\psi|[\hat{A}, \hat{B}]|\psi\rangle| \tag{7.4}$$

since $[\hat{A}', \hat{B}'] = [\hat{A}, \hat{B}]$.

For the quantities x and p_x, the value of the commutator is $\mathrm{i}\hbar$, and we recover

$$\Delta x \, \Delta p_x \geq \hbar/2 .$$

Equation (7.4) is the general form of the uncertainty relations for any two observables. If $\hat{A}$ and $\hat{B}$ do not commute, the two mean square deviations Δa and Δb cannot simultaneously be made as small as one wants (except for very special cases of states for which the expectation value of the commutator vanishes).

7.3 Ehrenfest's Theorem

Here, we calculate the time evolution of the expectation value of a quantity. By applying the result to the $\boldsymbol{r}$ and $\boldsymbol{p}$ variables, we shall find a form reminiscent of the classical equations of motion. We shall then understand how quantum mechanics and Newtonian mechanics can join together.

7.3.1 Evolution of the Expectation Value of an Observable

Consider a physical quantity A (which may depend explicitly on time) and its expectation value $\langle a\rangle = \langle\psi|\hat{A}|\psi\rangle$. We take the time derivative of this expression:

$$\frac{\mathrm{d}}{\mathrm{d}t}\langle a\rangle = \left(\frac{\mathrm{d}}{\mathrm{d}t}\langle\psi|\right)\hat{A}|\psi\rangle + \langle\psi|\left(\frac{\partial}{\partial t}\hat{A}\right)|\psi\rangle + \langle\psi|\hat{A}\left(\frac{\mathrm{d}}{\mathrm{d}t}|\psi\rangle\right).$$

Using the Schrödinger equation and its Hermitian conjugate,

$$\mathrm{i}\hbar\frac{\mathrm{d}|\psi\rangle}{\mathrm{d}t} = \hat{H}|\psi\rangle \quad \text{and} \quad -\mathrm{i}\hbar\frac{\mathrm{d}\langle\psi|}{\mathrm{d}t} = \langle\psi|\hat{H}. \tag{7.5}$$

we obtain

$$\frac{\mathrm{d}}{\mathrm{d}t}\langle a\rangle = \frac{1}{\mathrm{i}\hbar}\langle\psi|[\hat{A},\hat{H}]|\psi\rangle + \langle\psi|\frac{\partial\hat{A}}{\partial t}|\psi\rangle. \tag{7.6}$$

This formula, which was found by Dirac in 1925, is called the Ehrenfest theorem: it was in fact rederived and published in 1927 by Ehrenfest, as a step towards a more involved result. If the operator $\hat{A}$ does not depend explicitly on time, we obtain

$$\frac{\mathrm{d}}{\mathrm{d}t}\langle a\rangle = \frac{1}{\mathrm{i}\hbar}\langle\psi|[\hat{A},\hat{H}]|\psi\rangle. \tag{7.7}$$

We remark that the time evolution of the physical quantities of a system is governed by the Hamiltonian, i. e. the energy observable, through the *commutator* of each observable and this Hamiltonian. This is directly related to the fact that in the Schrödinger equation, the Hamiltonian governs the time evolution of the system.

In the above equations, we see the deep relation between the time evolution and the Hamiltonian of the system. This relation also appears in classical mechanics, as we shall see in Chap. 15. It is quite remarkable and intriguing that two physical concepts as fundamental and as mysterious as energy and time be so intimately related. We shall come back to this point when we study the time evolution of systems in Chap. 17.

7.3.2 Particle in a Potential $V(\boldsymbol{r})$

We denote by q_i the three position variables x, y, z and denote by p_i the coordinates of the momentum p_x, p_y, p_z $(i = 1, 2, 3)$. The operators $\hat{q}_i$ and $\hat{p}_i$ satisfy the commutation relations

$$[\hat{q}_i, \hat{q}_j] = 0 \,, \qquad [\hat{p}_i, \hat{p}_j] = 0 \,, \qquad [\hat{q}_j, \hat{p}_k] = \mathrm{i}\hbar\, \delta_{j,k} \,. \tag{7.8}$$

From these relations we obtain the commutation relations

$$[\hat{q}_j, \hat{p}_j^m] = m(\mathrm{i}\hbar)\hat{p}_j^{m-1} \,, \qquad [\hat{p}_j, \hat{q}_j^n] = -n(\mathrm{i}\hbar)\hat{q}_j^{n-1} \,, \tag{7.9}$$

which we can generalize to an arbitrary differentiable function $\hat{F} = F(\hat{q}_i, \hat{p}_i)$ of the operators $\hat{q}_i$ and $\hat{p}_i$:

$$[\hat{q}_j, \hat{F}] = \mathrm{i}\hbar \frac{\partial \hat{F}}{\partial \hat{p}_j} \quad \text{and} \quad [\hat{p}_j, \hat{F}] = -\mathrm{i}\hbar \frac{\partial \hat{F}}{\partial \hat{q}_j} \,. \tag{7.10}$$

We assume the Hamiltonian does not depend on time. Choosing $\hat{F} = \hat{H}$, we obtain the evolution equations

$$\frac{\mathrm{d}}{\mathrm{d}t}\langle q_j \rangle = \left\langle \frac{\partial \hat{H}}{\partial \hat{p}_j} \right\rangle , \quad \frac{\mathrm{d}}{\mathrm{d}t}\langle p_j \rangle = -\left\langle \frac{\partial \hat{H}}{\partial \hat{q}_j} \right\rangle . \tag{7.11}$$

In Chap. 15, we shall see how similar the structure of these equations is to the canonical Hamilton–Jacobi equations of analytical mechanics.

The Hamiltonian of a particle in a potential $V(\boldsymbol{r})$ is

$$\hat{H} = \frac{\hat{\boldsymbol{p}}^2}{2m} + V(\hat{\boldsymbol{r}}) \,. \tag{7.12}$$

Substituting in (7.11), we obtain

$$\frac{\mathrm{d}\langle \boldsymbol{r} \rangle}{\mathrm{d}t} = \frac{\langle \boldsymbol{p} \rangle}{m} \,, \tag{7.13}$$

$$\frac{\mathrm{d}\langle \boldsymbol{p} \rangle}{\mathrm{d}t} = -\langle \nabla V(\boldsymbol{r}) \rangle \,. \tag{7.14}$$

Equation (7.13) is the correct definition of the group velocity of a wave packet. It relates the expectation value of the momentum operator to the mean velocity, defined as the time derivative of the expectation value of the position. This relation between expectation values is identical to that found in classical physics. In contrast, (7.14) differs from the classical equation

$$\frac{\mathrm{d}\langle \boldsymbol{p} \rangle}{\mathrm{d}t} = -\nabla V(\boldsymbol{r})\Big|_{\boldsymbol{r}=\langle \boldsymbol{r} \rangle} \tag{7.15}$$

since, in general, $f(\langle \boldsymbol{r} \rangle) \neq \langle f(\boldsymbol{r}) \rangle$.

The Classical Limit. Suppose that the position distribution is peaked around some value $\boldsymbol{r}_0$. Then $\langle\nabla V(\boldsymbol{r})\rangle \simeq \nabla V(\boldsymbol{r}_0)$, and (7.14) and (7.15) are close to each other. In this case, (7.13) and (7.14), for the expectation values, are essentially the same as the classical equation of motion.[1] This observation constitutes the 1927 Ehrenfest theorem. It ensures, in particular, that one recovers classical dynamics for a macroscopic object. When the quantum uncertainties Δr and Δp are too small to be detected, one can safely consider that the wave packets are localized both in position and in momentum space. This justifies the correspondence principle, which guarantees that classical mechanics emerges as a limit of quantum mechanics.

To evaluate the validity criterion for the classical approximation, we can restrict ourselves to a one-dimensional case. We have

$$\frac{\mathrm{d}}{\mathrm{d}t}\langle p\rangle = \left\langle -\frac{\mathrm{d}V}{\mathrm{d}x}\right\rangle \neq -\left.\frac{\mathrm{d}V}{\mathrm{d}x}\right|_{x=\langle x\rangle} .$$

If we expand the function $f(x) = -\mathrm{d}V/\mathrm{d}x$ in the vicinity of $x = \langle x\rangle$, we obtain

$$f(x) = f(\langle x\rangle) + (x - \langle x\rangle)f'(\langle x\rangle) + \frac{1}{2}(x - \langle x\rangle)^2 f''(\langle x\rangle) + \ldots$$

and, by taking the expectation values,

$$\langle f\rangle = f(\langle x\rangle) + \frac{\Delta x^2}{2} f''(\langle x\rangle) + \ldots ,$$

where $\Delta x^2 = \langle(x - \langle x\rangle)^2\rangle$. The nonclassical term in the time evolution of the expectation value will be negligible if

$$|\Delta x^2 f''(\langle x\rangle)/f(\langle x\rangle)| \ll 1 ,$$

or, in terms of the potential V:

$$\left|\Delta x^2 \frac{\mathrm{d}^3 V}{\mathrm{d}x^3}\right| \ll \left|\frac{\mathrm{d}V}{\mathrm{d}x}\right| ,$$

i.e. if the potential is slowly varying over the extent of the wave packet (or over an interval of the order of the de Broglie wavelength). We notice that this condition is identically satisfied if $V(x)$ is a second-order polynomial.

7.3.3 Constants of Motion

Consider (7.7). It tells us under what conditions the quantity $\langle a\rangle$ remains constant as a function of time when the system evolves. In order for this to happen, it suffices that the observable $\hat{A}$ commutes with the Hamiltonian $\hat{H}$. In that case, for any state $|\psi\rangle$, we have $\mathrm{d}\langle a\rangle/\mathrm{d}t = 0$. Let us consider some important examples.

---

[1] A more detailed analysis relying on the Wigner representation of the density operator of the particle is presented in Appendix D.

Conservation of the Norm. If $\hat{A}$ is the identity operator $\hat{I}$, we obtain

$$\frac{\mathrm{d}\langle\psi|\psi\rangle}{\mathrm{d}t} = 0\,.$$

Notice that it is essential that the Hamiltonian be self-adjoint in order to have (7.5) and therefore the conservation of the norm.

Conservation of Energy for an Isolated System. In a time-independent problem, the choice $\hat{A} = \hat{H}$ yields

$$\frac{\mathrm{d}\langle E\rangle}{\mathrm{d}t} = 0\,.$$

Conservation of Momentum. Consider the motion of a free particle, of Hamiltonian $\hat{H} = \hat{\boldsymbol{p}}^2/2m$. The observables $\hat{p}_x, \hat{p}_y, \hat{p}_z$ commute with $\hat{H}$, and therefore

$$\frac{\mathrm{d}\langle p_i\rangle}{\mathrm{d}t} = 0\,, \qquad i = x, y, z\,.$$

As in classical physics, this is not true if the particle is placed in a potential $V(\boldsymbol{r})$, since the operators $\hat{p}_i$ no longer commute with $\hat{H}$.

Conservation of Angular Momentum. Consider the motion of a particle in a *central* potential $V(r)$. Classically, the angular momentum $\boldsymbol{L} = \boldsymbol{r} \times \boldsymbol{p}$ is a constant of the motion, as a result of rotational invariance. This remains true in quantum mechanics. One can check that

$$\left[\frac{\hat{\boldsymbol{p}}^2}{2m}, \hat{L}_i\right] = 0\,, \qquad \left[V(\hat{r}), \hat{L}_i\right] = 0\,, \qquad i = x, y, z\,,$$

which leads to

$$\frac{\mathrm{d}\langle L_i\rangle}{\mathrm{d}t} = 0\,, \qquad i = x, y, z\,.$$

This is no longer true if $V(\boldsymbol{r})$ is not central but depends on variables other than the modulus of $\boldsymbol{r}$, such as the polar and azimuthal angles θ and φ.

More generally, when a physical problem possesses a symmetry property (translation, rotation, etc.), this is reflected in the fact that the Hamiltonian commutes with the operators associated with this symmetry (the total momentum operator, the total angular-momentum operator, etc.). Owing to the Ehrenfest theorem, we know that the expectation value of that operator is a constant of the motion. In quantum mechanics, as in classical mechanics, it is essential to identify the symmetry properties of the system under consideration in order to exploit their consequences.

Remark If $|\psi\rangle$ is an eigenstate of $\hat{H}$, the expectation value $\langle a\rangle$ of *any* observable $\hat{A}$ is time independent, since the state $|\psi\rangle$ is stationary.

7.4 Commuting Observables

Observables which commute lead to interesting properties. This is particularly true if an observable commutes with the Hamiltonian.

7.4.1 Existence of a Common Eigenbasis for Commuting Observables

The following theorem is of great practical importance:

If two observables $\hat{A}$ and $\hat{B}$ commute, there exists an eigenbasis of $\mathcal{E}_{\mathrm{H}}$ formed from eigenvectors common to $\hat{A}$ and $\hat{B}$.

This theorem can be readily generalized to the case of several observables $\hat{A}$, $\hat{B}$, $\hat{C}$ which all commute with one another.

Consider as an example a two-dimensional isotropic harmonic oscillator. Finding the energy eigenfunctions seems, a priori, a difficult problem since it amounts to a second-order partial differential problem. However, the Hamiltonian can be split into the sum of two Hamiltonians acting on different variables:

$$\hat{H} = -\frac{\hbar^2}{2m}\frac{\partial^2}{\partial x^2} + \frac{1}{2}m\omega^2 x^2 - \frac{\hbar^2}{2m}\frac{\partial^2}{\partial y^2} + \frac{1}{2}m\omega^2 y^2 = \hat{H}_x + \hat{H}_y \,. \qquad (7.16)$$

The two operators $\hat{H}_x$ and $\hat{H}_y$, which are both one-dimensional harmonic oscillator Hamiltonians, obviously commute. Solving separately the eigenvalue problems for $\hat{H}_x$ and $\hat{H}_y$,

$$\hat{H}_x\phi_{n_1}(x) = E_{n_1}\phi_{n_1}(x)\,, \quad \hat{H}_y\phi_{n_2}(y) = E_{n_2}\phi_{n_2}(y)\,, \qquad (7.17)$$

we obtain the eigenvalues of $\hat{H}$ as the sums of the eigenvalues of $\hat{H}_x$ and $\hat{H}_y$, with eigenfunctions which are products of the corresponding one-dimensional eigenfunctions:

$$E_{n_1,n_2} = E_{n_1} + E_{n_2} = (n_1 + n_2 + 1)\hbar\omega\,,$$
$$\Phi_{n_1,n_2}(x,y) = \phi_{n_1}(x)\,\phi_{n_2}(y)\,.$$

7.4.2 Complete Set of Commuting Observables (CSCO)

A set of operators $\hat{A}, \hat{B}, \hat{C}, \ldots$ is said to form a complete set of commuting observables (CSCO) if their common eigenbasis is unique; in other words, to any set of eigenvalues $a_\alpha, b_\beta, c_\gamma, \ldots$ there corresponds a single eigenvector $|\alpha,\beta,\gamma,\ldots\rangle$ (up to a phase factor).

In general, for a given system, there exists an infinite number of CSCOs. In any given problem, one chooses the most convenient set. Neither the nature nor the number of observables which constitute a CSCO is fixed a priori.

Example. For a one-dimensional harmonic oscillator, the Hamiltonian

$$\hat{H}_x = \frac{\hat{p}_x^2}{2m} + \frac{1}{2}m\omega^2\hat{x}^2$$

is by itself a CSCO. There is only one eigenbasis of $\hat{H}_x$, made up of the Hermite functions $\phi_n(x)$. On the other hand, for the isotropic two-dimensional harmonic oscillator considered above, whose Hamiltonian is given in (7.16), this is no longer the case. A possible basis is the set of functions $\{\phi_{n_1}(x)\,\phi_{n_2}(y),\ n_1, n_2 \text{ integers}\}$, where $\phi_{n_1}(x)$ and $\phi_{n_2}(y)$ are the eigenstates of $\hat{H}_x$ and of $\hat{H}_y$, respectively. The eigenvalue corresponding to $\phi_{n_1}(x)\,\phi_{n_2}(y)$ is

$$E_{n_1,n_2} = \hbar\omega(n_1 + n_2 + 1)$$

and is degenerate, except in the case $n_1 = n_2 = 0$. This means that there are several different eigenbases of $\hat{H}$ (actually an infinite number). For instance, in the subspace corresponding to $2\hbar\omega$, two possible bases are

$$\{\phi_1(x)\,\phi_2(y),\ \phi_2(x)\,\phi_1(y)\}$$

and

$$\left\{\frac{1}{\sqrt{2}}[\phi_1(x)\,\phi_2(y) + \phi_2(x)\,\phi_1(y)],\ \frac{1}{\sqrt{2}}[\phi_1(x)\,\phi_2(y) - \phi_2(x)\,\phi_1(y)]\right\}.$$

Therefore $\hat{H}$ taken alone is not a CSCO in this case, while the set $\{\hat{H}_x, \hat{H}_y\}$ is a CSCO. Indeed, a knowledge of the two eigenvalues $\{E_{n_x} = (n_x + 1/2)\hbar\omega$, $E_{n_y} = (n_y + 1/2)\hbar\omega\}$ specifies the eigenvector uniquely.

7.4.3 Completely Prepared Quantum State

Why is the notion of a CSCO important physically? Suppose we want to specify the initial condition of an experiment. We need to know if this initial condition corresponds to a specific state, or if it is some more or less defined mixture of states. In the case of the isotropic two-dimensional oscillator, if we know the total energy $n\hbar\omega$, we know only that the state belongs to a subspace of dimension n, spanned by the n functions $\phi_{n_1}(x)\,\phi_{n_2}(y)$ with $n_1 + n_2 + 1 = n$. A measurement of the total energy is not sufficient to specify the state unambiguously. In contrast, if we measure simultaneously the energy of the motion along the x axis and that of the motion along the y axis (which is possible since the corresponding operators $\hat{H}_x$ and $\hat{H}_y$ commute), we can specify the state of the system completely. We shall say that we are dealing with a *completely prepared quantum system.*

More generally, we consider an isolated system in some unknown state $|\psi\rangle$, and we assume that $\{\hat{A}, \hat{B}, \dots, \hat{X}\}$ is a CSCO. If we measure successively all the physical quantities $A, B, \dots, X$ and obtain the results $a_\alpha, b_\beta, \dots, x_\xi$, the state of the system after this series of measurements is

$$|\psi_0\rangle = c\,\hat{P}_\xi \dots \hat{P}_\beta \hat{P}_\alpha |\psi\rangle\,, \tag{7.18}$$

where c is a normalization coefficient and where $\hat{P}_\alpha$, $\hat{P}_\beta$, ... project onto the eigensubspaces of $\hat{A}$, $\hat{B}$, ... associated with the eigenvalues a_α, b_β, ... By definition of a CSCO, the state $|\psi_0\rangle$ is

- an eigenstate of $\hat{A}$, $\hat{B}$, ... , $\hat{X}$,
- uniquely defined (up to an arbitrary phase factor).

Indeed, since $\hat{A}$ commutes with $\hat{B}, \dots, \hat{X}$, the projector $\hat{P}_\alpha$ also commutes with $\hat{P}_\beta, \dots, \hat{P}_\xi$. Therefore, using $\hat{P}_\alpha^2 = \hat{P}_\alpha$ and $\hat{A}\hat{P}_\alpha = a_\alpha \hat{P}_\alpha$, we find

$$\begin{aligned} \hat{A}|\psi_0\rangle &= c\,\hat{A}\left(\hat{P}_\xi \dots \hat{P}_\beta \hat{P}_\alpha^2\right)|\psi\rangle = c\,\hat{A}\hat{P}_\alpha\left(\hat{P}_\xi \dots \hat{P}_\beta \hat{P}_\alpha\right)|\psi\rangle \\ &= a_\alpha\, c\left(\hat{P}_\xi \dots \hat{P}_\beta \hat{P}_\alpha\right)|\psi\rangle = a_\alpha |\psi_0\rangle\,, \end{aligned}$$

which means that $|\psi_0\rangle$ is an eigenstate of $\hat{A}$ with eigenvalue a_α. In addition, the completeness of the set $\{\hat{A}, \hat{B}, \dots, \hat{X}\}$ ensures that there is only one state of the Hilbert space which is simultaneously an eigenstate of $\hat{A}$, $\hat{B}$, ... , $\hat{X}$ for the eigenvalues a_α, b_β, ... , x_ξ. This proves the uniqueness of $|\psi_0\rangle$, to within an arbitrary phase factor. We have obtained, by this series of measurements of all the physical quantities of a CSCO, a *completely prepared* quantum state.

Since $|\psi_0\rangle$ is an eigenstate of $\hat{A}$, $\hat{B}$, ... , any new measurement of $\hat{A}$, $\hat{B}$, ..., on this state will yield the same results a_α, b_β, ... This is also an important result. When two (or more) observables $\hat{A}$ and $\hat{B}$ commute, if one measures *successively* A, finding a result a_α, and B, finding a result b_β, this latter measurement does not change the value found previously for A. If we measure A again (provided the system has not evolved) we recover the result a_α with probability one.

Remarks

a. The order in which the measurements of $A, B, \dots, X$ are made is of no importance since $\hat{A}, \hat{B}, \dots, \hat{X}$ commute.
b. If the Hamiltonian of the system commutes with all the operators of the CSCO, the above statements are valid at any time. If not, all of the preceding results are valid only if all measurements of $A, B, \dots, X$ are performed within a time interval short compared with ω_0^{-1}, where ω_0 is a typical Bohr frequency of the Hamiltonian.

7.4.4 Symmetries of the Hamiltonian and Search of Its Eigenstates

The theorem presented at the beginning of this section plays an essential role when one is looking for the eigenstates of a given Hamiltonian. Let us give three examples.

Even Potential. Consider the one-dimensional motion of a particle in an even potential $V(x)$, i.e. $V(x) = V(-x)$. Let us introduce the (Hermitian) parity operator $\hat{P}$, defined by its action on any wave function $\psi(x)$ as follows:

$$\hat{P}\psi(x) = \psi(-x)\,.$$

One can check that if $V(x)$ is even, $\hat{P}$ commutes with the Hamiltonian $\hat{p}^2/2m + V(\hat{x})$. We can therefore look for a basis of eigenfunctions common to $\hat{H}$ and $\hat{P}$. The eigenvalues and eigenfunctions of $\hat{P}$ are very simple to determine. Since $\hat{P}^2 = 1$, the eigenvalues are ± 1. The corresponding eigenfunctions are the set of even wave functions (corresponding to the eigenvalue -1) and the set of odd wave functions (corresponding to the eigenvalue $+1$). We therefore know that we can search for the eigenfunctions of $\hat{H}$ as even or odd functions.

Rotation-Invariant Potential. Consider again the isotropic two-dimensional potential of Sect. 7.4.1. This problem may also be treated in polar coordinates (r, φ). The Hamiltonian has then the form

$$\hat{H} = -\frac{\hbar^2}{2m}\left(\frac{\partial^2}{\partial r^2} + \frac{1}{r}\frac{\partial}{\partial r} + \frac{1}{r^2}\frac{\partial^2}{\partial \varphi^2}\right) + \frac{1}{2}m\omega^2 r^2\,. \tag{7.19}$$

We can immediately check that, owing to the rotation invariance of this Hamiltonian around the z axis, it commutes with the z component of the angular momentum:

$$[\hat{H}, \hat{L}_z] = 0\,, \quad \text{where} \quad \hat{L}_z = \hat{x}\hat{p}_y - \hat{y}\hat{p}_x = \frac{\hbar}{\mathrm{i}}\left(x\frac{\partial}{\partial y} - y\frac{\partial}{\partial x}\right) = \frac{\hbar}{\mathrm{i}}\frac{\partial}{\partial \varphi}\,.$$

We can therefore diagonalize $\hat{H}$ and $\hat{L}_z$ simultaneously. The eigenfunctions of $\hat{L}_z$ are very simple in polar coordinates:

- First, these functions satisfy

  $$\hat{L}_z\psi(r,\varphi) = A\psi(r,\varphi) \qquad \Rightarrow \qquad \psi(r,\varphi) = f(r)\mathrm{e}^{\mathrm{i}A\varphi/\hbar}\,,$$

  where $f(r)$ is an arbitrary function.
- Secondly, since (r, φ) and $(r, \varphi + 2\pi)$ are two parameterizations of the same point of the plane, the only acceptable eigenvalues A are those such that $\psi(r, \varphi + 2\pi) = \psi(r, \varphi)$, in other words, $A/\hbar = n$, where n is an integer.

We are left with the determination of $f(r)$. Replacing $\psi(r,\varphi)$ by $f(r)\mathrm{e}^{\mathrm{i}n\varphi}$ in the eigenvalue equation of $\hat{H}$, we obtain

$$\left[-\frac{\hbar^2}{2m}\left(\frac{\mathrm{d}^2}{\mathrm{d}r^2}+\frac{1}{r}\frac{\mathrm{d}}{\mathrm{d}r}\right)+\frac{n^2\hbar^2}{2mr^2}+\frac{1}{2}m\omega^2r^2\right]f(r)=Ef(r)\,.$$

This is a simple linear differential equation whose solution can be found quite easily, and which determines, for each value of n, the energy levels of $\hat{H}$. We remark that this approach, which transforms a two-dimensional problem into a one-dimensional one, is not restricted to the harmonic oscillator. It applies for any *central* potential V, i. e. a potential which depends only on $r=\sqrt{x^2+y^2}$. If, on the other hand, V depends on both r and φ, $\hat{H}$ no longer commutes with $\hat{L}_z$ and this procedure does not apply.

The Case of a Periodic Potential. Consider the one-dimensional motion of a particle of mass m in a potential $V(x)$ which is periodic in space, of period a: $V(x+a)=V(x)$. Following the same arguments, we can prove some remarkable properties of the eigenfunctions $\psi(x)$ and of the energy levels that satisfy

$$-\frac{\hbar^2}{2m}\frac{\mathrm{d}^2\psi}{\mathrm{d}x^2}+V(x)\,\psi(x)=E\,\psi(x)\,. \qquad (7.20)$$

This problem has great practical importance. It is the basis of the treatment of electric conduction in crystals, where the periodic potential is created by the lattice of the atoms acting on an electron.

The translation symmetry of the Hamiltonian can be expressed as

$$[\hat{H},\hat{T}_a]=0\,,$$

where $\hat{T}_a$ is the translation operator; this acts on wave functions $\psi(x)$ such that $\hat{T}_a\psi(x)=\psi(x+a)$. The operator T_a is not Hermitian, but it is unitary ($\hat{T}_a^\dagger=\hat{T}_a^{-1}$). Therefore one can diagonalize $\hat{H}$ and $\hat{T}_a$ simultaneously.

Let $\psi(x)$ be an eigenfunction of $\hat{T}_a$ associated with a given (complex) eigenvalue λ. By definition, we have $\psi(x+na)=\lambda^n\psi(x)$, where n is a positive or negative integer. Since we know that exponentially increasing functions at $\pm\infty$ are not acceptable, this imposes the condition $|\lambda|=1$.

Consider a function $\psi(x)$ which is an eigenfunction of both $\hat{H}$ and $\hat{T}_a$. Since the eigenvalue λ of $\hat{T}_a$ is of unit modulus, we can write it as $\lambda=\mathrm{e}^{\mathrm{i}qa}$, with $-\pi\le qa<\pi$. Consequently, the function $\psi(x)$ can always be written as $\psi(x)=\mathrm{e}^{\mathrm{i}qx}\,u(x)$, where $u(x)$ is periodic with period a. This first important result is known as the *Bloch theorem.*

For a given potential $V(x)$ and for any value of q, one can then determine $u(x)$ by solving in one period (ranging from $x=0$ to $x=a$, for instance), the eigenvalue equation deduced from (7.20):

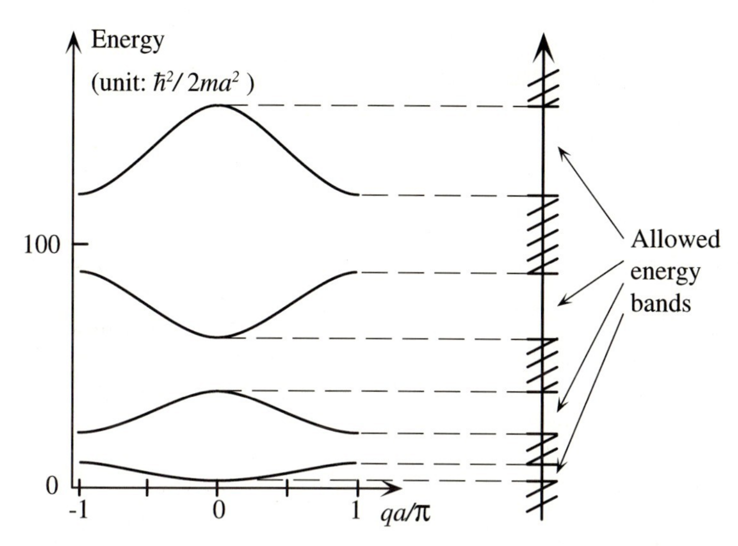

Fig. 7.1. Energy spectrum of a particle of mass m in the periodic potential $V(x) = (\mu\hbar^2/ma)\sum_n \delta(x - na)$. The figure, plotted for $\mu = 7$, represents the variation of the energies $E_n(q)$ with q; the spectrum consists of allowed energy bands separated by forbidden regions

$$\frac{1}{2m}\left(\frac{\hbar}{\mathrm{i}}\frac{\mathrm{d}}{\mathrm{d}x} + \hbar q\right)^2 u(x) + V(x)\,u(x) = E\,u(x)\,. \tag{7.21}$$

The boundary conditions result from the periodicity of $u(x)$:

$$u(a) = u(0)\,, \qquad u'(a) = u'(0)\,. \tag{7.22}$$

Mathematically, this problem with periodic boundary conditions is similar to the quantization of energies in an infinite square-well potential. One obtains a discrete set of eigenvalues $E_n(q)$, $n = 0, 1, \ldots$.

In order to obtain the spectrum of the initial Hamiltonian, we let q vary between $-\pi/a$ and π/a in order to obtain the entire spectrum of the initial Hamiltonian. In the case of a regular potential $V(x)$, each energy $E_n(q)$ is a continuous function of q. We therefore obtain a second important result concerning the spectrum: it consists of allowed energy bands, separated by *gaps* (or forbidden bands). An example of such a spectrum is given in Fig. 7.1.

This band structure of the spectrum plays an essential role in solid-state physics. Together with the Pauli principle, it allows one to predict whether the material under consideration will be a conductor, a semiconductor or an insulator. It is also of great practical interest for understanding the consequences of the presence of defects (which can be intentional, as in doped semiconductors) in the periodic structure.

7.5 Algebraic Solution of the Harmonic-Oscillator Problem

In order to illustrate how commutation relations function in practice, we give here the celebrated method, due to Dirac, for solving the harmonic-oscillator problem.

7.5.1 Reduced Variables

Consider the Hamiltonian

$$\hat{H} = \frac{\hat{p}^2}{2m} + \frac{1}{2}m\omega^2\hat{x}^2 . \tag{7.23}$$

With the change of observables

$$\hat{X} = \hat{x}\sqrt{\frac{m\omega}{\hbar}} , \qquad \hat{P} = \frac{\hat{p}}{\sqrt{m\hbar\omega}} , \tag{7.24}$$

we obtain

$$\hat{H} = \hbar\omega\hat{\mathcal{H}} , \qquad \text{where} \qquad \hat{\mathcal{H}} = \frac{1}{2}\left(\hat{X}^2 + \hat{P}^2\right) . \tag{7.25}$$

The commutation relation of $\hat{X}$ and $\hat{P}$ follows from that of $\hat{x}$ and $\hat{p}$:

$$[\hat{X}, \hat{P}] = \mathrm{i} . \tag{7.26}$$

Our goal is to derive the spectrum of the operator $\hat{\mathcal{H}}$ given in (7.25) by using only this commutation relation.

7.5.2 Annihilation and Creation Operators $\hat{a}$ and $\hat{a}^\dagger$

In order to solve the eigenvalue problem we introduce the following operators:

$$\hat{a} = \frac{1}{\sqrt{2}}(\hat{X} + \mathrm{i}\hat{P}) , \qquad \hat{a}^\dagger = \frac{1}{\sqrt{2}}(\hat{X} - \mathrm{i}\hat{P}) , \tag{7.27}$$

whose commutator is

$$[\hat{a}, \hat{a}^\dagger] = 1 . \tag{7.28}$$

These operators $\hat{a}$ and $\hat{a}^\dagger$ are called *annihilation* and *creation operators*, respectively, for reasons which will become clear in the following.

We also define the *number operator*

$$\hat{N} = \hat{a}^\dagger\hat{a} = \frac{1}{2}(\hat{X}^2 + \hat{P}^2 - 1) . \tag{7.29}$$

This Hermitian operator satisfies the commutation relations

$$[\hat{N}, \hat{a}] = -\hat{a} , \qquad [\hat{N}, \hat{a}^\dagger] = \hat{a}^\dagger , \tag{7.30}$$

and we have $\hat{\mathcal{H}} = \hat{N} + 1/2$, so that $\hat{\mathcal{H}}$ and $\hat{N}$ have the same eigenvectors. We want to show that the eigenvalues ν of $\hat{N}$ are the nonnegative integers and that these eigenvalues are not degenerate.

7.5.3 Eigenvalues of the Number Operator $\hat{N}$

The determination of the eigenvalues ν is performed by use of the following lemmas:

a. *The eigenvalues ν of the operator $\hat{N}$ are nonnegative.*
To show this, we consider an eigenvector $|\phi_\nu\rangle$ associated with the eigenvalue ν and calculate the square of the norm of the vector $\hat{a}|\phi_\nu\rangle$:

$$\|\hat{a}|\phi_\nu\rangle\|^2 = \langle\phi_\nu|\hat{a}^\dagger\hat{a}|\phi_\nu\rangle = \langle\phi_\nu|\hat{N}|\phi_\nu\rangle = \nu\,\langle\phi_\nu|\phi_\nu\rangle = \nu\;\||\phi_\nu\rangle\|^2 \,. \tag{7.31}$$

Therefore $\nu \geq 0$, and

$$\hat{a}|\phi_\nu\rangle = 0 \qquad \text{if and only if } \nu = 0\,. \tag{7.32}$$

b. *The vector $\hat{a}|\phi_\nu\rangle$ either is an eigenvector of $\hat{N}$, corresponding to the eigenvalue $\nu - 1$, or is equal to the null vector.*
Consider the vector $\hat{N}\hat{a}|\phi_\nu\rangle$. Using the commutation relation of $\hat{N}$ and $\hat{a}$, we obtain

$$\hat{N}(\hat{a}|\phi_\nu\rangle) = \hat{a}\hat{N}|\phi_\nu\rangle - \hat{a}|\phi_\nu\rangle = \nu\hat{a}|\phi_\nu\rangle - \hat{a}|\phi_\nu\rangle = (\nu - 1)(\hat{a}|\phi_\nu\rangle)\,.$$

Consequently,
(a) either $\hat{a}|\phi_\nu\rangle$ is different from the null vector, which means that $\nu - 1$ is an eigenvalue of $\hat{N}$ and $\hat{a}|\phi_\nu\rangle$ is a corresponding eigenvector;
(b) or $\hat{a}|\phi_\nu\rangle$ is the null vector.

c. *The vector $\hat{a}^\dagger|\phi_\nu\rangle$ is always an eigenvector of $\hat{N}$, corresponding to the eigenvalue $\nu + 1$.*
Using the commutation relation between $\hat{N}$ and $\hat{a}^\dagger$, we indeed obtain

$$\hat{N}(\hat{a}^\dagger|\phi_\nu\rangle) = \hat{a}^\dagger\hat{N}|\phi_\nu\rangle + \hat{a}|\phi_\nu\rangle = \nu\hat{a}^\dagger|\phi_\nu\rangle + \hat{a}^\dagger|\phi_\nu\rangle = (\nu + 1)(\hat{a}^\dagger|\phi_\nu\rangle)\,.$$

We notice that the vector $\hat{a}^\dagger|\phi_\nu\rangle$ cannot be equal to the null vector, since its norm is strictly positive:

$$\|\hat{a}^\dagger|\phi_\nu\rangle\|^2 = \langle\phi_\nu|\hat{a}\hat{a}^\dagger|\phi_\nu\rangle = \langle\phi_\nu|(\hat{N}+1)|\phi_\nu\rangle = (\nu+1)\||\phi_\nu\rangle\|^2\,. \tag{7.33}$$

Since ν is nonnegative, this implies that $\hat{a}^\dagger|\phi_\nu\rangle$ is an eigenvector of $\hat{N}$ with eigenvalue $\nu + 1$.

It is now simple to prove the following result:

The eigenvalues of $\hat{N}$ are the nonnegative integers only.

Consider a given eigenvalue ν of $\hat{N}$, with an associated eigenvector $|\phi_\nu\rangle$. We apply the operator $\hat{a}$ repeatedly to $|\phi_\nu\rangle$, generating thus a sequence of eigenvectors of $\hat{N}$, namely $|\phi_\nu\rangle$, $\hat{a}|\phi_\nu\rangle$, $\hat{a}^2|\phi_\nu\rangle$, ... , associated with the eigenvalues

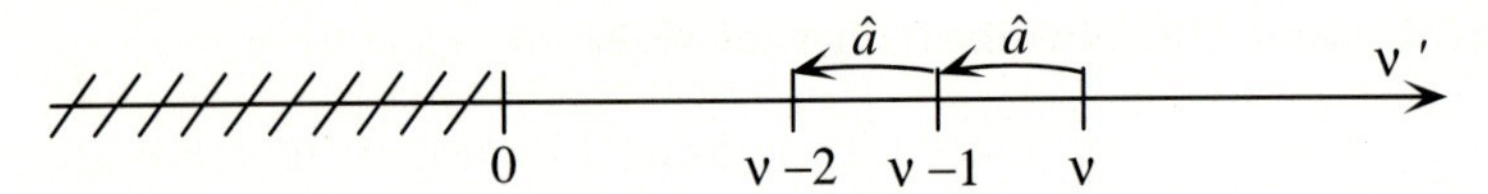

Fig. 7.2. By repeated action of the annihilation operator $\hat{a}$ on a given eigenstate $|\phi_\nu\rangle$ of $\hat{N}$, we construct a sequence of eigenstates associated with the eigenvalues $\nu - 1$, $\nu - 2$, ... Since the eigenvalues of $\hat{N}$ are positive, this sequence cannot be infinite, which implies that ν is an integer (see text)

ν, $\nu - 1$, $\nu - 2$, ... , respectively (see Fig. 7.2). Since the eigenvalues of $\hat{N}$ are nonnegative, one of the elements of the set $\{\nu, \nu-1, \nu-2, \dots\}$ is the *smallest*. Let us call this eigenvalue ν_{min}. Since ν_{min} is the smallest eigenvalue, $\nu_{\text{min}} - 1$ is *not* an eigenvalue. Therefore we deduce from lemma 2 that $\hat{a}|\phi_{\nu_{\text{min}}}\rangle$ is the null vector. Since we have established in (7.31) that $\|\hat{a}|\phi'_\nu\rangle\|^2 = \nu'\||\phi'_\nu\rangle\|^2$ for all ν', this implies that $\nu_{\text{min}} = 0$. Consequently, the initial eigenvalue ν, which by construction differs from ν_{min} by a integer, has to be a positive integer n. Coming back to the initial Hamiltonian $\hat{H} = \hbar\omega(\hat{N} + 1/2)$, we therefore recover the energy levels $(n + 1/2)\hbar\omega$ of the harmonic oscillator.

7.5.4 Eigenstates

Ground State. An eigenstate $|\phi_0\rangle$ associated with the eigenvalue $\nu = 0$, i. e. with an energy $\hbar\omega/2$, satisfies (7.32):

$$\hat{a}|\phi_0\rangle = 0 \qquad \Rightarrow \left(\hat{X} + \mathrm{i}\hat{P}\right)|\phi_0\rangle = 0\,. \tag{7.34}$$

In terms of wave functions, this relation becomes

$$\left(\frac{m\omega}{\hbar}x + \frac{\mathrm{d}}{\mathrm{d}x}\right)\phi_0(x) = 0\,. \tag{7.35}$$

This equation can be solved readily:

$$\phi_0(x) = C_0\; e^{-m\omega x^2/2\hbar}\,, \tag{7.36}$$

where C_0 is a normalization constant: we recover the result of Chap. 4. In particular, we observe that the ground-state is nondegenerate. To simplify the notation, we shall write $|\phi_0\rangle = |0\rangle$.

Nondegeneracy of Energy Levels. In order to prove that the energy levels are not degenerate, we proceed by induction. We have just seen that the ground state, associated with $n = 0$, is not degenerate. We suppose that the level $E_n = (n + 1/2)\hbar\omega$ is also not degenerate and we denote by $|n\rangle$ the corresponding eigenstate. We have to show now that the level E_{n+1} is not degenerate. We consider an eigenstate $|\psi_{n+1}\rangle$ of $\hat{N}$, associated with the eigenvalue $n + 1$: $\hat{N}|\psi_{n+1}\rangle = (n + 1)|\psi_{n+1}\rangle$. We know that $\hat{a}|\psi_{n+1}\rangle$ is an eigenstate of $\hat{N}$ with eigenvalue n. By assumption, the energy level E_n is not degenerate. Consequently, we must have $\hat{a}|\psi_{n+1}\rangle = \gamma|n\rangle$, where γ is a constant. If we multiply this equation by $\hat{a}^\dagger$, we obtain

$$\hat{a}^\dagger \hat{a}|\psi_{n+1}\rangle = \gamma \hat{a}^\dagger|n\rangle\,, \qquad \text{or, equivalently,} \quad |\psi_{n+1}\rangle = \frac{\gamma}{n+1}\hat{a}^\dagger|n\rangle\,.$$

This equation defines in a unique way (up to a phase factor) the eigenstate $|\psi_{n+1}\rangle$ associated with the eigenvalue $n+1$, which demonstrates that the energy levels of a one-dimensional harmonic oscillator are not degenerate.

Excited States. Assuming that the states $|n\rangle$ are normalized, owing to lemmas 2 and 3 given above and to (7.31) and (7.33), we obtain

$$\hat{a}|n\rangle = \sqrt{n}|n-1\rangle\,, \qquad \hat{a}^\dagger|n\rangle = \sqrt{n+1}|n+1\rangle\,. \tag{7.37}$$

Hence the names *annihilation* operator (for $\hat{a}$) and *creation* operator (for $\hat{a}^\dagger$), since these operators transform a state of energy $(n+1/2)\hbar\omega$ into a state $(n+1/2\mp1)\hbar\omega$. In other words, they annihilate or create a quantum of energy $\hbar\omega$. Similarly, the observable $\hat{N}$ is associated with the measurement of the number of quanta.

The sequence of states $|n\rangle$ is generated, starting from the ground state $|0\rangle$, by repeatedly applying the operator $\hat{a}^\dagger$:

$$|n\rangle = \frac{1}{\sqrt{n!}}(\hat{a}^\dagger)^n|0\rangle\,. \tag{7.38}$$

This allows us to find the wave function $\phi_n(x)$ of the nth excited state, starting from the ground state wave function:

$$\phi_n(x) = \frac{1}{\sqrt{n!}}\frac{1}{\sqrt{2^n}}\left(x\sqrt{\frac{m\omega}{\hbar}} - \sqrt{\frac{\hbar}{m\omega}}\frac{\mathrm{d}}{\mathrm{d}x}\right)^n \phi_0(x)\,. \tag{7.39}$$

This can be viewed as a compact formula for the Hermite functions defined in Chap. 4. In this example, we see the elegance and power of the algebraic method. This treatment of the harmonic oscillator and the operators $\hat{a}$, $\hat{a}^\dagger$ and $\hat{N}$ are basic tools in many branches of physics, such as quantum field theory, statistical mechanics and the many-body problem.

Further Reading

R.C. Hovis and H. Kragh, "P.A.M. Dirac and the beauty of physics", Sci. Am., May 1993, p. 62; M. Berry, "Paul Dirac, the purest soul in physics", Phys. World, February 1998, p. 36.

Exercises

7.1. Commutator algebra. Prove the following equalities:

$$[\hat{A}, \hat{B}\hat{C}] = [\hat{A}, \hat{B}]\hat{C} + \hat{B}[\hat{A}, \hat{C}]\,,$$

$$[\hat{A}, \hat{B}^n] = \sum_{s=0}^{n-1} \hat{B}^s[\hat{A}, \hat{B}]\hat{B}^{n-s-1}\,,$$

$$[\hat{A}, [\hat{B}, \hat{C}]] + [\hat{B}, [\hat{C}, \hat{A}]] + [\hat{C}, [\hat{A}, \hat{B}]] = 0 \quad \text{(Jacobi identity)}\,.$$

7.2. Glauber's formula. If two operators $\hat{A}$ and $\hat{B}$ do not commute, there is no simple relation between $\mathrm{e}^{\hat{A}}\mathrm{e}^{\hat{B}}$ and $\mathrm{e}^{\hat{A}+\hat{B}}$. Suppose here that $\hat{A}$ and $\hat{B}$ both commute with their commutator $[\hat{A}, \hat{B}]$. Prove Glauber's formula,

$$\mathrm{e}^{\hat{A}}\mathrm{e}^{\hat{B}} = \mathrm{e}^{\hat{A}+\hat{B}}\mathrm{e}^{[\hat{A},\hat{B}]/2}\,. \tag{7.40}$$

Hint: you can introduce the operator $\hat{F}(t) = \mathrm{e}^{t\hat{A}}\mathrm{e}^{t\hat{B}}$, where t is a dimensionless variable, and show that

$$\frac{\mathrm{d}F}{\mathrm{d}t} = \left(\hat{A} + \hat{B} + t[\hat{A}, \hat{B}]\right)\,\hat{F}(t).$$

You can then integrate this equation between $t = 0$ and $t = 1$.

Examples: $\hat{A} = \hat{x}/x_0, \hat{B} = \hat{p}/p_0$ (where x_0 and p_0 have the dimensions of a position and a momentum); $\hat{A} = \lambda\hat{a}, \hat{B} = \mu\hat{a}^\dagger$ (where $\hat{a}$ and $\hat{a}^\dagger$ are the annihilation and creation operators for a harmonic oscillator and λ and μ are two complex numbers).

7.3. Classical equations of motion for the harmonic oscillator. Show that for a harmonic oscillator with a potential $V(x) = m\omega^2 x^2/2$, the Ehrenfest theorem gives identically the classical equation of motion,

$$\frac{\mathrm{d}^2\langle x\rangle}{\mathrm{d}t^2} = -\omega^2\langle x\rangle\,.$$

7.4. Conservation law. Consider a system of two particles interacting through a potential $V(\boldsymbol{r}_1 - \boldsymbol{r}_2)$. Check that the total momentum $\boldsymbol{P} = \boldsymbol{p}_1 + \boldsymbol{p}_2$ is conserved. Show that this property can be extended to a system of n interacting particles.

7.5. Hermite functions. Prove, from (7.37) and the definition of $\hat{a}$ and $\hat{a}^\dagger$, the recursion relations for the following Hermite functions:

$$\hat{x}|n\rangle = \sqrt{\frac{\hbar}{2m\omega}}\left(\sqrt{n+1}|n+1\rangle + \sqrt{n}|n-1\rangle\right)\,, \tag{7.41}$$

$$\hat{p}|n\rangle = \mathrm{i}\sqrt{\frac{m\hbar\omega}{2}}\left(\sqrt{n+1}|n+1\rangle - \sqrt{n}|n-1\rangle\right)\,. \tag{7.42}$$

7.6. Generalized uncertainty relations.

a. Consider, in three dimensions, the radial variable $r = \sqrt{x^2 + y^2 + z^2}$ and a real function $f(r)$ of this variable. Show that the commutator of $\hat{p}_x$ with $f(\hat{r})$ is

$$[\hat{p}_x, \hat{f}] = -\mathrm{i}\hbar \frac{\hat{x}}{r} f'(\hat{r}) ,$$

where $f'(r)$ is the derivative of f.

b. Consider the operator $\hat{A}_x = \hat{p}_x - \mathrm{i}\lambda \hat{x} f(\hat{r})$, where λ is a real number.
- Calculate the square of the norm of $\hat{A}_x|\psi\rangle$ for an arbitrary vector $|\psi\rangle$.
- Add the analogous relations for $\hat{A}_y$ and $\hat{A}_z$, and derive an inequality relating $\langle p^2 \rangle$, $\langle r^2 f^2 \rangle$, $\langle f \rangle$ and $\langle r f' \rangle$ which holds for any function f and any state $|\psi\rangle$.

c. Considering the cases $f = 1$, $f = 1/r$ and $f = 1/r^2$, show that the following relations are satisfied in three dimensions:

$$\langle p^2 \rangle \langle r^2 \rangle \geq \frac{9}{4}\hbar^2 , \quad \langle p^2 \rangle \geq \hbar^2 \left\langle \frac{1}{r} \right\rangle^2 , \quad \langle p^2 \rangle \geq \frac{\hbar^2}{4} \left\langle \frac{1}{r^2} \right\rangle .$$

d. *Harmonic oscillator.* The Hamiltonian of a three-dimensional harmonic oscillator is $\hat{H} = \hat{p}^2/2m + m\omega^2 \hat{r}^2/2$.
- Using the first inequality, find a lower bound for the ground-state energy of this oscillator, and explain why this bound is equal to the ground-state energy.
- Write down the differential equation satisfied by the corresponding ground-state wave function and calculate this wave function.

e. *Hydrogen atom.* The Hamiltonian of the hydrogen atom is, considering the proton mass as very large compared with the electron mass,

$$\hat{H} = \frac{\hat{p}^2}{2m_e} - \frac{e^2}{r} ,$$

where, for simplicity, we set $e^2 = q^2/4\pi\epsilon_0$.
- Using the second inequality, find a lower bound for the ground-state energy of the hydrogen atom, and explain why this bound is equal to the ground-state energy.
- Write down the differential equation satisfied by the corresponding ground-state wave function $\phi(r)$ and calculate this wave function.

7.7. Quasi-classical states of the harmonic oscillator. We consider a one-dimensional harmonic oscillator of frequency ω and study the eigenstates $|\alpha\rangle$ of the annihilation operator, given by

$$\hat{a}|\alpha\rangle = \alpha|\alpha\rangle ,$$

where α is a complex number. We expand $|\alpha\rangle$ in the basis $\{|n\rangle\}$:

$$|\alpha\rangle = \sum_n C_n|n\rangle .$$

a. *Determination of α.*
   (i) Write down the recursion relation between the coefficients C_n.
   (ii) Express the C_n's in terms of the first coefficient C_0.
   (iii) Calculate the coefficients C_n by normalizing $|\alpha\rangle$, i.e. $\langle\alpha|\alpha\rangle = 1$.
   (iv) What are the allowed values of the number α?
   (v) In an energy measurement on the state $|\alpha\rangle$, what is the probability of finding the value $E_n = (n + 1/2)\hbar\omega$?

b. Consider a state $|\alpha\rangle$. Starting from the expression for the Hamiltonian and the definition of this state, do the following:
   (i) Calculate the expectation value $\langle E\rangle$.
   (ii) Calculate the expectation value of the square of the energy $\langle E^2\rangle$ (use the commutator of $\hat{a}$ and $\hat{a}^\dagger$).
   (iii) Deduce the value of the dispersion ΔE in this state.
   (iv) In what sense can one say that the energy is defined more and more accurately as $|\alpha|$ increases and becomes much greater than 1?

c. Calculate $\langle x\rangle, \Delta x, \langle p\rangle, \Delta p$ in the state $|\alpha\rangle$. In that state, what is the value of the product $\Delta x\, \Delta p$?

d. We assume that at $t = 0$, the oscillator is in the state $|\alpha\rangle$.
   (i) Write down the state $|\psi(t)\rangle$ of the system at time t.
   (ii) Show that the state $|\psi(t)\rangle$ is also an eigenstate of the operator $\hat{a}$ and give the corresponding eigenvalue.
   (iii) We set $\alpha = \alpha_0 e^{i\phi}$, where α_0 is real and positive. What are, at time t, the values of $\langle x\rangle$, $\langle p\rangle$ and $\Delta x\, \Delta p$?

e. We now determine the wave functions corresponding to $|\alpha\rangle$.
   (i) Check that the change of variables from x and p to X and P leads to the following expression for the operator $\hat{P}$ when it acts on wave functions $\psi(X,t)$:

$$\hat{P} = -i\frac{\partial}{\partial X} .$$

   Give the corresponding expression for the operator $\hat{X}$ when it acts on functions $\varphi(P,t)$.
   (ii) Calculate the wave function $\psi_\alpha(X)$ of the state $|\alpha\rangle$.
   (iii) Calculate the Fourier transform $\varphi_\alpha(P)$ of this wave function.
   (iv) Starting from the time dependence of $|\psi_\alpha(X,t)|^2$ and $|\varphi_\alpha(P,t)|^2$, explain the results obtained previously.

7.8. Time–energy uncertainty relation. Consider a state $|\psi\rangle$ of a system whose energy dispersion is ΔE, and an observable $\hat{A}$ whose expectation value and dispersion are $\langle a\rangle$ and Δa, respectively. Using the commutation relations, show that following inequality holds:

$$\Delta a\,\Delta E \geq \frac{\hbar}{2}\left|\frac{\mathrm{d}\langle a\rangle}{\mathrm{d}t}\right| .$$

Deduce from this that if the typical evolution timescale τ of the system is defined by $\tau = |\Delta a/(\mathrm{d}\langle a\rangle/\mathrm{d}t)|$, one has the inequality $\tau\Delta E \geq \hbar/2$.

7.9. Virial theorem. Consider a one-dimensional system with the Hamiltonian $\hat{H} = \hat{p}^2/2m + V(\hat{x})$, where $V(x) = \lambda x^n$.

a. Calculate the commutator $[\hat{H}, \hat{x}\hat{p}]$.
b. By taking the expectation value of this commutator, show that, for any eigenstate of $\hat{H}$, one has the relation

$$2\langle T\rangle = n\langle V\rangle ,$$

where $\hat{T} = \hat{p}^2/2m$ is the kinetic-energy operator. Check this relation on the harmonic oscillator.
c. Generalize this result to three dimensions by calculating $[\hat{H}, \hat{\boldsymbol{r}}\cdot\hat{\boldsymbol{p}}]$ and considering a potential $V(\boldsymbol{r})$ which is a homogeneous function of the variables x, y, z, of degree n. A homogeneous function of degree n satisfies $V(\alpha x, \alpha y, \alpha z) = \alpha^n V(x,y,z)$ and $\boldsymbol{r}\cdot\nabla V = nV$.
d. Show that, for an arbitrary potential $V(r)$, one has the general relation

$$2\langle T\rangle = \left\langle r\frac{\partial V}{\partial r}\right\rangle .$$

7.10. Benzene and cyclo-octatetraene molecules. Consider the states of an electron in a hexagonal C_6 molecule composed of six equally spaced atoms. The distance between two neighboring atoms is denoted by d. We denote by $|\xi_n\rangle$, $n = 1, \ldots, 6$, the states localized in the vicinity of the atoms $n = 1, \ldots, 6$, respectively. We assume that $\langle\xi_n|\xi_m\rangle = \delta_{n,m}$. The Hamiltonian $\hat{H}$ of this system is defined in the basis $\{|\xi_n\rangle\}$ by $\hat{H} = E_0\hat{I} + \hat{W}$, where

$$\hat{W}|\xi_n\rangle = -A(|\xi_{n+1}\rangle + |\xi_{n-1}\rangle)$$

and $A > 0$. We use here the cyclic conditions $|\xi_7\rangle \equiv |\xi_1\rangle$ and $|\xi_0\rangle \equiv |\xi_6\rangle$. We denote by $|\psi_n\rangle$ and E_n, $n = 1, \ldots, 6$, the eigenstates of $\hat{W}$ and the corresponding eigenvalues. For simplicity, we choose the origin of energy such that $E_0 = 0$.

We define the rotation operator $\hat{R}$ by $\hat{R}|\xi_n\rangle = |\xi_{n+1}\rangle$.

a. What are the eigenvalues λ_k, $k = 1, \dots, 6$ of $\hat{R}$?
b. The eigenvector corresponding to λ_k is denoted by $|\phi_k\rangle = \sum_{p=1}^{6} c_{k,p}|\xi_p\rangle$. Write down the recursion relation between the coefficients $c_{k,p}$ and determine these coefficients by normalizing $|\phi_k\rangle$.
c. Check that the vectors $|\phi_k\rangle$ form an orthonormal basis of the six-dimensional space under consideration.
d. Check that the same vectors $|\phi_k\rangle$ are eigenvectors of the operator $\hat{R}^{-1} = \hat{R}^\dagger$ defined by $\hat{R}^{-1}|\xi_n\rangle = |\xi_{n-1}\rangle$ and calculate the corresponding eigenvalues.
e. Show that $\hat{W}$ and $\hat{R}$ commute. What conclusions can we draw from that?
f. Express $\hat{W}$ in terms of $\hat{R}$ and $\hat{R}^{-1}$. Deduce the eigenstates of $\hat{W}$ and the corresponding eigenvalues. Discuss the degeneracies of the energy levels.
g. Consider now a regular eight-center chain of atoms closed into a ring (cyclo-octatetraene molecule).
   (i) Using a method similar to the preceding one, deduce the energy levels for an electron moving on this chain. Discuss the degeneracies of these levels.
   (ii) At time $t = 0$, the electron is assumed to be localized on the site $n = 1$, $|\psi(t = 0)\rangle = |\xi_1\rangle$. Calculate the probability $p_1(t)$ of finding the electron again on the site $n = 1$ at a later time t; set $\omega = A/\hbar$.
   (iii) Does there exist a time $t \neq 0$ for which $p_1(t) = 1$? Explain why. Is the propagation of an electron on the chain periodic?
h. Consider now an electron on a ring of N sites, located regularly on a circle with a distance d between two adjacent sites. The states localized in the vicinity of each center $n = 1, \dots, N$ are denoted by $|\xi_n\rangle$. The Hamiltonian is defined, as above, by $\hat{H} = E_0\hat{I} + \hat{W}$, where $\hat{W}|\xi_n\rangle = -A(|\xi_{n+1}\rangle + |\xi_{n-1}\rangle)$ and $A > 0$. By extending the argument above, calculate the energy levels and the corresponding eigenstates. What happens in the limit of a chain of infinite length closed into a ring?

8. The Stern–Gerlach Experiment

The capital things that have been said to mankind
have always been simple things.
Charles de Gaulle

In this chapter, we turn our attention to the celebrated 1922 experiment of Stern and Gerlach. We shall show, for this example of a highly "nonclassical" experimental situation, how one can construct phenomenologically the space of states and the relevant observables. We shall obtain a description of this experiment which, as a by-product, will provide us with a concrete way to discuss a measurement process in quantum mechanics.

8.1 Principle of the Experiment

A collimated beam of atoms is sent into a region where an inhomogeneous magnetic field is applied along the z direction, perpendicular to the initial velocity of the atoms (Fig. 8.1a). The possible deflection of the beam by the field gradient is then measured by observing the impacts of the atoms on a detection plate perpendicular to the initial direction of the beam.

8.1.1 Classical Analysis

We first analyze this experiment within classical mechanics. The atoms are neutral and are not subject to a magnetic Lorentz force. However, if they have a nonvanishing magnetic moment $\boldsymbol{\mu}$, a force

$$F_z = \mu_z \frac{\partial B_z}{\partial z} , \tag{8.1}$$

parallel to the z direction, acts on them and deflects their trajectory. The expression (8.1) is a well-known result of classical mechanics and magnetostatics. We first outline its derivation here. When a magnetic moment $\boldsymbol{\mu}$ is placed in a magnetic field $\boldsymbol{B}$, the magnetic interaction energy is

$$W = -\boldsymbol{\mu} \cdot \boldsymbol{B} , \tag{8.2}$$

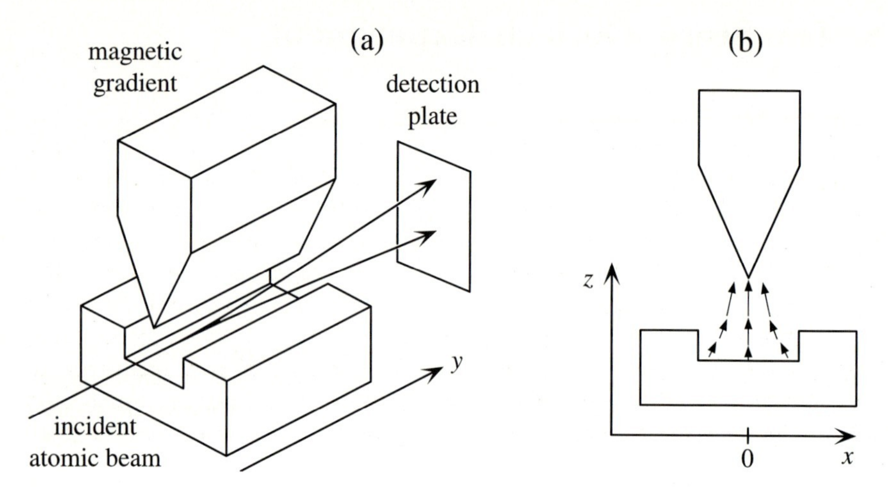

Fig. 8.1. (**a**) The Stern–Gerlach experiment: atoms from a collimated beam are deflected as they cross a region where an inhomogeneous magnetic field is applied. This experiment can be interpreted as a measurement of the component of the atomic magnetic moment along the direction of the field (z in the figure). (**b**) Magnetic gradient between the pole pieces of the magnet

and a torque:

$$\boldsymbol{\Gamma} = \boldsymbol{\mu} \times \boldsymbol{B} \tag{8.3}$$

is exerted on the magnetic moment. In addition, if the magnetic field is inhomogeneous, a force

$$\boldsymbol{F} = \boldsymbol{\nabla}(\boldsymbol{\mu} \cdot \boldsymbol{B}) = \sum_{i=x,y,z} \mu_i(t)\, \boldsymbol{\nabla} B_i \tag{8.4}$$

acts on the dipole.

We can make a classical model of an atom (let us say hydrogen for simplicity) by considering a particle with mass m_{e} and charge $-q$ (the electron), moving with a uniform velocity v in a circle of radius r centered on a charge $+q$. This positive fixed charge represents the nucleus and is supposed to be much heavier than the electron. The angular momentum of this system is

$$\boldsymbol{L} = \boldsymbol{r} \times \boldsymbol{p} = m_{\mathrm{e}} r v\, \boldsymbol{u}\,, \tag{8.5}$$

where $\boldsymbol{u}$ is the unit vector orthogonal to the orbital plane of the electron. The magnetic moment of this elementary current loop is

$$\boldsymbol{\mu} = I S\, \boldsymbol{u}\,, \tag{8.6}$$

where $I = -qv/(2\pi r)$ is the current in the loop, and $S = \pi r^2$ is the loop area. We then find a remarkably simple relation between the angular momentum and the magnetic moment of this classical system:

$$\boldsymbol{\mu} = \gamma_0 \boldsymbol{L}, \qquad \text{where} \qquad \gamma_0 = \frac{-q}{2m_e} \,. \tag{8.7}$$

Note that the proportionality coefficient, called the gyromagnetic ratio, does not depend on the radius r of the trajectory of the electron, nor on its velocity v. Strictly speaking, the presence of an external magnetic field perturbs the electron motion and modifies this very simple relation, but one can show that this perturbation is very weak for realistic fields, and we neglect it here.

From (8.3) one might naively expect that the magnetic moment of the atom would become aligned with the local magnetic field, as does the needle of a compass. However, the proportionality between the magnetic moment and the angular momentum gives rise to a radically different phenomenon, analogous to the gyroscopic effect. The evolution of the angular momentum is given by $\mathrm{d}\boldsymbol{L}/\mathrm{d}t = \boldsymbol{\Gamma}$. The proportionality between $\boldsymbol{L}$ and $\boldsymbol{\mu}$ then implies

$$\frac{\mathrm{d}\boldsymbol{\mu}}{\mathrm{d}t} = -\gamma_0 \, \boldsymbol{B} \times \boldsymbol{\mu} \,. \tag{8.8}$$

Consequently, for an atom at $\boldsymbol{r}$, the magnetic moment does not align with the axis of the local magnetic field $\boldsymbol{B}(\boldsymbol{r})$, but precesses around this axis with an angular frequency

$$\omega_0 = -\gamma_0 B(\boldsymbol{r}) \,. \tag{8.9}$$

The quantity ω_0 is called the Larmor frequency.

This precession phenomenon is very important in practice. It is a particular case of a general theorem[1] of electrodynamics proven by Larmor in 1897. This problem was considered independently the same year by Lorentz.

We assume that the classical trajectory of the atoms lies in the plane of symmetry $x = 0$ of the magnet (see Fig. 8.1b). Along this trajectory the magnetic field is always parallel to the z axis, so that the Larmor precession takes place around z. Also, owing to the symmetry of the device, the quantities $\partial B_z/\partial x$ and $\partial B_z/\partial y$ vanish along the atomic-beam trajectory (we neglect possible edge effects). If the displacement of the magnetic moment during a single precession period $2\pi/\omega_0$ is small compared with the typical scale of variation of the magnetic field, we can average the force (8.4) over the Larmor period. The contributions of μ_x and μ_y to (8.4) then vanish, and one is left only with the z component of the force $F_z = \mu_z(t)\, \partial B_z/\partial z$. In addition, we deduce from (8.8) that μ_z stays constant as the atom moves in the magnetic-field gradient, which justifies the result (8.1).

8.1.2 Experimental Results

In the absence of magnetic-field gradient one observes a single spot on the detecting plate, in the vicinity of $x = z = 0$ (Fig. 8.2a). The magnetic-field gradient provides a way to measure the z component of the magnetic

[1] See, e. g., J. D. Jackson, *Classical Electrodynamics*, Wiley, New York (1975).

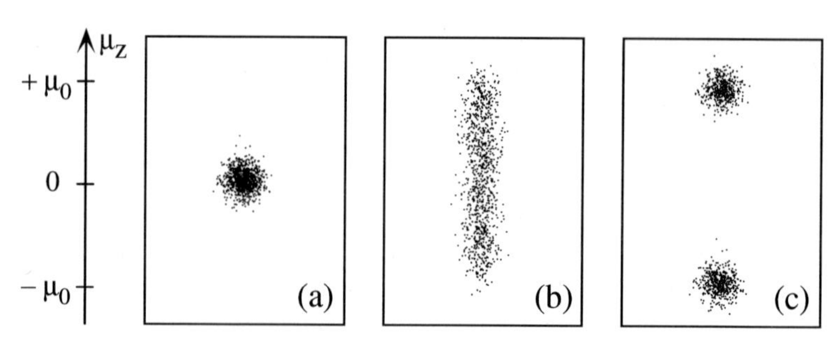

Fig. 8.2. Possible results of a Stern–Gerlach experiment. (**a**) In the absence of a magnetic-field gradient no deflection of the atomic trajectories occurs, and the atoms form a single spot around the point $x = z = 0$; each dot represents the impact of an atom on the detection screen. (**b**) Simulation of the result expected from classical mechanics, assuming that all atoms carry the same magnetic moment $\boldsymbol{\mu}_0$ with a random orientation; the distribution of the z component of the magnetic moment is uniform between $-\mu_0$ and $+\mu_0$. (**c**) Simulation of the result found experimentally with silver atoms: the experiment, which can be considered as a measurement of the z component of the magnetic moment, yields only the two results $+\mu_0$ and $-\mu_0$

moment of the atoms as they enter the field zone. Let us assume that all the atoms carry the same magnetic moment, of norm μ_0, and that this moment is oriented at random when an atom enters the field zone. Classically, this should produce some continuous distribution of μ_z between the two extreme values $-\mu_0$ and $+\mu_0$, so that one would expect that the impacts of the atoms on the screen would form an extended line parallel to z (Fig. 8.2b). The endpoints of the line correspond to atoms whose magnetic moments are oriented such that $\mu_z = +\mu_0$ and $\mu_z = -\mu_0$, respectively.

The experimental observation differs radically from this classical prediction. The set of impacts never forms a continuous line on the screen. For some atoms, such as silver, the impacts are grouped into two spots corresponding to $\mu_z = +\mu_0$ and $\mu_z = -\mu_0$, with $\mu_0 = 9.27 \times 10^{-24}\,\mathrm{J\,T^{-1}}$ (Fig. 8.2c). For other atoms, one may find three, four, etc. spots, which are always placed symmetrically with respect to the trajectory in the absence of a magnetic field. Some atoms, such as helium in its ground state, do not show any appreciable deviation. This latter case can obviously be interpreted as meaning that these atoms do not have a magnetic moment. We shall investigate in the next section how the quantum formalism can describe these results. We shall concentrate on the case where there are two spots, such as with silver atoms. This method can be extended to cases where there are 3, 4, ... spots.

To conclude this presentation, we remark that the order of magnitude of μ_0 is understandable. In fact, the only quantity with the dimensions of a magnetic moment that we can construct with fundamental and atomic constants is $\hbar q/m$, where q is the elementary charge (proton or electron) and m is a typical atomic mass. According to whether we choose for m the electron mass m_e or the proton mass m_p, we obtain results which differ by three orders of magnitude, ranging from 10^{-23} to 10^{-26} J T^{-1}. The result μ_0 of the Stern–Gerlach experiment is consistent with

$$\mu_0 = \hbar\,|\gamma_0| = \frac{\hbar q}{2m_e} , \tag{8.10}$$

which amounts to taking $L = \hbar$ in (8.7). The quantity (8.10) is the absolute value of the *Bohr magneton.*

Why would Stern and his colleague Gerlach want to do this experiment in 1921, five years before quantum mechanics as we now understand it was developed? Stern's goal was to test one of the mysterious aspects of the *old quantum theory,* which was called "space quantization". When theorists learned about the project of Stern and Gerlach, most of them were quite skeptical. For instance Born declared later, "It took me a time before I took this idea seriously. I thought always that direction [space] quantization was a kind of symbolic expression for something which you don't understand. But to take it literally like Stern did, this was his own idea... I tried to persuade Stern that there was no sense [in it], but then he told me that it was worth a try." The experiment was quite difficult, requiring at the same time a good vacuum system and a hot oven (1000°C) to produce an intense beam of silver atoms. When the result finally came out, it was at first considered as a very successful proof of the space quantization idea. However Einstein and Ehrenfest quickly pointed out crucial inconsistencies in the description of the experimental results within the old theory of quanta, and it was only when quantum mechanics was developed in 1926–27 (including the concept of spin) that a consistent description of this experiment became possible.

8.2 The Quantum Description of the Problem

The first stage of the quantum description consists in specifying the Hilbert space of the system, by examining the degrees of freedom of the atom. There are, a priori, two different sets of degrees of freedom involved in this experiment. First, the atoms can move in space, with the corresponding translational degrees of freedom along each of the three directions x, y, z. In addition, there is another degree of freedom corresponding to the internal magnetic moment of the atom.

In the case of ground-state helium atoms, for which no deviation occurs, the internal degree of freedom can be ignored. The state of the atom can be described by a wave function $\psi(\boldsymbol{r})$ whose evolution is given by the Schrödinger equation for a free particle (Chap. 2). For a silver atom, the internal degree of freedom associated with its magnetic moment plays a crucial role. This is what leads to the splitting of the trajectories of the atoms when they cross the inhomogeneous-field zone, giving rise to two spots.

The space of states that we have to consider, in order to explain the experiment, has the structure of a tensor product $\mathcal{E} = \mathcal{E}_{\text{external}} \otimes \mathcal{E}_{\text{internal}}$. The space corresponding to the translational degrees of freedom $\mathcal{E}_{\text{external}}$ is the space of wave functions seen in Chaps. 2 and 3: $\mathcal{E}_{\text{external}} = \mathcal{L}^2(R^3)$. In order to construct the space $\mathcal{E}_{\text{internal}}$ associated with the internal-magnetic-moment degree of freedom, we remark that the Stern–Gerlach experiment can be reinterpreted as *a measurement of the z component of the magnetic moment of the atom.* We denote the corresponding observable by $\hat{\mu}_z$.

The first experimental observation is that, whatever the state of the magnetic moment of the atom, a measurement of μ_z gives one of the results $+\mu_0$ and $-\mu_0$, and these values only. The dimension of the space $\mathcal{E}_{\text{internal}}$ is therefore at least 2, since there are at least two eigenstates of $\hat{\mu}_z$, with eigenvalues $+\mu_0$ and $-\mu_0$.

Obviously, there is nothing special about the z axis. The same remarks hold for the projections μ_x and μ_y of the magnetic moment on the x and y axes, and for the corresponding observables $\hat{\mu}_x$ and $\hat{\mu}_y$. There are at least two eigenstates of $\hat{\mu}_x$, with eigenvalues $+\mu_0$ and $-\mu_0$, and similarly for $\hat{\mu}_y$.

In which space $\mathcal{E}_{\text{internal}}$ should we describe the magnetic-moment states of the atom? The answer to this question is by no means obvious. Classically the magnetic moment $\boldsymbol{\mu}$ of a system is a vector quantity, characterized by its three components (μ_x, μ_y, μ_z) in a reference system. In quantum mechanics we must deal with a set of three observables $(\hat{\mu}_x, \hat{\mu}_y, \hat{\mu}_z)$ and we know that each of them has only two eigenvalues $+\mu_0$ and $-\mu_0$.

It is a remarkable fact that one can explain the experimental results under the "minimal" assumption that the Hilbert space $\mathcal{E}_{\text{internal}}$ associated with the magnetic moment is of *dimension 2.* As we shall see, this assumption is consistent, and it leads to an explanation of all phenomena related to the magnetic moment in the Stern–Gerlach experiment.[2]

Suppose therefore that $\mathcal{E}_{\text{internal}}$ is two-dimensional. A basis of this space is then provided by the two eigenstates of $\hat{\mu}_z$ corresponding to the two results $+\mu_0$ and $-\mu_0$. We write these states as $|+\rangle_z$ and $|-\rangle_z$. By assumption,

$$\hat{\mu}_z|+\rangle_z = \mu_0|+\rangle_z\,, \qquad \hat{\mu}_z|-\rangle_z = -\mu_0|-\rangle_z\,, \tag{8.11}$$

and any internal state $|\mu\rangle$ of the atom can be written as

$$|\mu\rangle = \alpha|+\rangle_z + \beta|-\rangle_z \tag{8.12}$$

where $|\alpha|^2 + |\beta|^2 = 1$. A measurement of the component μ_z of the magnetic moment then gives $+\mu_0$ (i. e. the atom is detected in the upper spot) with probability $|\alpha|^2$ and $-\mu_0$ (lower spot) with probability $|\beta|^2$.

Using a matrix representation, we have, in the basis $\{|\pm\rangle_z\}$,

$$|+\rangle_z = \begin{pmatrix} 1 \\ 0 \end{pmatrix}, \qquad |-\rangle_z = \begin{pmatrix} 0 \\ 1 \end{pmatrix}, \qquad |\mu\rangle = \begin{pmatrix} \alpha \\ \beta \end{pmatrix} \tag{8.13}$$

---

[2] As always in physics (be it quantum or classical), one can never prove that a given theoretical explanation is the only acceptable one. Theories can only be falsified. Our goal here is to propose a scheme that is as simple as possible and explains all observed phenomena.

and

$$\hat{\mu}_z = \mu_0 \begin{pmatrix} 1 & 0 \\ 0 & -1 \end{pmatrix} . \tag{8.14}$$

8.3 The Observables $\hat{\mu}_x$ and $\hat{\mu}_y$

Consider now the experimental situation shown in Fig. 8.3. We place two magnets consecutively in the beam. The first has a field gradient directed along z and splits the incident beam into two beams corresponding to the two internal states $|+\rangle_z$ and $|-\rangle_z$. When the beams leave the field zone, we stop the beam corresponding to the state $|-\rangle_z$ and keep only the beam in the state $|+\rangle_z$. This latter beam is then sent into another Stern–Gerlach device, whose axis is along the x axis, orthogonal to z. We therefore perform a *measurement of the x component* of the atomic magnetic moment, whose corresponding observable is $\hat{\mu}_x$. The result observed experimentally is that the beam is again split into two beams of *equal intensities*, corresponding to values of the magnetic moment along x equal to $+\mu_0$ and $-\mu_0$, respectively.

We want to find the form of the operator $\hat{\mu}_x$. By assumption, this operator acts in $\mathcal{E}_{\text{internal}}$ and is described by a 2×2 matrix in the basis $|\pm\rangle_z$:

$$\hat{\mu}_x = \mu_0 \begin{pmatrix} \alpha_x & \beta_x \\ \gamma_x & \delta_x \end{pmatrix} . \tag{8.15}$$

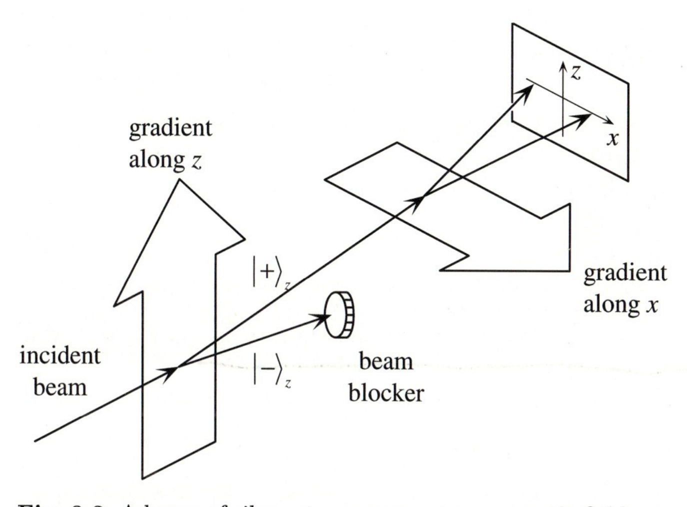

Fig. 8.3. A beam of silver atoms crosses two magnetic-field zones. The first creates a field gradient along z, the second a field gradient along x. After the first magnet, a shutter lets only the atoms in the internal state $|+\rangle_z$ continue. The second magnet allows one to perform a measurement of the x component of the magnetic moment. One finds the two results $+\mu_0$ and $-\mu_0$ with equal probabilities

There are several constraints on the four parameters $\alpha_x, \beta_x, \gamma_x, \delta_x$:

a. The operator $\hat{\mu}_x$ is Hermitian, therefore α_x and δ_x are real, and $\gamma_x = \beta_x^*$.
b. The possible results of a measurement of the x component of the magnetic moment are $+\mu_0$ and $-\mu_0$. These are the eigenvalues of the observable $\hat{\mu}_x$, which imposes the requirements that:

$$\text{the sum of eigenvalues} = \mathrm{Tr}(\hat{\mu}_x) \Rightarrow \alpha_x + \delta_x = 0\,, \tag{8.16}$$

$$\text{the product of eigenvalues} = \det(\hat{\mu}_x) \Rightarrow \alpha_x\delta_x - \beta_x\gamma_x = -1\,. \tag{8.17}$$

c. If the initial state is $|+\rangle_z$ and if we measure μ_x, we find the values $+\mu_0$ and $-\mu_0$ with equal probabilities. The expectation value of the results is therefore 0 for this initial state. This requires that

$$0 = {}_z\langle +|\hat{\mu}_x|+\rangle_z = \mu_0\alpha_x\,. \tag{8.18}$$

Combining this set of constraints, we deduce that the operator $\hat{\mu}_x$ is necessarily of the form

$$\hat{\mu}_x = \mu_0 \begin{pmatrix} 0 & \mathrm{e}^{-\mathrm{i}\phi_x} \\ \mathrm{e}^{\mathrm{i}\phi_x} & 0 \end{pmatrix}\,. \tag{8.19}$$

In the previous argument, we discussed the x axis. However, the same argument holds for any axis provided it is perpendicular to z. In particular, it is also true for the y axis, which is orthogonal to both x and z. We can repeat the argument and end up with an expression similar to (8.19):

$$\hat{\mu}_y = \mu_0 \begin{pmatrix} 0 & \mathrm{e}^{-\mathrm{i}\phi_y} \\ \mathrm{e}^{\mathrm{i}\phi_y} & 0 \end{pmatrix}\,. \tag{8.20}$$

The last step of our reasoning is to determine the relation between the phases ϕ_x and ϕ_y entering (8.19) and (8.20). In order to do that, consider an atomic beam prepared in the eigenstate of $\hat{\mu}_x$ with eigenvalue $+\mu_0$. This eigenstate, which we write as $|+\rangle_x$, is, in the basis $|\pm\rangle_z$,

$$|+\rangle_x = \frac{1}{\sqrt{2}}\left(|+\rangle_z + \mathrm{e}^{\mathrm{i}\phi_x}|-\rangle_z\right)\,. \tag{8.21}$$

If we measure the y component of the magnetic moment of atoms prepared in this state, a direct transcription of the above argument tells us that we shall find the two results $+\mu_0$ and $-\mu_0$ *with equal probabilities.* The expectation value of the result vanishes, and therefore

$$0 = {}_x\langle +|\hat{\mu}_y|+\rangle_x = \mu_0\cos(\phi_x - \phi_y)\,, \quad \text{i.e.} \quad \phi_y - \phi_x = \frac{\pi}{2}\,(\text{modulo}\,\pi)\,. \tag{8.22}$$

One can check that it is not possible to go further in this determination[3] of ϕ_x and ϕ_y. Any pair that satisfies (8.22) gives rise to operators $\hat{\mu}_x$ and $\hat{\mu}_y$

[3] The only extra ingredient one can impose is that the set of axes (x, y, z) be right-handed; the remaining arbitrary phase reflects the fact that the choice of the x and y axes in the plane orthogonal to z is arbitrary.

which can account for the set of experimental results. In order to simplify the notation, we choose the particular values $\phi_x = 0, \phi_y = \pi/2$. This leads to the following three operators $\hat{\mu}_x, \hat{\mu}_y, \hat{\mu}_z$ which describe the components of the magnetic moment along the three axes:

$$\hat{\mu}_x = \mu_0 \begin{pmatrix} 0 & 1 \\ 1 & 0 \end{pmatrix}, \quad \hat{\mu}_y = \mu_0 \begin{pmatrix} 0 & -\mathrm{i} \\ \mathrm{i} & 0 \end{pmatrix}, \quad \hat{\mu}_z = \mu_0 \begin{pmatrix} 1 & 0 \\ 0 & -1 \end{pmatrix}. \tag{8.23}$$

We recover, up to the coefficient μ_0, the Pauli matrices introduced in Chap. 6 (6.5). The eigenstates of $\hat{\mu}_x$ and $\hat{\mu}_y$ are

$$|\pm\rangle_x = \frac{1}{\sqrt{2}}\left(|+\rangle_z \pm |-\rangle_z\right), \qquad |\pm\rangle_y = \frac{1}{\sqrt{2}}\left(|+\rangle_z \pm \mathrm{i}|-\rangle_z\right). \tag{8.24}$$

8.4 Discussion

We now analyze our findings.

8.4.1 Incompatibility of Measurements Along Different Axes

The three operators $\hat{\mu}_x, \hat{\mu}_y, \hat{\mu}_z$ which we have just found do not commute. They obey the three cyclic commutation relations

$$[\hat{\mu}_x, \hat{\mu}_y] = 2\mathrm{i}\mu_0\, \hat{\mu}_z, \quad [\hat{\mu}_y, \hat{\mu}_z] = 2\mathrm{i}\mu_0\, \hat{\mu}_x, \quad [\hat{\mu}_z, \hat{\mu}_x] = 2\mathrm{i}\mu_0\, \hat{\mu}_y. \tag{8.25}$$

Physically, this means that one cannot know simultaneously two components of the magnetic moment of an atom. Suppose we start with atoms in the state $|+\rangle_z$ (the z component is known). If we measure the x component of the magnetic moment, the two results $+\mu_0$ and $-\mu_0$ are possible, with equal probabilities. Suppose we find $+\mu_0$ in this latter measurement. After the measurement of μ_x, the state of the system is the corresponding eigenstate of $\hat{\mu}_x$:

$$|+\rangle_x = \frac{1}{\sqrt{2}}(|-\rangle_z + |-\rangle_z).$$

A measurement of μ_x on this new state will again give the result $+\mu_0$. However, if we turn our attention back to the z axis and measure μ_z, the expression for $|+\rangle_x$ shows that the new measurement will give the results $+\mu_0$ and $-\mu_0$ with equal probabilities. Since we started initially with the state $|+\rangle_z$, for which the z component of the magnetic moment was well defined, we see that the intermediate measurement of μ_x has changed (or has perturbed) the state of the system.

In this simple example, we recover the paradoxical character of quantum "logic" as opposed to classical probabilistic logic. Suppose, for instance, that one performs a separation of red and blue objects, followed by a separation of, say, large and small objects. After the second separation, half of the "blue *and* large" objects would be ... red. In the example of the Stern–Gerlach experiment, sorting into the two categories $\mu_z = +\mu_0$ and $\mu_z = -\mu_0$ loses all its meaning if one attempts to sort the systems into subcategories $\mu_x = +\mu_0$ and $\mu_x = -\mu_0$.

8.4.2 Classical Versus Quantum Analysis

What would the argument lead to in classical mechanics? In a Stern–Gerlach apparatus oriented along the z axis, a magnetic moment precesses around this direction and, for realistic values of the parameters, it makes many rotations between the entrance and the exit. Therefore it seems that, just as in quantum mechanics, the final values of μ_x and μ_y should be completely uncorrelated with the initial values. However, nothing prevents us, at least in principle, from controlling sufficiently well the trajectories and the value of the field. Therefore, we can, to an arbitrary accuracy, make the precession angle equal to an even multiple of 2π. We then end up with a situation where we can measure μ_z without perturbing μ_x and μ_y.

Things become more involved if one performs a quantum description of the center of mass of the atom, still treating the magnetic moment classically. When the atom enters the Stern–Gerlach apparatus, the wave packet has some transverse extension Δz and some momentum dispersion Δp_z, with $\Delta z\, \Delta p_z \geq \hbar/2$. Let us denote by $b' = \partial B_z/\partial z$ the field gradient along the z axis, and denote by T the time it takes to cross the magnet. In order for the measurement of μ_z to be accurate, the momentum variation during the crossing of the field must be large compared with the initial dispersion, i. e.

$$\mu_0 b' T \gg \Delta p_z\,, \tag{8.26}$$

otherwise the spreading of the beam at the exit of the magnet will simply reflect the initial spreading. On the other hand, the angle of precession cannot be a constant, because the inhomogeneity of the field over the extension Δz induces a dispersion $\Delta\omega_0 = \gamma_0 b'\,\Delta z$ in the Larmor frequency (8.9). If we require that the values of μ_x and μ_y are not smeared out in the crossing of the field, the dispersion of the precession angle must be small compared with 2π:

$$T\,\Delta\omega_0 = T\,\gamma_0\, b'\,\Delta z \ll 2\pi\,. \tag{8.27}$$

Owing to the Heisenberg inequality and to the experimental result $\mu_0 \sim \hbar\gamma_0$ (see (8.10)), the two conditions (8.26) and (8.27) cannot be satisfied simultaneously; a quantum description of the center-of-mass motion of the atom suffices to make the measurements of μ_x, μ_y and μ_z "incompatible" with one another.

8.4.3 Measurement Along an Arbitrary Axis

Up to now, we have only considered measurements along the three axes x, y and z. Now we are interested in measuring the component of the magnetic moment along an arbitrary axis. This is shown in Fig. 8.4. We place a Stern–Gerlach apparatus along an arbitrary direction defined by the unit vector $\boldsymbol{u}_\theta$, such that

$$\boldsymbol{u}_\theta = \boldsymbol{u}_x \sin\theta + \boldsymbol{u}_z \cos\theta\,. \tag{8.28}$$

Classically, this corresponds to a measurement of the component μ_θ of the magnetic moment along $\boldsymbol{u}_\theta$, i. e. $\mu_\theta = \mu_x \sin\theta + \mu_z \cos\theta$. Using the correspondence principle, we assume that the corresponding observable is

$$\hat{\mu}_\theta = \hat{\mu}_x \sin\theta + \hat{\mu}_z \cos\theta = \mu_0 \begin{pmatrix} \cos\theta & \sin\theta \\ \sin\theta & -\cos\theta \end{pmatrix}. \tag{8.29}$$

This choice guarantees that the expectation values $\langle\mu_x\rangle, \langle\mu_y\rangle$ and $\langle\mu_z\rangle$ of the components of the magnetic moment transform as the components of a three-vector of the usual kind under rotations.

Just like $\hat{\mu}_x, \hat{\mu}_y, \hat{\mu}_z$, the operator $\hat{\mu}_\theta$ has the eigenvalues $+\mu_0$ and $-\mu_0$. Its eigenvectors are

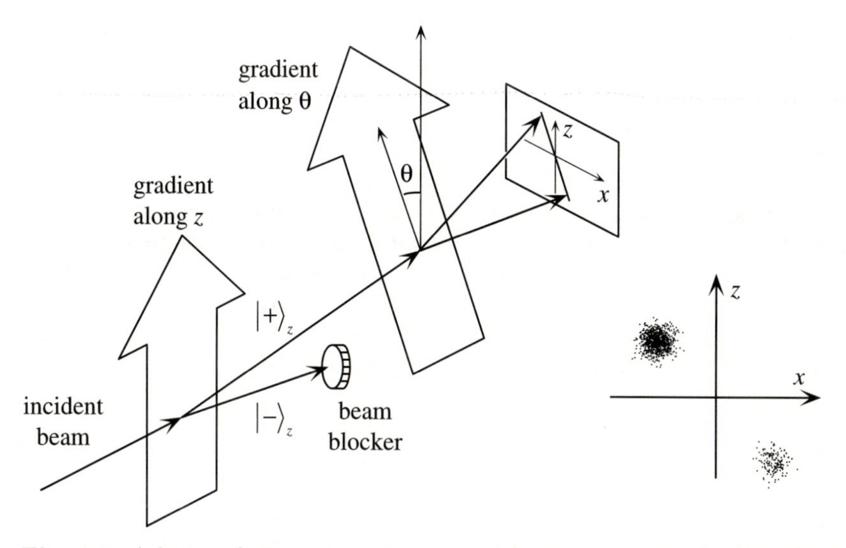

Fig. 8.4. A beam of silver atoms is prepared in the state $|+\rangle_z$. It then crosses a field gradient directed along $\boldsymbol{u}_\theta$. In this measurement of the component of the magnetic moment along $\boldsymbol{u}_\theta$, the two possible results are $+\mu_0$ and $-\mu_0$, with respective probabilities $\cos^2\theta/2$ and $\sin^2\theta/2$. The graph in the *lower right corner* shows a typical result for $\theta = \pi/4$

$$|+\rangle_\theta = \quad |+\rangle_z \cos(\theta/2) + |-\rangle_z \sin(\theta/2) = \begin{pmatrix} \cos(\theta/2) \\ \sin(\theta/2) \end{pmatrix}, \tag{8.30}$$

$$|-\rangle_\theta = -|+\rangle_z \sin(\theta/2) + |-\rangle_z \cos(\theta/2) = \begin{pmatrix} -\sin(\theta/2) \\ \cos(\theta/2) \end{pmatrix}. \tag{8.31}$$

The experimental observations are the following. If a beam prepared in the state $|+\rangle_z$ is sent into a field gradient directed along $\boldsymbol{u}_\theta$, one finds that this beam is split into two beams corresponding to magnetic moments along $\boldsymbol{u}_\theta$ equal to $+\mu_0$ and $-\mu_0$, with relative intensities $I_+(\theta) = I_+(0)\cos^2(\theta/2)$ and $I_-(\theta) = I_+(0)\sin^2(\theta/2)$.

In order to account for this result, we apply the principles of Chap. 5. A measurement of $\hat{\mu}_\theta$ can give two possible values, which are the eigenvalues $+\mu_0$ and $-\mu_0$; if the initial system is in the state $|+\rangle_z$, the respective probabilities for these two possible outcomes are

$$p_+ = |_\theta\langle +|+\rangle_z|^2 = \cos^2(\theta/2)\,, \tag{8.32}$$

$$p_- = |_\theta\langle -|+\rangle_z|^2 = \sin^2(\theta/2)\,. \tag{8.33}$$

Therefore, (8.29) explains why the experimental measurement, which involves a large number of atoms, gives two spots with relative intensities $\cos^2(\theta/2)$ and $\sin^2(\theta/2)$. The measurement gives a result with probability 1 only when θ is 0 or π, i. e. when the preparation axis $\boldsymbol{u}_z$ and the measurement axis $\boldsymbol{u}_\theta$ are parallel or antiparallel.

8.5 Complete Description of the Atom

We now address the question of giving a complete description of the state of the atom. The motion of the center of mass of the atom is described in the Hilbert space $\mathcal{E}_{\text{external}}$ of square-integrable functions $\mathcal{L}^2(R^3)$. The internal state corresponding to the degree of freedom associated with the magnetic moment is described in the two-dimensional space $\mathcal{E}_{\text{internal}}$.

8.5.1 Hilbert Space

The full Hilbert space is the tensor product of these two spaces:

$$\mathcal{E}_{\text{H}} = \mathcal{E}_{\text{external}} \otimes \mathcal{E}_{\text{internal}}\,.$$

Any element $|\psi\rangle$ of $\mathcal{E}_{\text{H}}$ is of the form

$$|\psi\rangle = |\psi_+\rangle \otimes |+\rangle + |\psi_-\rangle \otimes |-\rangle\,, \tag{8.34}$$

where $|\psi_+\rangle$ and $|\psi_-\rangle$ are vectors of $\mathcal{E}_{\text{external}}$, i. e. square-integrable functions of $\boldsymbol{r}$, and where $|+\rangle$ and $|-\rangle$ are the eigenstates of $\hat{\mu}_z$. For simplicity, in what follows we shall omit the subscript z when writing the kets $|\pm\rangle_z$. The space observables $\hat{A}_{\text{ext}}$ (for instance $\hat{x}$ or $\hat{p}$) and the magnetic-moment observables $\hat{\mu}_x, \hat{\mu}_y, \hat{\mu}_z$ act in different spaces and therefore *commute*. The product of two such observables is defined as

$$(\hat{A}_{\text{ext}} \otimes \hat{\mu}_x)(|\psi_\epsilon\rangle \otimes |\epsilon\rangle) = (\hat{A}_{\text{ext}}|\psi_\epsilon\rangle) \otimes (\hat{\mu}_x|\epsilon\rangle)\,, \quad \epsilon = \pm 1\,. \tag{8.35}$$

8.5.2 Representation of States and Observables

There are several possible representations of the states, with corresponding representations of the observables. We give two of these representations. The choice of one or the other, or a third one, is merely a matter of convenience.

(a) Mixed Representation. Any state $|\psi(t)\rangle$ can be represented by a vector of $\mathcal{E}_{\text{internal}}$ whose components are square-integrable functions:

$$\psi_+(\boldsymbol{r},t)|+\rangle + \psi_-(\boldsymbol{r},t)|-\rangle\,. \tag{8.36}$$

The scalar product of $|\psi(t)\rangle$ and $|\chi(t)\rangle$, represented by $\psi_+(\boldsymbol{r},t)|+\rangle + \psi_-(\boldsymbol{r},t)|-\rangle$ and by $\chi_+(\boldsymbol{r},t)|+\rangle + \chi_-(\boldsymbol{r},t)|-\rangle$, respectively, is

$$\langle\psi(t)|\chi(t)\rangle = \int \left(\psi_+^*(\boldsymbol{r},t)\chi_+(\boldsymbol{r},t) + \psi_-^*(\boldsymbol{r},t)\chi_-(\boldsymbol{r},t)\right)\,\mathrm{d}^3r\,.$$

The physical meaning of this representation follows from the fact that we are dealing with a pair of random variables $\boldsymbol{r}$ and μ_z. The probability law for the pair is the following:

$|\psi_+(\boldsymbol{r},t)|^2\,\mathrm{d}^3r$ *and* $|\psi_-(\boldsymbol{r},t)|^2\,\mathrm{d}^3r$ *are the probabilities of detecting the particle in a vicinity* d^3r *of the point* $\boldsymbol{r}$*, with the projection* μ_z *of its magnetic moment equal to* $+\mu_0$ *and* $-\mu_0$*, respectively.*

This results in the following properties:

a. Normalization:

$$\int \left(|\psi_+(\boldsymbol{r},t)|^2 + |\psi_-(\boldsymbol{r},t)|^2\right)\,\mathrm{d}^3r = 1\,. \tag{8.37}$$

b. Probability density of finding the particle at point $\boldsymbol{r}$ independently of the value of μ_z:

$$P(\boldsymbol{r},t) = |\psi_+(\boldsymbol{r},t)|^2 + |\psi_-(\boldsymbol{r},t)|^2\,. \tag{8.38}$$

c. *Conditional probabilities.* Knowing that the particle is at $\boldsymbol{r}$ (within d^3r), the probabilities that a measurement of μ_z yields the results $+\mu_0$ and $-\mu_0$ are

$$P_+(\boldsymbol{r},t) = \frac{|\psi_+(\boldsymbol{r},t)|^2}{P(\boldsymbol{r},t)}\,, \qquad P_-(\boldsymbol{r},t) = \frac{|\psi_-(\boldsymbol{r},t)|^2}{P(\boldsymbol{r},t)}\,, \tag{8.39}$$

respectively, where $P_+(\boldsymbol{r},t) + P_-(\boldsymbol{r},t) = 1$.

(b) Two-Component Wave Function or Spinor. It may also be convenient to use matrix representations for the states $|\psi(t)\rangle$ such as

$$\begin{pmatrix} \psi_+(\boldsymbol{r},t) \\ \psi_-(\boldsymbol{r},t) \end{pmatrix} , \tag{8.40}$$

and, for $\langle\psi(t)|$, the row matrix

$$\left(\psi_+^*(\boldsymbol{r},t), \psi_-^*(\boldsymbol{r},t)\right) . \tag{8.41}$$

The physical interpretation of ψ_+ and ψ_- as probability amplitudes for the pair of random variables $(\boldsymbol{r}, \mu_z)$ is the same as above.

Any observable which acts only on *space* variables is a 2×2 *scalar* matrix whose elements are operators acting in $\mathcal{L}^3(R)$. For instance, the kinetic energy operator can be written as

$$\frac{\hat{\boldsymbol{p}}^2}{2m} = \begin{pmatrix} -(\hbar^2/2m)\Delta & 0 \\ 0 & -(\hbar^2/2m)\Delta \end{pmatrix}$$

Any matrix acting only on magnetic-moment variables is a linear combination of the Pauli matrices (8.23) and of the identity. In this representation, the sum or the product of two observables is the sum or the product of the corresponding matrices.

8.5.3 Energy of the Atom in a Magnetic Field

If the atom is placed in a magnetic field $\boldsymbol{B}(\boldsymbol{r})$, the magnetic potential interaction energy is

$$\hat{W} = -\hat{\boldsymbol{\mu}} \cdot \boldsymbol{B}(\hat{\boldsymbol{r}}) . \tag{8.42}$$

In this formula, we have collected the set of three observables $\hat{\mu}_x, \hat{\mu}_y$ and $\hat{\mu}_z$ into the form of a vector operator $\hat{\boldsymbol{\mu}}$, and, by definition,

$$\hat{\boldsymbol{\mu}} \cdot \boldsymbol{B}(\hat{\boldsymbol{r}}) = \hat{\mu}_x B_x(\hat{\boldsymbol{r}}) + \hat{\mu}_y B_y(\hat{\boldsymbol{r}}) + \hat{\mu}_z B_z(\hat{\boldsymbol{r}}) . \tag{8.43}$$

8.6 Evolution of the Atom in a Magnetic Field

We now show how this theoretical model explains experimental facts.

8.6.1 Schrödinger Equation

Suppose the atom moves in space in a potential $V(\boldsymbol{r})$ and that, in addition, a magnetic field $\boldsymbol{B}$ acts on it. The Hamiltonian is the sum of two terms

$$\hat{H} = \hat{H}_{\text{ext}} \otimes \hat{I}_{\text{int}} + \hat{W} , \tag{8.44}$$

where

$$\hat{H}_{\text{ext}} = \frac{\hat{p}^2}{2m} + V(\hat{\boldsymbol{r}})$$

is of the same type as that which we studied in Chaps. 2–4. In particular, $\hat{H}_{\text{ext}}$ does not act on the internal magnetic moment variable. Conversely, $\hat{W}$ is given by (8.42). This operator acts in the space $\mathcal{E}_{\text{internal}}$ via the three operators $\hat{\mu}_x, \hat{\mu}_y, \hat{\mu}_z$; if the field is inhomogeneous, it also acts in the external Hilbert space through the three functions $B_x(\hat{\boldsymbol{r}}), B_y(\hat{\boldsymbol{r}}), B_z(\hat{\boldsymbol{r}})$.

The Schrödinger equation is

$$\mathrm{i}\hbar \frac{\mathrm{d}}{\mathrm{d}t}|\psi\rangle = \hat{H}|\psi\rangle . \tag{8.45}$$

Choosing the representation of states (8.36) and decomposing the vectors in the orthonormal basis $\{|+\rangle, |-\rangle\}$, we obtain the coupled differential system

$$\begin{aligned}
\mathrm{i}\hbar \frac{\partial}{\partial t}\psi_+(\boldsymbol{r},t) &= \left(-\frac{\hbar^2}{2m}\Delta + V(\boldsymbol{r})\right)\psi_+(\boldsymbol{r},t) \\
&\quad + \langle +|\hat{W}|+\rangle\, \psi_+(\boldsymbol{r},t) + \langle +|\hat{W}|-\rangle\, \psi_-(\boldsymbol{r},t) , \\
\mathrm{i}\hbar \frac{\partial}{\partial t}\psi_-(\boldsymbol{r},t) &= \left(-\frac{\hbar^2}{2m}\Delta + V(\boldsymbol{r})\right)\psi_-(\boldsymbol{r},t) \\
&\quad + \langle -|\hat{W}|+\rangle\, \psi_+(\boldsymbol{r},t) + \langle -|\hat{W}|-\rangle\, \psi_-(\boldsymbol{r},t) .
\end{aligned}$$

The matrix elements of $\hat{W}$ in the basis $\{|+\rangle, |-\rangle\}$ are functions of the external variables. They act in addition to the usual potential terms (diagonal terms) and couple the evolution equations of the components ψ_+ and ψ_-.

8.6.2 Evolution in a Uniform Magnetic Field

Consider a silver atom moving freely in space ($V(\boldsymbol{r}) = 0$), with a uniform applied magnetic field $\boldsymbol{B}$. We assume that at time $t = 0$ the complete atomic wave function (external and internal) is

$$\psi(\boldsymbol{r}, 0)(\alpha_0|+\rangle + \beta_0|-\rangle) , \tag{8.46}$$

i.e. it factorizes in the space and magnetic-moment variables. The total Hamiltonian contains both the kinetic energy of the atom and its interaction energy with the field $\boldsymbol{B}$:

$$\hat{H} = \frac{\hat{p}^2}{2m} - \hat{\boldsymbol{\mu}} \cdot \boldsymbol{B} , \tag{8.47}$$

where, from now on, we omit the identity operators $\hat{I}_{\text{int}}$ and $\hat{I}_{\text{ext}}$. At any later time t, the solution of the Schrödinger equation is also factorized:

$$\psi(\boldsymbol{r},t)(\alpha(t)|+\rangle + \beta(t)|-\rangle)\,, \tag{8.48}$$

where

$$i\hbar\frac{\partial\psi(\boldsymbol{r},t)}{\partial t} = -\frac{\hbar^2}{2m}\Delta\psi(\boldsymbol{r},t)\,, \tag{8.49}$$

$$i\hbar\frac{\mathrm{d}}{\mathrm{d}t}(\alpha(t)|+\rangle + \beta(t)|-\rangle) = -\hat{\boldsymbol{\mu}}\cdot\boldsymbol{B}\;(\alpha(t)|+\rangle + \beta(t)|-\rangle)\,. \tag{8.50}$$

Indeed, if we assume that the state is factorized as in (8.48), we can readily check that it satisfies the Schrödinger equation if (8.49) and (8.50) are satisfied. Since it coincides with the initial state (8.46) at $t = 0$, it therefore represents the solution of the time evolution equation.

We observe that the two types of degrees of freedom decouple from one another. The first equation (8.49) describes the motion of the particle in space (the decoupling remains valid even if a potential $V(\boldsymbol{r})$ is present). The second equation determines the evolution of the internal magnetic state of the atom. If $\boldsymbol{B}$ is parallel to z, (8.50) becomes

$$\begin{cases} i\hbar\dot{\alpha}(t) = -\mu_0 B\;\alpha(t) \\ i\hbar\dot{\beta}(t) = \;\;\mu_0 B\;\beta(t) \end{cases} \quad\Rightarrow\quad \begin{cases} \alpha(t) = \alpha_0\,\exp(-i\omega_0 t/2) \\ \beta(t) = \beta_0\,\exp(i\omega_0 t/2) \end{cases}, \tag{8.51}$$

where we have set $\omega_0 = -2\mu_0 B/\hbar$.

We can determine the expectation values M_x, M_y, M_z of the three components $\hat{\mu}_x, \hat{\mu}_y, \hat{\mu}_z$:

$$\begin{aligned} M_x(t) &= \langle\psi(t)|\hat{\mu}_x|\psi(t)\rangle = 2\mu_0\,\alpha_0\beta_0\,\cos\omega_0 t\,, \\ M_y(t) &= \langle\psi(t)|\hat{\mu}_y|\psi(t)\rangle = 2\mu_0\,\alpha_0\beta_0\,\sin\omega_0 t\,, \\ M_z(t) &= \langle\psi(t)|\hat{\mu}_z|\psi(t)\rangle = \mu_0\,(|\alpha_0|^2 - |\beta_0|^2)\,. \end{aligned} \tag{8.52}$$

Here we have assumed that α_0 and β_0 are real; the calculation can be generalized to complex coefficients with no difficulty.

As we could have expected from the Ehrenfest theorem, M_z is time independent, since $\hat{\mu}_z$ commutes with the Hamiltonian when $\boldsymbol{B}$ is along the z direction. On the contrary, M_x and M_y are not constants of the motion. In order to obtain a more intuitive picture of this evolution, we can rewrite these three equations in the form

$$\begin{cases} \dot{M}_x = -\omega_0 M_y \\ \dot{M}_y = \;\;\omega_0 M_x \\ \dot{M}_z = \;\;0 \end{cases} \quad \text{or} \quad \frac{\mathrm{d}\boldsymbol{M}}{\mathrm{d}t} = \boldsymbol{\Omega}\times\boldsymbol{M}\,, \tag{8.53}$$

where $\boldsymbol{\Omega} = \omega_0\boldsymbol{u}_z$. We recover the Larmor precession described in (8.8) (Fig. 8.5). We shall see later on that this precession can be observed experimentally and that the Larmor frequency can be measured very accurately, for instance in magnetic resonance experiments. This has numerous applications in physics, chemistry, biology and medicine.

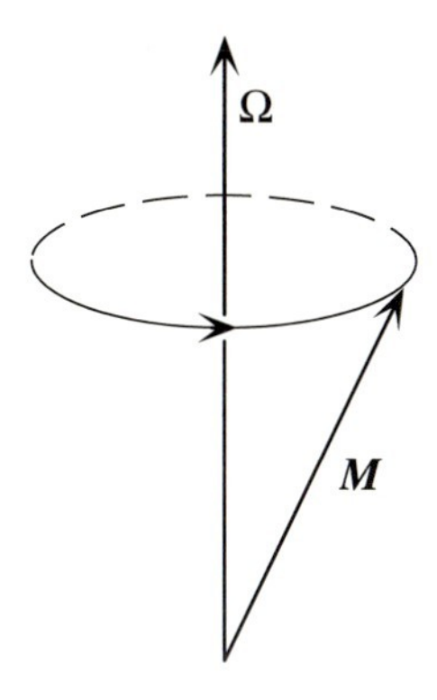

Fig. 8.5. Time evolution of the expectation values of the components of the magnetic moment of a silver atom placed in a magnetic field directed along z. One observes a gyroscopic motion identical to the Larmor precession of a classical magnetic moment placed in the same field

8.6.3 Explanation of the Stern–Gerlach Experiment

The last question we want to examine is whether the theoretical model that we have built can explain the observed spatial separation of the states $|\pm\rangle_z$. We consider an incident atomic beam propagating along y; each atom possesses a magnetic moment. In a region of length L, a magnetic field $\boldsymbol{B}$ parallel to z is applied with a gradient along z:

$$\boldsymbol{B}(\boldsymbol{r}) = B_z(\boldsymbol{r})\boldsymbol{u}_z\,, \qquad \text{where} \qquad B_z(\boldsymbol{r}) = B_0 + b'z\,. \tag{8.54}$$

In full rigor, (8.54) is incorrect since the field $\boldsymbol{B}(\boldsymbol{r})$ does not satisfy $\boldsymbol{\nabla}\cdot\boldsymbol{B} = 0$. A more realistic calculation can be done with a field $\boldsymbol{B} = B_0\boldsymbol{u}_z + b'(z\boldsymbol{u}_z - x\boldsymbol{u}_x)$, which satisfies Maxwell's equations. If the dominant part of the field $B_0\boldsymbol{e}_z$ is much larger than the transverse field $-b'x\boldsymbol{e}_x$ over the transverse extension Δx of the atomic wave packet (i.e. $B_0 \gg b'\Delta x$), the eigenstates of $-\hat{\boldsymbol{\mu}}\cdot\boldsymbol{B}$ remain practically equal to $|\pm\rangle_z$ and the present approach is valid.

Under these conditions, the Schrödinger equation (8.45) can be decoupled into two equations:

$$\mathrm{i}\hbar\frac{\partial}{\partial t}\psi_+(\boldsymbol{r},t) = \left(\frac{\hat{p}^2}{2m} - \mu_0 B\right)\psi_+(\boldsymbol{r},t)\,, \tag{8.55}$$

$$\mathrm{i}\hbar\frac{\partial}{\partial t}\psi_-(\boldsymbol{r},t) = \left(\frac{\hat{p}^2}{2m} + \mu_0 B\right)\psi_-(\boldsymbol{r},t)\,. \tag{8.56}$$

These two equations are both of the same type as the Schrödinger equation seen in Chap. 3, but the potential is not the same for ψ_+ and ψ_-. In order to proceed further, we set

$$\pi_\pm = \int |\psi_\pm(\boldsymbol{r},t)|^2\,\mathrm{d}^3r\,, \quad \pi_+ + \pi_- = 1\,, \tag{8.57}$$

where π_+ and π_- are the probabilities of finding $\mu_z = +\mu_0$ and $\mu_z = -\mu_0$. We deduce from (8.55) and (8.56) that

$$\frac{\mathrm{d}\pi_+}{\mathrm{d}t} = \frac{\mathrm{d}\pi_-}{\mathrm{d}t} = 0\,. \tag{8.58}$$

We define the functions

$$\phi_\pm(\boldsymbol{r},t) = \psi_\pm(\boldsymbol{r},t)/\sqrt{\pi_\pm}\,, \tag{8.59}$$

which are the *conditional* probability amplitudes of particles for which $\mu_z = \pm\mu_0$. These normalized functions also satisfy the Schrödinger-type equations (8.55) and (8.56).

We now define:

$$\langle \boldsymbol{r}_\pm \rangle = \int \boldsymbol{r}\, |\phi_\pm(\boldsymbol{r},t)|^2 \,\mathrm{d}^3r\,, \tag{8.60}$$

$$\langle \boldsymbol{p}_\pm \rangle = \int \phi^*_\pm(\boldsymbol{r},t)\, \frac{\hbar}{\mathrm{i}}\, \boldsymbol{\nabla}\phi_\pm(\boldsymbol{r},t)\; \mathrm{d}^3r\,, \tag{8.61}$$

where $\langle \boldsymbol{r}_+ \rangle$ and $\langle \boldsymbol{r}_- \rangle$ are the average positions of particles for which $\mu_z = +\mu_0$ and $\mu_z = -\mu_0$, respectively, and $\langle \boldsymbol{p}_\pm \rangle$ are their average momenta. A simple application of the Ehrenfest theorem gives

$$(\mathrm{d}/\mathrm{d}t)\langle \boldsymbol{r}_\pm \rangle = \langle \boldsymbol{p}_\pm \rangle/m\,, \tag{8.62}$$

$$(\mathrm{d}/\mathrm{d}t)\langle p_{x_\pm} \rangle = (d/dt)\langle p_{y_\pm} \rangle = 0\,, \tag{8.63}$$

$$(\mathrm{d}/\mathrm{d}t)\langle p_{z_\pm} \rangle = \pm\mu_0 b'\,. \tag{8.64}$$

At $t = 0$, we assume that

$$\langle \boldsymbol{r}_\pm \rangle = 0\,, \qquad \langle p_{x_\pm} \rangle = \langle p_{z_\pm} \rangle = 0\,, \qquad \langle p_{y_\pm} \rangle = mv\,.$$

We obtain, at time t,

$$\langle x_\pm \rangle = 0\,, \qquad \langle y_\pm \rangle = vt\,, \qquad \langle z_\pm \rangle = \pm\mu_0 b' t^2/2m\,. \tag{8.65}$$

Therefore there is a *spatial* separation along z of the initial beam into two beams. One beam corresponds to $\mu_z = +\mu_0$, the other to $\mu_z = -\mu_0$. When the beams leave the magnet of length L, their separation is

$$\delta z = \langle z_+ \rangle - \langle z_- \rangle = \frac{\mu_0 b'}{m}\,\frac{L^2}{v^2}\,. \tag{8.66}$$

If the field gradient is sufficiently strong that $\delta z > \Delta z$ (the separation is larger than the spatial extension of each wave packet), we obtain two well defined beams, one in the internal state $|+\rangle$, the other in the state $|-\rangle$. Therefore, the formalism we have developed in this section explains completely the Stern–Gerlach experiment and its results. It elicits two fundamental aspects of a measurement process in quantum mechanics:

- a measurement requires a finite spatial extension ($\delta z = 0$ in (8.66) if $L = 0$),
- a measurement is never instantaneous ($\delta z = 0$ if $T = L/v = 0$).

These two aspects were absent in the formulation of the principles of quantum mechanics presented in Chap. 5.

Finally a simple inspection of the evolution of the expectation value of the separation of the two spots, and of their dispersion at the exit of the magnet leads to the following result. Let $T = L/v$ be the time the atoms spend in the inhomogeneous magnetic field, and let us denote by $E_\perp = \langle p_z^2 \rangle / 2m$ the transverse energy communicated to the atom by the field gradient. In order for us to observe the splitting, the following condition must be satisfied:

$$TE_\perp \geq \hbar/2 \, .$$

This condition, where the value of the field gradient has disappeared, is one important aspect of the so-called *time–energy uncertainty relation*, which appears in any quantum measurement. We shall come back to this relation in Chap. 17.

8.7 Conclusion

In this chapter, we have proposed a quantum treatment which fully describes the phenomena encountered in Stern–Gerlach-type experiments with silver atoms (or, more generally, with "two-spot" atoms). We could generalize this approach to other classes of atoms or particles (three-spot atoms, four-spot atoms, etc.). In the case of three-spot atoms, for instance, there are three possible values $+\mu_0, 0, -\mu_0$ for the z component of the magnetic moment $\hat{\mu}_z$. By considering various combinations of measurements, one can construct the observables $\hat{\mu}_x$ and $\hat{\mu}_y$ in the basis $|+\rangle_z, |0\rangle_z, |-\rangle_z$, in other words, in a three-dimensional Hilbert space.

In fact, such a procedure becomes tedious if the dimension of the space $\mathcal{E}_{\text{internal}}$ is larger than 2. Later on, we shall use instead a more general approach which relies on the proportionality between magnetic moments and angular momenta. We shall see that the commutation relations (8.25), which we have found phenomenologically here, have a much larger scope and a much more fundamental character.

Further Reading

We strongly recommend any reader interested in the historical aspects of Stern and Gerlach's work to read "Space Quantization: Otto Stern's Luck Star", by B. Friedrich and D. Herschbach, Daedalus **127**, 165 (1998). See also M. Jammer, *The Conceptual Development of Quantum Mechanics*, McGraw-Hill, New York (1966); B.L. Van Der Waerden, *Sources of Quantum Mechanics*, North-Holland, Amsterdam (1967); J. Mehra and H. Rechenberg, *The Historical Development of Quantum Theory*, Springer, Berlin (1982).

Exercises

8.1. Determination of the magnetic state of a silver atom. Consider a silver atom in an arbitrary state of its magnetic moment

$$\alpha|+\rangle_z + \beta|-\rangle_z\,, \qquad \text{where} \qquad |\alpha|^2 + |\beta|^2 = 1\,. \tag{8.67}$$

a. Show that this is an eigenstate of $\boldsymbol{u}\cdot\hat{\boldsymbol{\mu}}$ with eigenvalue $+\mu_0$, where $\boldsymbol{u}$ is a unit vector, whose direction should be determined.
b. Alice sends to Bob *one* silver atom in the unknown state (8.67). Can Bob determine this state using Stern–Gerlach measurements?
c. Alice sends to Bob N ($\gg 1$) silver atoms, all prepared in the same unknown state (8.67). Give a possible strategy for Bob to determine this state (within statistical errors).

8.2. Results of repeated measurements: quantum Zeno paradox. The magnetic moment $\boldsymbol{\mu}$ of a neutron can be described in the same way as the magnetic moment of a silver atom in the Stern–Gerlach experiment. If a neutron is placed in a uniform magnetic field $\boldsymbol{B}$ parallel to the z axis, it can be considered as a two-state system for magnetic-moment measurements (disregarding space variables).

We denote by $|+\rangle$ and $|-\rangle$ the eigenstates of the observable $\hat{\mu}_z$. These eigenstates correspond to the two eigenvalues $+\mu_0$ and $-\mu_0$. The Hamiltonian of the system in the field $\boldsymbol{B}$ is $\hat{H} = -B\hat{\mu}_z$. We set $\omega = -2\mu_0 B/\hbar$.

a. Give the energy levels of the system.
b. At time $t = 0$ the neutron is prepared in the state $|\psi(0)\rangle = (|+\rangle + |-\rangle)/\sqrt{2}$. What results can be obtained by measuring μ_x on this state, with what probabilities?
c. Write down the state $|\psi(T)\rangle$ of the magnetic moment at a later time T.
d. We measure μ_x at time T. What is the probability of finding $+\mu_0$?
e. We now perform *on the same system* a sequence of N successive measurements at times $t_p = pT/N, \quad p = 1, 2, \ldots, N$. What is the probability that *all* these measurements give the result $\mu_x = +\mu_0$?
f. What does this probability become if $N \to \infty$? Interpret the result; do you think it makes sense physically?

9. Approximation Methods

Any necessary matter is by nature boring.
Aristotle, Metaphysics IV.5

In quantum mechanics the number of problems for which there exist analytical solutions is rather restricted, as it is in classical mechanics. We met some of these cases in Chap. 4 and we shall investigate the Coulomb problem in Chap. 11. In general one must resort to approximation methods. In this chapter, we present two of these methods: perturbation theory and the variational method.

9.1 Perturbation Theory

Perturbation theory consists of starting from a solvable problem and adding small modifications to the Hamiltonian.

9.1.1 Definition of the Problem

Consider the eigenvalue problem

$$\hat{H}|\psi\rangle = W|\psi\rangle \tag{9.1}$$

associated with the Hamiltonian $\hat{H}$. We assume that $\hat{H}$ can be cast into the form of a dominant term $\hat{H}_0$ plus a perturbation, which we write as $\lambda\hat{H}_1$, where λ is a real parameter:

$$\hat{H} = \hat{H}_0 + \lambda\hat{H}_1 \,. \tag{9.2}$$

We assume that we know the solution of the eigenvalue problem for $\hat{H}_0$,

$$\hat{H}_0|n,r\rangle = E_n|n,r\rangle \,, \qquad r = 1, 2, \dots, p_n \,, \tag{9.3}$$

where the degeneracy of the eigenvalue E_n is p_n, and where the p_n orthonormal eigenstates $|n,r\rangle$ with $r = 1, 2, \dots, p_n$ span the eigensubspace $\mathcal{E}_n$. We also assume that the term $\lambda\hat{H}_1$ is sufficiently weak to cause only small perturbations to the spectrum of $\hat{H}_0$.

Example. Consider the shift of the energy levels of a hydrogen atom in the presence of an external electric field (Stark effect). In the absence of the external field, the binding energies of the hydrogen atom are a few electron-volts and the size of the atom is of the order of a few angstroms. Therefore the electric field created by the proton and seen by the electron is of the order of $10^{10}\,\mathrm{V\,m^{-1}}$. This is enormous compared with any static field one deals with in a laboratory. Consequently the applied static field can be safely treated as a small perturbation to the Coulomb field. The parameter λ can be the intensity of the applied field, measured relative to some relevant reference, for instance the typical Coulomb field of $10^{10}\,\mathrm{V\,m^{-1}}$.

9.1.2 Power Expansion of Energies and Eigenstates

We assume that the energy levels W of $\hat{H}$ vary analytically with λ. Therefore, if λ is small enough, these levels and the corresponding states will be close to those of the nonperturbed Hamiltonian $\hat{H}_0$.

Perturbation theory consists in expanding $|\psi\rangle$ and W in powers of λ, i. e.

$$|\psi\rangle = |\psi^0\rangle + \lambda\, |\psi^1\rangle + \lambda^2\, |\psi^2\rangle + \dots\,, \tag{9.4}$$

$$W = W^{(0)} + \lambda\, W^{(1)} + \lambda^2\, W^{(2)} + \dots\,, \tag{9.5}$$

and calculating the coefficients of the expansion. We insert these expansions in the eigenvalue equation (9.1) to obtain

$$\begin{aligned}&\left(\hat{H}_0 + \lambda \hat{H}_1\right) \left(|\psi^0\rangle + \lambda\, |\psi^1\rangle + \dots\right) \\ &\quad = \left(W^{(0)} + \lambda W^{(1)} + \dots\right) \left(|\psi^0\rangle + \lambda |\psi^1\rangle + \dots\right),\end{aligned} \tag{9.6}$$

and identify each order in terms of powers of λ:

$$\hat{H}_0|\psi^0\rangle = W^{(0)}|\psi^0\rangle\,, \tag{9.7}$$

$$\hat{H}_0|\psi^1\rangle + \hat{H}_1|\psi^0\rangle = W^{(0)}|\psi^1\rangle + W^{(1)}|\psi^0\rangle\,, \tag{9.8}$$

$$\hat{H}_0|\psi^2\rangle + \hat{H}_1|\psi^1\rangle = W^{(0)}|\psi^2\rangle + W^{(1)}|\psi^1\rangle + W^{(2)}|\psi^0\rangle\,, \tag{9.9}$$

$$\dots = \dots\,.$$

We also have to take into account the normalization condition

$$1 = \langle\psi|\psi\rangle = \langle\psi^0|\psi^0\rangle + \lambda\left(\langle\psi^0|\psi^1\rangle + \langle\psi^1|\psi^0\rangle\right) + \dots \tag{9.10}$$

which yields

$$\langle\psi^0|\psi^0\rangle = 1\,, \tag{9.11}$$

$$\mathrm{Re}\langle\psi^0|\psi^1\rangle = 0\,, \tag{9.12}$$

$$\dots = 0\,.$$

Since the description of the perturbed state is given in the same Hilbert space as for the unperturbed state, each term $|\psi^i\rangle$ can be expanded in the original eigenbasis of H_0:

$$|\psi^i\rangle = \sum_n \sum_{r=1}^{p_n} \gamma^i_{n,r} |n,r\rangle \,. \tag{9.13}$$

The series of equations (9.7)–(9.12) provides recursion relations for calculating all terms $|\psi^i\rangle$ and $W^{(i)}$. At any given order, we obtain an approximation to the exact solution.

We note that (9.7) implies that $|\psi^0\rangle$ is an eigenvector of H_0 and that $W^{(0)}$ is an eigenvalue of H_0. Therefore

$$W^{(0)} = E_n \tag{9.14}$$

and $|\psi^0\rangle$ is a vector of the corresponding eigensubspace $\mathcal{E}_n$.

9.1.3 First-Order Perturbation in the Nondegenerate Case

If the level E_n is not degenerate, we simply denote the corresponding eigenvector by $|n\rangle$. The solution to first order is particularly simple. Let $|\psi_n\rangle = |\psi^0_n\rangle + \lambda|\psi^1_n\rangle + \dots$ be the perturbed state and $W_n = W_n^{(0)} + \lambda W_n^{(1)} + \dots$ the corresponding energy level. Equation (9.7) then implies

$$|\psi^0_n\rangle = |n\rangle \,, \qquad W_n^{(0)} = E_n \,, \tag{9.15}$$

which expresses the fact that the perturbed state and energy level are close to the unperturbed ones. We take the scalar product of (9.8) with the vector $\langle n|$. Taking into account (9.15) and the fact that $\langle n|\hat{H}_0 = E_n\langle n|$, we obtain, by setting $\Delta E_n^{(1)} = \lambda W_n^{(1)}$,

$$\Delta E_n^{(1)} = \langle n|\lambda \hat{H}_1|n\rangle \,. \tag{9.16}$$

To first order, the energy shift ΔE_n of the level E_n is equal to the *expectation value* of the perturbing Hamiltonian for the unperturbed state $|n\rangle$.

9.1.4 First-Order Perturbation in the Degenerate Case

Suppose that the level E_n of $\hat{H}_0$ has a p_n-fold degeneracy. We denote by $|n,r\rangle$, $r = 1,\dots,p_n$, an orthonormal basis of the corresponding eigensubspace. In general the perturbation $\lambda\hat{H}_1$ will lift the degeneracy and the level E_n will be split into p_n sublevels $E_n + \lambda W^{(1)}_{n,q}$, $q = 1,\dots,p_n$. We denote by $|\psi_{n,q}\rangle$ the corresponding eigenstates and $|\psi^0_{n,q}\rangle$ the zeroth order in λ of each of these eigenstates. As stated at the end of Sect. 9.1.2, we know that each $|\psi^0_{n,q}\rangle$ belongs to the eigensubspace $\mathcal{E}_n$. Note that there is no reason why $|\psi^0_{n,q}\rangle$ should coincide with one of the basis vectors $|n,r\rangle$, since these have been chosen arbitrarily. In other words, we have, in general,

$$|\psi^0_{n,q}\rangle = \sum_{r=1}^{p_n} C_{q,r}|n,r\rangle \tag{9.17}$$

and we want to determine the coefficients $C_{q,r}$.

We multiply (9.8) on the left by $\langle n,r'|$ and obtain

$$\sum_{r=1}^{p_n}\langle n,r'|\lambda\hat{H}_1|n,r\rangle\, C_{q,r} = \lambda W^{(1)}_{n,q}\, C_{q,r'}\,. \tag{9.18}$$

For any given value of q, this is nothing but the eigenvalue problem for the $p_n \times p_n$ matrix $\langle n,r'|\lambda\hat{H}_1|n,r\rangle$. The p_n shifts $\Delta E^{(1)}_{n,q} = \lambda W^{(1)}_{n,q}$ of the level E_n are given by the solutions of the *secular equation*[1]

$$\begin{vmatrix} \langle n,1|\lambda\hat{H}_1|n,1\rangle - \Delta E & \dots & \langle n,1|\lambda\hat{H}_1|n,p_n\rangle \\ \vdots & \langle n,r|\lambda\hat{H}_1|n,r\rangle - \Delta E & \vdots \\ \langle n,p_n|\lambda\hat{H}_1|n,1\rangle & \dots & \langle n,p_n|\lambda\hat{H}_1|n,p_n\rangle - \Delta E \end{vmatrix} = 0$$

We also obtain the $C_{q,r}$ and therefore the eigenstates to zeroth order in λ corresponding to these eigenvalues.

Summary. *In all cases, degenerate or not, the first-order energy shift of a level E_n is obtained by* **diagonalizing the restriction** *of the perturbing Hamiltonian to the corresponding subspace.*

9.1.5 First-Order Perturbation to the Eigenstates

Consider the nondegenerate case. Using (9.8) and taking the scalar product with the eigenstate $|k\rangle$ for $k \neq n$, we obtain

$$(E_n - E_k)\,\langle k|\psi^1_n\rangle = \langle k|\hat{H}_1|n\rangle\,.$$

Therefore we can write $|\psi^1_n\rangle$ as

$$|\psi^1_n\rangle = |n\rangle\,\langle n|\psi^1_n\rangle + \sum_{k\neq n}\frac{\langle k|\hat{H}_1|n\rangle}{E_n - E_k}\,|k\rangle\,. \tag{9.19}$$

Equation (9.11) implies $\mathrm{Re}(\langle n|\psi^1_n\rangle) = 0$. By making the change of phase $|\psi_n\rangle \to \mathrm{e}^{\mathrm{i}\alpha}|\psi_n\rangle$ in (9.4), we can choose α such that $\mathrm{Im}(\langle n|\psi^1_n\rangle) = 0$ without

[1] Perturbation theory was first used in celestial mechanics, by Laplace and Lagrange. The initial purpose was to calculate the perturbations of the motions of planets around the sun (dominant term) due to the gravitational field of the other planets (perturbation). Poisson and Cauchy showed that the problem was basically an eigenvalue problem (6×6 matrices for Saturn, up to 8×8 for Neptune).

loss of generality. The first-order perturbation $|\psi_n^1\rangle$ to the state vector is then completely determined as

$$|\psi_n^1\rangle = \sum_{k\neq n} \frac{\langle k|\hat{H}_1|n\rangle}{E_n - E_k} |k\rangle . \tag{9.20}$$

9.1.6 Second-Order Perturbation to the Energy Levels

We consider the nondegenerate case for simplicity. Using the above result for the first-order perturbation to the state vector, and taking the scalar product of (9.9) with the eigenstate $|n\rangle$, we obtain the second-order correction to the energy of the eigenstates:

$$\Delta E_n^{(2)} = \lambda^2 W_n^{(2)} = \lambda^2 \sum_{k\neq n} \frac{|\langle k|\hat{H}_1|n\rangle|^2}{E_n - E_k} . \tag{9.21}$$

9.1.7 Examples

Harmonic Potential with a Modified Spring Constant. We consider first a harmonic oscillator which is perturbed by a complementary harmonic term. We write $\hat{H} = \hat{H}_0 + \lambda\hat{H}_1$, where

$$\hat{H}_0 = \frac{\hat{p}^2}{2m} + \frac{1}{2}m\omega^2\hat{x}^2 , \qquad \lambda\hat{H}_1 = \frac{\lambda}{2}m\omega^2\hat{x}^2 . \tag{9.22}$$

The energy levels of $\hat{H}_0$ are well known: $E_n = (n+1/2)\,\hbar\omega$. Since the Hamiltonian $\hat{H}$ still corresponds to a harmonic oscillator, we know exactly its energy levels as long as $\lambda > -1$:

$$W_n = \left(n + \frac{1}{2}\right) \hbar\omega \sqrt{1+\lambda} . \tag{9.23}$$

Perturbation theory gives, to first order (see (9.16)),

$$\Delta E_n^{(1)} = \langle n|\frac{\lambda}{2}m\omega^2\hat{x}^2|n\rangle . \tag{9.24}$$

This energy shift can be calculated easily by using the expression for $\hat{x}$ in terms of creation and annihilation operators,

$$\hat{x} = \sqrt{\frac{\hbar}{2m\omega}}\,(\hat{a} + \hat{a}^\dagger) \tag{9.25}$$

and we obtain

$$\Delta E_n^{(1)} = \left(n + \frac{1}{2}\right) \hbar\omega\, \frac{\lambda}{2} . \tag{9.26}$$

As expected, this coincides with the first-order term in the expansion in powers of λ of the exact result (9.23).

Anharmonic Potential. Consider a harmonic potential which is perturbed by a quartic potential:

$$\hat{H}_0 = \frac{\hat{p}^2}{2m} + \frac{1}{2}m\omega^2\hat{x}^2 , \qquad \lambda\hat{H}_1 = \lambda\frac{m^2\omega^3}{\hbar}\hat{x}^4 , \tag{9.27}$$

where λ is a dimensionless real parameter. Using again the expression for $\hat{x}$ in terms of annihilation and creation operators, we find the shift of the energy level $E_n = (n + 1/2)\,\hbar\omega$ to first order in λ to be

$$\Delta E_n^{(1)} = \lambda\frac{m^2\omega^3}{\hbar}\,\langle n|\hat{x}^4|n\rangle = \frac{3\lambda}{4}\,\hbar\omega\,(2n^2 + 2n + 1)\,. \tag{9.28}$$

9.1.8 Remarks on the Convergence of Perturbation Theory

In using the expansions (9.4) and (9.5), we implicitly assumed that the solution could be expanded in a power series in λ, and therefore that it is analytic in the vicinity of $\lambda = 0$ and that the series converges for λ sufficiently small.

In the first example given above, corresponding to a harmonic oscillator with a modified spring constant, the exact result is known: $W_n = (n + 1/2)\,\hbar\omega\,\sqrt{1+\lambda}$. We see that the series converges for $-1 < \lambda \leq 1$. This is physically quite reasonable: for $\lambda < -1$ the potential $m\omega^2(1+\lambda)x^2/2$ is repulsive and there are no bound states. For $\lambda > 1$ the "dominant" term $m\omega^2x^2/2$ is actually "small" compared with $\lambda\, m\omega^2x^2/2$, and it is that term which should be treated as a perturbation.

The case of the anharmonic potential (second example) is somewhat pathological in the sense that one can prove that the power expansion in λ never converges: the power series has a vanishing radius of convergence! Nevertheless, the result (9.28) is a good approximation as long as the correction to the unperturbed term $(n + 1/2)\hbar\omega$ is small. For a fixed value of λ (small compared with unity), this will only occur for values of n smaller than some value $n_{\max}(\lambda)$, since the correction increases as n^2. This can be understood physically since:

- The term proportional to x^4 is only small if the extension of the wave function is not too large; it becomes dominant as soon as $\langle x^2\rangle$ is large.
- For $\lambda \geq 0$ the potential $m\omega^2\hat{x}^2/2 + \lambda\hat{H}_1$ has bound states, while for $\lambda < 0$ (even arbitrarily small) the force becomes *repulsive* for x sufficiently large. The Hamiltonian is no longer bounded from below and there are no bound states. Therefore when one crosses the value $\lambda = 0$, the physical nature of the problem changes dramatically. This is reflected in the mathematical properties of the solution; there is a singularity at $\lambda = 0$ and the power series expansion around the origin has a vanishing radius of convergence.

A well-known example of a series which does not converge but whose first terms give an excellent approximation to the exact answer is Stirling's formula, used to approximate Euler's gamma function:

$$\Gamma(x) = \sqrt{\frac{2\pi}{x}} \left(\frac{x}{e}\right)^x \left(1 + \frac{1}{12\,x} + \frac{1}{288\,x^2} + \dots\right).$$

This is called an asymptotic series; it can be used safely in computers, although it is not convergent.

9.2 The Variational Method

We briefly describe the variational method, which is very convenient for estimating the approximate value of energy levels (mostly the ground state) and which is frequently used in quantum chemistry.

9.2.1 The Ground State

The first use of the variational method is to derive an upper bound on the ground-state energy of a quantum system. It is based on the following theorem:

Let $|\psi\rangle$ be any normalized state; the expectation value of a Hamiltonian $\hat{H}$ in this state is always greater than or equal to the ground-state energy E_0 of this Hamiltonian, i. e.

$$\langle\psi|\hat{H}|\psi\rangle \geq E_0 \qquad \text{for any } |\psi\rangle\,. \tag{9.29}$$

To prove this result, we expand $|\psi\rangle$ in an eigenbasis of $\hat{H}$:

$$|\psi\rangle = \sum_n C_n|n\rangle\,, \qquad \sum_n C_n C_n^* = 1\,,$$

where $\hat{H}|n\rangle = E_n|n\rangle$ and, by definition, $E_0 \leq E_n$. Calculating $\langle\psi|\hat{H}|\psi\rangle - E_0$, we obtain

$$\langle\psi|\hat{H}|\psi\rangle - E_0 = \sum_n E_n C_n C_n^* - E_0 \sum_n C_n C_n^* = \sum_n (E_n - E_0)|C_n|^2 \geq 0\,,$$

which proves (9.29). Alternatively, one may simply observe that if the spectrum of an operator is bounded from below, the expectation value of this operator is necessarily greater than or equal to the lower bound of the spectrum.

In practice, this result is used in the following way. We choose a state $|\psi\rangle$ which depends on some parameters and we calculate $\langle E\rangle$ in this state. The minimum value that we find by varying the parameters gives an approximation to the ground-state energy, which is, furthermore, an upper bound on this energy level.

Example. Consider the harmonic oscillator $\hat{H} = \hat{p}^2/2m + m\omega^2\hat{x}^2/2$ and the normalized test function

$$\psi_a(x) = \sqrt{\frac{2a^3}{\pi}} \frac{1}{x^2 + a^2} \, .$$

In this case there is a single variational parameter a, and we obtain

$$E(a) = \langle \psi_a | \hat{H} | \psi_a \rangle = \int \psi_a(x) \left(-\frac{\hbar^2}{2m} \frac{\mathrm{d}^2}{\mathrm{d}x^2} + \frac{1}{2} m\omega^2 x^2 \right) \psi_a(x) \, \mathrm{d}x \, .$$

We can compute $E(a)$ by using

$$\int_{-\infty}^{+\infty} \frac{\mathrm{d}x}{x^2 + a^2} = \frac{\pi}{a}$$

and its derivatives with respect to a. We obtain

$$E(a) = \frac{\hbar^2}{4ma^2} + \frac{1}{2} m\omega^2 a^2 \, ,$$

which has a minimum for $a^2 = \hbar/(m\omega\sqrt{2})$. Hence

$$E_{\text{min}} = \frac{\hbar\omega}{\sqrt{2}} \, .$$

This gives an upper bound on the exact result $\hbar\omega/2$. The difference between the exact result and the value derived from the variational method can be further reduced by choosing more elaborate test functions with several variational parameters. Had we chosen Gaussian functions as the set of test functions, we would have obtained the exact result of course, since the true ground state of $\hat{H}$ would have been an element of this set.

9.2.2 Other Levels

One can generalize the variational method to other states by using the following theorem:

The function

$$|\psi\rangle \longrightarrow E_\psi = \frac{\langle \psi | \hat{H} | \psi \rangle}{\langle \psi | \psi \rangle}$$

is stationary with respect to $|\psi\rangle$ if and only if $|\psi\rangle$ is an eigenstate of $\hat{H}$.

To prove this result we consider a variation $|\delta\psi\rangle$ of $|\psi\rangle$, i. e. $|\psi\rangle \to |\psi\rangle + |\delta\psi\rangle$. Expanding the above formula to first order, we find

$$\langle \psi | \psi \rangle \, \delta E_\psi = \langle \delta\psi | (\hat{H} - E_\psi) | \psi \rangle + \langle \psi | (\hat{H} - E_\psi) | \delta\psi \rangle \, .$$

If $|\psi\rangle$ is an eigenstate of $\hat{H}$ with eigenvalue E, then $E_\psi = E$ and $(\hat{H} - E_\psi)|\psi\rangle = 0$. Consequently $\delta E_\psi = 0$ whatever the infinitesimal variation $|\delta\psi\rangle$.

Conversely, if $\delta E_\psi = 0$ whatever the variation $|\delta\psi\rangle$, we must have

$$\langle\delta\psi|(\hat{H} - E_\psi)|\psi\rangle + \langle\psi|(\hat{H} - E_\psi)|\delta\psi\rangle = 0\,.$$

In particular, this must happen if we make the choice

$$|\delta\psi\rangle = \eta\,(\hat{H} - E_\psi)|\psi\rangle\,,$$

where η is an infinitesimal number. Inserting this in the above formula, we obtain

$$\langle\psi|(\hat{H} - E_\psi)^2|\psi\rangle = 0\,.$$

The norm of the vector $(\hat{H} - E_\psi)|\psi\rangle$ vanishes, and therefore

$$(\hat{H} - E_\psi)|\psi\rangle = 0\,.$$

This means that $|\psi\rangle$ is an eigenvector of $\hat{H}$ with eigenvalue E_ψ.

In practice, we can use this result in the following way. We choose a set of wave functions (or state vectors) which depend on a set of parameters, which we call α collectively. We calculate the expectation value of the energy $E(\alpha)$ for these wave functions. All the extrema of $E(\alpha)$ with respect to the variations of α will be approximations to the energy levels. Of course, these extrema will not in general be exact solutions, since the choice of test wave functions does not cover the entire Hilbert space.

9.2.3 Examples of Applications of the Variational Method

Calculations of Energy Levels. Consider a particle of mass m placed in an isotropic 3D potential $V(r) \propto r^\beta$. We choose the normalized Gaussian test function

$$\psi_a(\boldsymbol{r}) = (a/\pi)^{3/4}\exp(-ar^2/2)\,. \tag{9.30}$$

We find, in this state,

$$\langle p^2\rangle = \frac{3}{2}a\hbar^2\,, \qquad \langle r^\beta\rangle = a^{-\beta/2}\frac{\Gamma(3/2+\beta/2)}{\Gamma(3/2)}\,.$$

This gives an upper bound on the ground state for the following potentials:

- The harmonic potential ($\beta = 2$), for which we recover the exact result.
- The Coulomb potential $V(r) = -e^2/r$. We find

$$E_0 = -\frac{4}{3\pi}\frac{me^4}{\hbar^2}\,, \quad \text{to be compared with the exact result} \quad -\frac{1}{2}\frac{me^4}{\hbar^2}\,.$$

- The linear potential $V(r) = gr$. We find

$$E_0 = \left(\frac{81}{2\pi}\right)^{1/3} \left(\frac{\hbar^2 g^2}{2m}\right)^{1/3} \simeq 2.345 \left(\frac{\hbar^2 g^2}{2m}\right)^{1/3},$$

to be compared with the coefficient 2.338 of the exact result.

Relation to Perturbation Theory. We have the following result.

The first order of perturbation theory provides an upper bound on the ground-state energy.

Indeed, in first-order perturbation theory, the ground-state energy is

$$W_0 = \langle\psi_0|\,(\hat{H}_0 + \lambda\hat{H}_1)\,|\psi_0\rangle\,,$$

where $|\psi_0\rangle$ is the ground-state wave function of H_0. Because of the theorem (9.29), W_0 is an upper bound on the ground-state energy of $H_0 + \lambda H_1$.

Uncertainty Relations. Using the inequality (9.29) for systems whose ground state is known, we can derive uncertainty relations between $\langle p^2\rangle$ and $\langle r^\alpha\rangle$, where α is a given exponent.

a. *The $\langle r^2\rangle\,\langle p^2\rangle$ uncertainty relation.* Consider a one-dimensional harmonic oscillator, whose ground-state energy is $\hbar\omega/2$. Whatever the state $|\psi\rangle$, we have

$$\frac{\langle p^2\rangle}{2m} + \frac{1}{2}m\omega^2\langle x^2\rangle \geq \frac{\hbar\omega}{2} \quad \Rightarrow \quad \langle p^2\rangle + m^2\omega^2\langle x^2\rangle - \hbar m\omega \geq 0\,.$$

We recognize a second-degree polynomial inequality in the variable $m\omega$. The necessary and sufficient condition for this to hold for all values of $m\omega$ is

$$\langle p^2\rangle\langle x^2\rangle \geq \frac{\hbar^2}{4}\,. \tag{9.31}$$

In three dimensions, using the notation $r^2 = x^2 + y^2 + z^2$, we obtain, in the same manner

$$\langle p^2\rangle\langle r^2\rangle \geq \frac{9\hbar^2}{4}\,. \tag{9.32}$$

b. *The $\langle 1/r\rangle\,\langle p^2\rangle$ uncertainty relation.* The hydrogen atom Hamiltonian is $H = \hat{p}^2/2m - e^2/\hat{r}$ and its ground-state energy is $E_0 = -me^4/(2\hbar^2)$ (see Chap. 11). Consequently, we have for all $|\psi\rangle$

$$\frac{\langle p^2\rangle}{2m} - e^2\left\langle\frac{1}{r}\right\rangle \geq -\frac{me^4}{2\hbar^2}\,.$$

We have again a second-degree polynomial in the variable me^2 which is always positive, from which we deduce

$$\langle p^2\rangle \geq \hbar^2\left\langle\frac{1}{r}\right\rangle^2\,. \tag{9.33}$$

Exercises

9.1. Perturbed harmonic oscillator. Using the results (9.19) and (9.21), calculate the second-order energy shift of the perturbed harmonic oscillator (9.22), and compare this with the power expansion in λ of the exact result (9.23).

9.2. Comparison of the ground states of two potentials. Consider two potentials $V_1(\boldsymbol{r})$ and $V_2(\boldsymbol{r})$ such that $V_1(\boldsymbol{r}) < V_2(\boldsymbol{r})$ at all points $\boldsymbol{r}$. Show that the energy of the ground state of a particle moving in the potential V_1 is always lower than the energy of the ground state of the same particle moving in V_2.

9.3. Existence of a bound state in a potential well. Consider a particle moving in one dimension in a potential $V(x)$ which tends to zero at $\pm\infty$ and which is such that $V(x) \leq 0$ for all x. Show that there is always at least one bound state for this motion. Is this result still valid in three dimensions?

9.4. Generalized Heisenberg inequalities. Consider the Hamiltonian $\hat{H} = p^2/2m + gr^\alpha$, where g and α have the same sign and where $\alpha > -2$. The energy levels E_n of $\hat{H}$ can be derived from the eigenvalues ε_n of the operator $(-\Delta_\rho + \eta\rho^\alpha)$ (where ρ is a dimensionless variable and $\eta = |\alpha|/\alpha$) by use of the scaling law

$$E_n = \varepsilon_n \, |g|^{2/(\alpha+2)} \left(\frac{\hbar^2}{2m}\right)^{\alpha/(\alpha+2)} ,$$

as one can check directly by applying the scaling $r = \rho\left(\hbar^2/(2m|g|)\right)^{1/(\alpha+2)}$.

Show, using the variational method, that the following general relation holds:

$$\langle p^2 \rangle \, \langle r^\alpha \rangle^{2/\alpha} \geq \kappa\, \hbar^2 \qquad \text{where} \qquad \kappa = |\alpha| \, 2^{2/\alpha} \left(\frac{|\varepsilon_0|}{\alpha+2}\right)^{(\alpha+2)/\alpha} .$$

Here ε_0 is the smallest eigenvalue of the operator $-\Delta_\rho + \eta\rho^\alpha$.

10. Angular Momentum

Done like a Frenchman; turn and turn again!
William Shakespeare, Henry VI

Angular momentum plays a central role in physics. It is a constant of the motion in rotation invariant problems. It is also essential in the interpretation of physical phenomena such as magnetism, which is conceived classically as originating from the motion of charges. Ferromagnetism, however, cannot be explained with classical ideas. Instead, it arises from the intrinsic magnetic moment of electrons related to their spin, i. e. their intrinsic angular momentum, whose origin and description are purely quantum mechanical.

Starting from the classical definition of the orbital angular momentum $\boldsymbol{L} = \boldsymbol{r} \times \boldsymbol{p}$ of a particle, we shall first write down the commutation relations of the corresponding observable $\hat{\boldsymbol{L}}$. These commutation relations will then become the definition of *all* angular-momentum observables $\hat{\boldsymbol{J}}$, including cases where classical analogs do not exist. We shall study the general form of the eigenstates and eigenvalues of such observables.

We shall then come back to orbital angular momenta in the wave function formalism and we shall determine the corresponding eigenfunctions, called spherical harmonics. Among their many applications, these functions will be a useful tool when we study the hydrogen atom in the next chapter.

The fundamental proportionality relation between the angular momentum and the magnetic moment of a microscopic system will provide us with an experimental means to verify the quantization of angular momenta. We shall return to the experimental results which we analyzed phenomenologically in Chap. 8, i. e. the Stern–Gerlach experiment. The foundations of this analysis will then be seen to be much deeper and more general. We shall see how experiment can prove that there exist in nature angular momenta which have no classical analog, namely the spins of particles. The spin-1/2 formalism and the complete description of a particle including its intrinsic spin variables will be treated in Chap. 12.

10.1 Orbital Angular Momentum and the Commutation Relations

In classical mechanics, the angular momentum $\boldsymbol{L}$, with respect to the origin, of a particle of momentum $\boldsymbol{p}$ located at position $\boldsymbol{r}$ is

$$\boldsymbol{L} = \boldsymbol{r} \times \boldsymbol{p} . \tag{10.1}$$

Our starting point will be to assume, according to the correspondence principle, that the angular-momentum observable is

$$\hat{\boldsymbol{L}} = \hat{\boldsymbol{r}} \times \hat{\boldsymbol{p}} . \tag{10.2}$$

The three components $\hat{L}_x$, $\hat{L}_y$, $\hat{L}_z$ of this (vector) observable do not commute. One finds, after a simple calculation,

$$[\hat{L}_x, \hat{L}_y] = \mathrm{i}\hbar \hat{L}_z , \quad [\hat{L}_y, \hat{L}_z] = \mathrm{i}\hbar \hat{L}_x , \quad [\hat{L}_z, \hat{L}_x] = \mathrm{i}\hbar \hat{L}_y , \tag{10.3}$$

which we can summarize as

$$\hat{\boldsymbol{L}} \times \hat{\boldsymbol{L}} = \mathrm{i}\hbar \, \hat{\boldsymbol{L}} . \tag{10.4}$$

The angular momentum with respect to an arbitrary point $\boldsymbol{r}_0$ is $\boldsymbol{L} = (\boldsymbol{r} - \boldsymbol{r}_0) \times \boldsymbol{p}$. It is straightforward to check that this observable also satisfies the above commutation relations.

Consider a system of N particles with position and momentum operators $\hat{\boldsymbol{r}}_i, \hat{\boldsymbol{p}}_i$, $i = 1, \dots, N$. The total angular-momentum operator is

$$\hat{\boldsymbol{L}}^{(\mathrm{tot})} = \sum_{i=1}^{N} \hat{\boldsymbol{L}}_i = \sum_{i=1}^{N} \hat{\boldsymbol{r}}_i \times \hat{\boldsymbol{p}}_i .$$

One can check that $\hat{\boldsymbol{L}}^{(\mathrm{tot})}$ satisfies the three commutation relations (10.3), since an operator $\hat{\boldsymbol{L}}_i$ commutes with all of the other $\hat{\boldsymbol{L}}_j$ $(j \neq i)$. We shall therefore take as the definition of an angular-momentum (vector) observable $\hat{\boldsymbol{J}}$ the following fundamental relation between its components, directly inspired by (10.4):

$$\hat{\boldsymbol{J}} \times \hat{\boldsymbol{J}} = \mathrm{i}\hbar \, \hat{\boldsymbol{J}} . \tag{10.5}$$

10.2 Eigenvalues of Angular Momentum

The commutation relation (10.5) defines an angular-momentum observable. In group theory this relation is the Lie algebra of the rotation group (notice that, by defining the dimensionless observable $\hat{\boldsymbol{K}} = \hat{\boldsymbol{J}}/\hbar$, we obtain $[\hat{K}_x, \hat{K}_y] = \mathrm{i}\hat{K}_z$, in which $\hbar$ has disappeared). The quantization of angular momenta was actually derived by Elie Cartan as early as 1914 in his analysis of Lie groups, long before quantum mechanics was developed.

10.2.1 The Observables $\hat{J}^2$ and $\hat{J}_z$ and the Basis States $|j, m\rangle$

The observable $\hat{J}^2 = \hat{J}_x^2 + \hat{J}_y^2 + \hat{J}_z^2$, which is associated with the square of the angular momentum, commutes with each component of $\hat{\boldsymbol{J}}$:

$$[\hat{J}^2, \hat{\boldsymbol{J}}] = 0\,. \tag{10.6}$$

To show this, consider for instance the component $\hat{J}_x$:

$$[\hat{J}_x, \hat{J}^2] = [\hat{J}_x, \hat{J}_y^2 + \hat{J}_z^2] = \mathrm{i}\hbar\left(\hat{J}_y\hat{J}_z + \hat{J}_z\hat{J}_y\right) - \mathrm{i}\hbar\left(\hat{J}_y\hat{J}_z + \hat{J}_z\hat{J}_y\right) = 0\,.$$

As a consequence, starting from the three operators $\hat{J}_x$, $\hat{J}_y$ and $\hat{J}_z$ and functions of only these three operators, one can construct a CSCO made up of the square of the angular momentum $\hat{J}^2$ and one of the components of $\hat{\boldsymbol{J}}$. By convention, we choose the CSCO $\{\hat{J}^2, \hat{J}_z\}$. The eigenvectors common to these two operators are denoted by $|j, m\rangle$. The dimensionless quantum numbers j and m are defined such that the eigenvalues of $\hat{J}^2$ and $\hat{J}_z$ are $j(j+1)\hbar^2$ and $m\hbar$, respectively. In other words, we set

$$\hat{J}^2|j, m\rangle = j(j+1)\hbar^2|j, m\rangle\,, \tag{10.7}$$

$$\hat{J}_z|j, m\rangle = m\hbar|j, m\rangle\,. \tag{10.8}$$

We can always choose $j \geq 0$. Indeed, all eigenvalues of $\hat{J}^2$ are nonnegative since $\langle\psi|\hat{J}^2|\psi\rangle \geq 0$ for all $|\psi\rangle$, and any nonnegative real number can always be written $j(j+1)$, where j is also nonnegative. For the moment, there is no other restriction on the possible values of j and m. We assume these eigenvectors are orthonormal:

$$\langle j, m|j', m'\rangle = \delta_{j,j'}\,\delta_{m,m'}\,.$$

Since $\hat{J}^2$ and $\hat{J}_z$ form a CSCO, the vector $|j, m\rangle$ is *unique* for given values of j and m.

Some systems are such that the only observables are angular-momentum observables and functions of these observables. This is, for instance, the case for the free motion of a particle on a sphere. However, in general, a system will have other degrees of freedom. The term "CSCO" used above is then an abuse of language, since a true CSCO will contain other observables $\hat{A}$, $\hat{B}$, etc., and the corresponding eigenbasis will depend on other quantum numbers $|\alpha, \beta, \ldots, j, m\rangle$. Fortunately, the existence of these extra quantum numbers does not affect the diagonalization of $\hat{J}^2$ and $\hat{J}_z$, which is of interest here. Once this diagonalization is performed, one will have to diagonalize the other relevant observables of the CSCO. For instance, when we consider the hydrogen atom in the next chapter, we shall use a CSCO made up of the Hamiltonian $\hat{H}$, $\hat{L}^2$ and $\hat{L}_z$, $\hat{\boldsymbol{L}}$ being the orbital angular momentum. In such a case, it is straightforward to write the complete form of (10.7) and (10.8). Assume for instance that $\{\hat{A}, \hat{J}^2, \hat{J}_z\}$ form a true CSCO for a given system; the common eigenbasis $|\alpha, j, m\rangle$ is unique and we have

$$\begin{aligned}
\hat{A}|\alpha,j,m\rangle &= a_\alpha|\alpha,j,m\rangle\,, \\
\hat{J}^2|\alpha,j,m\rangle &= j(j+1)\hbar^2|\alpha,j,m\rangle\,, \\
\hat{J}_z|\alpha,j,m\rangle &= m\hbar|\alpha,j,m\rangle\,, \\
\langle\alpha,j,m|\alpha',j',m'\rangle &= \delta_{\alpha,\alpha'}\delta_{j,j'}\delta_{m,m'}\,.
\end{aligned}$$

One can verify that the arguments developed below are unchanged but with a more complicated notation, owing to the presence of the index α, which is unaffected and remains a "spectator" index in the derivation.

The physical significance of the choice of the CSCO $\{\hat{J}^2, \hat{J}_z\}$ and of the quantum numbers (10.7) corresponds to the following questions:

a. What are the possible results of a measurement of the square of the angular momentum?
b. Once the square of the angular momentum is fixed, what are the possible results of a measurement of the projection of this angular momentum on an axis, here the z axis?

The method we shall follow in order to determine the quantum numbers j and m is similar to the algebraic technique developed for the harmonic oscillator in Chap. 7.

10.2.2 The Operators $\hat{J}_\pm$

We first introduce the two operators $\hat{J}_+$ and $\hat{J}_-$,

$$\hat{J}_+ = \hat{J}_x + \mathrm{i}\hat{J}_y \qquad \text{and} \qquad \hat{J}_- = \hat{J}_x - \mathrm{i}\hat{J}_y\,, \tag{10.9}$$

which are Hermitian conjugates of one another: $\hat{J}_+^\dagger = \hat{J}_-$, $\hat{J}_-^\dagger = \hat{J}_+$. Since $\hat{J}_\pm$ are linear combinations of $\hat{J}_x$ and $\hat{J}_y$, which commute with $\hat{J}^2$, these operators $\hat{J}_\pm$ also commute with $\hat{J}^2$:

$$[\hat{J}^2, \hat{J}_\pm] = 0\,. \tag{10.10}$$

On the other hand, $\hat{J}_+$ and $\hat{J}_-$ do not commute with $\hat{J}_z$. Using the relations (10.5), we find

$$\begin{aligned}
[\hat{J}_z, \hat{J}_\pm] &= [\hat{J}_z, \hat{J}_x] \pm \mathrm{i}[\hat{J}_z, \hat{J}_y] = \mathrm{i}\hbar\hat{J}_y \pm \mathrm{i}(-\mathrm{i}\hbar\hat{J}_x) \\
&= \pm\hbar\hat{J}_\pm\,.
\end{aligned} \tag{10.11}$$

10.2.3 Action of $\hat{J}_\pm$ on the States $|j,m\rangle$

Consider a given state $|j,m\rangle$ and the two vectors $\hat{J}_\pm|j,m\rangle$. Using the definitions (10.7) and (10.8) and the commutation relations we have just established, we find that

$$\hat{J}^2 \hat{J}_\pm |j,m\rangle = \hat{J}_\pm \hat{J}^2 |j,m\rangle = j(j+1)\,\hbar^2 \hat{J}_\pm |j,m\rangle\,, \tag{10.12}$$
$$\hat{J}_z \hat{J}_\pm |j,m\rangle = (\hat{J}_\pm \hat{J}_z \pm \hbar \hat{J}_\pm)|j,m\rangle = (m\pm 1)\hbar\, \hat{J}_\pm |j,m\rangle\,. \tag{10.13}$$

From these two relations we deduce that:

- *The vector $\hat{J}_+|j,m\rangle$ is an eigenvector of $\hat{J}^2$ and $\hat{J}_z$, corresponding to the eigenvalues $j(j+1)\hbar^2$ and $(m+1)\hbar$. Otherwise it is equal to the null vector.*
- *The vector $\hat{J}_-|j,m\rangle$ is an eigenvector of $\hat{J}^2$ and $\hat{J}_z$, corresponding to the eigenvalues $j(j+1)\hbar^2$ and $(m-1)\hbar$. Otherwise it is equal to the null vector.*

In other words, starting from a vector $|j,m\rangle$, the repeated action of the operators $\hat{J}_+$ and $\hat{J}_-$ generates a whole series of vectors in the same eigen-subspace of $\hat{J}^2$, corresponding to eigenvalues of $\hat{J}_z$ which differ from m by positive or negative integers (Fig. 10.1). However, we expect intuitively that the number of such vectors should be limited, since the projection $m\hbar$ of the angular momentum along a given axis should not exceed the modulus $\sqrt{j(j+1)}\hbar$ of the angular momentum. In order to make this statement more quantitative, we consider the square of the norm of $\hat{J}_\pm|j,m\rangle$:

$$\|\hat{J}_\pm |j,m\rangle\|^2 = \langle j,m|\hat{J}_\pm^\dagger \hat{J}_\pm|j,m\rangle = \langle j,m|\hat{J}_\mp \hat{J}_\pm|j,m\rangle\,.$$

Using:

$$\hat{J}_\mp \hat{J}_\pm = (\hat{J}_x \mp \mathrm{i}\hat{J}_y)(\hat{J}_x \pm \mathrm{i}\hat{J}_y) = \hat{J}_x^2 + \hat{J}_y^2 \pm \mathrm{i}[\hat{J}_x, \hat{J}_y] = \hat{J}^2 - \hat{J}_z^2 \mp \hbar \hat{J}_z\,,$$

we obtain:

$$\|\hat{J}_\pm |j,m\rangle\|^2 = \big[j(j+1) - m(m\pm 1)\big]\hbar^2\,. \tag{10.14}$$

In order for these two quantities to be positive, we must have

$$-j \le m \le j\,. \tag{10.15}$$

This inequality relates the projection of the angular momentum on the z axis to its modulus. The forbidden region $|m| > j$ of the j,m plane corresponds to the hatched areas in Fig. 10.1.

10.2.4 Quantization of j and m

We can now obtain the main result of this section, i. e. the quantization of the values of j and m. Let us start from a given eigenstate $|j,m\rangle$ of $\hat{J}^2$ and $\hat{J}_z$. By applying $\hat{J}_+$ repeatedly to this state, we generate a series of eigenvectors of $\hat{J}^2$ and $\hat{J}_z$ proportional to $|j,m+1\rangle$, $|j,m+2\rangle$, … . Because of the inequality (10.15), this series cannot be infinite. Consider the maximum value m_{max} that can be reached: by definition $\hat{J}_+|j,m_{\mathrm{max}}\rangle$ is not an eigenvector of $\hat{J}^2$ and $\hat{J}_z$. Consequently, it is equal to the null vector and its norm is zero. From the combination of (10.14) and (10.15), we find that this is possible if

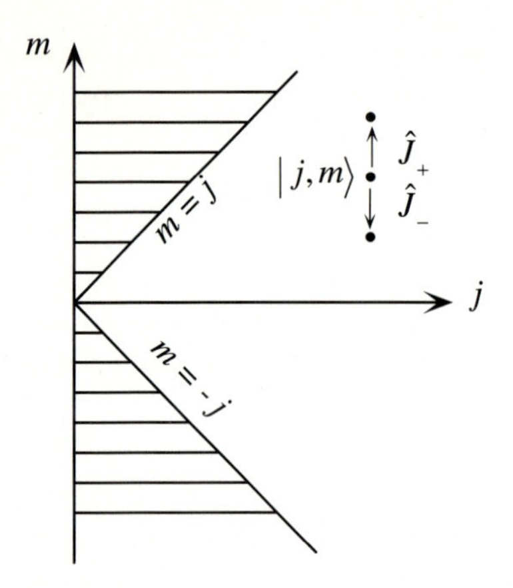

Fig. 10.1. Geometric representation of the action of the operators $\hat{J}_{\pm}$ on a ket $|j,m\rangle$. The areas with horizontal hatching, which correspond to $|m| > j$, are forbidden

and only if $m_{\max} = j$. We therefore obtain an initial result: there exists an integer N such that

$$m + N = j$$

In other words, the two real numbers j and m always differ by an integer.

Consider now the state $|j, m_{\max} = j\rangle$ and apply repeatedly the operator $\hat{J}_-$ to this vector. We generate in this way a second series of eigenvectors of $\hat{J}^2$ and $\hat{J}_z$ proportional to $|j, j-1\rangle$, $|j, j-2\rangle$, ... Again, because of the inequality (10.15), this series cannot be infinite and there exists a minimum value $m_{\min}$ that can be reached. Consequently the vector $\hat{J}_-|j, m_{\min}\rangle$ is not an eigenvector of $\hat{J}^2$ and $\hat{J}_z$, and it is necessarily equal to the null vector. Its norm is zero, which, owing to (10.14) and (10.15), implies $m_{\min} = -j$. Consequently, there exists an integer N' such that

$$j - N' = -j \, .$$

In other words, the eigenvalues of the square of the angular momentum (10.7) are such that the number j is integer or half-integer:

$$j = N'/2 \, . \tag{10.16}$$

Summary. *If $\hat{\boldsymbol{J}}$ is an observable such that $\hat{\boldsymbol{J}} \times \hat{\boldsymbol{J}} = \mathrm{i}\hbar\hat{\boldsymbol{J}}$, the eigenvalues of the observable $\hat{J}^2 = \hat{J}_x^2 + \hat{J}_y^2 + \hat{J}_z^2$ are of the form $j(j+1)\hbar^2$, where j is a positive (or zero) integer or half integer. The eigenvalues of the observable $\hat{J}_z$ are of the form $m\hbar$, where m is an integer or a half integer (Fig. 10.2).*

If a system is in an eigenstate of $\hat{J}^2$ corresponding to the value j, the only possible values of m are the $2j+1$ numbers $m = -j, -j+1, \ldots, j-1, j$.

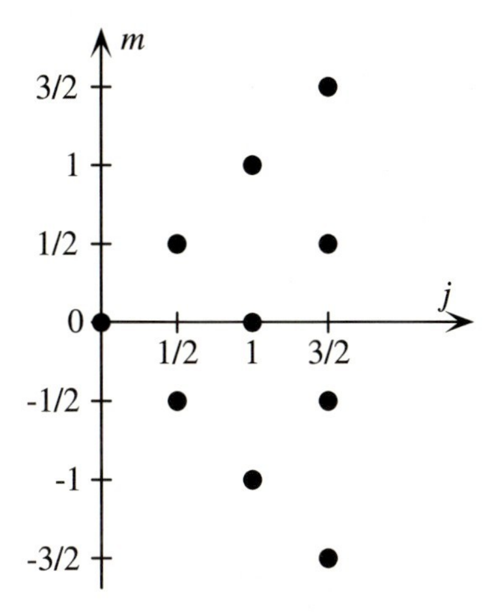

Fig. 10.2. Allowed values of the pairs (j, m) for an arbitrary angular-momentum observable

Remark. We have already emphasized that for $m \leq j - 1$, the vector $\hat{J}_+|j,m\rangle$, is proportional to $|j, m+1\rangle$, and for $m \geq -j+1$, $\hat{J}_-|j,m\rangle$ is proportional to $|j, m-1\rangle$. In all of what follows, we shall choose the phases of the vectors $|j,m\rangle$ such that (10.14) becomes

$$\hat{J}_\pm|j,m\rangle = \sqrt{j(j+1) - m(m\pm 1)}\,\hbar\,|j, m\pm 1\rangle\,. \tag{10.17}$$

For $m = j$ and $m = -j$ we recall

$$\hat{J}_+|j,j\rangle = 0\,, \qquad \hat{J}_-|j,-j\rangle = 0\,. \tag{10.18}$$

10.2.5 Measurement of $\hat{J}_x$ and $\hat{J}_y$

We can calculate the expectation value and the uncertainty of the result of a measurement of the transverse components of the angular momentum when the state of the system is $|j,m\rangle$. Of course, the results derived above for the z axis can be transposed to the x and y axes. Therefore the only possible results of a measurement of J_x or J_y are $m'\hbar$, m' being one of the values $-j, -j+1, \ldots, j-1, j$.

Expectation Values of J_x and J_y. Using $\hat{J}_x = (\hat{J}_+ + \hat{J}_-)/2$ and $\hat{J}_y = (\hat{J}_+ - \hat{J}_-)/2\mathrm{i}$, we see that the expectation values of these operators in the state $|j,m\rangle$ vanish identically. We obtain for $\hat{J}_x$, for instance:

$$\langle j,m|\hat{J}_x|j,m\rangle = \frac{1}{2}\langle j,m|\hat{J}_+|j,m\rangle + \frac{1}{2}\langle j,m|\hat{J}_-|j,m\rangle = 0$$

since the states $|j,m\rangle$ and $|j,m\pm 1\rangle$ are orthogonal.

If the system is prepared in an eigenstate $|j, m\rangle$ of $\hat{J}^2$ and $\hat{J}_z$, the expectation values of J_x and J_y vanish.

Mean Square Deviation of J_x and J_y. We can calculate the expectation values of J_x^2 and J_y^2 as follows:

$$\langle j, m|\hat{J}_x^2 + \hat{J}_y^2|j, m\rangle = \langle j, m|\hat{J}^2 - \hat{J}_z^2|j, m\rangle = [j(j+1) - m^2]\hbar^2 .$$

The mean square deviation ΔJ_x, which is equal to ΔJ_y for symmetry reasons, is therefore

$$\Delta J_x = \Delta J_y = \hbar \sqrt{[j(j+1) - m^2]/2} .$$

The uncertainty in the measurements of J_x and J_y vanishes if and only if $j = 0$. We recover in this example the general uncertainty relation

$$\Delta J_x \, \Delta J_y \geq \frac{\hbar}{2} |\langle J_z \rangle| ,$$

since

$$|m| \leq j \quad \Rightarrow \quad \hbar^2 \left[j(j+1) - m^2\right]/2 \geq \hbar^2 |m|/2 .$$

10.3 Orbital Angular Momentum

Here we consider the orbital angular momentum of a particle with respect to the origin, $\hat{\boldsymbol{L}} = \hat{\boldsymbol{r}} \times \hat{\boldsymbol{p}}$. According to the previous results, if $\ell(\ell+1)\hbar^2$ are the eigenvalues of $\hat{L}^2$ ($\ell \geq 0$) and $m\hbar$ those of $\hat{L}_z$, then 2ℓ and $2m$ are integers. In the specific case of orbital angular momenta, we shall see that ℓ and m are integers.

10.3.1 The Quantum Numbers m and ℓ are Integers

m is an Integer. When dealing with an orbital angular momentum, we consider a particle moving in space, whose state can be described by a wave function. In the wave function formalism the operator $\hat{L}_z$ has the form

$$\hat{L}_z = \hat{x}\hat{p}_y - \hat{y}\hat{p}_x = -\mathrm{i}\hbar \left(x\frac{\partial}{\partial y} - y\frac{\partial}{\partial x} \right) .$$

In order to solve the eigenvalue problem for the angular momentum, it is convenient to switch to spherical coordinates. We give the corresponding useful formulas in Sect. 10.3.2 below. If z is the polar axis, θ the colatitude ($0 \leq \theta \leq \pi$) and φ the azimuthal angle ($0 \leq \varphi < 2\pi$), the operator $\hat{L}_z$ has the very simple form

$$\hat{L}_z = \frac{\hbar}{\mathrm{i}} \frac{\partial}{\partial \varphi} \,. \tag{10.19}$$

Consider a state of the particle which is an eigenstate of $\hat{L}_z$ with eigenvalue $m\hbar$. The corresponding wave function $\psi_m(\boldsymbol{r})$ therefore satisfies

$$\hat{L}_z \, \psi_m(\boldsymbol{r}) = m\hbar \, \psi_m(\boldsymbol{r}) \,.$$

The form (10.19) of $\hat{L}_z$ gives us the following very simple φ dependence of the wave function:

$$\psi_m(\boldsymbol{r}) = \Phi_m(r,\theta) \, \mathrm{e}^{\mathrm{i}m\varphi} \,,$$

where $\Phi_m(r,\theta)$ is arbitrary at this stage. In the change $\varphi \to \varphi + 2\pi$, we notice that x, y and z *do not change*, and the function $\psi_m(\boldsymbol{r})$ keeps the same value. It must therefore be periodic in φ with period 2π. Consequently,

$$\mathrm{e}^{\mathrm{i}m\varphi} = \mathrm{e}^{\mathrm{i}m(\varphi+2\pi)} \Rightarrow \mathrm{e}^{\mathrm{i}2\pi m} = 1 \,,$$

which shows that m must be an integer in the case of an orbital angular momentum.

ℓ is an Integer. In the general analysis performed in Sect. 10.2, we have seen that m and j differ by an integer. Therefore, in the case of an orbital angular momentum, since m is an integer, ℓ is also an integer.

10.3.2 Spherical Coordinates

We give here some formulas in spherical coordinates which will be useful in what follows. We have already seen the expression for $\hat{L}_z$:

$$\hat{L}_z = \frac{\hbar}{\mathrm{i}} \frac{\partial}{\partial \varphi} \,.$$

We shall use the following expression for the operator $\hat{L}^2 = \hat{L}_x^2 + \hat{L}_y^2 + \hat{L}_z^2$:

$$\hat{L}^2 = -\hbar^2 \left(\frac{1}{\sin\theta} \frac{\partial}{\partial\theta} \sin\theta \frac{\partial}{\partial\theta} + \frac{1}{\sin^2\theta} \frac{\partial^2}{\partial\varphi^2} \right) \,. \tag{10.20}$$

The operators $\hat{L}_\pm$, corresponding to the operators $\hat{J}_\pm$ introduced in Sect. 10.2, have the form:

$$\hat{L}_\pm = \hat{L}_x \pm \mathrm{i}\hat{L}_y = \hbar \mathrm{e}^{\pm\mathrm{i}\varphi} \left(\pm \frac{\partial}{\partial\theta} + \mathrm{i}\cot\theta \frac{\partial}{\partial\varphi} \right) \,. \tag{10.21}$$

Notice that all the angular momentum operators are scale invariant: they involve only the angular variables θ and φ and do not depend on the radial coordinate.

We shall frequently make use of the following expression for the Laplacian operator Δ:

$$\Delta = \frac{1}{r} \frac{\partial^2}{\partial r^2} r - \frac{1}{r^2\hbar^2} \hat{L}^2 \,. \tag{10.22}$$

10.3.3 Eigenfunctions of $\hat{L}^2$ and $\hat{L}_z$: the Spherical Harmonics

The eigenfunctions common to the observables $\hat{L}^2$ and $\hat{L}_z$ are called the spherical harmonics and are denoted $Y_{\ell,m}(\theta,\varphi)$, the eigenvalues being $\ell(\ell+1)\hbar^2$ and $m\hbar$, respectively:

$$\hat{L}^2\, Y_{\ell,m}(\theta,\varphi) = \ell(\ell+1)\hbar^2\, Y_{\ell,m}(\theta,\varphi)\,, \tag{10.23}$$

$$\hat{L}_z\, Y_{\ell,m}(\theta,\varphi) = m\hbar\, Y_{\ell,m}(\theta,\varphi)\,. \tag{10.24}$$

The spherical harmonics form a basis for the square integrable functions on the unit sphere. They are completely defined as follows:

a. They are normalized to unity:

$$\iint Y^*_{\ell,m}(\theta,\varphi)\, Y_{\ell',m'}(\theta,\varphi)\, \sin\theta\, \mathrm{d}\theta\, \mathrm{d}\varphi = \delta_{\ell,\ell'}\, \delta_{m,m'}\,.$$

b. Their phases are such that the recursion relation (10.17), which we rewrite below, is satisfied, and that $Y_{\ell,0}(0,0)$ is real and positive:

$$\hat{L}_\pm Y_{\ell,m}(\theta,\varphi) = \sqrt{\ell(\ell+1)-m(m\pm 1)}\;\hbar\, Y_{\ell,m\pm 1}(\theta,\varphi)\,. \tag{10.25}$$

c. As seen in Sect. 10.3.1, the fact that $Y_{\ell,m}(\theta,\varphi)$ is an eigenfunction of $\hat{L}_z$ leads to a very simple dependence on φ:

$$Y_{\ell,m}(\theta,\varphi) = F_{\ell,m}(\theta)\, \mathrm{e}^{\mathrm{i}m\varphi}\,. \tag{10.26}$$

d. Starting from the relation

$$\hat{L}_+ Y_{\ell,\ell}(\theta,\varphi) = 0\,, \tag{10.27}$$

we obtain, using (10.21) and (10.26),

$$Y_{\ell,\ell}(\theta,\varphi) = C\, (\sin\theta)^\ell \mathrm{e}^{\mathrm{i}\ell\varphi}\,, \tag{10.28}$$

where the modulus and phase of the normalization constant C are determined by the previous conditions.

Summary. *For a particle moving in space, the orbital angular-momentum-operator is $\hat{\boldsymbol{L}} = \hat{\boldsymbol{r}} \times \hat{\boldsymbol{p}}$. The eigenvalues of the observable $\hat{L}^2 = \hat{L}_x^2 + \hat{L}_y^2 + \hat{L}_z^2$ are of the form $\hbar^2\ell(\ell+1)$, where ℓ is an integer ≥ 0.*

If the system is in an eigenstate of $\hat{L}^2$ corresponding to the quantum number ℓ, the $2\ell+1$ possible eigenvalues of the observable $\hat{L}_z$ are $m\hbar$, with m integer such that $-\ell \leq m \leq \ell$.

The corresponding eigenfunctions $\psi_{\ell,m}(\boldsymbol{r})$ are

$$\psi_{\ell,m}(\boldsymbol{r}) = R_{\ell,m}(r)\, Y_{\ell,m}(\theta,\varphi)\,.$$

The radial dependence of these functions, which is contained in the function $R_{\ell,m}(r)$, is a priori arbitrary, since this variable does not appear in the angular momentum-operators $\hat{L}^2$ and $\hat{L}_z$.

10.3.4 Examples of Spherical Harmonics

The spherical harmonics play a central role in atomic and molecular physics. Together with their linear combinations, they form the *atomic orbitals* of external electrons in monovalent atoms, in particular of the hydrogen atom, which we shall consider in the next chapter. The first few spherical harmonics are

$$\ell = 0 \qquad Y_{0,0}(\theta, \varphi) = \frac{1}{\sqrt{4\pi}}\,, \tag{10.29}$$

$$\ell = 1 \qquad Y_{1,1}(\theta, \varphi) = -\sqrt{\frac{3}{8\pi}}\sin\theta\, \mathrm{e}^{\mathrm{i}\varphi}\,, \tag{10.30}$$

$$Y_{1,0}(\theta, \varphi) = \sqrt{\frac{3}{4\pi}}\cos\theta\,, \tag{10.31}$$

$$Y_{1,-1}(\theta, \varphi) = \sqrt{\frac{3}{8\pi}}\sin\theta\, \mathrm{e}^{-i\varphi}\,. \tag{10.32}$$

We have drawn in Fig. 10.3 the functions $|Y_{\ell,m}(\theta,\varphi)|^2 = |F_{\ell,m}(\theta)|^2$ for the lowest values of ℓ and m.

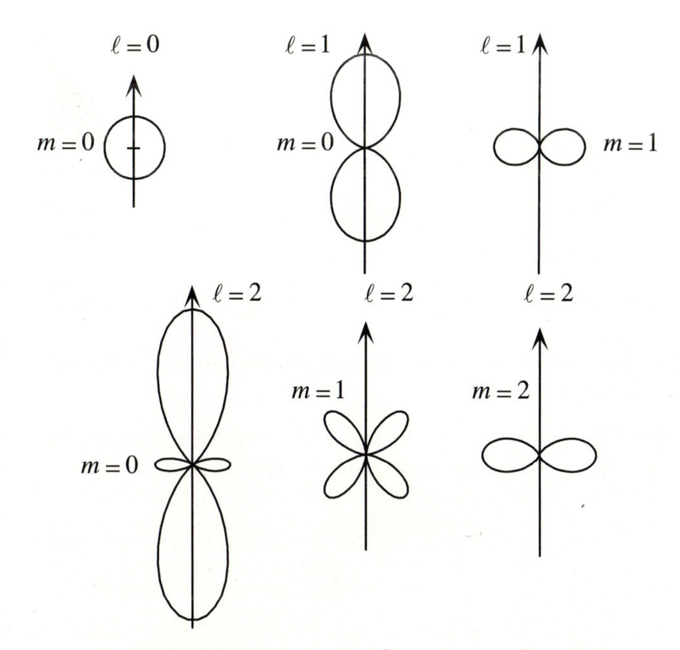

Fig. 10.3. Graphs of $|Y_{\ell,m}(\theta,\varphi)|^2 = |F_{\ell,m}(\theta)|^2$ in terms of the polar angle θ, for $\ell = 0, 1, 2$ and for $|m| \leq \ell$

10.3.5 Example: Rotational Energy of a Diatomic Molecule

A simple illustration of the quantization of the values of $\hat{L}^2$ can be obtained through the rotational energy spectrum of a molecule. Such a spectrum is presented in Fig. 10.4 for the diatomic cesium molecule Cs_2. The spectrum was obtained[1] by measuring the frequency of the photons needed to ionize Cs_2 molecules formed in a very cold atomic vapor of cesium atoms (temperature $\sim 100\,\mu$K). The data of Fig. 10.4, which represents only a small fraction of the total spectrum, exhibits a series of peaks characteristic of a quantized rotational energy.

One can visualize a diatomic molecule formed by two atoms of mass M separated by a distance R as a two-body system bound by a potential (Sect. 4.2.3). Classically, if the interatomic distance R is at its equilibrium value, the molecule has a rotational energy

$$E_{\rm rot} = \frac{L^2}{2I}, \qquad (10.33)$$

where $I = MR^2/2$ is the moment of inertia of the system and L is its angular momentum with respect to its center of mass. In quantum mechanics, this result transposes into

$$E_{\rm rot}(\ell) = \frac{\hbar^2 \ell(\ell+1)}{2I}, \qquad (10.34)$$

which shows that the rotational energy is quantized. The formula (10.34) accounts very well for the series of peaks in Fig. 10.4. The distance between two consecutive peaks increases linearly with the peak index, as expected from

$$E_{\rm rot}(\ell) - E_{\rm rot}(\ell-1) = \frac{\hbar^2}{I}\ell .$$

The moment of inertia deduced from this spectrum corresponds to a distance $R = 1.3\,$nm between the two cesium atoms. This distance, much larger than the usual interatomic spacing in diatomic molecules, indicates that the Cs_2 dimer has actually been prepared in a long-range molecular state.

If one investigates the absorption spectrum of the cold molecular gas over a much wider range, one finds several series of lines such as the one in Fig. 10.4. Each series corresponds to a particular vibrational state of the molecule. The moments of inertia associated with these series differ slightly from one another: this is a consequence of the variation of the average distance between the two atoms in the various vibrational states of the molecule.

Remark. The study of rotational excitations of molecules is an important field of research in physics, chemistry and astrophysics. If we choose x, y and

[1] These data, corresponding to the 17th excited vibrational state, have been extracted from A. Fioretti et al., Eur. Phys. J. D **5**, 389 (1999).

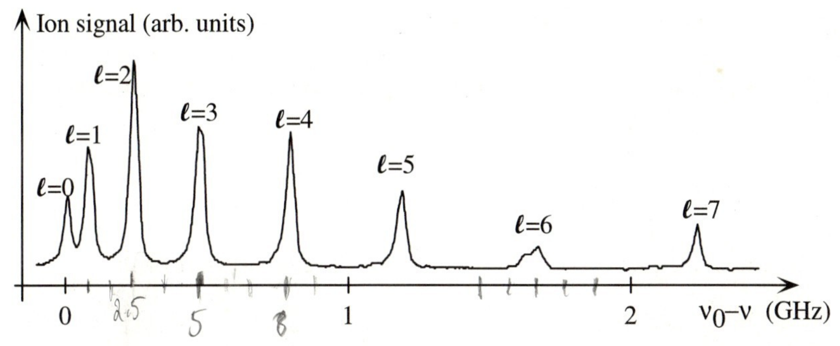

Fig. 10.4. Rotational spectrum of cold Cs_2 molecules, showing the quantization of $\hat{L}^2$. This spectrum was obtained by measuring the number of molecular ions produced by a laser beam crossing the assembly of cold molecules as a function of the laser frequency ν (ν_0 is a reference frequency corresponding to $\ell = 0$). The height of each peak is proportional to the population of the corresponding rotational level ℓ

z to be along the principal axes of inertia of a rigid rotator and denote by I_x, I_y, and I_z the corresponding moments of inertia, the rotational energy spectrum arises from the Hamiltonian

$$\hat{H}_{\mathrm{R}} = \frac{\hat{L}_x^2}{2I_x} + \frac{\hat{L}_y^2}{2I_y} + \frac{\hat{L}_z^2}{2I_z} \, .$$

where we neglect the vibrational energies (there exist subtleties in this problem since $\hat{L}_x$, $\hat{L}_y$, $\hat{L}_z$ refer to a body-fixed reference system and not to a space-fixed one). If the three moments of inertia are all different, the diagonalization of such an operator cannot be written in closed form, except for low values of $\hat{\boldsymbol{L}}^2$. If two of the moments of inertia are equal, for instance $I_x = I_y \equiv I$, the spectrum is simple since $\hat{H}_{\mathrm{R}} = \left(\hat{L}^2 - \hat{L}_z^2\right)/(2I) + \hat{L}_z^2/(2I_z)$ whose eigenvalues are

$$E_{l,m} = \hbar^2 \left(\frac{l(l+1) - m^2}{2I} + \frac{m^2}{2I_z} \right) \, .$$

From this point of view, a diatomic molecule can be considered as a rigid rotator for which $I_x = I_y = MR^2/2$ and $I_z \simeq 0$. Therefore the excitation energies of the z term are very large, and we can restrict ourselves to the ground state $m = 0$, thus ending up with the form (10.34).

10.4 Angular Momentum and Magnetic Moment

To a large extent, the experimental evidence for the quantization of angular momenta relies on the fact that when a charged particle possesses an angular momentum, it also possesses a magnetic moment. We have already illustrated this with a simple classical model in Sect. 8.1.1. Using the quantum analog of this analysis, we reinterpret in this section simple experimental results such as the Stern–Gerlach experiment, which have led to the discovery of *half-integer* angular momenta, i.e. angular momenta which correspond to half-integer values of j and m.

10.4.1 Orbital Angular Momentum and Magnetic Moment

In Chap. 8, by considering a very simple classical model of a hydrogen-like atom, we derived the proportionality relation between the magnetic moment $\boldsymbol{\mu}$ of the current loop formed by the electron and its angular momentum $\boldsymbol{L}$:

$$\boldsymbol{\mu} = \gamma_0 \boldsymbol{L}\,, \qquad \text{where} \qquad \gamma_0 = \frac{-q}{2m_\mathrm{e}}\,. \tag{10.35}$$

In this chapter we have established the quantum properties of the observable $\hat{\boldsymbol{L}}$. Consider now a particle with orbital angular momentum. We postulate that if the particle has a magnetic moment $\boldsymbol{\mu}$ associated with this orbital motion, the corresponding observable $\hat{\boldsymbol{\mu}}$ is proportional to $\hat{\boldsymbol{L}}$. In particular, for the magnetic moment associated with the orbital angular momentum of an electron, we set

$$\hat{\boldsymbol{\mu}} = \gamma_0 \hat{\boldsymbol{L}} \qquad \text{and} \qquad \hat{H}_\mathrm{M} = -\hat{\boldsymbol{\mu}} \cdot \boldsymbol{B}\,, \tag{10.36}$$

where $\hat{H}_\mathrm{M}$ is the magnetic-energy observable of the system in a magnetic field $\boldsymbol{B}$.

From the properties of $\hat{\boldsymbol{L}}$, we deduce the following results, anticipating the conclusions of the next chapter.

1. Consider an electron moving in a central potential. We suppose that the electron is in a given energy level E_n and in an eigenstate of the orbital angular momentum with eigenvalue $\ell(\ell+1)\hbar^2$. As a consequence of rotation invariance, in the absence of an external magnetic field, the $2\ell+1$ states corresponding to $m = -\ell, \dots, +\ell$ have the *same* energy E_n. Let us denote these states by $|n,\ell,m\rangle$; they are eigenstates of $\hat{L}_z$ with eigenvalues $m\hbar$. From the assumption (10.36), the state $|n,\ell,m\rangle$ is also an eigenstate of $\hat{\mu}_z$. The corresponding eigenvalue is $\mu_z = \gamma_0 m\hbar$. The negative quantity

$$\mu_\mathrm{B} = \gamma_0 \hbar = \frac{-q\hbar}{2m_\mathrm{e}} \sim -9.27 \times 10^{-24}\,\mathrm{J\,T^{-1}} \tag{10.37}$$

   is called the *Bohr magneton.*
2. If we place the system in a magnetic field $\boldsymbol{B}$ parallel to z, the degeneracy is lifted. The state $|n,\ell,m\rangle$ is an eigenstate of the observable $\hat{H}_\mathrm{M}$, with the eigenvalue

$$W_m = -m\mu_{\mathrm{B}}B\,.$$

We therefore expect to observe a splitting of the atomic energy level E_n into $2\ell+1$ sublevels, equally spaced by an interval $\Delta E = -\mu_{\mathrm{B}}B$. This is called the *Zeeman effect*. It can be observed in a transition $E_n \to E_{n'}$. In the absence of a magnetic field, this transition occurs at a single frequency $(E_n - E_{n'})/2\pi\hbar$. If we apply a field B, several lines appear. The number of such lines is directly related to the angular momenta ℓ and ℓ' of the initial and final levels.

10.4.2 Generalization to Other Angular Momenta

These arguments developed for the case of an orbital electron can be generalized to any microscopic system (an atom, an electron, a nucleus, etc.). For a microscopic quantum system which is in an eigenstate of the square of the angular momentum $\hat{J}^2$, with eigenvalue $j(j+1)\hbar^2$, we postulate that the magnetic moment $\hat{\boldsymbol{\mu}}$ and the angular momentum $\hat{\boldsymbol{J}}$ are proportional:

$$\hat{\boldsymbol{\mu}} = \gamma\hat{\boldsymbol{J}}\,, \tag{10.38}$$

where γ is the gyromagnetic ratio of the system in that state. If we apply a magnetic field $\boldsymbol{B}$, this system acquires a magnetic energy corresponding to the observable

$$\hat{H}_{\mathrm{M}} = -\hat{\boldsymbol{\mu}}\cdot\boldsymbol{B}\,. \tag{10.39}$$

In general, a complex system, such as an atom or a nucleus, has a whole series of energy levels, each of which is in an eigensubspace of $\hat{J}^2$. The constant γ then depends on the level under consideration.

The above relations are a *theoretical conjecture*, i. e. we have not proved them. In order to check this theory, we must submit it to experiment. A direct verification of (10.38) would consist in measuring $\boldsymbol{\mu}$ and $\boldsymbol{J}$ separately and observing that the proportionality holds. However, although direct measurements of angular momenta are perfectly feasible, they fall outside the scope of this book. On the other hand, it is easy to imagine magnetic-moment measurements. For instance, the conclusions of Sect. 10.4.1 concerning the energy spectrum of a system with orbital angular momentum placed in a magnetic field can be generalized to any $\hat{\boldsymbol{J}}$.

Another consequence of this proportionality relation between $\hat{\boldsymbol{J}}$ and $\hat{\boldsymbol{\mu}}$ is the *Larmor precession* phenomenon, which takes place at the quantum level for the expectation values $\langle\boldsymbol{\mu}\rangle$. The Ehrenfest theorem yields

$$\frac{\mathrm{d}}{\mathrm{d}t}\langle\boldsymbol{\mu}\rangle = \frac{1}{\mathrm{i}\hbar}\langle[\hat{\boldsymbol{\mu}}, \hat{H}_{\mathrm{M}}]\rangle\,,$$

where we assume that only the term $\hat{H}_{\mathrm{M}}$ in the Hamiltonian does not commute with $\hat{\boldsymbol{\mu}}$. Indeed, the other terms of the Hamiltonian are supposed to be rotationally invariant. Hence they commute with $\hat{\boldsymbol{J}}$ and with $\hat{\boldsymbol{\mu}}$. Owing to (10.5), the commutation relations of $\hat{\boldsymbol{\mu}}$ are

$$\hat{\boldsymbol{\mu}} \times \hat{\boldsymbol{\mu}} = \mathrm{i}\hbar\gamma\hat{\boldsymbol{\mu}} \,.$$

Therefore, a simple calculation yields

$$\frac{\mathrm{d}}{\mathrm{d}t}\langle\boldsymbol{\mu}\rangle = -\gamma\boldsymbol{B} \times \langle\boldsymbol{\mu}\rangle \,.$$

The expectation value $\langle\boldsymbol{\mu}\rangle$ satisfies the equations of motion found in Chap. 8 for the classical quantity (cf. (8.8)). This comes from the fact that the Hamiltonian is linear in $\hat{\boldsymbol{\mu}}$. The measurement of the Larmor precession frequency provides a direct determination of the gyromagnetic ratio γ and a consistency test of the results. We therefore have an experimental means to check that (10.38) and (10.39) are verified, and, after this has been checked, we can measure angular momenta via measurements of magnetic moments.

10.4.3 What Should We Think about Half-Integer Values of j and m ?

To conclude this chapter, we come back to the half-integer values of j and m which we found in the general derivation of the eigenvalues of angular momenta. In relation to orbital angular momentum, we did not accept such values (Sect. 10.3.1). Nevertheless, one may wonder whether these values appear in nature, or whether they are simply a mathematical artifact.

If we recall the Stern–Gerlach experiment and the proportionality relation between $\boldsymbol{J}$ and $\boldsymbol{\mu}$, we see a simple and natural answer to this question. We have seen in Chap. 8 that this experiment can be visualized as a means to measure μ_z. If the angular momentum is an integer, μ_z can only take an *odd* number of values:

$$\mu_z = -\hbar\gamma j, \;\; -\hbar\gamma(j-1), \;\; \ldots\,, 0, \;\; \ldots\,, \;\; \hbar\gamma(j-1), \;\; \hbar\gamma j \,.$$

This means that we should observe an odd number of spots on the screen located after the field gradient. Conversely, if j is half-integer, μ_z has an *even* number of possible values, corresponding to an even number of spots.

We know that two spots are observed for some atoms, such as silver. This is one of the many decisive proofs that there exist in nature half-integer angular momenta. For the moment, we know nothing about their nature except that they cannot correspond to orbital angular momenta $\boldsymbol{L} = \boldsymbol{r}\times\boldsymbol{p}$. We shall study such half integer angular momenta in Chap. 12, for the particular example of a spin 1/2.

Further Reading

- A.R. Edmonds, *Angular Momentum in Quantum Mechanics*, Princeton University Press, Princeton (1950).

- For a discussion of the rotational properties of molecules, see for example L. Landau and E. Lifshitz, *Quantum Mechanics*, chapters XI and XIII, Pergamon Press, Oxford (1965); C. Cohen-Tannoudji, B. Diu, and F. Laloë, *Quantum Mechanics*, Chap. VI, Wiley, New-York (1977); C.H. Townes and A.L. Schawlow, *Microwave Spectroscopy*, Chaps. 1 to 4, McGraw-Hill, New York (1955); G. Herzberg, *Molecular Spectra and Molecular Structure*, vol. I, D. Van Nostrand, Princeton (1963).

Exercises

10.1. Operator invariant under rotation. Show that if an operator $\hat{A}$ commutes with two components of the angular momentum (e. g. $\hat{J}_x$ and $\hat{J}_y$), it also commutes with the third component (e. g. $\hat{J}_z$).

10.2. Commutation relations for $\hat{\boldsymbol{r}}$ and $\hat{\boldsymbol{p}}$. Prove the following commutation relations:

$$[\hat{L}_j, \hat{x}_k] = \mathrm{i}\hbar\varepsilon_{jk\ell}\hat{x}_\ell\,, \qquad [\hat{L}_j, \hat{p}_k] = \mathrm{i}\hbar\varepsilon_{jk\ell}\hat{p}_\ell\,,$$

where $\varepsilon_{ijk} = 1$ if (i,j,k) is an even permutation of (x,y,z), $\varepsilon_{ijk} = -1$ if (i,j,k) is an odd permutation, and $\varepsilon_{ijk} = 0$ otherwise. Deduce the following identity:

$$[\hat{\boldsymbol{L}}, \hat{p}^2] = [\hat{\boldsymbol{L}}, \hat{r}^2] = 0\,.$$

10.3. Rotation-invariant potential. Consider a particle in a potential $V(\boldsymbol{r})$. What is the condition on $V(\boldsymbol{r})$ in order for $\boldsymbol{L}$ to be a constant of the motion?

10.4. Unit angular momentum. Consider a system in an eigenstate of $\hat{L}^2$ with eigenvalue $2\hbar^2$, i. e. $\ell = 1$.

a. Starting from the action of the operators $\hat{L}_+$ and $\hat{L}_-$ on the basis states $\{|\ell, m\rangle\}$ common to $\hat{L}^2$ and $\hat{L}_z$, find the matrices which represent $\hat{L}_x$, $\hat{L}_y$ and $\hat{L}_z$.
b. Give, in terms of the angles θ and φ, the probability density for a system in the eigenstate of $\hat{L}^2$ and $\hat{L}_x$ corresponding to the eigenvalues $\ell = 1$ and $m_x = 1$.

10.5. Commutation relations for $\hat{J}_x^2$, $\hat{J}_y^2$ and $\hat{J}_z^2$.

a. Show that $[\hat{J}_x^2, \hat{J}_y^2] = [\hat{J}_y^2, \hat{J}_z^2] = [\hat{J}_z^2, \hat{J}_x^2]$.

b. Show that these three commutators vanish in states where $j = 0$ or $j = 1/2$, for example

$$\langle j, m_1 | [\hat{J}_z^2, \hat{J}_x^2] | j, m_2 \rangle = 0\,,$$

for any relevant pair m_1, m_2 in the cases $j = 0$ and $j = 1/2$.

c. Show that these commutators also vanish in states where $j = 1$. Find the common eigenbasis of $\hat{J}_x^2$, $\hat{J}_y^2$ of $\hat{J}_z^2$ in this case.

11. Initial Description of Atoms

However much we inflate our ideas
beyond conceivable spaces,
we only give birth to atoms,
at the expense of the truth of things.
Blaise Pascal

At the end of the 19th century, understanding atomic spectroscopy was a challenge for the physics community. The explanation of spectroscopic data was one of the first great victories of quantum theory. In modern science and technology, the mastery of atomic physics has been responsible for decisive progress ranging from laser technology to the exploration of the cosmos.

The particular case of the hydrogen atom is perhaps the most striking. Its particularly simple spectrum delivered the first clues to the quantum laws. It has been used as a test bed for the development of quantum theory. Its hyperfine structure is responsible both for the hydrogen maser (Chap. 13) and for a revolution in astrophysics, since the corresponding 21 cm line has been extensively studied in radio astronomy to probe the structure of the interstellar and intergalactic media. Furthermore, the hydrogen atom is probably the physical system which is known with the greatest accuracy. It can be calculated "completely" in the sense that the accuracy of the present experimental results is the same as the accuracy of theoretical computer calculations.

The successive approximations one makes in atomic physics are of various origins. One starts, as in this chapter, by treating the problem of a nonrelativistic spinless electron in the Coulomb field of the proton. The problem therefore consists in finding the energy levels of the Hamiltonian

$$\hat{H} = \frac{\hat{\boldsymbol{p}}_\mathrm{e}{}^2}{2m_\mathrm{e}} + \frac{\hat{\boldsymbol{p}}_\mathrm{p}{}^2}{2m_\mathrm{p}} - \frac{q^2}{4\pi\varepsilon_0|\hat{\boldsymbol{r}}_\mathrm{e} - \hat{\boldsymbol{r}}_\mathrm{p}|} ,$$

where $\hat{\boldsymbol{p}}_\mathrm{e}, \hat{\boldsymbol{p}}_\mathrm{p}, \hat{\boldsymbol{r}}_\mathrm{e}, \hat{\boldsymbol{r}}_\mathrm{p}$ are the momentum and position operators of the electron and of the proton. Taking the effects of relativistic kinematics and spin effects of the electron into account requires a formalism which is not covered in this book: the Dirac equation. The resulting corrections are small compared with the leading terms. Up to this point, the problem can be solved analytically. Other fine-structure effects, such as the Lamb shift, require the more elaborate formalism of quantum field theory.

The theoretical treatment of complex atoms (i. e. atoms with more than one electron) involves serious computational problems, even at the nonrelativistic stage. The helium atom, with its two electrons, can only be calculated numerically. Actually, this calculation was considered as the first true test of quantum mechanics, since the much simpler case of hydrogen could be treated successfully by several other approaches derived from the "old" Bohr–Sommerfeld quantum theory. Owing to the accuracy of the present numerical calculations, the helium atom is considered to be known exactly.[1]

In this chapter we shall first consider, in Sect. 11.1, the problem of the interaction of two particles via a potential which depends only on their relative coordinates. We shall see how this reduces to the problem of the *relative* motion of the two particles. In Sect. 11.2 we shall restrict ourselves to the case of a central potential, which depends only on the distance between the particles. We shall use the invariance properties of the problem in order to choose a CSCO constructed from the Hamiltonian and the angular momentum, i. e. $\hat{H}, \hat{L}^2$ and $\hat{L}_z$, and we shall see how the traditional quantum numbers used in atomic physics show up. In Sect. 11.3 we shall study the Coulomb potential and we shall calculate the bound-state energies of hydrogen in the nonrelativistic approximation. In Sect. 11.4 we shall extend the results to hydrogen-like atoms. Finally, in Sect. 11.5, we shall give a qualitative interpretation of the spectra of alkali atoms. The problems of complex atoms, of the Mendeleev classification and of some subtle effects due to spin will be treated in Chapts. 13 and 16.

11.1 The Two-Body Problem; Relative Motion

Consider a system of two particles, of masses M_1 and M_2, at positions $\boldsymbol{r}_1$ and $\boldsymbol{r}_2$, whose mutual interaction is given by a potential $V(\boldsymbol{r}_1-\boldsymbol{r}_2)$. The potential depends only on the relative position of the particles. The Hamiltonian is

$$\hat{H} = \frac{\hat{\boldsymbol{p}}_1{}^2}{2M_1} + \frac{\hat{\boldsymbol{p}}_2{}^2}{2M_2} + V(\hat{\boldsymbol{r}}_1 - \hat{\boldsymbol{r}}_2)\,, \tag{11.1}$$

and the system is described by wave functions $\Psi(\boldsymbol{r}_1, \boldsymbol{r}_2)$.

We can separate the *global* motion of the center of mass of the system and the *relative* motion of the two particles. We introduce the position and momentum operators of the center of mass,

$$\hat{\boldsymbol{R}} = \frac{M_1\hat{\boldsymbol{r}}_1 + M_2\hat{\boldsymbol{r}}_2}{M_1 + M_2}\,, \qquad \hat{\boldsymbol{P}} = \hat{\boldsymbol{p}}_1 + \hat{\boldsymbol{p}}_2\,, \tag{11.2}$$

and the *relative* position and momentum operators,

[1] T. Kinoshita, "Ground state of the helium atom", Phys. Rev. **105**, 1490 (1957).

$$\hat{\boldsymbol{r}} = \hat{\boldsymbol{r}}_1 - \hat{\boldsymbol{r}}_2\,, \qquad \hat{\boldsymbol{p}} = \frac{M_2\hat{\boldsymbol{p}}_1 - M_1\hat{\boldsymbol{p}}_2}{M_1 + M_2}\,. \tag{11.3}$$

We can rewrite the Hamiltonian as

$$\hat{H} = \hat{H}_{\text{c.m.}} + \hat{H}_{\text{rel}}\,, \tag{11.4}$$

where

$$\hat{H}_{\text{c.m.}} = \frac{\hat{\boldsymbol{P}}^2}{2M}\,, \qquad \hat{H}_{\text{rel}} = \frac{\hat{\boldsymbol{p}}^2}{2\mu} + V(\hat{\boldsymbol{r}})\,. \tag{11.5}$$

Here we have introduced the total mass M and the reduced mass μ:

$$M = M_1 + M_2\,, \qquad \mu = \frac{M_1 M_2}{M_1 + M_2}\,. \tag{11.6}$$

Just as in classical mechanics, the Hamiltonian $\hat{H}$ separates into the sum of (i) the Hamiltonian $\hat{H}_{\text{c.m.}}$, describing the free motion of the center of mass (momentum $\boldsymbol{P}$, total mass M), and (ii) the Hamiltonian $\hat{H}_{\text{rel}}$, which describes the relative motion of the two particles in the potential $V(\boldsymbol{r})$ (momentum $\boldsymbol{p}$, reduced mass μ).

Let $\{\hat{X}_i\}$ and $\{\hat{P}_i\}$ be the components of $\hat{\boldsymbol{R}}$ and $\hat{\boldsymbol{P}}$, and $\{\hat{x}_i\}$ and $\{\hat{p}_i\}$ those of $\hat{\boldsymbol{r}}$ and $\hat{\boldsymbol{p}}$. The commutation relations are

$$[\hat{X}_j, \hat{P}_k] = \mathrm{i}\hbar\delta_{jk}\,, \qquad [\hat{x}_j, \hat{p}_k] = \mathrm{i}\hbar\delta_{jk}\,, \tag{11.7}$$

and

$$[\hat{X}_j, \hat{p}_k] = 0\,, \qquad [\hat{x}_j, \hat{P}_k] = 0\,. \tag{11.8}$$

In other words, the position and momentum operators of the center of mass and of the relative motion obey the canonical commutation relations (11.7), while any variable associated with the center-of-mass motion commutes with any variable associated with the relative motion (11.8).

These commutation relations imply

$$[\hat{\boldsymbol{P}}, \hat{H}_{\text{rel}}] = 0\,, \qquad [\hat{\boldsymbol{P}}, \hat{H}] = 0\,, \qquad [\hat{H}, \hat{H}_{\text{rel}}] = 0\,. \tag{11.9}$$

Consequently, we can look for a basis of eigenfunctions of $\hat{H}$ which will be simultaneously eigenfunctions of $\hat{\boldsymbol{P}}$ and $\hat{H}_{\text{rel}}$. The eigenfunctions of $\hat{\boldsymbol{P}}$ are the plane waves $\mathrm{e}^{\mathrm{i}\boldsymbol{K}.\boldsymbol{R}}$, where $\boldsymbol{K}$ is an arbitrary wave vector. Consequently, the desired basis of eigenfunctions of $\hat{H}$ has the form

$$\Psi(\boldsymbol{R}, \boldsymbol{r}) = \mathrm{e}^{\mathrm{i}\boldsymbol{K}.\boldsymbol{R}}\,\psi(\boldsymbol{r})\,,$$

where $\psi(\boldsymbol{r})$ is an eigenfunction of $\hat{H}_{\text{rel}}$:

$$\hat{H}_{\text{rel}}\,\psi(\boldsymbol{r}) = E\,\psi(\boldsymbol{r})\,. \tag{11.10}$$

The eigenvalues E_{tot} of $\hat{H}$ are

$$E_{\text{tot}} = \frac{\hbar^2 K^2}{2M} + E\,, \tag{11.11}$$

i. e. the sum of the *kinetic energy* of the global system ($\hat{H}_{\text{c.m.}}$) and the *internal energy* ($\hat{H}_{\text{rel}}$).

According to the Ehrenfest theorem, the relation $[\hat{H}, \hat{\boldsymbol{P}}] = 0$ implies the conservation of the total momentum, $\mathrm{d}\langle \boldsymbol{P}\rangle/\mathrm{d}t = \mathbf{0}$. This is due to the fact that the potential depends only on the relative variable $\boldsymbol{r} = \boldsymbol{r}_1 - \boldsymbol{r}_2$, in other words, the Hamiltonian of the system is translation invariant.

In what follows, we are interested only in the *relative* motion of the two particles, corresponding to the eigenvalue problem (11.10). This amounts to studying the quantum motion of a particle of mass μ in a potential $V(\boldsymbol{r})$, since $\hat{\boldsymbol{p}}$ and $\hat{\boldsymbol{r}}$ satisfy the canonical commutation relations of momentum and position operators. For an atomic system made up of an electron ($M_1 = m_{\text{e}}$) and the rest of the atom (M_2), we have $M_2 \gg m_{\text{e}}$. Therefore we neglect the small difference between the reduced mass μ and the electron mass m_{e}, remembering that it is easy to correct for reduced-mass effects if necessary.

11.2 Motion in a Central Potential

Consider a particle of mass m_{e} moving in a *central* potential. This means that the potential $V(r)$ depends only on the distance $r = |\boldsymbol{r}|$ and not on the orientation of $\boldsymbol{r}$.

11.2.1 Spherical Coordinates

Spherical coordinates are obviously well adapted to our problem. Equation (11.10) is then written, using the expression (10.22) for the Laplacian, as

$$\left(-\frac{\hbar^2}{2m_{\text{e}}}\frac{1}{r}\frac{\partial^2}{\partial r^2}r + \frac{\hat{L}^2}{2m_{\text{e}}r^2} + V(r)\right)\psi(\boldsymbol{r}) = E\,\psi(\boldsymbol{r})\,. \tag{11.12}$$

As already noticed in Sect. 7.3.3, the Hamiltonian $\hat{H}_{\text{rel}}$ commutes with the three angular-momentum operators $\hat{L}_i$, $i = x, y, z$. Each $\hat{L}_i$ commutes with $\hat{L}^2$. In addition, $\hat{L}_i$ acts only on the variables θ and φ, and it commutes with r, $\partial/\partial r$ and $V(r)$. In other words, the Hamiltonian $\hat{H}_{\text{rel}}$, which from now on will be denoted by $\hat{H}$ for simplicity, commutes with the angular momentum:

$$[\hat{H}, \hat{\boldsymbol{L}}] = 0\,.$$

Consequently $\hat{H}, \hat{L}^2$ and a given component of $\hat{\boldsymbol{L}}$, e. g. $\hat{L}_z$, form a set of commuting observables. We shall verify a posteriori that this set is *complete* by checking that the corresponding common eigenfunction basis is unique.

According to the Ehrenfest theorem, the relation $[\hat{H}, \hat{\boldsymbol{L}}] = 0$ implies the conservation of angular momentum, $\mathrm{d}\langle \boldsymbol{L} \rangle / \mathrm{d}t = \boldsymbol{0}$. This is due to the fact that the potential depends only on the radial variable $r = |\boldsymbol{r}_1 - \boldsymbol{r}_2|$; in other words, the Hamiltonian of the system is rotation invariant.

11.2.2 Eigenfunctions Common to $\hat{H}$, $\hat{L}^2$ and $\hat{L}_z$

Separation of the Angular Variables. Part of the eigenvalue problem (11.12) is already solved since we know the form of the eigenfunctions common to $\hat{L}^2$ and $\hat{L}_z$. These are the spherical harmonics. We separate the variables in the following way:

$$\psi_{\ell,m}(\boldsymbol{r}) = R_\ell(r)\, Y_{\ell,m}(\theta, \varphi)\,, \tag{11.13}$$

$$\hat{L}^2 \psi_{\ell,m}(\boldsymbol{r}) = \ell(\ell+1)\hbar^2\, \psi_{\ell,m}(\boldsymbol{r})\,, \tag{11.14}$$

$$\hat{L}_z \psi_{\ell,m}(\boldsymbol{r}) = m\hbar\, \psi_{\ell,m}(\boldsymbol{r})\,, \tag{11.15}$$

where ℓ and m are integers, with $|m| \leq \ell$. Substituting in (11.12), the eigenvalue equation becomes

$$\left(-\frac{\hbar^2}{2m_\mathrm{e}} \frac{1}{r} \frac{\mathrm{d}^2}{\mathrm{d}r^2} r + \frac{\ell(\ell+1)\hbar^2}{2m_\mathrm{e} r^2} + V(r) \right) R_\ell(r) = E\, R_\ell(r)\,. \tag{11.16}$$

This equation is independent of the quantum number m. That is why we have not put an index m on the unknown function $R_\ell(r)$ in (11.13). This differential equation is the *radial equation* and $R_\ell(r)$ is called the *radial wave function*.

The boundary condition – i.e. normalizability of the wave function – which we must impose in finding bound states is $\int |\psi(\boldsymbol{r})|^2\, \mathrm{d}^3 r = 1$, i.e., in spherical coordinates,

$$\int \mathrm{d}^2\Omega \int_0^\infty \mathrm{d}r\; r^2\, |\psi(r, \theta, \varphi)|^2 = 1\,.$$

Here Ω is the solid angle, where $\mathrm{d}^2\Omega = \sin\theta\, \mathrm{d}\theta\, \mathrm{d}\varphi$. Since the spherical harmonics are normalized, we obtain

$$\int_0^\infty \mathrm{d}r\; r^2\, |R_\ell(r)|^2 = 1 \tag{11.17}$$

for the radial wave function $R_\ell(r)$.

If we introduce the *reduced wave function* $u_\ell(r) = r\, R_\ell(r)$, the Schrödinger equation becomes

$$\left(-\frac{\hbar^2}{2m_\mathrm{e}} \frac{\mathrm{d}^2}{\mathrm{d}r^2} + \frac{\ell(\ell+1)\hbar^2}{2m_\mathrm{e} r^2} + V(r) \right) u_\ell(r) = E\, u_\ell(r)\,, \tag{11.18}$$

where $\int_0^\infty |u_\ell(r)|^2\, \mathrm{d}r = 1$. One can prove that any normalizable solution $R_\ell(r)$ is bounded at the origin, and therefore $u_\ell(0) = 0$. This equation has the

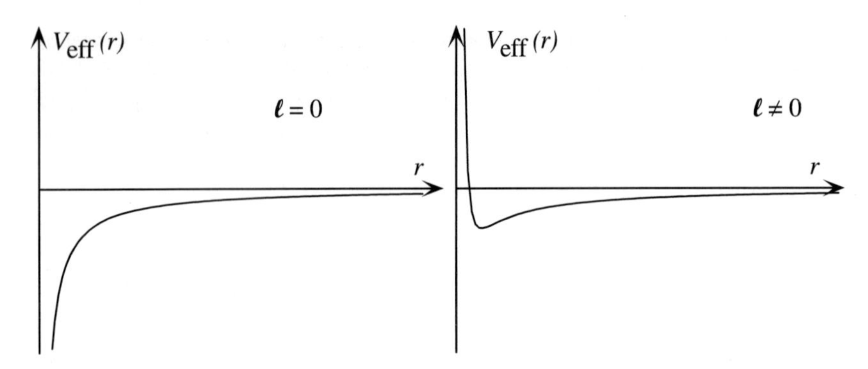

Fig. 11.1. Effective potential entering the one-dimensional Schrödinger equation for the reduced radial wave function $u_\ell(r)$. For $\ell = 0$ (*left*) the motion occurs in the "bare" potential $V(r)$; for $\ell \neq 0$ (*right*) the effective potential is the superposition of $V(r)$ and of the centrifugal barrier $\ell(\ell+1)\hbar^2/(2m_{\rm e}r^2)$. The figure is drawn for a Coulomb potential $V(r) \propto 1/r$

structure of the Schrödinger equation describing the one-dimensional motion of a particle of mass $m_{\rm e}$ in the following potential:

$$V_{\rm eff}(r) = V(r) + \frac{\ell(\ell+1)\hbar^2}{2m_{\rm e}r^2} . \qquad (11.19)$$

This effective potential is a superposition of the interaction potential between the two particles 1 and 2, and a *centrifugal barrier* term which is repulsive and increases as the angular momentum increases (Fig. 11.1).

The Radial Quantum Number n'. The radial equation depends on the parameter ℓ. For each ℓ, corresponding to a given value of the square of the angular momentum, we deal with a one-dimensional problem of the same type as that which we studied in Chap. 4. The bound-state energy levels ($E < 0$) correspond to solutions $R_\ell(r)$ which satisfy (11.17).

For a given ℓ, we can arrange the possible values of the bound-state energies in an increasing sequence, which we label by an integer n' ($n' = 0, 1, 2, \ldots$), the state $n' = 0$ being the most strongly bound. Depending on the potential, this sequence may be finite (as for a square-well potential) or infinite (as for the Coulomb potential).

The general mathematical properties of the differential equation (11.16), together with the conditions that $R_\ell(0)$ is finite and that $R_\ell(r)$ can be normalized as in (11.17), show that this number n' corresponds to the number of *nodes* of the radial wave function, i.e. the number of times it vanishes between $r = 0$ and $r = \infty$. This is independent of the form of the potential $V(r)$ (provided it is not too pathological).

The quantum number n' is called the *radial quantum number*. A radial wave function, defined by the two quantum numbers ℓ and n' and normalized to unity, is unique (up to a phase factor). The eigenvalues of the Hamiltonian are therefore labeled in general by the two quantum numbers n' and ℓ. They do not depend on the quantum number m, as a consequence of the rotation invariance of the system. This means that the $2\ell+1$ states corresponding to given values of n' and ℓ and to different values of m have the same energy and are *degenerate*.

These general considerations apply to any two-body system with a central potential: this includes the hydrogen atom, and also to a certain extent alkali atoms, diatomic molecules, the deuteron and quark systems.

The Principal Quantum Number n. In Sect. 11.3, we shall solve (11.18) exactly in the case of a Coulomb potential $V(r) = -q^2/4\pi\varepsilon_0 r$. In this particular case, the energy levels depend only on the quantity $n' + \ell + 1$. It is therefore customary to label atomic levels with the three quantum numbers ℓ, m and the positive integer n, called the *principal quantum number*, defined by the relation

$$n = n' + \ell + 1 .$$

The energy eigenstates are then classified by their values of n ($n = 1, 2, 3, \ldots$). The classification of atomic states by the three integers (n, ℓ, m) is just a redefinition of a catalog made in terms of (n', ℓ, m). For a given value of n, there are only n possible values of ℓ: $\ell = 0, 1, \ldots, n-1$. For each value of ℓ, there are $2\ell+1$ possible values of m. The wave function of an energy eigenstate is labeled by the three corresponding quantum numbers ($\psi_{n,\ell,m}(\boldsymbol{r})$), and the corresponding energy is denoted by $E_{n,\ell}$.

Spectroscopic Notation ($s, p, d, f, \ldots$ states). The measurement of the energy levels of an atom is often performed by the observation of the wavelengths of its spectral lines. We show in Fig. 11.2 the energies $E_{n,\ell}$ of the valence electron of sodium and some of the observed transitions. Each horizontal line represents a state; the number on the left is the value of the principal quantum number n. Each column corresponds to a given value of ℓ. The energy of the state is given on the vertical axis (for instance, $E_{3,0} = -5.13\,\text{eV}$). On the right, we give the energy levels E_n of hydrogen, which, as we shall see, depend only on n.

The quantum theory of the emission of a photon by an excited atom imposes *selection rules* (Chap. 17). For a transition from a state (n, ℓ) to a state (n_0, ℓ_0) by emission of a photon of energy $\hbar\omega = E_{n,\ell} - E_{n_0,\ell_0}$, all transitions are not allowed. Only the transitions for which $\ell = \ell_0 \pm 1$ are intense.

Experimental observations in the 19th century showed that one can group the lines into *series*, which were given names according to their aspect. In the case of sodium, after the theory had been understood, it turned out that these series correspond to the following transitions:

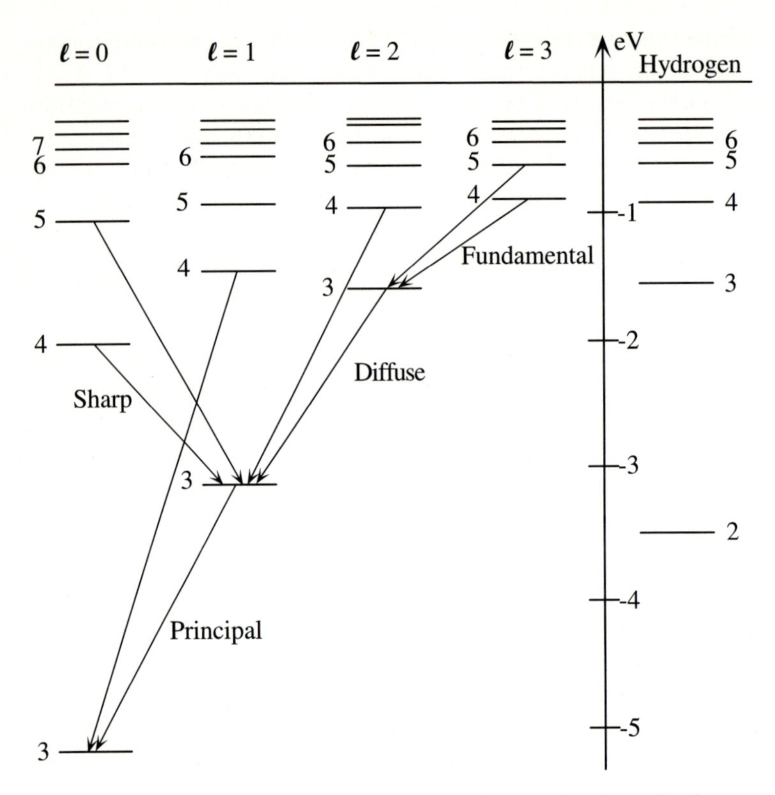

Fig. 11.2. Energy levels of the external electron of sodium (*left*) and energy levels of hydrogen (*right*)

the *sharp* series	$\hbar\omega = E_{n,\ell=0} - E_{3,1}$,
the *principal* series	$\hbar\omega = E_{n,\ell=1} - E_{3,0}$,
the *diffuse* series	$\hbar\omega = E_{n,\ell=2} - E_{3,1}$,
the *fundamental*series	$\hbar\omega = E_{n,\ell=3} - E_{3,2}$.

Each of these four series corresponds to transitions from a state of given ℓ (and various values of n) to a well defined state. Consequently, tradition associates each value of ℓ with the initial letter of the corresponding series (spectroscopic notation):

Symbolic letter:	s	p	d	f	g	h
Corresponding value of ℓ :	0	1	2	3	4	5 .

A state of well defined energy is then noted by a number (the value of n) followed by a letter (corresponding to the value of ℓ):

$$n = 1, \quad \ell = 0 : \quad \text{state 1s} ; \qquad n = 3, \quad \ell = 2 : \quad \text{state 3d} .$$

11.3 The Hydrogen Atom

The hydrogen atom is the simplest atomic system. Here, we consider the problem in its first approximation, where we neglect spin effects. We consider the problem of a particle of mass $m_{\rm e}$ in the Coulomb field of the proton, which is considered as infinitely massive (the reduced-mass correction is straightforward):

$$V(r) = -\frac{q^2}{4\pi\varepsilon_0 r} = -\frac{e^2}{r} \, .$$

Here q is the elementary charge and we set $e^2 = q^2/4\pi\varepsilon_0$. The radial equation is

$$\left(-\frac{\hbar^2}{2m_{\rm e}} \frac{1}{r} \frac{{\rm d}^2}{{\rm d}r^2} r + \frac{\ell(\ell+1)\hbar^2}{2m_{\rm e} r^2} - \frac{e^2}{r} \right) R_\ell(r) = E \, R_\ell(r) \, . \qquad (11.20)$$

11.3.1 Orders of Magnitude: Appropriate Units in Atomic Physics

The above equation involves three constants: $\hbar$ (action), $m_{\rm e}$ (mass), and e^2 (a product of an energy and a length). It is useful to form units of length and energy relevant to our problem using these three constants. We can then reformulate the eigenvalue equation (11.20) in terms of dimensionless quantities.

What are the appropriate units in atomic physics? In other words, what are the orders of magnitude of the expected results? We first note that $e^2/\hbar$ has the dimensions of a velocity. Unless the differential equation has pathologies (which fortunately is not the case), $e^2/\hbar$ must represent the typical velocity v of the electron in the lowest energy levels of the hydrogen atom. This velocity has to be compared with the velocity of light c, which is the absolute velocity standard in physics. The ratio between these two velocities forms a *dimensionless constant* α, which is a combination of the fundamental constants $q, \hbar$ and c:

$$\alpha = \frac{e^2}{\hbar c} = \frac{q^2}{4\pi\varepsilon_0 \hbar c} \sim \frac{1}{137} \, .$$

The smallness of this constant α guarantees that the nonrelativistic approximation is acceptable up to effects of the order of $v^2/c^2 \sim 10^{-4}$. The constant α is called, for (unfortunate) historical reasons, the *fine structure constant.* A more appropriate terminology would have been "fundamental constant of electromagnetic interactions".

Any charge Q is an integer multiple of the elementary charge $Q = Zq$ (or an integer multiple of $q/3$ if one includes quarks). Therefore the fundamental form of Coulomb's law between two charges $Q = Zq$ and $Q' = Z'q$ is $V(r) = \alpha ZZ'(\hbar c/r)$, where Z and Z' are integers, which involves only mechanical quantities. The introduction of electric units and of ε_0 is only a convenient device to describe macroscopic cases, where Z and Z' are very large. The experimental determination of the fundamental constant α is a keypoint in physics: the accepted value corresponds to $1/\alpha = 137.0359779(32)$.

The length unit of the problem is the *Bohr radius*

$$a_1 = \frac{\hbar^2}{m_e e^2} = \frac{1}{\alpha} \frac{\hbar}{m_e c} \sim 0.53\,\text{Å}\,,$$

where $\hbar/m_e c$ is the Compton wavelength of the electron. The Bohr radius is the typical size of an atom.

The energy unit relevant to the hydrogen atom is

$$E_{\mathrm{I}} = \frac{m_e e^4}{2\hbar^2} = \frac{1}{2} m_e c^2 \alpha^2 \sim 13.6\,\mathrm{eV}\,,$$

which, as we shall see, is the ionization energy of the atom. The electron-volt is a typical energy of external atomic electrons.

The atomic timescale is $2\pi\hbar^3/(m_e e^4) \sim 1.5 \times 10^{-16}$ s. It represents the period of the classical circular motion of an electron around a proton for the energy $-E_{\mathrm{I}}$.

11.3.2 The Dimensionless Radial Equation

Having identified the relevant length and energy scales of the problem, we introduce the dimensionless quantities $\rho = r/a_1$ and $\varepsilon = -E/E_{\mathrm{I}}$. We define ε with a minus sign so that this quantity is positive if we are dealing with a bound state, whose energy is negative. We obtain the following dimensionless equation:

$$\left(\frac{1}{\rho} \frac{\mathrm{d}^2}{\mathrm{d}\rho^2} \rho - \frac{\ell(\ell+1)}{\rho^2} + \frac{2}{\rho} - \varepsilon \right) R_\ell(\rho) = 0\,. \tag{11.21}$$

This equation is well known to mathematicians, and the following properties can be demonstrated rigorously.

a. For each value of ℓ, we obtain an infinite set of normalizable solutions, labeled by an integer $n' = 0, 1, \ldots$:

$$R(\rho) = \mathrm{e}^{-\sqrt{\varepsilon}\rho}\, \rho^\ell\, Q_{n',\ell}(\rho)\,. \tag{11.22}$$

where $Q_{n',\ell}(\rho) = C_0 + C_1\rho + \ldots + C_{n'}\rho^{n'}$ is called a Laguerre polynomial of degree n'. It has n' real zeros between $\rho = 0$ and $\rho = +\infty$.

b. These normalizable solutions correspond to particular values of ε, i.e.

$$\varepsilon = \frac{1}{(n' + \ell + 1)^2}\,. \tag{11.23}$$

Table 11.1. Radial wave functions $R_{n,\ell}(\rho)$ for the Coulomb problem, for $n = 1, 2, 3$

$n=1$	$\ell=0$	$2\,\mathrm{e}^{-\rho}$
$n=2$	$\ell=0$	$\frac{1}{\sqrt{2}}\left(1-\frac{\rho}{2}\right)\,\mathrm{e}^{-\rho/2}$
	$\ell=1$	$\frac{1}{2\sqrt{6}}\,\rho\,\mathrm{e}^{-\rho/2}$
$n=3$	$\ell=0$	$\frac{2}{3^{3/2}}\left(1-\frac{2}{3}\rho+\frac{2}{27}\rho^2\right)\,\mathrm{e}^{-\rho/3}$
	$\ell=1$	$\frac{2^{5/2}}{3^{7/2}}\,\rho\left(1-\frac{\rho}{6}\right)\,\mathrm{e}^{-\rho/3}$
	$\ell=2$	$\frac{2^{3/2}}{3^{9/2}\sqrt{5}}\,\rho^2\,\mathrm{e}^{-\rho/3}$

As already mentioned in Sect. 11.2, the integer n' gives the number of nodes of the radial wave function and is called the radial quantum number. The principal quantum number is the integer $n = n' + \ell + 1$. The first few radial wave functions $R_{n,\ell}(\rho)$ are given in Table 11.1. We remark that $\varepsilon_n = 1/n^2$ is an eigenvalue of *all* radial equations corresponding to values of ℓ smaller than n: $\ell = 0, 1, \ldots, n-1$.

Although we shall not give here a rigorous proof that the normalizable solutions of (11.21) can indeed be cast into the form (11.22), we can check that these solutions make sense both around the origin and at infinity.

Around $\rho = 0$. The Coulomb term $1/\rho$ and the constant term ϵ are negligible compared with the centrifugal term $\ell(\ell+1)/\rho^2$ (for $\ell \neq 0$). Assuming a power-law dependence $R_{n,\ell}(\rho) \propto \rho^s$ around $\rho = 0$, we find that the only possible exponent s compatible with a normalizable solution is $s = \ell$ ($s = -\ell - 1$ is not square integrable for $\ell \geq 1$). This corresponds to the expansion of (11.22) around $\rho = 0$. Notice that in the case of an s wave ($\ell = 0$), a solution behaving as $1/r$ is square integrable; however, it does not satisfy the Schrödinger equation, since $\Delta(1/r) = -4\pi\delta(\boldsymbol{r})$.

At Infinity. Keeping only the leading terms in the expansion in $R_{n,\ell}$, we obtain

$$R_{n,\ell}(\rho) \sim \mathrm{e}^{-\sqrt{\varepsilon}\rho}\,\left(C_{n'}\rho^{n-1} + C_{n'-1}\rho^{n-2} + \ldots\right) .$$

If we insert this expansion into the differential equation (11.21), we can immediately check that the term in $\mathrm{e}^{-\sqrt{\varepsilon}\rho}\rho^{n-1}$ always cancels in the equation, while the coefficient of the next term, $\mathrm{e}^{-\sqrt{\varepsilon}\rho}\rho^{n-2}$, is proportional to $C_{n'}(1 - n\sqrt{\varepsilon})$. We can check that this term also cancels for the particular choice $\varepsilon = 1/n^2$. The subsequent terms of the expansion, which depend on the centrifugal barrier, allow the determination of the coefficients $C_{n'}, C_{n'-1}, \ldots, C_0$.

Coming back to the initial variables for length and energy, we can summarize the above results as follows.

Each solution of the Schrödinger equation (11.12) corresponding to a bound state for the Coulomb problem can be labeled by three integers (or quantum numbers):

$$n = 1, 2, \ldots, \qquad \ell = 0, 1, \ldots, n-1, \qquad m = -\ell, \ldots, \ell\,.$$

The energy of a solution depends only on the principal quantum number n:

$$E_n = -\frac{E_{\mathrm{I}}}{n^2}, \qquad \text{where} \qquad E_{\mathrm{I}} = \frac{m_{\mathrm{e}} e^4}{2\hbar^2} \sim 13.6\,\mathrm{eV}\,.$$

To each energy level there correspond several possible values of the angular momentum. The total degeneracy (with respect to ℓ and m) of a level with given n is

$$\sum_{\ell=0}^{n-1} (2\ell + 1) = n^2\,.$$

The wave function corresponding to a given set n, ℓ, m is unique (up to a phase factor) and reads

$$\psi_{n,\ell,m}(\boldsymbol{r}) = Y_{\ell,m}(\theta,\varphi)\,\mathrm{e}^{-r/(n\,a_1)}\left(\frac{r}{a_1}\right)^{\ell} \times \left(C_0 + C_1\frac{r}{a_1} + \ldots + C_{n-\ell-1}\left(\frac{r}{a_1}\right)^{n-\ell-1}\right), \tag{11.24}$$

where the C_k's ($k = 0, \ldots, n-\ell-1$) are the coefficients of the Laguerre polynomials and where $a_1 = \hbar^2/(m_{\mathrm{e}} e^2) \sim 0.53$ Å.

Remark. The degeneracy with respect to ℓ is a specific property of the $1/r$ potential. For central potentials different from the Coulomb ($1/r$) and the harmonic (r^2) potentials, it is not the case that two series of energy levels corresponding to two different values ℓ and ℓ' of the angular momentum overlap, and one has to use the two quantum numbers n and ℓ to specify the energy levels (i. e. $E_{n,\ell}$). This degeneracy in the case of the Coulomb problem is a signature of an extra symmetry, called a dynamical symmetry. This symmetry, which can be represented by an O(4) or SU(2) × SU(2) Lie group, was used by Pauli in his 1925 derivation of the hydrogen spectrum. The existence of this "hidden symmetry", which is additional to the rotational invariance of the problem, is also present in classical mechanics. In addition to the angular momentum, a second independent quantity is conserved for the Coulomb potential: the Lenz vector. A direct consequence of this additional constant of motion is the fact that in a $1/r$ potential, all trajectories with a negative total energy are closed, which is not true for other central potentials (except for the harmonic case).

11.3.3 Spectrum of Hydrogen

In Fig. 11.3 we represent the energies E_n of the hydrogen atom. Each line represents an energy level, the number to the right of the line is the value of n, each column corresponds to a given value of ℓ, and we give the value of the energy on the vertical axis.

The selection rule given for the observable spectral lines of the sodium atom $\ell = \ell_0 \pm 1$ still holds. The most famous series is the Balmer series. It corresponds to transitions from states ns to the state 2p:

$$\hbar\omega = E_n - E_2 = 13.6\,\frac{n^2-4}{4n^2}\ \mathrm{eV}\,.$$

The first few lines of the Balmer series are in the visible part of the spectrum ($\hbar\omega \sim 2$ to $3\,\mathrm{eV}$; $\lambda \sim 0.5\,\mu\mathrm{m}$). The Lyman series, corresponding to transitions to the ground state, lies in the ultraviolet ($\lambda \leq 121.5\,\mathrm{nm}$).

Integers played an important role in science in the 19th century. Examples can be found in chemical reactions, atomic theory, and the classification and evolution of species in zoology and in botany. It was by chance that, in 1885, Balmer, who was a high-school teacher in Basel and was fascinated by numerology, learned about the positions of the first four lines of hydrogen. He realized that the wavelengths

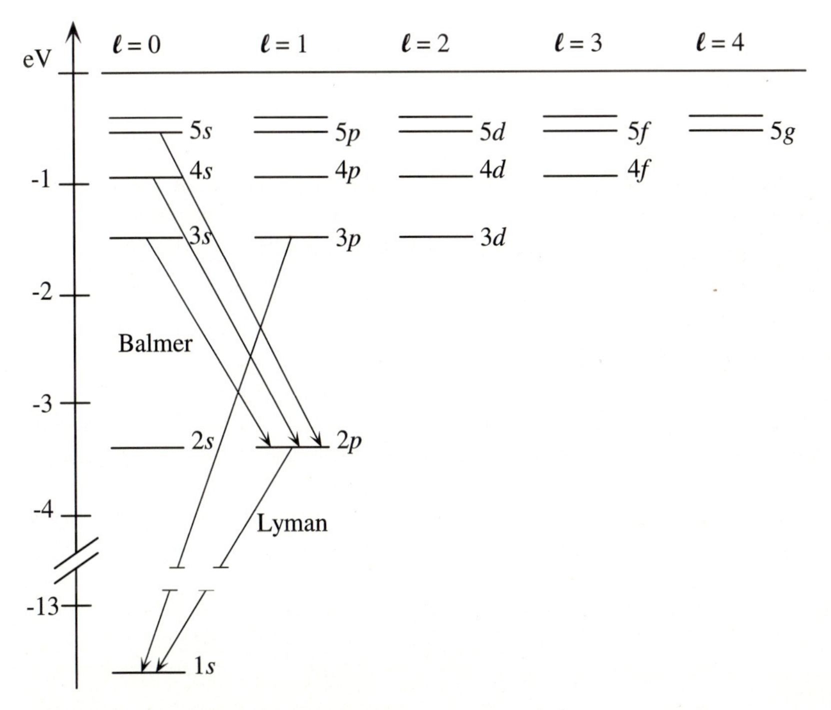

Fig. 11.3. Energy levels of hydrogen

of the lines could be represented to an accuracy of 10^{-3} by a formula involving integers: $1/\lambda \propto (n^2 - 4)/n^2, n \geq 3$; this same formula still applied accurately when Huggins found eight more lines of hydrogen in stellar spectra. Although he was not a physicist, Balmer found the simplicity of the formula quite striking. In his 1885 paper, he wrote: "It appears to me that hydrogen ... more than any other substance is destined to open new paths to the knowledge of the structure of matter and its properties".

In 1912, Niels Bohr, who was 27, was working with Rutherford on an atomic model. He was not aware of Balmer's formula or of the analogous results obtained by Rydberg for alkali atoms. One day, by chance, he learned of the existence of Balmer's formula; it took him only a few weeks to construct his celebrated model of the hydrogen atom, which was one of the turning points of quantum physics.

11.3.4 Stationary States of the Hydrogen Atom

The Ground State (1s). The ground state corresponds to $n = 1$, and therefore $\ell = 0$ and $m = 0$ (the 1s state in spectroscopic language). Since the spherical harmonic $Y_{0,0}(\theta, \varphi)$ is a constant equal to $1/\sqrt{4\pi}$, the normalized wave function of this state is

$$\psi_{1,0,0}(\boldsymbol{r}) = \frac{\mathrm{e}^{-r/a_1}}{\sqrt{\pi a_1^3}} .$$

The probability of finding the electron in a spherical shell of thickness $\mathrm{d}r$, represented in Fig. 11.4, is

$$P(r)\,\mathrm{d}r = |\psi_{1,0,0}(\boldsymbol{r})|^2\, 4\pi r^2\, \mathrm{d}r .$$

The probability density per unit volume is proportional to the exponential function e^{-2r/a_1}, and is maximum at $r = 0$. The most probable distance between the electron and the proton is the Bohr radius $a_1 = 0.53\,\text{Å}$.

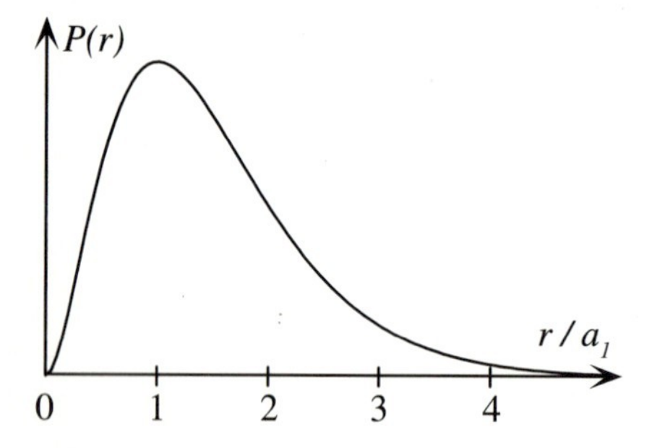

Fig. 11.4. Radial probability density $P(r)$, giving the probability of finding the electron between r and $r + \mathrm{d}r$ in a hydrogen atom prepared in its ground state

Other States. Figure 11.5 represents the radial probability density $P_{n,\ell}(r) = r^2 |R_{n,\ell}(r)|^2$ for various states n, ℓ. We note the reduction of the number of nodes of the radial wave function as ℓ increases for a given n. For a level n, l, the function $P_{n,\ell}(r)$ has $n' = n - \ell - 1$ zeros, where n' is the degree of the corresponding Laguerre polynomial. In particular, for $\ell = n - 1$, we remark that $P(r)$ has a single maximum, located at a distance $r = n^2 a_1$ (see (11.24)).

Figure 11.6 represents some spatial probability densities $|\psi_{n,\ell,m}(\boldsymbol{r})|^2$ in the plane $y = 0$ (these are axially symmetric functions around the z axis). For large quantum numbers $n \gg 1$, we notice that one gets closer to "classical" situations, corresponding to a well-localized particle.

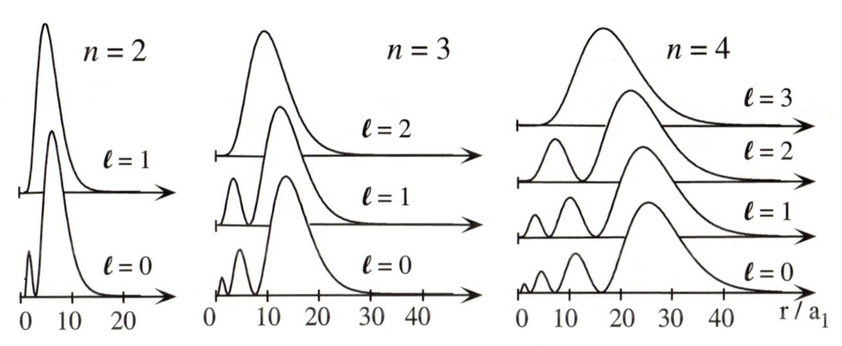

Fig. 11.5. Radial probability density $r^2 |R_{n,\ell}(r)|^2$ of the states $n = 2, 3, 4$ of hydrogen

11.3.5 Dimensions and Orders of Magnitude

Consider a hydrogen atom prepared in a stationary state $|n, \ell, m\rangle$. Using the virial theorem (Chap. 7, exercise 7.9), one can show that the classical relation between the kinetic energy and the potential energy still holds for the expectation values of these quantities:

$$E_n^{(\mathrm{kin})} = \left\langle \frac{p^2}{2m_\mathrm{e}} \right\rangle = -E_n = \frac{E_\mathrm{I}}{n^2} , \tag{11.25}$$

$$E_n^{(\mathrm{pot})} = \left\langle \frac{-e^2}{r} \right\rangle = 2\,E_n = -\frac{2E_\mathrm{I}}{n^2} . \tag{11.26}$$

Using the properties of the Laguerre polynomials, one finds that the mean radius has the following variation with n and ℓ:

$$\langle r \rangle = \frac{a_1}{2} \left[3n^2 - \ell(\ell + 1)) \right] . \tag{11.27}$$

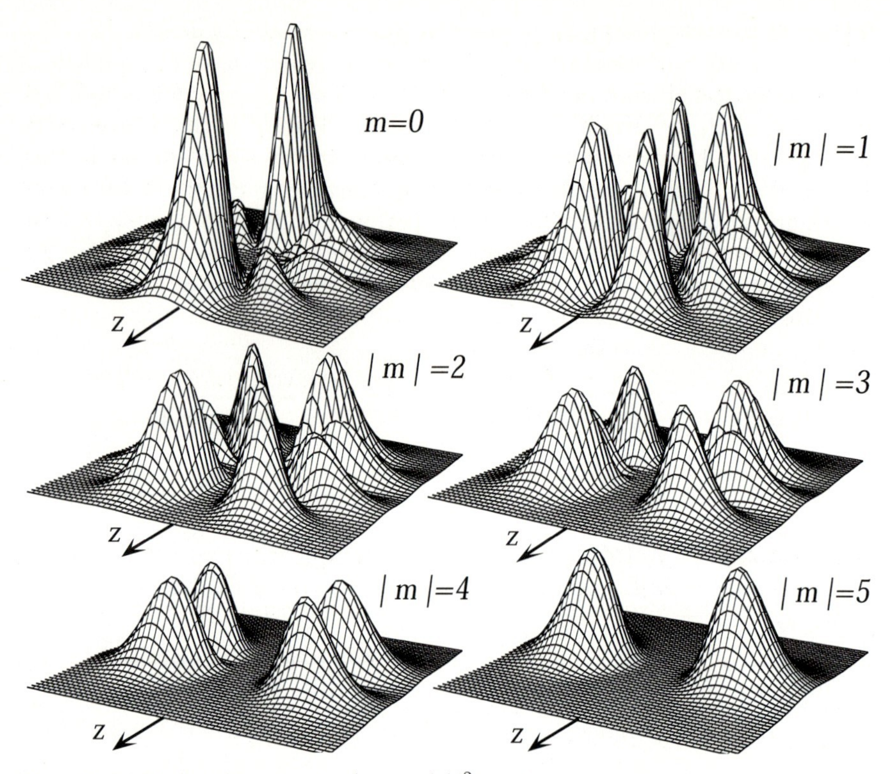

Fig. 11.6. Probability density $|\psi_{n,\ell,m}(\boldsymbol{r})|^2$ in the $y = 0$ plane for $n = 6$, $\ell = 5$ (mesh size $60\,a_1 \times 60\,a_1$). For $m = 0$, the particle is localized in the vicinity of the z axis. For large $|m|$ (in particular, $m = \pm 5$), the particle is localized in the plane $z = 0$, in the vicinity of a circle centered on the origin, of radius $r = 30\,a_1$ (circular state). The vertical scale of the surface for $m = 0$ has been reduced by a factor of 2 with respect to the five other surfaces in order to improve the visibility

One can also show that

$$\left\langle \frac{1}{r} \right\rangle = \frac{1}{n^2 a_1}\,, \qquad \left\langle \frac{1}{r^2} \right\rangle = \frac{2}{n^3\,(2\ell+1)\,a_1^2}$$

$$\langle r^2 \rangle = \frac{n^2\,a_1^2}{2}\left[5n^2 + 1 - 3\ell(\ell+1)\right] .$$

Also, setting $\rho = r/a_1$, one obtains the following result for $p > -2\ell - 1$:

$$\frac{p+1}{n^2}\langle \rho^p \rangle \;-\; (2p+1)\langle \rho^{p-1} \rangle \;+\; \frac{p}{4}\left[(2\ell+1)^2 - p^2\right]\langle \rho^{p-2} \rangle = 0 .$$

Remark. Because of the n^2 quadratic variation of the mean atomic radius, the maximal probability density for a given radial wave function decreases as $1/n^4$. For this reason, a readjustment of scales is necessary in order to visualize Fig. 11.5 properly.

11.3.6 Time Evolution of States of Low Energies

We can now briefly discuss the nature of the motion of the electron in the hydrogen atom when it is prepared in a linear superposition of some of its lowest-energy states. In one-dimensional problems such as the harmonic oscillator or the inversion of the NH_3 molecule, we have seen that superpositions of stationary states oscillate periodically. How does the wave function of the electron evolve in a three-dimensional problem such as the hydrogen atom?

Consider the superposition of wave functions of different energies, with a maximum value $m = \ell$ of the azimuthal quantum number. For simplicity we restrict our analysis to the equatorial plane $\theta = \pi/2$, perpendicular to the z axis. The spherical harmonics $Y_{\ell,\ell}(\theta, \varphi)$ have a maximum in this plane and vary as $\exp(\mathrm{i}\ell\varphi)$. The wave function is therefore

$$\psi(r, \pi/2, \varphi, t) = \sum_{n,\ell} \gamma_{n,\ell}\, \mathrm{e}^{-\mathrm{i}E_n t/\hbar}\, R_{n,\ell}(r)\, \mathrm{e}^{\mathrm{i}\ell\varphi} \,.$$

The probability density $|\psi|^2$ in this equatorial plane is a function of r and φ, and it varies with time. We investigate the nature of this variation in two examples.

- Consider first a superposition with equal weight of the 1s ground state and the 2s first excited state, both of zero angular momentum. The wave function does not have any angular dependence:

$$|\psi(r, t)|^2 = \frac{1}{2}\, |R_{1,0}(r) + \mathrm{e}^{-3i\omega t/4} R_{2,0}(r)|^2 \,,$$

  where $\hbar\omega = E_{\mathrm{I}}$. The radial variation of $R_{2,0}(r)$ has a zero and changes sign. Therefore, the interference between the two radial functions is alternately destructive and constructive around the origin, and the system exhibits a radial beat between $r = a_1$ and $r = 4a_1$ (breathing mode).
- Consider now a superposition, still with equal weights, of the 1s ground state and the 2p first excited state ($\ell = 1$):

$$|\psi(r, \varphi, t)|^2 = \frac{1}{2}\, |R_{1,0}(r) + \mathrm{e}^{-\mathrm{i}(3\omega t/4 - \varphi)} R_{2,1}(r)|^2 \,.$$

  At time $t = 0$, this distribution shows an asymmetry in φ. The interference between the two wave functions $R_{1,0}(r)$ and $R_{2,1}(r)$ is constructive at $\varphi = 0$ and destructive at $\varphi = \pi$. As a function of time, this asymmetry displays a rotation around the polar axis with a constant angular velocity $3\omega/4$.

More complicated superpositions of states produce combinations of these two basic fundamental motions of radial and angular pulsations, with various frequencies. This motion is a relatively simple wave phenomenon, but one has little intuitive understanding of it since it concerns waves which are *attracted* by a center. This is a much less frequent situation than waves confined by walls, in a waveguide for instance. Another situation where an attractive situation occurs is gravitation, where there are (classical) density waves in galactic clouds. However, in that case, the timescales are too large to allow the direct observation of such motions.

11.4 Hydrogen-Like Atoms

The eigenstates of the Hamiltonian describing an atom of atomic number Z, ionized $Z-1$ times, are readily obtained from the previous results. We just have to replace the potential $-e^2/r$ by

$$V(r) = -\frac{Ze^2}{r}\,.$$

If we make this change in the radial equation, we recover the same equation as for hydrogen. What changes physically is the length and energy scales. Hydrogen-like atoms have the same wave functions as the hydrogen atom, but distances are *reduced* by a factor Z, and energies are *multiplied* by a factor Z^2:

$$a_1^{(Z)} = \frac{\hbar^2}{Zm_e e^2}\,, \qquad E_n^{(Z)} = -\frac{Z^2 m_e e^4}{2n^2\hbar^2}\,. \tag{11.28}$$

This also applies, to a first approximation, to *internal* electrons of an atom of large Z, which can be thought of as evolving in the field of the nucleus only. In lead, for instance ($Z = 82$), an internal electron is on average at a distance $a_1/Z \simeq 6 \times 10^{-13}$ m, with an energy $-E_I Z^2 \sim -10^5$ eV. There exist sizeable corrections to the nonrelativistic approximation in such cases since the average velocity of the electron, of order $Z\alpha c$, becomes close to the velocity of light.

11.5 Muonic Atoms

The μ lepton, or muon, discovered in 1937, has the physical properties of a heavy electron. Like the electron, it is elementary or point-like, it has the same electric charge and the same spin, but it is 200 times more massive: $m_\mu = 206.8 m_e$. It is unstable and decays into an electron and two neutrinos, $\mu \to e + \bar{\nu}_e + \nu_\mu$, with a lifetime of 2×10^{-6} s.

With particle accelerators, one can produce muons, slow them down and have them captured by atoms, where they form hydrogen-like systems. In a complex atom, the muon is not constrained with respect to electrons by the Pauli principle. The muon expels electrons, cascades from one level to another and eventually falls into the vicinity of the nucleus at a distance from it of $a_\mu = \hbar^2/Zm_\mu e^2$, which is 200 times smaller than the corresponding distance for the internal electrons. Therefore, it forms a hydrogen-like atom, since it does not feel the electric field of the electrons. The lifetime of the muon is much larger than the total time for the cascades ($\sim 10^{-14}$ s). It is also much larger than the typical atomic time $\hbar^3/m_\mu e^4 \sim 10^{-19}$ s. The muon can therefore be considered as stable on these timescales.

The Bohr radius of a muonic atom is of the same order as a nuclear radius. Consider again lead ($Z = 82$), whose nuclear radius is $R \approx 8.5$ fm. We find $a_\mu \approx 3.1$ fm, which means that the μ penetrates the nucleus noticeably. In fact, in the ground state, it has a 90% probability of being inside the nucleus. The description of the nucleus as a point particle creating a Coulomb potential is inadequate, and one has to turn to a more elaborate model of the electrostatic potential created by this nucleus. Consequently, the spectra of muonic atoms provide information on the structure of nuclei, in particular concerning their charge distribution, i. e. the distribution of protons inside these nuclei.

For a spherical nucleus, the potential is harmonic inside the nucleus (assuming a constant charge density), and Coulomb-like outside the nucleus. If the nucleus is deformed, i. e. flattened or cigar-shaped, the spherical symmetry is broken, and the levels will no longer be degenerate with respect to the magnetic quantum number m. This results in a splitting of the spectral lines.

Figure 11.7, obtained at CERN, shows the spectra of muonic atoms in the cases of gold ($Z = 79$), which has a spherical nucleus, and of uranium ($Z = 92$), which has a deformed nucleus. We notice the more complicated structure of the higher-energy line for uranium. This is a very accurate method to determine the deformations of nuclei.

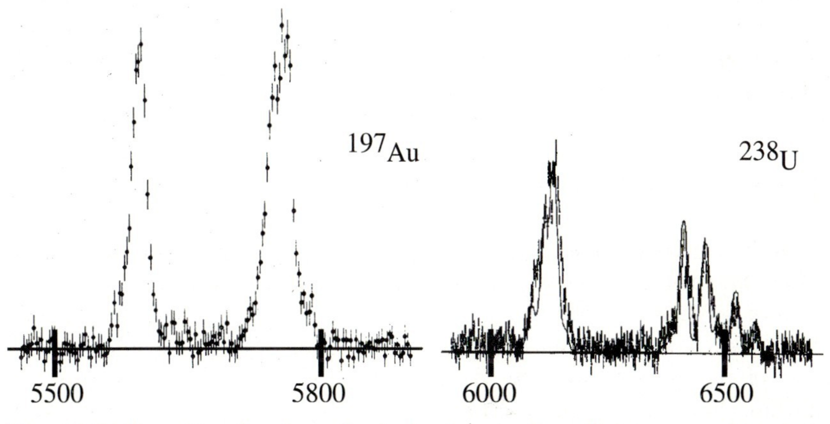

Fig. 11.7. Transition line from the 2p level (actually split into two sublevels $2p_{1/2}$ and $2p_{3/2}$; see Sect. 13.2) to the 1s level in muonic atoms of gold ($Z = 79, A = 197$) and uranium ($Z = 92, A = 238$) (horizontal scale in keV). The gold nucleus is spherical, and the corresponding spectrum has a simple shape; the uranium nucleus is deformed, and the upper peak is split into four lines (CERN document)

The existence of the muon has been a mystery for more than 40 years. When it was discovered, Rabi said, "Who ordered that?" Why a heavy electron? All matter around us that we know about can be built starting with protons, neutrons, electrons and neutrinos, or, in terms of fundamental constituents, with the family of quarks and leptons {u, d, e, ν}. So why a heavy electron, with which one can imagine a Gulliver universe: atoms, molecules, chemistry and biology 200 times smaller than the matter we know? There are many applications of the muon, such as probing nuclei, crystals and pyramids, but why is it there? What role is it supposed to play?

In 1974, with the discovery of a new quark, the charm quark c, it was realized that the muon forms, together with its neutrino, the charm quark and the strange quark s (constituent of the strange particles discovered in the 1940s), a new family of quarks and leptons (c, s, μ, ν_μ). This family generates, at higher scales, a new atomic and nuclear physics, but its members are unstable. In 1975–76, with the discovery of a new lepton τ, another new quark was discovered, the b quark (beautiful or bottom), and, in 1995, the top quark t, hence a third family (t, b, τ, ν_τ).[2]

In 1989, at the LEP electron collider, it was proven that the (light) fundamental constituents of matter belong to only these three families. Present ideas are that the Big Bang could not have happened so nicely, if these two extra ("useless") families of quarks and leptons did not exist. They are of importance in order to create the world! However, we still do not understand the masses of these quarks and leptons (and therefore their stability). The mass problem is one of the great problems of modern physics.

11.6 Spectra of Alkali Atoms

The wave functions of the hydrogen atom enable us to understand qualitatively some characteristics of the spectra of alkali atoms, in particular Fig. 11.2 above.

The alkali atoms (Li, Na, K, Rb, Cs, Fr) are monovalent, i. e. they have one more electron than the number corresponding to a set of *complete shells*. We shall explain this expression in Chap. 16, when we study the Pauli principle and complex atoms. Sodium, for instance, has 11 electrons. Ten of them are localized close to the nucleus, the last one, the valence electron, is less bound than the others, and can be excited more easily.

If sodium is heated to a high temperature, the electron cloud will be excited, but this excitation will mostly affect the valence electron. In Fig. 11.2 we have indicated the energy levels by using only the quantum numbers n and ℓ of the valence electron, starting from the value $n = 3$, since the levels $n = 1$ and $n = 2$ are filled (Pauli principle).

For alkali atoms, one may consider, to a good approximation, that the nucleus of charge $+Ze$ and the $Z - 1$ internal electrons form a spherically symmetric charge distribution. The potential $V(r)$ felt by the valence electron

[2] T.M. Liss and P.L. Lipton, "The discovery of the top quark", Sci. Am., September 1997, p. 36; M. Perl, "The leptons after 100 years", Phys. Today, October 1997, p. 34.

can therefore be described to a first approximation by a function of the type

$$V(r) = -\frac{e^2}{r} Z_{\text{eff}}(r) \, .$$

The boundary values of $Z_{\text{eff}}(r)$ are Z for $r \to 0$ and 1 for $r \to \infty$. With this definition of $V(r)$, the radial equation becomes

$$\left(-\frac{\hbar^2}{2m_{\text{e}}} \frac{1}{r} \frac{\mathrm{d}^2}{\mathrm{d}r^2} r + \frac{\ell(\ell+1)\hbar^2}{2m_{\text{e}} r^2} - \frac{e^2}{r} Z_{\text{eff}}(r) \right) R_\ell(r) = E R_\ell(r) \, .$$

Referring to the form of the hydrogen wave functions, one can understand the following:

- If the probability density of the electron around the center ($r = 0$) is large, then Z_{eff} is noticeably larger than 1, and the electron is more strongly bound than in hydrogen.
- Conversely, if the electron's mean distance from the center is large, $Z_{\text{eff}} \sim 1$.

More precisely for a given n, the value of the wave function at the origin becomes smaller as the quantum number ℓ increases, because of the term $(r/a_1)^\ell$ originating from the centrifugal potential. Therefore we expect that, for a given n, the levels of sodium should be lower than those of hydrogen, and that they should get closer to the hydrogen case as ℓ increases. For increasing values of n and ℓ, the average distance between the electron and the nucleus also increases. The potential resembles more and more the Coulomb potential of a hydrogen atom, and the energy levels tend to those of hydrogen. This can be checked in Fig. 11.2.

Further Reading

- The richness of the physics involved in the understanding of the hydrogen atom is described in T.W. Hänsch, A.L. Schawlow and G.W. Series, "The spectrum of atomic hydrogen", Sci. Am., March 1979; D. Kleppner, M.G. Littman and M.L. Zimmerman, "Highly Excited Atoms", Sci. Am., May 1981; M. Nauenberg, C. Stroud and J. Yeazell, "The classical limit of an atom", Sci. Am., June 1994, p. 24.
- Once the physics related to Coulomb interactions has been understood, atoms can also be used as microscopic laboratories to test other theories, such as electroweak interactions which manifest themselves as a parity-violating effect. See, for example, M.A. Bouchiat and L. Pottier, "Atomic preference between right and left", Sci. Am., June 1984; S.C. Bennett and C.E. Wieman, Phys. Rev. Lett. **82**, 2484 (1999).
- For more elaborate treatments of atoms see, for example, W. Thirring, *Quantum Mechanics of Atoms and Molecules*, Chap. 4.3, Springer, New York (1981).

Exercises

11.1. Expectation value of r for the Coulomb problem. Consider the dimensionless radial equation for the hydrogen atom,

$$\left(\frac{\mathrm{d}^2}{\mathrm{d}\rho^2} - \frac{\ell(\ell+1)}{\rho^2} + \frac{2}{\rho}\right) u_{n,\ell}(\rho) = \varepsilon\, u_{n,\ell}(\rho)\,, \tag{11.29}$$

where $u_{n,\ell}(\rho) = \rho R_{n,\ell}(\rho)$ is the reduced wave function and satisfies the conditions $\int_0^\infty |u_{n,\ell}(\rho)|^2\,\mathrm{d}\rho = 1$ and $u_{n,\ell}(0) = 0$.

a. By multiplying this equation by $\rho\, u_{n,\ell}(\rho)$ and integrating over ρ, show that

$$\frac{\langle\rho\rangle}{n^2} + \frac{\ell(\ell+1)}{n^2} - 2 = \int_0^{+\infty} \rho\, u_{n,\ell}(\rho)\, u''_{n,\ell}(\rho)\,\mathrm{d}r\,.$$

You can use the result $\langle 1/\rho\rangle = 1/n^2$ deduced from the virial theorem (11.26).

b. By multiplying the Schrödinger equation by $\rho^2 u'_{n,\ell}(\rho)$ and integrating over ρ, show that

$$\frac{\langle\rho\rangle}{n^2} - 1 = -\int_0^{+\infty} \rho\, u_{n,\ell}(\rho)\, u''_{n,\ell}(\rho)\,\mathrm{d}r\,.$$

c. Deduce from the above results that $\langle\rho\rangle = [3n^2 - \ell(\ell+1)]/2$.

11.2. Three-dimensional harmonic oscillator in spherical coordinates. We treat the three-dimensional harmonic-oscillator problem as a central-potential problem. Consider the Hamiltonian

$$\hat{H} = \frac{\hat{p}^2}{2m} + \frac{1}{2}m\omega^2\hat{r}^2\,,$$

where $\hat{r}^2 = \hat{x}^2 + \hat{y}^2 + \hat{z}^2$.

a. Introduce the dimensionless quantities $\rho = r\sqrt{m\omega/\hbar}$ and $\epsilon = E/\hbar\omega$. Show that the radial equation (11.16) becomes

$$\left(-\frac{1}{\rho}\frac{\mathrm{d}^2}{\mathrm{d}\rho^2}\rho + \frac{\ell(\ell+1)}{\rho^2} + \rho^2 - 2\epsilon\right) R_\ell(\rho) = 0\,. \tag{11.30}$$

b. It can be proved that the normalizable solutions of (11.30) can be labeled by an integer n':

$$R_{n',\ell}(\rho) = \rho^\ell\, P_{n',\ell}(\rho^2)\, \mathrm{e}^{-\rho^2/2}\,,$$

where $P_{n',\ell}$ is a polynomial of degree n'. These solutions correspond to particular values of ϵ:

$$\epsilon = 2n' + \ell + 3/2\,.$$

In the following we set $n = 2n' + \ell$. Show that one recovers the levels $E = \hbar\omega(n_1 + n_2 + n_3 + 3/2)$ (n_i integer ≥ 0) that were obtained in Cartesian coordinates in Chap. 4 (exercise 4.3) and are associated with the eigenstates $|n_1; n_2; n_3\rangle$. To what values of the angular momentum do the energy levels E_n correspond?

c. Give the explicit correspondence between the states $|n_1; n_2; n_3\rangle$ and $|n, \ell, m\rangle$ for $n = n_1 + n_2 + n_3 = 1$.

11.3. Relation between the Coulomb problem and the harmonic oscillator. Consider the three-dimensional harmonic-oscillator problem treated in the previous exercise, but writing the potential as $V(r) = K^2 m\omega^2 r^2/2$, where K is dimensionless, in order to keep track of the parameters. Using the dimensionless variable ρ of the previous exercise, the radial equation for the reduced wave function $u = \rho R(\rho)$ is

$$\left(\frac{\mathrm{d}^2}{\mathrm{d}\rho^2} - \frac{\ell(\ell+1)}{\rho^2} - K^2\rho^2 + 2\frac{E}{\hbar\omega}\right) u(\rho) = 0\,. \tag{11.31}$$

Similarly, consider the Coulomb problem with a potential $V(r) = -Ze^2/r$, where Z is dimensionless. The radial equation for the variable $\rho = r/a_1$ is

$$\left(\frac{\mathrm{d}^2}{\mathrm{d}\rho^2} - \frac{\ell(\ell+1)}{\rho^2} + \frac{2Z}{\rho} + \frac{E}{E_{\mathrm{I}}}\right) u(\rho) = 0\,. \tag{11.32}$$

a. Show that, under the transformation $u(\rho) = x^\alpha f(x)$, where $x = \sqrt{\rho}$, and with an appropriate choice of α, one can cast the hydrogen problem (11.32) into the same form as the harmonic-oscillator problem (11.31).
b. Discuss the correspondence between the parameters of the two problems.
c. Recalling the results of the previous exercise, find the energy levels of the hydrogen atom.

11.4. Confirm or invalidate the following assertions.

a. If $[\hat{H}, \hat{\boldsymbol{L}}] = \boldsymbol{0}$, the energy levels do not depend on m (i. e. on the eigenvalues of the projection of one of the components of the angular momentum $\hat{\boldsymbol{L}}$).
b. If $[\hat{H}, \hat{L}^2] = 0$, the energy levels do not depend on ℓ.

11.5. Centrifugal-barrier effects. Consider a central potential. We denote by E_ℓ the lowest energy level for a given ℓ. Show that E_ℓ increases with ℓ.

11.6. Algebraic method for the hydrogen atom. We consider the radial dimensionless equation for the Coulomb problem (11.29) and we introduce the operators:

$$A_\ell^- = \frac{\mathrm{d}}{\mathrm{d}\rho} + \frac{\ell+1}{\rho} - \frac{1}{\ell+1}, \qquad A_\ell^+ = \frac{\mathrm{d}}{\mathrm{d}\rho} - \frac{\ell+1}{\rho} + \frac{1}{\ell+1}.$$

a. Calculate $A_\ell^- A_\ell^+$. Show that (11.29) can be written

$$\left(A_\ell^- A_\ell^+\right) u_\ell = \left(\epsilon - \frac{1}{(\ell+1)^2}\right) u_\ell . \tag{11.33}$$

b. Show that

$$A_\ell^+ A_\ell^- = A_{\ell+1}^- A_{\ell+1}^+ - \frac{1}{(\ell+2)^2} + \frac{1}{(\ell+1)^2} .$$

By multiplying (11.33) by A_ℓ^+, show that $A_\ell^+ u_\ell(\rho)$ satisfies the radial equation, with the same eigenvalue ε but with an angular momentum $\ell' = \ell + 1$.

c. Similarly, show that $A_{\ell-1}^- u_\ell(\rho)$ satisfies the radial equation with the same eigenvalue ε but an angular momentum $\ell' = \ell - 1$.
d. Calculate the expectation value of $A_\ell^- A_\ell^+$ for the radial function $u_\ell(\rho)$, and show that $\varepsilon \leq 1/(\ell+1)^2$.
e. Show that, for a given value of ε, there exists a maximum value ℓ_{max} of the angular momentum such that $\varepsilon = 1/n^2$, where we have set $n = \ell_{\mathrm{max}} + 1$. Show that the corresponding radial wave function $u_{\ell_{\mathrm{max}}}(\rho)$ satisfies the differential equation

$$\left(\frac{\mathrm{d}}{\mathrm{d}\rho} - \frac{n}{\rho} + \frac{1}{n}\right) u_{\ell_{\mathrm{max}}}(\rho) = 0 .$$

f. Deduce from these results the energy levels and the corresponding wave functions of the hydrogen atom.

11.7. Molecular potential. Consider a central potential of the form

$$V(r) = A/r^2 - B/r \qquad (A, B > 0) .$$

We want to calculate the energy levels of a particle of mass m_{e} in this potential.

a. Write down the radial equation.
b. By a change of notation, reduce this equation to an eigenvalue problem which is formally identical to the Kepler problem. Check that one can solve this equation by using the same arguments as for the hydrogen atom.
c. Give the explicit values of the energy levels in terms of A and B.

12. Spin 1/2 and Magnetic Resonance

The object must be presentable.
Charles Perrault, The Ducks and the Little Water Spaniel

The elaboration of the concept of spin was certainly the most difficult step of all of quantum theory during the first quarter of the 20th century. After the triumph of the Bohr model in 1913, the "old theory of quanta" of Bohr and Sommerfeld, which was based on the idea of quantum restrictions on classical quantities, had accumulated successes. People contemplated reaching a unified theory of spectroscopic data. Alas! the catalog of more and more ad hoc and complicated recipes could also be considered as reflecting a scholarly form of ignorance. The accumulation of successes was followed by a similar accumulation of unexplained facts, and even paradoxes. The anomalous Zeeman effect, the splitting of spectral lines, electron shells in complex atoms and the Stern–Gerlach experiment seemed to be challenges to the scientific community. Nobody imagined that these effects had a common origin.

The explanation was simple, but it was revolutionary. For the first time people were facing a purely quantum effect, with no classical analog. Nearly all the physical world depends on this concept, that of spin 1/2.

We have already mentioned experimental arguments which show that there exist in nature half-integer angular momenta. It is not possible to account for various effects in atomic-physics experiments if one assumes that the electron, which is a *point-like* particle down to distances of 10^{-18} m, has only the three degrees of freedom corresponding to translations in space. A series of experimental and theoretical arguments show that it must have an *internal* degree of freedom, its *intrinsic angular momentum.* This quantity does not have a classical analog. Any attempt to make a semiclassical model for this intrinsic angular momentum immediately leads to inconsistencies. In other words, the electron is both a point-like and a spinning object; spin is a purely quantum property.

We shall refer mainly to the case of the electron, but our considerations extend to many other particles or systems. The proton, the neutron, quarks and neutrinos have the same spin as the electron. In relativistic quantum mechanics, the structure of the Lorentz group shows that the spin of a particle characterizes this particle as much its mass and its electric charge do.

Fig. 12.1. Three physicists discussing the optimal way to measure spin effects in proton collisions at the Argonne ZGS accelerator (U.S.A.) (CERN document)

Here, we are interested in a spin of 1/2, i. e. an intrinsic angular momentum which corresponds to eigenvalues $j = 1/2, m = \pm 1/2$. In general, one uses the term *spin* for the intrinsic angular momentum of a particle, as opposed to its orbital angular momentum. The spin can then take any of the values seen in Chap. 10: $j = 0$ for the π meson, $j = 1$ for the photon and the deuteron, etc.

Our first goal here is to become familiar with the concept of spin 1/2. It is a purely quantum physical quantity, and the intuitive representation each of us constructs for it is a personal matter, as shown in Fig. 12.1. We shall then come back to the relation between angular momenta and magnetic moments in order to describe, in the final section, a phenomenon of great practical importance: magnetic resonance.

The hypothesis of spin for the electron was due to Uhlenbeck and Goudsmit in 1925. Both of them were quite young, since Uhlenbeck was still hesitating between a career in physics and in history, while Goudsmit had not yet passed his final exams. As soon as they realized that their hypothesis could explain many experimental facts, they discussed it with their professor, Ehrenfest, who encouraged them to publish their work. This idea was received with very varied feelings by the community. Bohr was very enthusiastic about it, while Pauli and Lorentz had several objections. One of these objections was connected with relativity. If one models the electron as an extended charged sphere whose electrostatic energy is equal to its mass energy $m_e c^2$, the radius of the sphere is of the order of $e^2/(m_e c^2)$ and the equatorial velocity of the sphere has to be much larger than the speed of light in order to account for the angular momentum $\hbar/2$ (one actually obtains $v_{\rm eq} \sim c/\alpha = 137c$). We know now that this objection is not relevant; it must be interpreted as an argument showing that a classical representation of the electron's intrinsic angular momentum is impossible. Spin is a fully quantum concept.

12.1 The Hilbert Space of Spin 1/2

The degree of freedom associated with the spin of a particle manifests itself in the measurement of particular physical quantities. These are the projections of the intrinsic angular momentum on three orthogonal axes x, y, z, and all functions of these quantities. The fundamental property of a spin-1/2 particle is that in a measurement of the projection of its spin along any axis, the only outcomes that one can observe are the two values $+\hbar/2$ and $-\hbar/2$.

From this experimental result, it follows that if we measure the square of any component of the spin, we find a single value $\hbar^2/4$, with probability one. Therefore a measurement of the square of the spin $S^2 = S_x^2 + S_y^2 + S_y^2$ always gives the result $S^2 = 3\hbar^2/4$, whatever the spin state of the particle. Any spin state is a linear superposition of two basis states and the degree of freedom corresponding to spin is described in a *two-dimensional* Hilbert space, $\mathcal{E}_{\mathrm{spin}}$.

12.1.1 Spin Observables

Let $\hat{\boldsymbol{S}}$ be the vector spin observable, i. e. a set of three observables $\{\hat{S}_x, \hat{S}_y, \hat{S}_z\}$. These three observables have the commutation relations of an angular momentum:

$$\hat{\boldsymbol{S}} \times \hat{\boldsymbol{S}} = \mathrm{i}\hbar\hat{\boldsymbol{S}} . \tag{12.1}$$

Each of the observables $\hat{S}_x, \hat{S}_y$ and $\hat{S}_z$ has eigenvalues $\pm\hbar/2$. The observable $\hat{S}^2 = \hat{S}_x^2 + \hat{S}_y^2 + \hat{S}_z^2$ is proportional to the identity in $\mathcal{E}_{\mathrm{spin}}$, with eigenvalue $3\hbar^2/4$.

12.1.2 Representation in a Particular Basis

We choose a basis of states in which both $\hat{S}^2$ and $\hat{S}_z$ are diagonal, which we denote by $\{|+\rangle, |-\rangle\}$:

$$\hat{S}_z|+\rangle = \frac{\hbar}{2}|+\rangle , \qquad \hat{S}_z|-\rangle = -\frac{\hbar}{2}|-\rangle , \qquad \hat{S}^2|\pm\rangle = \frac{3\hbar^2}{4}|\pm\rangle . \tag{12.2}$$

Using the notation of Chap. 10, the states $|\pm\rangle$ would be $|j = 1/2, m = \pm 1/2\rangle$. The action of $\hat{S}_x$ and $\hat{S}_y$ on the elements of this basis is written as (see (10.17))

$$\hat{S}_x|+\rangle = \hbar/2|-\rangle , \qquad \hat{S}_x|-\rangle = \hbar/2|+\rangle , \tag{12.3}$$

$$\hat{S}_y|+\rangle = \mathrm{i}\hbar/2|-\rangle , \qquad \hat{S}_y|-\rangle = -\mathrm{i}\hbar/2|+\rangle . \tag{12.4}$$

An arbitrary spin state $|\Sigma\rangle$ can be written as

$$|\Sigma\rangle = \alpha\,|+\rangle \,+\, \beta\,|-\rangle , \qquad |\alpha|^2 + |\beta|^2 = 1 . \tag{12.5}$$

The probabilities of finding $+\hbar/2$ and $-\hbar/2$ in a measurement of S_z on this state are $P(+\hbar/2) = |\alpha_+|^2$, $P(-\hbar/2) = |\alpha_-|^2$, respectively.

12.1.3 Matrix Representation

It is convenient to use the following matrix representations for the states and the operators:

$$|+\rangle = \begin{pmatrix} 1 \\ 0 \end{pmatrix}, \quad |-\rangle = \begin{pmatrix} 0 \\ 1 \end{pmatrix}, \quad |\Sigma\rangle = \begin{pmatrix} \alpha_+ \\ \alpha_- \end{pmatrix}. \tag{12.6}$$

We can use the Pauli matrices $\hat{\boldsymbol{\sigma}} \equiv \{\hat{\sigma}_x, \hat{\sigma}_y, \hat{\sigma}_z\}$ introduced in Chap. 6:

$$\hat{\sigma}_x = \begin{pmatrix} 0 & 1 \\ 1 & 0 \end{pmatrix}, \qquad \hat{\sigma}_y = \begin{pmatrix} 0 & -i \\ i & 0 \end{pmatrix}, \qquad \hat{\sigma}_z = \begin{pmatrix} 1 & 0 \\ 0 & -1 \end{pmatrix}. \tag{12.7}$$

These satisfy the commutation relations

$$\hat{\boldsymbol{\sigma}} \times \hat{\boldsymbol{\sigma}} = 2\mathrm{i}\,\hat{\boldsymbol{\sigma}}\,. \tag{12.8}$$

The spin observables are represented as

$$\hat{\boldsymbol{S}} = \frac{\hbar}{2}\,\hat{\boldsymbol{\sigma}}\,. \tag{12.9}$$

In this basis, the eigenstates $|\pm\rangle_x$ of $\hat{S}_x$ and $|\pm\rangle_y$ of $\hat{S}_y$ are

$$|\pm\rangle_x = \frac{1}{\sqrt{2}}\begin{pmatrix} 1 \\ \pm 1 \end{pmatrix}, \qquad |\pm\rangle_y = \frac{1}{\sqrt{2}}\begin{pmatrix} 1 \\ \pm \mathrm{i} \end{pmatrix}. \tag{12.10}$$

12.1.4 Arbitrary Spin State

Consider the most general spin state $|\Sigma\rangle$. Up to a phase factor, this state can always be written as

$$|\Sigma\rangle = \mathrm{e}^{-\mathrm{i}\varphi/2}\cos(\theta/2)\,|+\rangle + \mathrm{e}^{\mathrm{i}\varphi/2}\sin(\theta/2)\,|-\rangle\,,$$

where $0 \le \theta \le \pi$ and $0 \le \varphi < 2\pi$. One can then check that $|\Sigma\rangle$ is the eigenstate with eigenvalue $\hbar/2$ of the operator $\hat{S}_{\boldsymbol{u}} = \boldsymbol{u} \cdot \hat{\boldsymbol{S}}$, i. e. the projection of the spin along an axis of unit vector $\boldsymbol{u}$, with polar angle θ and azimuthal angle φ. In fact, we have

$$\boldsymbol{u} = \sin\theta\,\cos\varphi\;\boldsymbol{e}_x + \sin\theta\,\sin\varphi\;\boldsymbol{e}_y + \cos\theta\;\boldsymbol{e}_z\,,$$

i. e., in matrix notation,

$$\hat{S}_{\boldsymbol{u}} = \frac{\hbar}{2}\begin{pmatrix} \cos\theta & \sin\theta\,\mathrm{e}^{-\mathrm{i}\varphi} \\ \sin\theta\,\mathrm{e}^{\mathrm{i}\varphi} & -\cos\theta \end{pmatrix}.$$

In other words, for any spin-1/2 state, there is always a direction $\boldsymbol{u}$ for which this state is an eigenstate of the operator $\boldsymbol{u} \cdot \hat{\boldsymbol{S}}$, the projection of the spin along this direction. This remarkable property does not generalize to higher spins.

12.2 Complete Description of a Spin-1/2 Particle

The preceding discussion is similar to the phenomenological analysis of the magnetic moment in Chap. 8. The state in three-dimensional Euclidean space of a spin-1/2 particle is described in the space $\mathcal{E}_{\text{external}}$ of square-integrable functions $\mathcal{L}^2(R^3)$, and the spin state is described in the space $\mathcal{E}_{\text{spin}}$ introduced above.

12.2.1 Hilbert Space

The complete Hilbert space is the tensor product of these two spaces:

$$\mathcal{E}_{\text{H}} = \mathcal{E}_{\text{external}} \otimes \mathcal{E}_{\text{spin}} \, . \tag{12.11}$$

Any element $|\psi\rangle$ of $\mathcal{E}_{\text{H}}$ can be written as

$$|\psi\rangle = |\psi_+\rangle \otimes |+\rangle + |\psi_-\rangle \otimes |-\rangle \, , \tag{12.12}$$

where $|\psi_+\rangle$ and $|\psi-\rangle$ are elements of $\mathcal{E}_{\text{external}}$.

We notice that the *space* observables $\hat{A}_{\text{ex}}$ ($\hat{x}, \hat{p}$, etc.) and the *spin* observables $\hat{B}_{\text{sp}}$ ($\hat{S}_x, \hat{S}_y$, etc.) act in different spaces, and therefore they commute. The (tensor) product of two such observables is defined as

$$\left(\hat{A}_{\text{ex}} \otimes \hat{B}_{\text{sp}}\right) (|\psi_\sigma\rangle \otimes |\sigma\rangle) = \left(\hat{A}_{\text{ex}}|\psi_\sigma\rangle\right) \otimes \left(\hat{B}_{\text{sp}}|\sigma\rangle\right) \quad \sigma = \pm \, . \tag{12.13}$$

12.2.2 Representation of States and Observables

There are several possible representations of the states and of the observables, each of which has its advantages according to the problem under consideration.

Mixed Representation. The state is represented by a vector of $\mathcal{E}_{\text{spin}}$ whose components are square-integrable functions:

$$\psi_+(\boldsymbol{r},t) \, |+\rangle + \psi_-(\boldsymbol{r},t) \, |-\rangle \, . \tag{12.14}$$

We recall the physical interpretation of this representation: $|\psi_+(\boldsymbol{r},t)|^2 \, \mathrm{d}^3r$ and $|\psi_-(\boldsymbol{r},t)|^2 \, \mathrm{d}^3r$ are the probabilities of finding the particle in a volume d^3r around the point $\boldsymbol{r}$, with a spin component $+\hbar/2$ and $-\hbar/2$, respectively, along z.

An operator in $\mathcal{E}_{\text{external}}$ acts on the functions $\psi_\pm(\boldsymbol{r},t)$, a spin operator acts on the vectors $|+\rangle$ and $|-\rangle$, and the products of operators $\hat{A}_{\text{ex}} \otimes \hat{B}_{\text{sp}}$ can be deduced from (12.13).

Two-Component Wave Function. The state vector is represented in the form

$$\begin{pmatrix} \psi_+(\boldsymbol{r},t) \\ \psi_-(\boldsymbol{r},t) \end{pmatrix} . \tag{12.15}$$

The physical interpretation of ψ_+ and ψ_- as probability amplitudes for the pair of random variables $(\boldsymbol{r}, S_z)$ is the same as above.

Atomic States. In many problems of atomic physics, it is useful to use the quantum numbers n, ℓ, m to classify the states $|n,\ell,m\rangle$, which form a basis of $\mathcal{E}_{\text{external}}$. The introduction of spin is done in the space spanned by the family $\{|n,\ell,m\rangle \otimes |\sigma\rangle\}$, where the spin quantum number can take the two values ± 1. It is convenient to use the compact notation

$$|n,\ell,m,\sigma\rangle \equiv |n,\ell,m\rangle \otimes |\sigma\rangle , \tag{12.16}$$

where the states of an electron are described by *four* quantum numbers. The action of space operators on the states $|n,\ell,m\rangle$ is known (see Chap. 11), and therefore the action of general operators on the states $|n,\ell,m,\sigma\rangle$ can readily be inferred from the considerations above.

12.3 Spin Magnetic Moment

We have already emphasized in Chap. 10 that the connection between the angular momentum of a system and its magnetic moment allows a quantitative test of the angular-momentum theory. This connection remains valid for the spin degree of freedom and gives direct evidence for the existence of half-integer angular momenta.

12.3.1 The Stern–Gerlach Experiment

To the spin of a particle, there corresponds a magnetic moment which is proportional to it:

$$\hat{\boldsymbol{\mu}} = \gamma \hat{\boldsymbol{S}} = \mu_0 \hat{\boldsymbol{\sigma}} , \tag{12.17}$$

where $\mu_0 = \gamma\hbar/2$. This proportionality relation is fundamental. It implies the commutation relations between the components of the magnetic moment that we found in Chap. 8 by analyzing the Stern–Gerlach experiment. We recall that this experiment gives direct access to the nature of the angular momentum of the particle under consideration. The deviation of the beams is proportional to μ_z, and therefore to J_z. If the magnetic moment of the atom is due to an orbital angular momentum, we expect an odd number of spots. The observation of an even number of spots, two for monovalent atoms such as silver, is proof that half-integer angular momenta exist in nature.

12.3.2 Anomalous Zeeman Effect

We place an atom, prepared in a level of energy E and angular momentum j, in a magnetic field $\boldsymbol{B}$ parallel to z. The magnetic energy is

$$\hat{W} = -\hat{\boldsymbol{\mu}} \cdot \boldsymbol{B} . \tag{12.18}$$

The corresponding level is split into $2j+1$ sublevels with energies

$$E - \gamma\hbar B_0 m , \quad m = -j, \ldots , j$$

A corresponding splitting of each line is observed in the spectrum. If all angular momenta are orbital angular momenta (i. e. they have a classical interpretation), j must be an integer. In that case $2j+1$ is odd and we expect a splitting into an odd number of levels. The splitting of spectral lines in a magnetic field, first observed by Zeeman in 1896–1903, shows that in many cases, in particular for alkali atoms, this is not true. There is a splitting into an *even* number of levels.

Faraday was convinced as early as 1845 that there was a deep connection between optical and magnetic phenomena. In one of the last experiments of his life, in 1862, he attempted to investigate the influence of magnetic fields on radiation. Many technical problems prevented him from obtaining a positive answer. It was only in 1896 that these experiments were redone successfully by Zeeman. At that time, theorists, in particular Lorentz, had already predicted that one should observe a splitting into an odd number of lines (one line, i. e. no splitting, or three). Zeeman first confirmed this result on the spectra of cadmium and zinc. The discovery, in the particular case of sodium, of what was to be called the "anomalous Zeeman effect", i. e. an even number of lines, remained a real challenge to the scientific community, who were totally confused by this phenomenon, for more than 25 years. It was only in the years 1925–1926, after much struggle, with the ideas of Pauli, Uhlenbeck and Goudsmit, that the introduction of the notion of spin completely solved the problem. The "anomalous Zeeman effect" then appeared as a natural phenomenon.

12.3.3 Magnetic Moment of Elementary Particles

The electron, the proton and the neutron are spin-1/2 particles. The corresponding spin magnetic moment is related to the spin $\boldsymbol{S}$ by the relation $\hat{\boldsymbol{\mu}} = \gamma\hat{\boldsymbol{S}}$. Experiments give the following values of the gyromagnetic ratios:

$$\begin{aligned} \text{electron} &\quad \gamma \simeq 2\gamma_0 = -q/m_{\mathrm{e}} , \\ \text{proton} &\quad \gamma \simeq +2.79\, q/m_{\mathrm{p}} , \\ \text{neutron} &\quad \gamma \simeq -1.91\, q/m_{\mathrm{p}} . \end{aligned}$$

The possible results of a measurement of the component of these magnetic moments along a given axis are therefore

$$\begin{aligned} \text{electron} &\quad \mu_z = \pm\mu_{\mathrm{B}} = \mp q\hbar/2m_{\mathrm{e}} , \\ \text{proton} &\quad \mu_z = \pm 2.79\, q\hbar/2m_{\mathrm{p}} , \\ \text{neutron} &\quad \mu_z = \pm 1.91\, q\hbar/2m_{\mathrm{p}} . \end{aligned}$$

The quantity $\mu_B = -9.274 \times 10^{-24}\,\mathrm{J\,T^{-1}}$ is called the *Bohr magneton*. The quantity $\mu_N = q\hbar/2m_p = 5.051 \times 10^{-27}\,\mathrm{J\,T^{-1}}$ is called the *nuclear magneton.*

Dirac's relativistic theory of the electron predicts the value of the electron magnetic moment, i. e.

$$\hat{\boldsymbol{\mu}} = g_e \left(\frac{q}{2m_e}\right) \hat{\boldsymbol{S}}\,, \quad \text{where} \quad g_e = 2\,.$$

The value measured experimentally for the gyromagnetic factor g_e nearly coincides with this prediction. One can account for the slight difference between the experimental result and Dirac's prediction by taking into account the coupling of the electron to the quantized electromagnetic field (quantum electrodynamics). This constitutes one of the most spectacular successes of fundamental physics. The experimental and theoretical values of the quantity g_e coincide within the limits of the accuracy of experiments and of computer calculations. At present, we have for the electron, setting $g_e = 2(1+a)$,

$$a^{\text{theo.}} = 0.001\,159\,652\,\underline{200}\,(40)\,, \tag{12.19}$$

$$a^{\text{exp.}} = 0.001\,159\,652\,\underline{193}\,(10)\,; \tag{12.20}$$

the errors in parentheses relate to the two last digits.

The coefficients $+2.79$ and -1.91 for the proton and the neutron are due to the internal structure of these particles. They can be measured with great accuracy by magnetic-resonance experiments: $\mu_p/\mu_N = 2.792\,847\,3\underline{86}\,(63)$ and $\mu_n/\mu_N = -1.913\,042\,\underline{75}\,(45)$; they can be calculated with a 10% accuracy by using the quark model.

12.4 Uncorrelated Space and Spin Variables

In most physical situations, such as the Stern–Gerlach experiments, the space and the spin variables are correlated. For instance, in Chap. 11 we studied only a first approximation to the hydrogen atom, where we neglected spin effects. If we include the spin degree of freedom in this approximation, we find that the states $|n, \ell, m, +\rangle$ and $|n, \ell, m, -\rangle$ are degenerate. Actually, there exist corrections to this approximation, such as the *fine structure* of the hydrogen atom, which we shall describe in Chap. 13 and which is due to the interaction between the spin magnetic moment and the electromagnetic field created by the proton. The degeneracy is then partially lifted and the new eigenstates are combinations of the initial states $|n, \ell, m, \sigma\rangle$. In other words, in a given eigenstate of the total Hamiltonian, the spatial wave function of an electron depends on its spin state, and the two random variables $\boldsymbol{r}$ and S_z are correlated.

This correlation is sometimes extremely weak. In such cases, the two variables $\boldsymbol{r}$ and S_z can be considered as independent and their probability law can be factorized. Such a physical situation can be represented by a factorized state vector,

$$\Phi(\boldsymbol{r},t)\begin{pmatrix}\alpha_+(t)\\ \alpha_-(t)\end{pmatrix}. \tag{12.21}$$

If one performs spin measurements in this case, the results are independent of the position of the particle. The only observables which are relevant are 2×2 Hermitian matrices with numerical coefficients (which can depend on time).

Such cases happen in practice, in particular in magnetic-resonance experiments. We then use the term "spin state of the proton" instead of "state of the proton" since the position of the proton in space does not play any role in the experiment under consideration.

12.5 Magnetic Resonance

In the fundamental proportionality relation between the spin and the associated magnetic moment $\boldsymbol{\mu} = \gamma \boldsymbol{S}$, the determination of the gyromagnetic coefficient γ is an important issue. In the case of objects such as the electron or the proton, it provides a test of fundamental interactions. For nuclei in molecules, the value of γ gives precise information about the electronic environment and the chemical bonding in these molecules.

We describe below how one can perform very accurate measurements of γ. As often in physics, this consists in using a resonance phenomenon. We shall then present some applications of magnetic resonance, electronic and nuclear.

12.5.1 Larmor Precession in a Fixed Magnetic Field $\boldsymbol{B}_0$

We choose the z axis to be parallel to the field $\boldsymbol{B}_0$. Ignoring space variables (see Sect. 12.4), the Hamiltonian is

$$\hat{H} = -\hat{\boldsymbol{\mu}} \cdot \boldsymbol{B}_0 = -\mu_0 B_0\, \hat{\sigma}_z\,. \tag{12.22}$$

We set

$$-\mu_0 B_0/\hbar = \omega_0/2\,, \quad \text{i.e.} \quad \omega_0 = -\gamma B_0\,. \tag{12.23}$$

The eigenstates of $\hat{H}$ are the eigenstates $|+\rangle$ and $|-\rangle$ of $\hat{\sigma}_z$.

Consider an arbitrary state $|\psi(t)\rangle$ such that $|\psi(0)\rangle = \alpha|+\rangle + \beta|-\rangle$, where $|\alpha|^2 + |\beta|^2 = 1$. Its time evolution is

$$|\psi(t)\rangle = \alpha\, \mathrm{e}^{-\mathrm{i}\omega_0 t/2}\, |+\rangle + \beta\, \mathrm{e}^{\mathrm{i}\omega_0 t/2}\, |-\rangle\,. \tag{12.24}$$

The expectation value $\langle \boldsymbol{\mu} \rangle$ reads

$$\langle \mu_x \rangle = 2\mu_0 \, \mathrm{Re}\left(\alpha^* \beta \, \mathrm{e}^{\mathrm{i}\omega_0 t}\right) = C \, \cos(\omega_0 t + \varphi) \,, \tag{12.25}$$

$$\langle \mu_y \rangle = 2\mu_0 \, \mathrm{Im}\left(\alpha^* \beta \, \mathrm{e}^{\mathrm{i}\omega_0 t}\right) = C \, \sin(\omega_0 t + \varphi) \,, \tag{12.26}$$

$$\langle \mu_z \rangle = \mu_0 \, \left(|\alpha|^2 - |\beta|^2\right) \,, \tag{12.27}$$

where C and φ are the modulus and phase, respectively, of the complex number $\mu_0 \alpha^* \beta$. We recover the Larmor precession that we derived for an arbitrary angular momentum in Sect. 10.4.2. The projection $\langle \mu_z \rangle$ of the magnetic moment along the field is time independent, and the component of $\langle \boldsymbol{\mu} \rangle$ perpendicular to $\boldsymbol{B}$ rotates with an angular velocity ω_0. The fact that μ_z is a constant of motion is a consequence of the commutation relation $[\hat{H}, \hat{\mu}_z] = 0$ and of the Ehrenfest theorem.

This provides a simple method to measure the angular frequency ω_0. We place a coil in a plane parallel to $\boldsymbol{B}_0$ and prepare a macroscopic quantity of spins all in the same spin state $|\psi(0)\rangle$. The precession of $\langle \boldsymbol{\mu} \rangle$ at the frequency ω_0 causes a periodic variation of the magnetic flux in the coil, and this induces an electric current at the same frequency. This method is, however, not as accurate as the resonance experiment we present below.

12.5.2 Superposition of a Fixed Field and a Rotating Field

A technique invented by Rabi in the 1930s allows one to perform a very accurate measurement of ω_0 by means of a resonance phenomenon. We place the magnetic moment in a known field B_0, on which we superimpose a weak field B_1 which rotates at a variable angular velocity ω in the xy plane. Such a field can be obtained using two coils oriented along the x and y axes, with alternating currents at frequency $\omega/2\pi$ with a phase difference equal to $\pi/2$ (radio frequencies are used in experiments of this type). We show below that, at resonance, i.e. for $\omega = \omega_0$, the spin flips between its two possible states $|\pm\rangle$. Notice that this calculation, which is characteristic of a driven two-state system, is similar to that performed for the ammonia maser in Chap. 6.

The form of the Hamiltonian is

$$\hat{H} = -\hat{\boldsymbol{\mu}} \cdot \boldsymbol{B} = -\mu_0 B_0 \hat{\sigma}_z - \mu_0 B_1 \cos \omega t \, \hat{\sigma}_x - \mu_0 B_1 \sin \omega t \, \hat{\sigma}_y \,. \tag{12.28}$$

We set

$$|\psi(t)\rangle = a_+(t)|+\rangle + a_-(t)|-\rangle \,. \tag{12.29}$$

The Schrödinger equation yields the differential system

$$\mathrm{i}\dot{a}_+ = \frac{\omega_0}{2} a_+ + \frac{\omega_1}{2} \mathrm{e}^{-\mathrm{i}\omega t} \, a_- \,, \tag{12.30}$$

$$\mathrm{i}\dot{a}_- = \frac{\omega_1}{2} \mathrm{e}^{\mathrm{i}\omega t} \, a_+ - \frac{\omega_0}{2} a_- \,, \tag{12.31}$$

where we have defined $\mu_0 B_0/\hbar = -\omega_0/2$, $\mu_0 B_1/\hbar = -\omega_1/2$.

A change of functions $b_\pm(t) = \exp(\pm i\omega t/2) a_\pm(t)$ leads to

$$i\dot{b}_+ = -\frac{\omega - \omega_0}{2}\, b_+ + \frac{\omega_1}{2}\, b_- \,, \tag{12.32}$$

$$i\dot{b}_- = \frac{\omega_1}{2}\, b_+ + \frac{\omega - \omega_0}{2}\, b_- \,. \tag{12.33}$$

The above transformation is the quantum form of a change of reference frame. It transforms from the laboratory frame to a frame rotating, like the magnetic field, with an angular velocity ω around the z axis. With this change of reference frame, the basis of the Hilbert space is time dependent, whereas the Hamiltonian is time independent:

$$\hat{\tilde{H}} = \frac{\hbar}{2}\begin{pmatrix} \omega_0 - \omega & \omega_1 \\ \omega_1 & \omega - \omega_0 \end{pmatrix} = -\frac{\hbar}{2}(\omega - \omega_0)\hat{\sigma}_z + \frac{\hbar}{2}\omega_1 \hat{\sigma}_x \,.$$

One can check that the equations (12.32), (12.33) imply $\ddot{b}_\pm + (\Omega/2)^2\, b_\pm = 0$, where

$$\Omega^2 = (\omega - \omega_0)^2 + \omega_1^2 \,. \tag{12.34}$$

Suppose that the spin is initially in the state $|+\rangle$, i.e. $b_-(0) = 0$. One finds

$$b_-(t) = -\frac{i\omega_1}{\Omega} \sin\left(\frac{\Omega t}{2}\right) , \tag{12.35}$$

$$b_+(t) = \cos\left(\frac{\Omega t}{2}\right) + i\frac{\omega - \omega_0}{\Omega} \sin\left(\frac{\Omega t}{2}\right) . \tag{12.36}$$

The probability that a measurement of S_z at time t gives the result $-\hbar/2$ is

$$\begin{aligned}\mathcal{P}_{+\to-}(t) &= |\langle -|\psi(t)\rangle|^2 = |a_-(t)|^2 = |b_-(t)|^2 \\ &= \left(\frac{\omega_1}{\Omega}\right)^2 \sin^2\left(\frac{\Omega t}{2}\right) .\end{aligned} \tag{12.37}$$

This formula, which is due to Rabi, exhibits a resonance phenomenon:

- If the frequency ω of the rotating field is noticeably different from the frequency ω_0 we want to measure, more precisely if $|\omega - \omega_0| \gg \omega_1$, the probability that the spin flips, i.e. that we measure $S_z = -\hbar/2$, is very small for all t.
- If we choose $\omega = \omega_0$, then the probability of a spin flip is equal to one at times $t_n = (2n+1)\pi/\omega_1$ (n integer) even if the amplitude of the rotating field $\boldsymbol{B}_1$ is very small.
- For $|\omega - \omega_0| \sim \omega_1$, the probability amplitude oscillates with an appreciable amplitude, smaller than one.

In Fig. 12.2, we have drawn the time oscillation of the probability $\mathcal{P}_{+\to-}$ off resonance and at resonance. For a typical magnetic field of 1 T, the resonance frequency is $\omega_e/2\pi \sim 28$ GHz for an electron, and 2.79 $\omega_N/2\pi \sim 43$ MHz for a proton. These frequencies correspond to decameter waves in the nuclear case and centimeter waves in the electronic case.

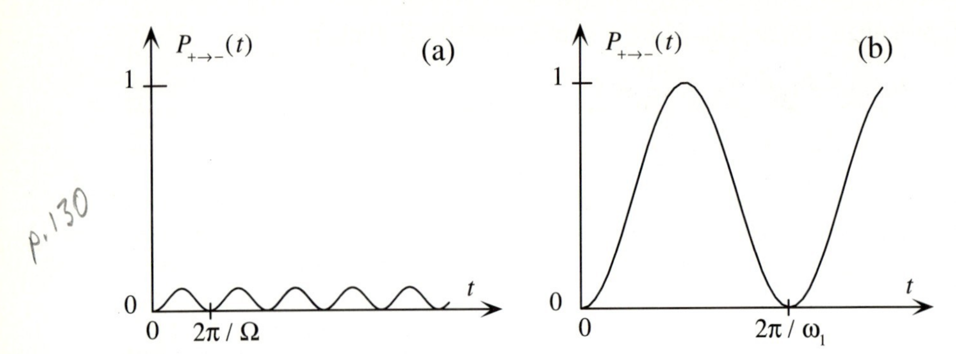

Fig. 12.2. Rabi oscillations: (**a**) slightly off resonance, $\omega - \omega_0 = 3\omega_1$; (**b**) at resonance, $\omega = \omega_0$

12.5.3 Rabi's Experiment

The resonance effect described above was understood by Rabi in 1939. It provides a very accurate measurement of a magnetic moment. The device used by Rabi consists of a source, two Stern–Gerlach deflectors with magnetic fields in opposite directions, and a detector (Fig. 12.3). Between the two Stern–Gerlach magnets, a zone containing a superposition of a uniform field $\boldsymbol{B}_0$ and a rotating field $\boldsymbol{B}_1$, as described above, is placed.

Consider first the effect of the two Stern–Gerlach magnets in the absence of the fields $\boldsymbol{B}_0$ and $\boldsymbol{B}_1$. A particle emitted by the source in the spin state $|+\rangle$ undergoes two successive deflections in two opposite directions and reaches the detector. When the fields $\boldsymbol{B}_0$ and $\boldsymbol{B}_1$ are present, this is not true anymore. If the frequency ω of the rotating field is close to the Larmor frequency ω_0, the resonance phenomenon will change the component μ_z of the particle. When such a spin flip occurs between the two Stern–Gerlach magnets, the two deflections have the same direction (upwards in the case of Fig. 12.3), and the particle misses the detector. The signal registered on the detector as a

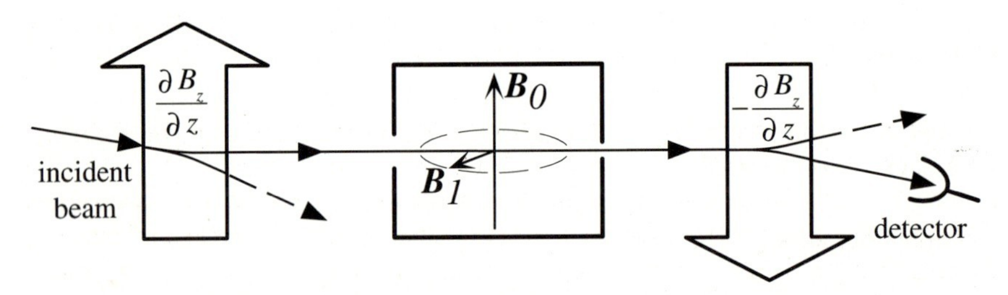

Fig. 12.3. Apparatus developed by Rabi for observing the magnetic resonance effect. In the absence of magnetic resonance, all particles emitted in the state $|+\rangle$ reach the detector. If the resonance occurs, the spins of the particles flip between the two magnets and the signal drops

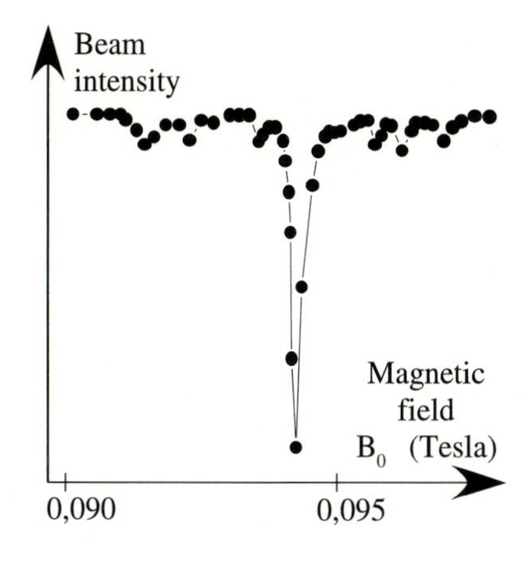

Fig. 12.4. Signal (obtained by Rabi) recorded on the detector of Fig. 12.3 with a beam of HD molecules, as a function of the field B_0 ($B_1 = 10^{-4}\,\mathrm{T}, \omega/2\pi = 4\,\mathrm{MHz}$)

function of the frequency of the rotating field $\boldsymbol{B}_1$ undergoes a sharp drop for $\omega = \omega_0$ (Fig. 12.4). This leads to a measurement of the ratio $|\mu|/j = \hbar\omega_0/B_0$ for a particle of angular momentum j. Actually, this measurement is so precise that the main source of error comes from the determination of B_0. In practice, as shown in Fig. 12.4, the frequency ω is fixed and one varies the magnitude of the field $\boldsymbol{B}_0$, or, equivalently, the frequency ω_0.

In 1933 Stern, with his apparatus, had measured the proton magnetic moment with a 10% accuracy. That was a very difficult experiment, since nuclear magnetic moments are 1000 times smaller than electronic ones. One must operate with H_2 or HD molecules, where the pairing cancels the effects due to the magnetic moment of the electrons. In 1939, with his resonance apparatus, Rabi gained a factor of 1000 in the accuracy of the measurement. The resonance is very selective in frequency, and the presence of other magnetic moments causes no problem. Rabi's result made an impression on the minds of people. It was greeted as a great achievement. Stern remarked that Rabi had attained the theoretical accuracy of the measurement, which is fixed by the uncertainty relations. When Hulthén announced that Rabi had been awarded the Nobel prize on December 10, 1944 on Stockholm radio, he said that, "By this method Rabi has literally established radio relations with the most subtile particles of matter, with the world of the electrons and of the atomic nucleus."

The great breakthrough in the application of nuclear magnetic resonance (N.M.R.) came with the work of Felix Bloch at Stanford and of Edward Purcell at MIT, in 1945. Owing to the development of radio wave technologies during the Second World War, due to the development of radar, Bloch and Purcell were able to operate on condensed matter, and not on molecular beams. Here one uses macroscopic numbers of spins, thereby obtaining much more intense signals and more manageable experiments. The resonance is observed by measuring, for instance, the absorption of the wave generating the rotating field $\boldsymbol{B}_1$. The imbalance between the populations of the two states

$|+\rangle$ and $|-\rangle$, which is necessary in order for a signal to be obtained, results from the conditions of thermal equilibrium. In a field $B_0 = 1\,\mathrm{T}$, the magnetic energy for a proton is $2.79\,\mu_N B \sim 10^{-7}\,\mathrm{eV}$ and the relative population difference between the two spin states due to the Boltzmann factor at room temperature is $\pi_+ - \pi_- \sim 4\times10^{-6}$. This relative difference is small, but quite sufficient for one to observe a significant signal since one deals with samples containing a macroscopic number of spins (typically 10^{23}).

12.5.4 Applications of Magnetic Resonance

The applications of magnetic resonance, in domains ranging from solid-state physics and low temperatures to chemistry, biology and medicine, are numerous. Because of its magnetic effects, spin can play the role of a local probe inside matter. NMR has transformed chemical analysis and the determination of the structure of molecules (see Fig. 12.5). It has become an invaluable tool in molecular biology. Since 1980, NMR has also caused a revolution in medical diagnosis and physiology. It allows one to measure and visualize, in three dimensions and with a spatial precision better than a millimeter, the concentration of water in "soft matter" (muscles, brain, etc.), which, in contrast to bones, is difficult to observe with X rays. One can study in this way the structure and metabolism of living tissues, and one can detect internal injuries, tumors, etc. The nuclear spin, which was a curiosity to some visionary physicists of the 1940s and 1950s, has become one of the great hopes of modern medicine.

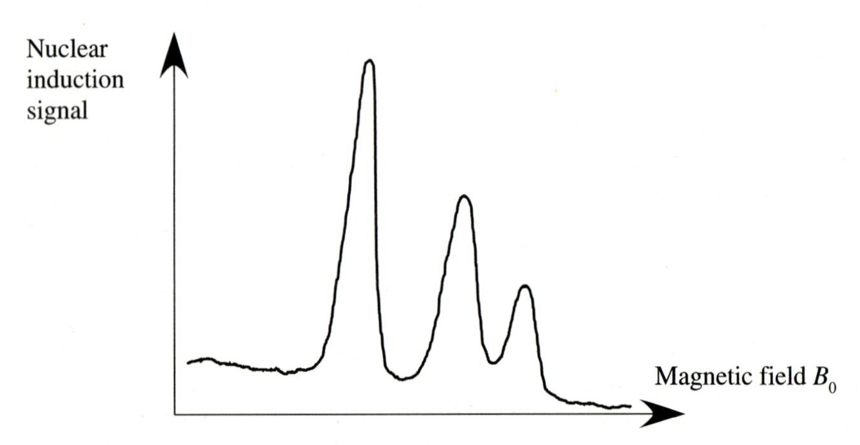

Fig. 12.5. One of the first examples of nuclear magnetic resonance applied to chemistry: the resonance signal obtained from the protons of the ethanol molecule CH_3CH_2OH consists of a three-peak structure. These peaks are associated with the three protons of the CH_3 group, the two protons of the CH_2 group and the single proton of the OH group. The magnetic field B_0 was $\sim 0.8\,\mathrm{T}$, and the total trace is $7.5\,\mu\mathrm{T}$ wide

One can use NMR to visualize the activity of the brain in real time. It is possible to localize and register the response of the visual cortex to a stimulation. The following step, after submitting a volunteer to a sequence of such excitations, is to ask him/her to *think* about the signal. The NMR response of the brain is *the same* as that obtained with an external stimulation! This may be considered a direct proof that we think, which is somewhat comforting for the mind.

12.5.5 Rotation of a Spin 1/2 Particle by 2π

It seems obvious and a matter of common geometric sense that a rotation by 2π of a system around a fixed axis is equivalent to the identity. However, strictly speaking, this is not true for a spin-1/2 particle.

Let us come back to the calculation of section Sect. 12.5.1 and suppose that, at time $t = 0$, the state of the system is $|+x\rangle$:

$$|\psi(t=0)\rangle = \frac{1}{\sqrt{2}}\left(|+\rangle + |-\rangle\right) .$$

The mean value of the magnetic moment, given by (12.25), (12.26) and (12.27), is $\langle \boldsymbol{\mu} \rangle = \mu_0\, \boldsymbol{u}_x$. Equation (12.24) gives the evolution of this state. After a time $t = 2\pi/\omega_0$, classically, the system has precessed by an angle 2π around $\boldsymbol{B}$. In quantum mechanics one can verify that the expectation value is back to its initial value $\langle \boldsymbol{\mu} \rangle = \mu_0\, \boldsymbol{u}_x$. What can we say, however, about the state vector? We can check that $|\psi(t)\rangle$ is still an eigenvector of $\hat{S}_x$ (or of $\hat{\mu}_x$) with an eigenvalue $+\hbar/2$. However, we notice that, quite surprisingly, the state vector has *changed sign*:

$$|\psi(t=2\pi/\omega_0)\rangle = -\frac{1}{\sqrt{2}}\left(|+\rangle + |-\rangle\right) = -|\psi(0)\rangle .$$

A rotation by 2π is therefore not equivalent to the identity for a spin-1/2 particle. Only rotations of $4n\pi$ give back the initial state identically. This property can also be guessed from the dependence $\mathrm{e}^{im\varphi}$ for orbital angular momenta: using the same formula for $m = 1/2$ and $\varphi = 2\pi$ would give $\mathrm{e}^{\mathrm{i}\pi} = -1$.

This peculiarity was understood as soon as spin 1/2 was discovered, in 1926. It remained a controversial point for more than 50 years. Does the phase of the state vector after a rotation of 2π have a physical meaning? A positive experimental answer was only obtained in the 1980s in a series of remarkable experiments.[1] The spin-1/2 particles were sent into a two-channel interferometer. In one of them a magnetic field rotated the spins by multiples of 2π. The change in sign of the wave function was observed by means of the displacement of the interference fringes, and the experimental signal confirmed that a rotation of 4π was needed to recover the fringe pattern

[1] A.W. Overhauser, A.R. Collela and S.A. Werner, Phys. Rev. Lett. **33**, 1237 (1974); Phys. Rev. Lett. **34**, 1472 (1975); Phys. Rev. Lett. **35**, 1053 (1975).

which was measured in the absence of rotation. The fact that a rotation of a spin-1/2 particle by 2π is not equivalent to the identity, contrary to common sense, is another manifestation that spin 1/2 is basically quantum mechanical.

This property reflects an important mathematical structure which relates the two groups SO(3) and SU(2). Let us first see how these two groups appear in this context. Rotations in Euclidean space R^3 form the well-known group SO(3). Each rotation $\mathcal{R}$ of R^3 can be parameterized by a unit vector $\boldsymbol{u}$ (the axis of rotation) and an angle of rotation φ $(0 \leq \varphi \leq \pi)$ around this axis; with each rotation $\mathcal{R} \equiv (\boldsymbol{u}, \varphi)$, one can associate a *rotation operator* $\hat{M}(\mathcal{R})$ of $\mathcal{E}_{\text{spin}}$:

$$\hat{M}(\mathcal{R}) = \cos(\varphi/2)\, \hat{I} - \mathrm{i} \sin(\varphi/2)\, \boldsymbol{u} \cdot \hat{\boldsymbol{\sigma}}\, .$$

This rotation operator is unitary $(\hat{M}\hat{M}^\dagger = \hat{M}^\dagger \hat{M} = \hat{I})$ and of determinant 1, and it gives the transform $|\psi'\rangle$ of a state vector $|\psi\rangle$ under a rotation $\mathcal{R}$: $|\psi'\rangle = \hat{M}(\mathcal{R})\, |\psi\rangle$. On can check this property immediately for the result (12.24) in the particular case $\boldsymbol{u} = \boldsymbol{u}_z$ and $\varphi = \omega_0 t$. One says that the group SU(2) formed by the matrices $\hat{M}(\mathcal{R})$ is a *representation* of the rotation group.

The minus sign found previously arises as a consequence of a general property of this representation: it is a *projective* representation, i. e. for two arbitrary rotations $\mathcal{R}$ and $\mathcal{R}'$, $\hat{M}(\mathcal{R}\mathcal{R}')$ is not equal, strictly speaking, to $\hat{M}(\mathcal{R})\, \hat{M}(\mathcal{R}')$, but can differ from it by a phase factor. In particular, if one chooses $\mathcal{R} = \mathcal{R}' \equiv (\boldsymbol{u}_z, \pi)$, one finds

$$\mathcal{R}\mathcal{R}' = I_{R^3} \qquad \text{but} \qquad \hat{M}(\mathcal{R})\, \hat{M}(\mathcal{R}') = (-\mathrm{i}\hat{\sigma}_z)\, (-\mathrm{i}\hat{\sigma}_z) = -\hat{I}\, .$$

Mathematically, one says that there is a local isomorphism between the Lie algebras of the two groups SO(3) and SU(2), but that these two groups are not globally isomorphic. This formalism, which is called *spinor theory*, was developed in the early 20th century by the mathematician Élie Cartan.

Further Reading

- For the development of the ideas which led to the concept of spin, see, for example M. Jammer, *The Conceptual Development of Quantum Mechanics*, Chap. 3 (McGraw-Hill, New York, 1956); G. Uhlenbeck, "Fifty years of spin", Phys. Today, June 1976, p. 43.
- R.L. Jaffe, "Where does the proton really get its spin?", Phys. Today, September 1995, p. 24.
- I.L. Pykett, "Medical applications of NMR", Sci. Am., May 1982; "Images of pain in the brain", Sci. Am., 1991,
- E. Cartan, *The Theory of Spinors*, Hermann, Paris (1966).

Exercises

12.1. Products of Pauli matrices. Show that

$$\hat{\sigma}_j \, \hat{\sigma}_k = \delta_{j,k} + \mathrm{i}\varepsilon_{j,k,\ell} \, \hat{\sigma}_l \, , \tag{12.38}$$

where $\varepsilon_{j,k,\ell} = 1$ or -1 if (j, k, ℓ) is an even or odd permutation, respectively, of (x, y, z), and $\varepsilon_{j,k,\ell} = 0$ otherwise.

12.2. Algebra with Pauli matrices. Consider the Pauli matrices $\hat{\boldsymbol{\sigma}}$, and two vectors $\boldsymbol{A}$ and $\boldsymbol{B}$. Show that

$$(\hat{\boldsymbol{\sigma}} \cdot \boldsymbol{A})(\hat{\boldsymbol{\sigma}} \cdot \boldsymbol{B}) = \boldsymbol{A} \cdot \boldsymbol{B} + \mathrm{i}\hat{\boldsymbol{\sigma}} \cdot (\boldsymbol{A} \wedge \boldsymbol{B}) \, .$$

12.3. Spin and orbital angular momentum. Consider a spin-1/2 particle whose state is $|\psi\rangle = \psi_+(\boldsymbol{r})|+\rangle + \psi_-(\boldsymbol{r})|-\rangle$. Let $\hat{\boldsymbol{S}}$ be the spin observable and $\hat{\boldsymbol{L}}$ the orbital angular momentum. We assume that

$$\psi_+(\boldsymbol{r}) = R(r)\left(Y_{0,0}(\theta,\varphi) + \frac{1}{\sqrt{3}} Y_{1,0}(\theta,\varphi)\right) \, ,$$

$$\psi_-(\boldsymbol{r}) = \frac{R(r)}{\sqrt{3}} \left[Y_{1,1}(\theta,\varphi) - Y_{1,0}(\theta,\varphi)\right] \, .$$

a. What is the normalization condition on $R(r)$?
b. What are the probabilities of finding $\pm\hbar/2$ in measurements of S_z or S_x?
c. What are the possible results of a measurement of L_z? Give the corresponding probabilities.

12.4. Geometric origin of the commutation relations of $\hat{\boldsymbol{J}}$. The general definition of the angular momentum $\hat{\boldsymbol{J}} = (\hat{J}_x, \hat{J}_y, \hat{J}_z)$ of a quantum system is based on the transformation of the state of this system under a rotation. More specifically, we assume that in an infinitesimal rotation by an angle $\mathrm{d}\varphi$ around the axis $\boldsymbol{u}$, the state vector of the system $|\psi\rangle$ is changed in accordance with

$$|\psi\rangle \longrightarrow \left(1 - \mathrm{i}\frac{\mathrm{d}\varphi}{\hbar}\, \boldsymbol{u} \cdot \hat{\boldsymbol{J}} + \ldots\right) |\psi\rangle \, , \tag{12.39}$$

where we neglect terms of order $\mathrm{d}\varphi^2$ or higher (see exercise 5.1 of Chap. 5). We know that in a geometric rotation of $\mathrm{d}\varphi$ around $\boldsymbol{u}$ in three-dimensional Euclidean space, a vector $\boldsymbol{V}$ transforms in accordance with

$$\boldsymbol{V} \longrightarrow \boldsymbol{V} + \mathrm{d}\varphi\, \boldsymbol{u} \times \boldsymbol{V} + \ldots \, .$$

a. We consider the following four successive rotations:
   - rotation around the x axis by an angle $\mathrm{d}\alpha$,
   - rotation around the y axis by an angle $\mathrm{d}\beta$,

- rotation around the x axis by an angle $-\mathrm{d}\alpha$,
- rotation around the y axis by an angle $-\mathrm{d}\beta$.

We are interested in the resulting geometric transformation to second order in $\mathrm{d}\alpha$ and $\mathrm{d}\beta$.
(i) Justify that the terms in $\mathrm{d}\alpha^2$ and $\mathrm{d}\beta^2$ do not contribute.
(ii) Calculate the term in $\mathrm{d}\alpha\,\mathrm{d}\beta$.
(iii) Show that the resulting transformation after these four infinitesimal rotations is a rotation of $-\mathrm{d}\alpha\,\mathrm{d}\beta$ around the z axis.

b. We apply the previous four successive rotations to a quantum system.
(i) Write the final state of the system in terms of the initial state $|\psi\rangle$, $\hat{\boldsymbol{J}}$ and the angles $\mathrm{d}\alpha$ and $\mathrm{d}\beta$ up to second order in these angles.
(ii) Deduce that one is led to $[\hat{J}_x, \hat{J}_y] = \mathrm{i}\hbar\hat{J}_z$ in order to ensure the consistency of the definition (12.39).

c. Consider a spin-1/2 system which we rotate around the z axis.
(i) How does the state vector $|\psi\rangle = \alpha|+\rangle + \beta|-\rangle$ transform under a rotation of angle $\mathrm{d}\varphi$?
(ii) Generalize the result to an arbitrary angle φ.
(iii) Relate this to the Larmor precession phenomenon discussed in Sect. 12.5.1.

13. Addition of Angular Momenta, Fine and Hyperfine Structure of Atomic Spectra

It is not obviously true that two and two are four,
assuming **four** *means three and one.*
Gottfried Wilhelm Leibniz

When atomic spectral lines are observed with sufficient resolution, they appear in general to have a complex structure, each line being in fact a group of nearby components. The *fine* and *hyperfine* splittings of atomic levels are of particular importance, both from the fundamental point of view and in terms of applications. The origin of such structures lies in the magnetic interactions of the electron inside the atom. In this chapter we shall study some examples of these interactions and the ensuing effects.

In order to do this, we need a technical tool: the addition of angular momenta in quantum mechanics and the notion of the total angular momentum of a system. This notion is useful in many physical problems and we shall give its basic elements in Sect. 13.1. In Sect. 13.2 we shall describe the *spin–orbit* interaction of the electron spin magnetic moment with the magnetic field originating from the orbital motion of the electron around the nucleus. A well-known example of the resulting splitting is the yellow line (D line) of sodium. In Sect. 13.3 we shall describe the *hyperfine* interaction between the spin magnetic moments of the electron and of the proton in the ground state of atomic hydrogen. This interaction produces a splitting which is responsible for the 21 cm line of hydrogen, of considerable interest in astrophysics.

13.1 Addition of Angular Momenta

We now consider the total angular momentum of a complex system. We shall start with the simple case of two spin-1/2 subsystems and then generalize the procedure.

13.1.1 The Total-Angular Momentum Operator

In classical mechanics, one defines the *total* angular momentum of a system of two (or n) particles as the sum

$$L_{\text{tot}} = L_1 + L_2$$

of the individual angular momenta of these two (or n) particles.

Consider now, in quantum mechanics, two angular-momentum observables $\hat{\boldsymbol{J}}_1$ and $\hat{\boldsymbol{J}}_2$, which, by definition, act in two different Hilbert spaces $\mathcal{E}_1$ and $\mathcal{E}_2$. For instance, this may concern a system of two particles: $\mathcal{E}_1$ and $\mathcal{E}_2$ are then the spaces $\mathcal{L}^2(R^3)$ of square-integrable functions of $\boldsymbol{r}_1$ and $\boldsymbol{r}_2$, respectively. A similar situation can also arise for a particle moving in space ($\mathcal{E}_1 = \mathcal{L}^2(R^3)$) which possesses an intrinsic angular momentum ($\mathcal{E}_2 = \mathcal{E}_{\text{spin}}$).

The Hilbert space of the total system is the tensor product

$$\mathcal{E} = \mathcal{E}_1 \otimes \mathcal{E}_2 .$$

By *definition*, the *total-angular-momentum* observable of the system is

$$\hat{\boldsymbol{J}} \equiv \hat{\boldsymbol{J}}_1 \otimes \hat{I}_2 + \hat{I}_1 \otimes \hat{\boldsymbol{J}}_2 = \hat{\boldsymbol{J}}_1 + \hat{\boldsymbol{J}}_2 , \tag{13.1}$$

where $\hat{I}_1$ and $\hat{I}_2$ are the identity operators in $\mathcal{E}_1$ and $\mathcal{E}_2$, respectively. This observable acts in $\mathcal{E}$ and is an angular-momentum observable. Indeed, it satisfies the commutation relations

$$\hat{\boldsymbol{J}} \times \hat{\boldsymbol{J}} = \mathrm{i}\hbar \hat{\boldsymbol{J}} , \tag{13.2}$$

since $\hat{\boldsymbol{J}}_1$ and $\hat{\boldsymbol{J}}_2$ commute. Therefore we know that we can diagonalize $\hat{J}^2$ and $\hat{J}_z$ simultaneously. We also know the set of their possible eigenvalues: $\hbar^2 j(j+1)$ with $2j$ integer for $\hat{J}^2$, and $\hbar m$ with $m = -j, \dots, j-1, j$ for $\hat{J}_z$ for a given j.

One can check that the four angular-momentum observables

$$\hat{J}_1^2, \hat{J}_2^2, \hat{J}^2, \hat{J}_z$$

commute. Moreover, we shall see in Sect. 13.1.4 that this set forms a CSCO in the sense of Sect. 10.2. Their common eigenbasis is therefore unique. We write their common eigenvectors as $|j_1, j_2; j, m\rangle$. We have, by definition,

$$\hat{J}_1^2 |j_1, j_2; j, m\rangle = j_1(j_1+1)\hbar^2 \, |j_1, j_2; j, m\rangle , \tag{13.3}$$

$$\hat{J}_2^2 |j_1, j_2; j, m\rangle = j_2(j_2+1)\hbar^2 \, |j_1, j_2; j, m\rangle , \tag{13.4}$$

$$\hat{J}^2 |j_1, j_2; j, m\rangle = j(j+1)\hbar^2 \, |j_1, j_2; j, m\rangle , \tag{13.5}$$

$$\hat{J}_z |j_1, j_2; j, m\rangle = m\hbar \, |j_1, j_2; j, m\rangle . \tag{13.6}$$

As in Chap. 10, we omit the presence of other possible quantum numbers which would be cumbersome to write and are irrelevant to the present discussion.

13.1.2 Uncoupled and Coupled Bases

The Hilbert space $\mathcal{E}$ corresponding to the degrees of freedom associated with the angular momentum is generated by the family of factorized states

$$\{|j_1, m_1\rangle \otimes |j_2, m_2\rangle\} \equiv \{|j_1, m_1; j_2, m_2\rangle\} .$$

In this basis, the observables $\hat{J}_1^2, \hat{J}_{1z}, \hat{J}_2^2, \hat{J}_{2z}$ are diagonal. Consider the eigensubspace of the two observables $\hat{J}_1^2$ and $\hat{J}_2^2$ corresponding to given values of j_1 and j_2. The dimension of this subspace is $(2j_1+1)(2j_2+1)$ and we ask the following question:

What are, in this subspace, the eigenvectors of $\hat{J}^2$ and $\hat{J}_z$, and the corresponding eigenvalues $j(j+1)\hbar^2$ and $m\hbar$?

In other words, we want to perform in each eigensubspace of $\hat{J}_1^2$ and $\hat{J}_2^2$ a change of basis to go from the *uncoupled* eigenbasis common to $\{\hat{J}_1^2, J_{1z}, \hat{J}_2^2, \hat{J}_{2z}\}$ to the *coupled* eigenbasis common to $\{\hat{J}_1^2, \hat{J}_2^2, \hat{J}^2, \hat{J}_z\}$. The eigenvalues of $\hat{J}^2$ and $\hat{J}_z$ will be expressed as functions of j_1, j_2, m_1 and m_2. Once the determination of the values of j has been performed, we shall express the eigenstates $|j_1, j_2; j, m\rangle$ in terms of the states $|j_1, m_1; j_2, m_2\rangle$:

$$|j_1, j_2; j, m\rangle = \sum_{m_1 m_2} C^{j,m}_{j_1,m_1;j_2,m_2} |j_1, m_1; j_2, m_2\rangle\,, \tag{13.7}$$

$$C^{j,m}_{j_1,m_1;j_2,m_2} = \langle j_1, m_1; j_2, m_2 | j_1, j_2; j, m\rangle\,. \tag{13.8}$$

The coefficients $C^{j,m}_{j_1,m_1;j_2,m_2}$ of the change of basis (13.7) are called the *Clebsch–Gordan coefficients.*

13.1.3 A Simple Case: the Addition of Two Spins of 1/2

The case of two spin-1/2 particles will be of particular interest in the following (21 cm line of hydrogen, Pauli principle, etc.). We shall first treat this case in an elementary way, before considering the general problem of the coupling of two arbitrary angular momenta.

The Hilbert Space of the Problem. Consider a system of two spin-1/2 particles, for instance the electron and the proton in a hydrogen atom or the two electrons of a helium atom. We denote the particles by 1 and 2. The Hilbert space of the system is

$$\mathcal{E}_{\mathrm{H}} = \mathcal{E}^1_{\mathrm{external}} \otimes \mathcal{E}^1_{\mathrm{spin}} \otimes \mathcal{E}^2_{\mathrm{external}} \otimes \mathcal{E}^2_{\mathrm{spin}}\,.$$

We denote by $\mathcal{E}_{\mathrm{s}}$ the tensor product of the two spin spaces:

$$\mathcal{E}_{\mathrm{s}} = \mathcal{E}^1_{\mathrm{spin}} \otimes \mathcal{E}^2_{\mathrm{spin}}\,. \tag{13.9}$$

$\mathcal{E}_{\mathrm{s}}$ is a four-dimensional space generated by the family $\{|\sigma_1\rangle \otimes |\sigma_2\rangle\}, \sigma_1 = \pm, \sigma_2 = \pm$, which we write in the simpler form

$$\{|+\,;\,+\rangle\,,\ |+\,;\,-\rangle\,,\ |-\,;\,+\rangle\,,\ |-\,;\,-\rangle\}\,, \tag{13.10}$$

by setting $|\sigma_1\rangle \otimes |\sigma_2\rangle \equiv |\sigma_1\,;\,\sigma_2\rangle$. The total spin operator is

$$\hat{\boldsymbol{S}} = \hat{\boldsymbol{S}}_1 + \hat{\boldsymbol{S}}_2\,.$$

The most general state (space + spin) $|\psi\rangle$ of this system of two spin-1/2 particles can be written as

$$|\psi\rangle = \psi_{++}(\boldsymbol{r}_1, \boldsymbol{r}_2)|+\,;\,+\rangle + \psi_{+-}(\boldsymbol{r}_1, \boldsymbol{r}_2)|+\,;\,-\rangle$$
$$+ \psi_{-+}(\boldsymbol{r}_1, \boldsymbol{r}_2)|-\,;\,+\rangle + \psi_{--}(\boldsymbol{r}_1, \boldsymbol{r}_2)|-\,;\,-\rangle\,. \tag{13.11}$$

Matrix Representation. We can use a matrix representation of the spin states and spin operators for this system. In the basis (13.10), a state is represented by a four-component vector. The observables $\hat{\boldsymbol{S}}_1$ and $\hat{\boldsymbol{S}}_2$ (extended to the tensor product space) can easily be written in the following way, using the Pauli matrices and a block 2×2 notation for the 4×4 matrices:

$$\hat{S}_{1x} = \frac{\hbar}{2}\begin{pmatrix} 0 & \vdots & \hat{I} \\ \cdots & \cdots & \cdots \\ \hat{I} & \vdots & 0 \end{pmatrix}, \qquad \hat{S}_{2x} = \frac{\hbar}{2}\begin{pmatrix} \hat{\sigma}_x & \vdots & 0 \\ \cdots & \cdots & \cdots \\ 0 & \vdots & \hat{\sigma}_x \end{pmatrix},$$

$$\hat{S}_{1y} = \frac{\hbar}{2}\begin{pmatrix} 0 & \vdots & -\mathrm{i}\hat{I} \\ \cdots & \cdots & \cdots \\ \mathrm{i}\hat{I} & \vdots & 0 \end{pmatrix}, \qquad \hat{S}_{2y} = \frac{\hbar}{2}\begin{pmatrix} \hat{\sigma}_y & \vdots & 0 \\ \cdots & \cdots & \cdots \\ 0 & \vdots & \hat{\sigma}_y \end{pmatrix},$$

$$\hat{S}_{1z} = \frac{\hbar}{2}\begin{pmatrix} \hat{I} & \vdots & 0 \\ \cdots & \cdots & \cdots \\ 0 & \vdots & -\hat{I} \end{pmatrix}, \qquad \hat{S}_{2z} = \frac{\hbar}{2}\begin{pmatrix} \hat{\sigma}_z & \vdots & 0 \\ \cdots & \cdots & \cdots \\ 0 & \vdots & \hat{\sigma}_z \end{pmatrix},$$

where $\hat{I}$ stands for the 2×2 identity matrix.

Total Spin States. We consider $\mathcal{E}_\mathrm{s}$ and we denote by $|S, M\rangle$ the eigenstates of $\hat{S}^2$ and $\hat{S}_z$ with eigenvalues $S(S+1)\hbar^2$ and $M\hbar$, respectively. Since $\hat{S}_z = \hat{S}_{1z} + \hat{S}_{2z}$, the largest possible value of M is $1/2 + 1/2 = 1$. The corresponding state is unique; it is the state $|+\,;\,+\rangle$. Similarly, the smallest possible value of M is $-1/2 - 1/2 = -1$, and the corresponding eigenstate is $|-\,;\,-\rangle$.

Let us calculate the action of the square of the total spin on these two vectors:

$$\begin{aligned}\hat{S}^2|+\,;\,+\rangle &= \left(\hat{S}_1^2 + \hat{S}_2^2 + 2\hat{\boldsymbol{S}}_1\cdot\hat{\boldsymbol{S}}_2\right)|+\,;\,+\rangle \\ &= \left(\frac{3}{4}\hbar^2 + \frac{3}{4}\hbar^2 + \frac{\hbar^2}{2}\left(\hat{\sigma}_{1x}\hat{\sigma}_{2x} + \hat{\sigma}_{1y}\hat{\sigma}_{2y} + \hat{\sigma}_{1z}\hat{\sigma}_{2z}\right)\right)|+\,;\,+\rangle \\ &= 2\hbar^2|+\,;\,+\rangle\,.\end{aligned}$$

Similarly,

$$\hat{S}^2|-\,;\,-\rangle = 2\hbar^2|-\,;\,-\rangle\,.$$

The two states $|+\,;\,+\rangle$ and $|-\,;\,-\rangle$ are therefore eigenstates of $\hat{S}^2$ with an eigenvalue of $2\hbar^2$, which corresponds to an angular momentum equal to 1. In the notation of Sect. 13.1.2, we have, therefore,

$$|s_1 = 1/2,\ m_1 = 1/2; s_2 = 1/2,\ m_2 = 1/2\rangle$$
$$= |s_1 = 1/2,\ s_2 = 1/2; S = 1,\ M = 1\rangle$$

and:

$$|s_1 = 1/2,\ m_1 = -1/2; s_2 = 1/2,\ m_2 = -1/2\rangle$$
$$= |s_1 = 1/2,\ s_2 = 1/2; S = 1,\ M = -1\rangle\,.$$

Since we have recognized two states $|S = 1, M = \pm 1\rangle$ of angular momentum 1, we now look for the third one $|S = 1, M = 0\rangle$. In order to do this, we use the general relation found in Chap. 10, $\hat{S}_-|j,m\rangle \propto |j,m-1\rangle$, and we obtain

$$\hat{S}_-|S = 1, M = 1\rangle = \left(\hat{S}_{1-} + \hat{S}_{2-}\right)|+\,;\,+\rangle \quad \propto \quad |-\,;\,+\rangle + |+\,;\,-\rangle\,.$$

After normalization, we obtain the state

$$|S = 1, M = 0\rangle = \frac{1}{\sqrt{2}}\left(|+\,;\,-\rangle + |-\,;\,+\rangle\right)\,.$$

One can check that this state is indeed an eigenstate of $\hat{S}^2$ and $\hat{S}_z$, with eigenvalues $2\hbar^2$ and 0, respectively.

We have identified a three-dimensional subspace in $\mathcal{E}_\mathrm{s}$ corresponding to a total angular momentum equal to 1. The orthogonal subspace, of dimension 1, is generated by the vector

$$\frac{1}{\sqrt{2}}\left(|+\,;\,-\rangle - |-\,;\,+\rangle\right)\,.$$

One can readily verify that this vector is an eigenvector of $\hat{S}^2$ and of $\hat{S}_z$ with both eigenvalues equal to zero.

To summarize, the total spin in the particular case $j_1 = j_2 = 1/2$ corresponds to

$$S = 1 \quad \text{or} \quad S = 0\,,$$

and the four corresponding eigenstates, which form a basis of $\mathcal{E}_\mathrm{H}$, are

$$|1, M\rangle : \begin{cases} |1,\ 1\rangle = |+\,;\,+\rangle \\ |1,\ 0\rangle = (|+\,;\,-\rangle + |-\,;\,+\rangle)/\sqrt{2} \\ |1,\ -1\rangle = |-\,;\,-\rangle \end{cases}, \tag{13.12}$$

$$|0,0\rangle : \qquad |0,\ 0\rangle = (|+\,;\,-\rangle - |-\,;\,+\rangle)/\sqrt{2}\,. \tag{13.13}$$

In this particular case of two spin-1/2 particles, we have solved the problem of Sect. 13.1.2 by decomposing the $4 = 2 \times 2$-dimensional space $\mathcal{E}_\mathrm{s}$ (the tensor product of the two two-dimensional spaces) into a *direct sum* of a one-dimensional space ($S = 0$) and a three-dimensional space ($S = 1$), i. e. $2 \times 2 = 1 + 3$.

Symmetry Properties. The following symmetry properties will be important when we consider identical particles and the Pauli principle.

The three states $|1, M\rangle$ are called collectively the *triplet state* of the two-spin system. They are *symmetric* with respect to the interchange of the z projections of the spins of the two particles, σ_1 and σ_2. The state $|0, 0\rangle$ is called the *singlet state* and is *antisymmetric* with respect to the same exchange. In mathematical terms, if we define a permutation operator $\hat{P}^s_{12}$ in $\mathcal{E}_s$ by the relation

$$\hat{P}^s_{12}|\sigma_1 \; ; \; \sigma_2\rangle = |\sigma_2 \; ; \; \sigma_1\rangle \,, \tag{13.14}$$

the triplet and singlet states are eigenvectors of this operator:

$$\hat{P}^s_{12}|1, M\rangle = |1, M\rangle \,, \qquad \hat{P}^s_{12}|0, 0\rangle = -|0, 0\rangle \,. \tag{13.15}$$

13.1.4 Addition of Two Arbitrary Angular Momenta

We now want to establish the following result:

Consider two angular-momentum observables $\hat{\boldsymbol{J}}_1$ and $\hat{\boldsymbol{J}}_2$. In the subspace corresponding to given values of j_1 and j_2, the possible values for the quantum number j associated with the total angular momentum $\hat{\boldsymbol{J}}$ are

$$j = |j_1 - j_2| \;,\; |j_1 - j_2| + 1 \;,\; \ldots \;,\; j_1 + j_2 - 1 \;,\; j_1 + j_2 \,.$$

Construction of the States such that $j = j_1 + j_2$. We first remark that any vector $|j_1, m_1; j_2, m_2\rangle$ is an eigenvector of $\hat{J}_z = \hat{J}_{1z} + \hat{J}_{2z}$ with eigenvalue $m\hbar$, and $m = m_1 + m_2$. One therefore deduces the following result:

The vector $|j_1, j_2; j, m\rangle$ corresponding to $m = j_1 + j_2$ exists and is unique.

Indeed, the maximum values of m_1 and m_2 are j_1 and j_2, and therefore the maximum value of m is $m_{\max} = j_1 + j_2$. We deduce that the maximum value of j is also $j_{\max} = j_1 + j_2$ since the index m can take all the values $m = -j \;,\; -j + 1 \ldots ,j$ in a given eigensubspace of $\hat{J}^2$.

There is only one normalized vector in the Hilbert space which fulfills the condition $m = m_{\max}$ (up to a phase factor):

$$|j_1, m_1 = j_1; j_2, m_2 = j_2\rangle \,.$$

This vector is also an eigenstate of $\hat{J}^2$, with eigenvalue $j(j+1)\hbar^2$, and $j = j_1 + j_2$, as can be checked directly using the following expression:

$$\hat{J}^2 = \hat{J}_1^2 + \hat{J}_2^2 + \hat{J}_{1+}\hat{J}_{2-} + \hat{J}_{1-}\hat{J}_{2+} + 2\hat{J}_{1z}\hat{J}_{2z} \,.$$

Consequently, we can write

$$|j = j_1 + j_2, m = j_1 + j_2\rangle = |m_1 = j_1; m_2 = j_2\rangle \,. \tag{13.16}$$

Remark. In what follows, we omit the indices j_1 and j_2 in the left- and right-hand sides of (13.16); these are implicit in the forms $|j,m\rangle \equiv |j_1,j_2;j,m\rangle$ and $|m_1;m_2\rangle \equiv |j_1,m_1;j_2,m_2\rangle$.

We now define, as in Chap. 10, the raising and lowering operators as follows:

$$\hat{J}_+ = \hat{J}_{1+} + \hat{J}_{2+}\,, \qquad \hat{J}_- = \hat{J}_{1-} + \hat{J}_{2-}\,.$$

We have

$$\hat{J}_\pm |j,m\rangle \propto |j,m\pm 1\rangle\,,$$

$$\hat{J}_{1\pm}|m_1;m_2\rangle \propto |m_1 \pm 1;m_2\rangle\,, \qquad \hat{J}_{2\pm}|m_1;m_2\rangle \propto |m_1;m_2 \pm 1\rangle\,,$$

where the proportionality coefficients are given in (10.17). Starting with $|j = j_1+j_2, m = j_1+j_2\rangle$, we can generate a series of states $|j = j_1+j_2, m'\rangle$ for $m' = j_1+j_2-1, \dots, -(j_1+j_2)$ by applying repeatedly the operator $\hat{J}_-$. For instance, using the normalization coefficients given in (10.17), we find

$$\begin{aligned}|\psi_\mathrm{a}\rangle &= |j = j_1+j_2, m = j_1+j_2-1\rangle \\ &\propto \sqrt{j_1}\,|m_1 = j_1-1; m_2 = j_2\rangle + \sqrt{j_2}\,|m_1 = j_1; m_2 = j_2-1\rangle\,.\end{aligned} \tag{13.17}$$

Eigensubspaces of $\hat{J}_z$. A graphical representation of the uncoupled basis states $|m_1;m_2\rangle$ is given in Fig. 13.1. Each dot in the m_1, m_2 plane represents a basis state. A fixed $m = m_1 + m_2$, corresponding to an eigensubspace $\mathcal{E}(m)$ of $\hat{J}_z$, is represented by a straight dashed line. The dot in the upper right corner corresponds to the state (13.16). As we have already noted, the dimension of this particular eigensubspace $\mathcal{E}(j_1+j_2)$ is 1. The next dashed line corresponds to $m = j_1+j_2-1$, and the corresponding eigensubspace $\mathcal{E}(j_1+j_2-1)$ has dimension 2, with the following possible basis:

$$|m_1 = j_1-1; m_2 = j_2\rangle\,, \qquad |m_1 = j_1; m_2 = j_2-1\rangle\,. \tag{13.18}$$

In general the eigenvalue $m\hbar$ of $\hat{J}_z$ has some degeneracy, except for $m = \pm(j_1+j_2)$.

By construction, each eigensubspace $\mathcal{E}(m)$ of $\hat{J}_z$ is invariant under the action of the Hermitian operators $\hat{J}_+\hat{J}_-$ and $\hat{J}_-\hat{J}_+$. Indeed $\hat{J}_+$ and $\hat{J}_-$ globally increase and decrease, respectively, the value of m_1+m_2 by 1. From the expression

$$\hat{J}^2 = \frac{1}{2}\left(\hat{J}_+\hat{J}_- + \hat{J}_-\hat{J}_+\right) + \left(\hat{J}_{1z} + \hat{J}_{2z}\right)^2\,,$$

it follows that $\mathcal{E}(m)$ is also globally invariant under the action of $\hat{J}^2$.

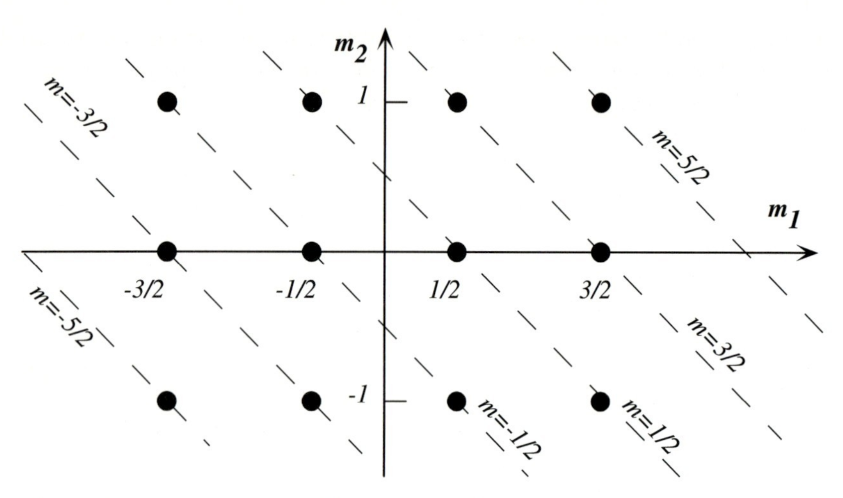

Fig. 13.1. Representation of the uncoupled basis states $|m_1; m_2\rangle$. The *dashed lines* represent the eigensubspaces $\mathcal{E}(m)$ of $\hat{J}_z$. These subspaces are globally invariant under the action of $\hat{J}^2$. This figure corresponds to $j_1 = 3/2$ and $j_2 = 1$

Construction of all States of the Coupled Basis. The total dimension of the Hilbert space, is $(2j_1 + 1)(2j_2 + 1)$. Inside this space we have already identified the $2j + 1$ vectors of the coupled basis with $j = j_1 + j_2$. We now describe the principle by which all remaining states of the coupled basis can be determined.

Consider the subspace $\mathcal{E}(j_1 + j_2 - 1)$ of $\hat{J}_z$, a possible basis of which is given in (13.18). Inside this subspace, we have already identified the vector $|\psi_\mathrm{a}\rangle$ given in (13.17), which is, by construction, an eigenvector of $\hat{J}_z$ and $\hat{J}^2$, with eigenvalues $(j_1 + j_2 - 1)\hbar$ and $(j_1 + j_2)(j_1 + j_2 + 1)\hbar^2$. Consider the vector of $\mathcal{E}(j_1 + j_2 - 1)$ orthogonal to $|\psi_\mathrm{a}\rangle$:

$$|\psi_\mathrm{b}\rangle = \sqrt{j_2}\,|m_1 = j_1 - 1; m_2 = j_2\rangle - \sqrt{j_1}\,|m_1 = j_1; m_2 = j_2 - 1\rangle\,.$$

Since $\mathcal{E}(j_1+j_2-1)$ is globally invariant under the action of $\hat{J}^2$, we can diagonalize this operator inside $\mathcal{E}(j_1 + j_2 - 1)$, and the corresponding eigenbasis is orthogonal. We know that $|\psi_\mathrm{a}\rangle$ is an eigenvector of $\hat{J}^2$. Therefore $|\psi_\mathrm{b}\rangle$, which is orthogonal to $|\psi_\mathrm{a}\rangle$, is also an eigenvector of $\hat{J}^2$, i.e.

$$\hat{J}^2\,|\psi_\mathrm{b}\rangle = j(j+1)\hbar^2|\psi_\mathrm{b}\rangle\,, \tag{13.19}$$

and we want to determine the value of j. On one hand we have $\hat{J}_z|\psi_\mathrm{b}\rangle = m\hbar|\psi_\mathrm{b}\rangle$, with $m = j_1 + j_2 - 1$; therefore $j \geq j_1 + j_2 - 1$, since one always has $j \geq m$. On the other hand, we cannot have $j = j_1 + j_2$, since this would mean that there existed two independent vectors ($|\psi_\mathrm{a}\rangle$ and $|\psi_\mathrm{b}\rangle$) corresponding to the same values of j and m (i.e. $j_1 + j_2$ and $j_1 + j_2 - 1$, respectively). This cannot be true, since (i) there is only one vector corresponding to $j = m =$

j_1+j_2 and (ii) there is a one-to-one correspondence, through the action of $\hat{J}_\pm$, between states associated with (j,m) and states associated with $(j,m\pm 1)$. Consequently, we must have $j = j_1+j_2-1$ in (13.19):

$$|\psi_{\mathrm{b}}\rangle \propto |j = j_1+j_2-1, m = j_1+j_2-1\rangle\,.$$

By applying repeatedly the operator $\hat{J}_-$ to $|\psi_{\mathrm{b}}\rangle$, we generate a new series of states, which are labeled $|j = j_1+j_2-1, m'\rangle$.

We have now identified all vectors in the two subspaces $\mathcal{E}(j_1+j_2)$ and $\mathcal{E}(j_1+j_2-1)$. We can repeat the same operation for the subspace $\mathcal{E}(j_1+j_2-2)$ (whose dimension is 3, with two vectors already identified), etc., until all the eigenstates $|m_1;m_2\rangle$ have been used. This occurs when we reach a quantum number m such that the dimension of $\mathcal{E}(m)$ is less than or equal to the dimension of $\mathcal{E}(m+1)$ ($m=-1/2$ for the values chosen in Fig. 13.1). Altogether, we find $2j_{\min}+1$ series of states, where $j_{\min} = \min(j_1,j_2)$. The possible values for j are therefore

$$j = j_1+j_2\,,\quad j = j_1+j_2-1\,,\;\ldots,\quad j = j_1+j_2-2j_{\min} = |j_1-j_2|\,.$$

We can check that the numbers of states of the coupled and uncoupled bases coincide, since

$$(2j_1+1)(2j_2+1) = 2(j_1+j_2)+1+2(j_1+j_2)-1+\ldots+2|j_1-j_2|+1\,.$$

Mathematically, we have decomposed the $(2j_1+1)\times(2j_2+1)$-dimensional space (the tensor product of a $(2j_1+1)$- and a $(2j_2+1)$-dimensional space) into the direct sum of spaces of $2(j_1+j_2)+1$, $2(j_1+j_2)-1$, ... , etc. dimensions.

Using this general procedure, one can determine the coefficients which relate the vectors of the uncoupled basis to those of the coupled basis (the Clebsch–Gordan coefficients, defined in (13.7)). The general expression for a Clebsch–Gordan coefficient is quite involved.[1] Here we shall give two examples which are useful in numerous problems. Consider first the case where $j_2 = 1/2$. One finds in this case

$$\left|j=j_1+\frac{1}{2},m\right\rangle = \cos\theta_m\left|m+\frac{1}{2};-\frac{1}{2}\right\rangle - \sin\theta_m\left|m-\frac{1}{2};\frac{1}{2}\right\rangle\,,$$

$$\left|j=j_1-\frac{1}{2},m\right\rangle = \sin\theta_m\left|m+\frac{1}{2};-\frac{1}{2}\right\rangle + \cos\theta_m\left|m-\frac{1}{2};\frac{1}{2}\right\rangle\,,$$

where

$$\cos\theta_m = \sqrt{\frac{j_1-m+1/2}{2j_1+1}}\,,\qquad \sin\theta_m = (-1)^{2(j_1+m)}\sqrt{\frac{j_1+m+1/2}{2j_1+1}}\,.$$

As a particular example, for $j_1 = 1/2$, we recover the triplet and singlet states introduced in Sect. 13.1.3.

[1] See, e.g., exercise 13.4 and A.R. Edmonds, *Angular Momentum in Quantum Mechanics*, Princeton University Press, Princeton (1950).

Another case of practical interest is the addition of two angular momenta equal to 1: $j_1 = 1$ and $j_2 = 1$. The possible values of j are $2, 1, 0$, and the corresponding vectors are, for $j = 2$ (with $\epsilon = \pm 1$),

$$
\begin{aligned}
|j=2, m=2\epsilon\rangle &= |\epsilon;\epsilon\rangle\,, \\
\sqrt{2}\,|j=2, m=\epsilon\rangle &= |\epsilon;0\rangle + |0;\epsilon\rangle\,, \\
\sqrt{6}\,|j=2, m=0\rangle &= |+1;-1\rangle + 2\,|0;0\rangle + |-1;+1\rangle\,.
\end{aligned}
$$

For $j = 1$, one finds

$$
\begin{aligned}
\sqrt{2}\,|j=1, m=\epsilon\rangle &= |\epsilon;0\rangle - |0;\epsilon\rangle\,, \\
\sqrt{2}\,|j=1, m=0\rangle &= |+1;-1\rangle - |-1;+1\rangle\,,
\end{aligned}
$$

and, finally, the $j = 0$ state is given by

$$
\sqrt{3}\,|j=0, m=0\rangle = |+1;-1\rangle - |0;0\rangle + |-1;+1\rangle\,.
$$

13.1.5 One-Electron Atoms, Spectroscopic Notations

In Chap. 11, we neglected spin effects in the hydrogen atom. If one takes spin into account, the classification of atomic states requires four quantum numbers: $|n, \ell, m, \sigma\rangle$, $\sigma = \pm$. The states $\sigma = \pm$ are degenerate in energy in the Coulomb approximation. The *spin–orbit* interaction, which we shall discuss later on, lifts this degeneracy, and we shall see that the energy eigenstates are the eigenstates $|n, \ell, j, m_j\rangle$ of the *total angular momentum* $\boldsymbol{J} = \boldsymbol{L} + \boldsymbol{S}$. Their energies do not depend on the quantum number m_j giving the projection on z of $\boldsymbol{J}$, since the Hamiltonian is rotation invariant. In the case of an electron of orbital angular momentum ℓ and spin 1/2, the values of j are therefore

$$
j = \ell \pm 1/2\,,
$$

except for $\ell = 0$ in which case $j = 1/2$.

One can classify the states according to the above quantum numbers. In the spectroscopic notation, one adds, on the right of the symbol $(n\ell)$ described in Chap. 11, the value of j. For instance,

$$
\begin{aligned}
2\mathrm{p}_{3/2} \quad &\Leftrightarrow \quad n=2\,,\ \ell=1\,,\ j=\frac{3}{2}=\ell+\frac{1}{2}\,; \\
3\mathrm{d}_{3/2} \quad &\Leftrightarrow \quad n=3\,,\ \ell=2\,,\ j=\frac{3}{2}=\ell-\frac{1}{2}\,.
\end{aligned}
$$

13.2 Fine Structure of Monovalent Atoms

The resonance lines of monovalent atoms are split into two components. One example is the yellow line of sodium, corresponding to the transition 3p $\to$ 3s. This line is split into two lines, called D_1 and D_2, of wavelengths $\lambda_1 \simeq$

589.6 nm and $\lambda_2 \simeq 589.0$ nm, respectively. The same effect is observed in the hydrogen atom: the Lyman α line, corresponding to the transition 2p $\to$ 1s, is also split into two components.

This splitting is due to the spin–orbit interaction: the first excited level, which has an orbital angular momentum $\ell = 1$ (p state), is split into two sublevels because of this interaction. One level corresponds to $j = 3/2$, and the other to $j = 1/2$. The splitting is weak compared with the main effect, i. e. the energy difference between the initial levels (1s and 2p for hydrogen). The $2\mathrm{p}_{3/2}$–$2\mathrm{p}_{1/2}$ energy difference for hydrogen is roughly 4.5×10^{-5} eV, corresponding to a frequency of 10 GHz; the $3\mathrm{p}_{3/2}$–$3\mathrm{p}_{1/2}$ splitting in sodium is $\sim 2 \times 10^{-3}$ eV (500 GHz).

The physical origin of the spin–orbit coupling can be understood from a classical argument. Suppose we model the hydrogen atom as an electron orbiting with a velocity $\boldsymbol{v}$ around a proton. The proton is much heavier than the electron and is assumed to be at rest in the laboratory frame. It creates an electrostatic field acting on the electron,

$$\boldsymbol{E} = \frac{q}{4\pi\varepsilon_0 r^3}\,\boldsymbol{r}\,. \tag{13.20}$$

In the rest frame of the electron, the proton moves at velocity $-\boldsymbol{v}$ and this gives rise, in addition to the electric field (13.20), to a magnetic field

$$\boldsymbol{B} = -\boldsymbol{v} \times \boldsymbol{E}/c^2 = \frac{q}{4\pi\varepsilon_0 m_\mathrm{e} c^2 r^3}\,\boldsymbol{L}\,. \tag{13.21}$$

Here $\boldsymbol{L} = m_\mathrm{e}\boldsymbol{r} \times \boldsymbol{v}$ stands for the angular momentum of the electron in the laboratory frame. In order to derive (13.21), we have assumed that $|\boldsymbol{v}| \ll c$ and have considered only the dominant terms in v/c. The spin magnetic moment of the electron $\hat{\boldsymbol{\mu}}_\mathrm{s} = -(q/m_\mathrm{e})\hat{\boldsymbol{S}}$ interacts with this magnetic field, and this gives rise to a magnetic energy, which can be written in the laboratory frame as

$$W_{\mathrm{s.o.}} = \frac{1}{2}\,\frac{e^2}{m_\mathrm{e}^2 c^2}\,\frac{1}{r^3}\,\boldsymbol{L}\cdot\boldsymbol{S}\,. \tag{13.22}$$

The associated quantum Hamiltonian is obtained by replacing r, $\boldsymbol{L}$ and $\boldsymbol{S}$ with the corresponding operators. We can rewrite this expression using the natural atomic units, i. e. the Bohr radius a_1 and the ionization energy E_I, together with the fine structure constant α:

$$\hat{W}_{\mathrm{s.o.}} = \alpha^2 E_\mathrm{I}\left(\frac{a_1}{\hat{r}}\right)^3 \frac{\hat{\boldsymbol{L}}\cdot\hat{\boldsymbol{S}}}{\hbar^2}\,. \tag{13.23}$$

Remarks

a. The spin–orbit coupling is a *relativistic* effect. We notice in (13.23) that, since a_1/r and $\hat{\boldsymbol{L}}\cdot\hat{\boldsymbol{S}}/\hbar^2$ are of the order of 1, the spin–orbit coupling with of order $\alpha^2 \simeq (1/137)^2$ compared with the main effect. This is indeed of order v^2/c^2, since $v/c \sim \alpha$.

b. For the s states ($\ell = 0$), the term (13.23) vanishes. However, one can show that there exists a relativistic shift of the levels, called the Darwin term, whose value is

$$W_{\mathrm{D}} = \frac{\pi^2 e^2 \hbar^2}{2m_{\mathrm{e}}^2 c^2}\, |\psi(0)|^2 \,.$$

This term vanishes for $\ell \neq 0$, since $\psi(0) = 0$ in that case. This term only affects s waves.

All the above terms can be obtained directly and exactly in the framework of the relativistic Dirac equation. When one solves this equation for the Coulomb potential, one finds that the states $2\mathrm{s}_{1/2}$ and $2\mathrm{p}_{1/2}$ of hydrogen are degenerate, and that the state $2\mathrm{p}_{3/2}$ lies 10 GHz above these two states. Experimentally, a splitting between the two states $2\mathrm{s}_{1/2}$ and $2\mathrm{p}_{1/2}$, called the *Lamb shift* after its discoverer, is observed. This splitting, of the order of 1 GHz, is due to the coupling of the electron to the quantized electromagnetic field. The calculation of the Lamb shift was the first spectacular success of quantum electrodynamics.

The spin–orbit splitting is small, and we can calculate it using perturbation theory (Chap. 9). For instance, for the level $n = 2$ of hydrogen, we have to diagonalize the restriction of $W_{\mathrm{s.o.}}$ (13.23) to the subspace generated by the six states $|n = 2, \ell = 1, m, \sigma\rangle$. This coupling $\hat{W}_{\mathrm{s.o.}}$ involves the scalar product $\hat{\boldsymbol{L}} \cdot \hat{\boldsymbol{S}}$, which is diagonal in the basis $|n, \ell, j, m_j\rangle$ of the eigenstates of the *total angular momentum*. We have the equality

$$\hat{\boldsymbol{L}} \cdot \hat{\boldsymbol{S}} = \frac{1}{2}\left((\hat{\boldsymbol{L}} + \hat{\boldsymbol{S}})^2 - \hat{L}^2 - \hat{S}^2\right) = \frac{1}{2}\left(\hat{J}^2 - \hat{L}^2 - \hat{S}^2\right),$$

with eigenvalues $[j(j+1) - \ell(\ell+1) - 3/4]\hbar^2/2$. Using (13.23), one therefore obtains the splitting between the $j = \ell + 1/2$ and $j = \ell - 1/2$ states (for instance $2\mathrm{p}_{3/2}$ and $2\mathrm{p}_{1/2}$) as follows:

$$\Delta E(n, \ell) \equiv E(j = \ell + 1/2) - E(j = \ell - 1/2) = (\ell + 1/2) A_{n,\ell}\,,$$

where

$$A_{n,\ell} = \alpha^2 E_{\mathrm{I}} \int |\psi_{n,\ell,m}(\boldsymbol{r})|^2 \left(\frac{a_1}{r}\right)^3 \, \mathrm{d}^3 r \,.$$

One can check easily that this quantity is independent of m and that its numerical value coincides with the experimental result for the $2\mathrm{p}_{1/2}$–$2\mathrm{p}_{3/2}$ splitting of hydrogen. For other atoms, one may observe more complicated effects. For instance, in sodium, there is an *inversion* of the spin–orbit effect: $E(3\mathrm{d}_{3/2}) > E(3\mathrm{d}_{5/2})$. This comes from an effect of the core of internal electrons.

The origin of this *doubling* of the p levels, like the existence of an *even* number of levels in the "anomalous" Zeeman effect and of even numbers is the Mendeleev classification (an *even* number of electrons in a closed shell), is due to the electron spin, entangled with the Pauli principle in the latter case. However, in the early days of quantum mechanics, it was nearly impossible to guess that all these manifestations of the number 2 had a common origin. The discovery of spin and of the Pauli principle was truly one of the most difficult steps of quantum mechanics. It was so difficult that it took some time for the physics community to really appreciate the significance and importance of these notions. Pauli was awarded the Nobel prize as late as 1945, whereas his contemporaries Heisenberg, Dirac and Schrödinger had received the prize in the early 1930s. Note that a direct calculation of the spin–orbit coupling, with the usual formulations of special relativity, gives a value *twice* as large as (13.22), and therefore a fine-structure splitting twice too large. This is why Pauli, at the end of 1925, did not believe in the idea of spin, and called it a "heresy" in a letter to Niels Bohr. However, in March 1926, L.H. Thomas remarked that the rest frame of the electron is not an inertial frame, and that a correct calculation introduces a factor of 1/2 in the formula (the Thomas precession[2]). This convinced Pauli of the validity of the spin-1/2 concept.

Finally, one can understand the origin of the name *fine structure constant* for α, which governs the order of magnitude of fine structure effects. The name was introduced in 1920 by Sommerfeld, who had calculated the fine structure of hydrogen in the framework of the old quantum theory, by considering the relativistic effects due to the eccentricity of the orbits. Sommerfeld's calculation gave the correct result, but this was simply another awkward coincidence due to the particular symmetries of the hydrogen problem and to the ensuing degeneracies with respect to ℓ.

13.3 Hyperfine Structure; the 21 cm Line of Hydrogen

An even smaller effect (a splitting of the order of 6×10^{-6} eV) has very important practical applications. This effect comes from the magnetic interaction between the spin magnetic moments of the electron and the proton:

$$\hat{\boldsymbol{\mu}}_{\mathrm{e}} = \gamma_{\mathrm{e}} \hat{\boldsymbol{S}}_{\mathrm{e}}\,, \qquad \gamma_{\mathrm{e}} = -q/m_{\mathrm{e}}\,; \tag{13.24}$$

$$\hat{\boldsymbol{\mu}}_{\mathrm{p}} = \gamma_{\mathrm{p}} \hat{\boldsymbol{S}}_{\mathrm{p}}\,, \qquad \gamma_{\mathrm{p}} \simeq 2.79\, q/m_{\mathrm{p}}\,. \tag{13.25}$$

This interaction is called the spin–spin, or hyperfine, interaction. We shall consider only its effect on the ground state of hydrogen, $n=1, \ell=0$.

13.3.1 Interaction Energy

We neglect here effects due to the internal structure of the proton and treat it as a point-like particle. The calculation of the magnetic field created at a point $\boldsymbol{r}$ by a magnetic dipole $\boldsymbol{\mu}_{\mathrm{p}}$ located at the origin is a well-known problem in magnetostatics.[3] The result can be written

$$\boldsymbol{B}(\boldsymbol{r}) = -\frac{\mu_0}{4\pi r^3}\left(\boldsymbol{\mu}_{\mathrm{p}} - \frac{3(\boldsymbol{\mu}_{\mathrm{p}}\cdot\boldsymbol{r})\,\boldsymbol{r}}{r^2}\right) + \frac{2\mu_0}{3}\,\boldsymbol{\mu}_{\mathrm{p}}\,\delta(\boldsymbol{r})\,. \tag{13.26}$$

[2] See, e. g., J.D. Jackson, *Classical Electrodynamics*, Sect. 11.8, Wiley, New York (1975).

[3] See, e. g., J.D. Jackson, *Classical Electrodynamics*, Sect. 5.6, Wiley, New York (1975).

The interaction Hamiltonian between the magnetic moment $\boldsymbol{\mu}_{\mathrm{e}}$ of the electron and this magnetic field reads

$$\hat{W} = -\hat{\boldsymbol{\mu}}_{\mathrm{e}} \cdot \hat{\boldsymbol{B}} .$$

For $\boldsymbol{r} \neq 0$, $\hat{W}$ reduces to the usual dipole–dipole interaction:

$$r \neq 0 , \qquad \hat{W}_{\mathrm{dip}} = \frac{\mu_0}{4\pi \hat{r}^3} \left(\hat{\boldsymbol{\mu}}_{\mathrm{e}} \cdot \hat{\boldsymbol{\mu}}_{\mathrm{p}} - \frac{3(\hat{\boldsymbol{\mu}}_{\mathrm{e}} \cdot \hat{\boldsymbol{r}})(\hat{\boldsymbol{\mu}}_{\mathrm{p}} \cdot \hat{\boldsymbol{r}})}{\hat{r}^2} \right) .$$

This interaction will not contribute to our calculation, because of the following mathematical property. For any function $g(r), r = |\boldsymbol{r}|$, that is regular at $r = 0$, an angular integration yields

$$\int g(r)\, W_{\mathrm{dip}}(\boldsymbol{r})\, \mathrm{d}^3 r = 0 . \tag{13.27}$$

At $\boldsymbol{r} = 0$, the field (13.26) is singular because of the contribution of the term proportional to $\delta(\boldsymbol{r})$. This leads to a *contact interaction*

$$\hat{W}_{\mathrm{cont}} = -\frac{2\mu_0}{3} \hat{\boldsymbol{\mu}}_{\mathrm{e}} \cdot \hat{\boldsymbol{\mu}}_{\mathrm{p}}\, \delta(\hat{\boldsymbol{r}}) .$$

The origin of the singularity at $\boldsymbol{r} = 0$ is the point-like nature of the proton that we have assumed in our analysis. This implies that all field lines converge to the same point. A calculation taking into account the finite size of the proton and the corresponding modification to the field leads to essentially the same result, because the size of the proton is very small compared with the size of the probability distribution of the electron in the 1s state. Note that this point-like model is strictly valid for positronium, which is an atom consisting of an electron and a positron, both being point-like objects.

Here we have taken into account the magnetic interaction between the proton and electron spins. There is also a contribution from the magnetic interaction between the proton spin and the magnetic moment associated with the current loop formed by the electron, which is proportional to its orbital angular momentum $\boldsymbol{L}$. In the following we are interested in the properties of the ground state, for which $\ell = 0$, and this additional term does not contribute.

13.3.2 Perturbation Theory

The observable $\hat{W}$ acts on space and spin variables. We consider the orbital ground state of the hydrogen atom, which, owing to the spin variables, is a four-state system. An arbitrary state of this four-dimensional subspace can be written as

$$|\psi\rangle = \psi_{100}(\boldsymbol{r})\, |\Sigma\rangle , \tag{13.28}$$

where $\psi_{100}(\boldsymbol{r})$ is the ground-state wave function found in Chap. 11: $\psi_{100}(\boldsymbol{r}) = \mathrm{e}^{-r/a_1}/\sqrt{\pi a_1^3}$. First-order perturbation theory requires that we diagonalize the restriction of $\hat{W}$ to this subspace.

We shall proceed in two steps. First we treat the space variables, which results in an operator acting only on spin variables, and then we diagonalize this latter operator. Consider

$$\hat{H}_1 = \int \psi_{100}^*(\boldsymbol{r})\, \hat{W}\, \psi_{100}(\boldsymbol{r})\, \mathrm{d}^3r\,.$$

The probability density for the ground-state level $|\psi_{100}(\boldsymbol{r})|^2$ is isotropic. As a consequence of (13.27), $\hat{W}_{\mathrm{dip}}$ does not contribute to $\hat{H}_1$. The contact term is readily evaluated as

$$\hat{H}_1 = -\frac{2\mu_0}{3}\hat{\boldsymbol{\mu}}_{\mathrm{e}}\cdot\hat{\boldsymbol{\mu}}_{\mathrm{p}}\,|\psi_{100}(0)|^2\,. \tag{13.29}$$

$\hat{H}_1$ is an operator which acts only on spin states. It can be cast into the form

$$\hat{H}_1 = \frac{A}{\hbar^2}\hat{\boldsymbol{S}}_{\mathrm{e}}\cdot\hat{\boldsymbol{S}}_{\mathrm{p}}\,, \tag{13.30}$$

where the constant A can be inferred from the values of $\gamma_{\mathrm{e}}, \gamma_{\mathrm{p}}$ and $\psi_{100}(0)$:

$$A = -\frac{2}{3}\frac{\mu_0}{4\pi}\frac{4}{a_1^3}\gamma_{\mathrm{e}}\gamma_{\mathrm{p}}\hbar^2 = \frac{16}{3}\times 2.79\,\frac{m_{\mathrm{e}}}{m_{\mathrm{p}}}\alpha^2 E_{\mathrm{I}}\,.$$

One obtains

$$A \simeq 5.87\times10^{-6}\,\mathrm{eV}\,,\qquad \nu = \frac{A}{h} \simeq 1417\,\mathrm{MHz}\,,\qquad \lambda = \frac{c}{\nu} \simeq 21\,\mathrm{cm}\,. \tag{13.31}$$

13.3.3 Diagonalization of $\hat{H}_1$

The diagonalization of $\hat{H}_1$ in the Hilbert space of spin states is simple. Considering the total spin $\hat{\boldsymbol{S}} = \hat{\boldsymbol{S}}_{\mathrm{e}} + \hat{\boldsymbol{S}}_{\mathrm{p}}$, one has

$$\hat{\boldsymbol{S}}_{\mathrm{e}}\cdot\hat{\boldsymbol{S}}_{\mathrm{p}} = \frac{1}{2}\left(\hat{S}^2 - \hat{S}_{\mathrm{e}}^2 - \hat{S}_{\mathrm{p}}^2\right)\,,$$

which is diagonal in the basis of the eigenstates $|S, M\rangle$ of the total spin, with the following eigenvalues:

$$\frac{\hbar^2}{2}\,[S(S+1) - 3/2]\,,\qquad \text{where}\quad S = 0 \quad\text{or}\quad S = 1\,.$$

The ground state $E_0 = -E_{\mathrm{I}}$ of the hydrogen atom is therefore split by the hyperfine interaction into two sublevels corresponding to the triplet $|1, M\rangle$ and singlet $|0, 0\rangle$ states:

$$
\begin{aligned}
E_+ &= E_0 + A/4\,, \qquad \text{triplet state} \quad |1, M\rangle\,; \\
E_- &= E_0 - 3A/4\,, \qquad \text{singlet state} \quad |0, 0\rangle\,.
\end{aligned}
\tag{13.32}
$$

The difference between these two energies is equal to A, i.e. 5.87×10^{-6} eV; it corresponds to the characteristic line of hydrogen at a wavelength $\lambda \sim 21$ cm.

Remarks

(1) In its ground state, the hydrogen atom constitutes a four-level system with two energy levels. By a method whose principle is similar to that discussed in Chap. 6, it is possible (but technically more complicated) to devise a hydrogen maser.[4] Among other things, this allows one to measure the constant A, or, equivalently, the frequency $\nu = A/h$ with an impressive accuracy:

$$
\nu = \underbrace{14}_{\text{A}}\,\underbrace{20}_{\text{B}}\,\underbrace{40}_{\text{C}}\,\underbrace{5751}_{\text{D}}\,.\,\underbrace{7684}_{\text{E}} \pm 0.0017\ \text{Hz}\,.
$$

In this result, we have labeled several groups of digits. The first two digits (A) were obtained by Fermi in 1930; they correspond to the contact term considered above. The following two digits (B) can be calculated using the Dirac equation and the experimental value for the anomalous magnetic moment of the electron (a deviation of the order of 10^{-3}). Other corrections can account for the next two digits (C): relativistic vacuum polarization corrections, the finite size of the nucleus, polarization of the nucleus, etc. The set (D, E) is out of range for theorists at present.

Such an accuracy has, in particular, provided a means to test the predictions of general relativity.[5] A hydrogen maser was sent in a rocket to an altitude of 10 000 km, and the variation of its frequency as the gravitational field and the velocity varied was measured. Despite numerous difficulties, it was possible to check the predictions of relativity with an accuracy of 7×10^{-5}; the result is still one of the most accurate verifications of the theory (more precisely, of the equivalence principle).

(2) The hyperfine splitting of alkali atoms has the same origin as that of hydrogen, although it is more difficult to calculate theoretically. The frequencies listed in Table 13.1 are observed. These splittings have been used in masers and atomic clocks. One of the many applications is the definition of the time standard on the basis of the hyperfine effect in the cesium-133 isotope in its ground state ($\Delta E \sim 3.8 \times 10^{-5}$ eV). One second is defined as being equal to 9 192 631 770 periods of the corresponding line. The relative accuracy of the practical realization of this definition is 10^{-15}. Such impressive precision has been made possible by the use of laser-cooled atoms, whose residual temperature is of the order of only 1 μK.

[4] H.M. Goldenberg, D. Kleppner and N.F. Ramsey, Phys. Rev. Lett. **8**, 361 (1960).
[5] R. Vessot et al., Phys. Rev. Lett. **45**, 2081 (1980).

Table 13.1. Frequencies associated with hyperfine splittings of alkali atoms

Atom	Frequency (GHz)	State
^{7}Li	0.83	2s
^{23}Na	1.77	3s
^{39}K	0.46	4s
^{85}Rb	3.04	5s
^{87}Rb	6.83	5s
^{133}Cs	9.19	6s

13.3.4 The Effect of an External Magnetic Field

If we place the hydrogen atom in an external magnetic field, the magnetic Hamiltonian becomes

$$\hat{H}_{\mathrm{M}} = \frac{A}{\hbar^2}\hat{\boldsymbol{S}}_{\mathrm{e}} \cdot \hat{\boldsymbol{S}}_{\mathrm{p}} - \hat{\boldsymbol{\mu}}_{\mathrm{e}} \cdot \boldsymbol{B}_0 - \hat{\boldsymbol{\mu}}_{\mathrm{p}} \cdot \boldsymbol{B}_0 \,. \tag{13.33}$$

Here we do not take into account space variables, and we assume that the Zeeman splitting is small enough that first-order perturbation theory is valid.

The nuclear magneton μ_{N} is much smaller than the Bohr magneton μ_{B}. Therefore, we can neglect the last term on the right-hand side of (13.33). In this approximation, the diagonalization of H_{M} is simple. We set $\eta = q\hbar B_0/(2m_{\mathrm{e}})$ and $\tan 2\theta = 2\eta/A$, and we obtain the following splitting:

$$\begin{aligned}
(A/4) + \eta &\to |1,\ 1\rangle \,, \\
(A/4) - \eta &\to |1, -1\rangle \,, \\
-(A/4) + \sqrt{A^2/4 + \eta^2} &\to \cos\theta\,|1,0\rangle + \sin\theta\,|0,0\rangle \,, \\
-(A/4) - \sqrt{A^2/4 + \eta^2} &\to -\sin\theta\,|1,0\rangle + \cos\theta\,|0,0\rangle \,.
\end{aligned}$$

The levels are represented in Fig. 13.2.

One observes, as for NH_3, a competition between the hyperfine coupling and the presence of the field. For weak fields, the states $|1,\ 0\rangle$ and $|0,\ 0\rangle$ are unaffected, whereas the energies of the states $|1,\ 1\rangle$ and $|1, -1\rangle$ vary linearly with B. There is a splitting of the 21 cm line into three components. For strong fields, the eigenstates are the factorized states $|\sigma_{\mathrm{e}}\,;\,\sigma_{\mathrm{p}}\rangle$. The transition region ($\eta \sim A$) is around $B \sim 0.1\,\mathrm{T}$.

13.3.5 The 21 cm Line in Astrophysics

In galaxies, matter exists in two main forms. The first, which is directly visible, is condensed matter: stars at various stages of their evolution and planets, which are now being discovered in solar systems other than ours. However, there also exists a diffuse interstellar medium, composed mainly of

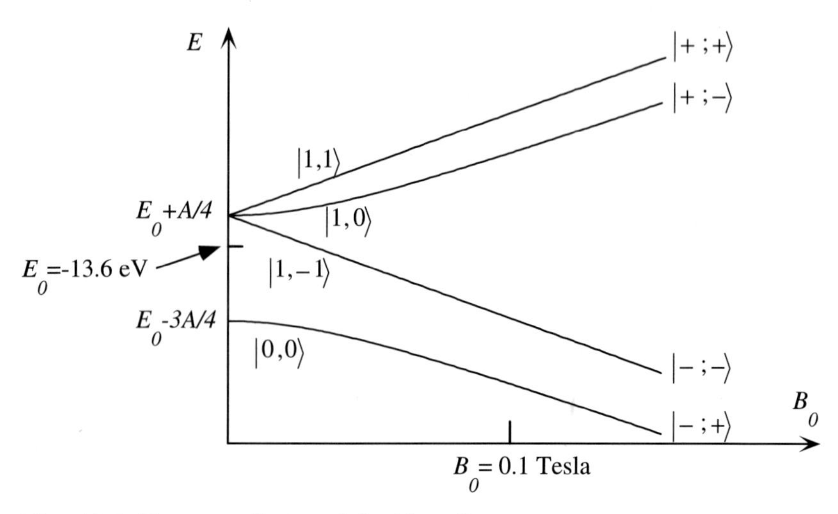

Fig. 13.2. Zeeman splitting of the 21 cm line

atomic hydrogen, whose total mass is quite important (from 10 to 50% of the total visible galactic mass).

The temperature of these interstellar clouds is typically 100 K. Since the corresponding thermal energy $kT \sim 10^{-2}\,\text{eV}$ is much smaller than E_{I}, hydrogen atoms cannot be appreciably excited by thermal collisions from the 1s ground state to the other states of the Lyman series. However, transitions between the two hyperfine states $S = 1$ and $S = 0$ can occur easily. The emission of 21 cm radiation corresponds to a spontaneous transition from the $S = 1$ state to the $S = 0$ state. This emission is very weak, because the lifetime of the triplet $S = 1$ state is extremely long: $\tau \sim 3.5 \times 10^{14}\,\text{s} \sim 10^7$ years.[6] Nevertheless, the amounts of atomic hydrogen in the interstellar medium are so large that an appreciable signal is emitted.

The observation of this line of hydrogen has deeply modified our understanding of the interstellar medium. Measurements of the intensity of the line give the mass distribution of the amount of hydrogen. The Doppler shift allows us to measure the velocities of the hydrogen clouds. The splitting of the line and its polarization provide a measurement of the magnetic field inside the interstellar medium. By analyzing the structure of our galaxy, the Milky Way (which is difficult to observe because we are in its plane), it has been possible to show that it is a spiral galaxy, of radius 50 000 light-years, and that we are 30 000 light-years from the center (see Fig. 13.3). One can also measure the density of the interstellar medium (0.3 atoms cm^{-3} on average), its temperature (20 K to 100 K), its structure (roughly one interstellar cloud

[6] This very long lifetime is due to a combination of two facts: the energy difference is very small, and the emission proceeds through a magnetic dipole transition (while atomic resonance lines correspond to electric dipole transitions).

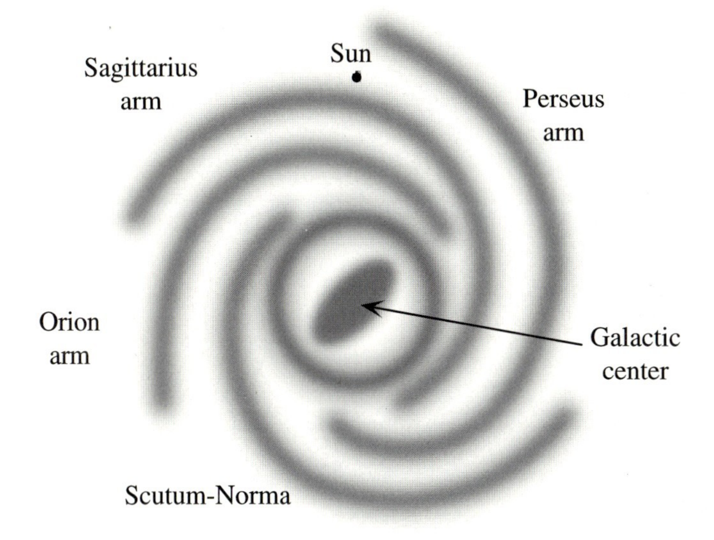

Fig. 13.3. Spiral structure of the Milky Way, reconstructed from radio-astronomical observations at a wavelength of 21 cm (By courtesy of Mr. Frederic Zantonio)

every 1000 light-years along a line of sight) and its extension outside the plane of the galaxy (roughly 1000 light-years).

Finally, this 21 cm line, which comes from the ground state of the simplest and most abundant element in the universe, provides a coding procedure for communicating with possible extraterrestrial civilizations, who, obviously, see the same universe, with the same dominant element, whose ground state has the same hyperfine structure. On the few space probes which have now left the solar system, NASA scientists have put a plate which carries a message from our civilization and which uses this key.

The plate on the Pioneer 10 space probe, represented in Fig. 13.4, is an example. At the top, two atoms with parallel and antiparallel spins are represented, meaning that the hyperfine transition of wavelength 21 cm takes place between these two states. The transition is symbolized by a line between the two atoms, which is used as a length unit (21 cm) and a time unit (the inverse of the frequency). The length unit may be checked by comparing the sketch of the Pioneer probe, which would be recovered with the plate. Two human beings are represented on the same scale.

The spider-like diagram on the left gives the directions and frequencies of the main known pulsars that can be observed from the earth at present. A given configuration of this kind happens only at one time and at one location in our galaxy. Knowing the history of pulsars, it is therefore possible to find the few stars which were roughly in the right place at the right time. The extraterrestrial beings (whom, by definition, we know nothing about, except that if they exist they are intelligent) will therefore be able to localize both

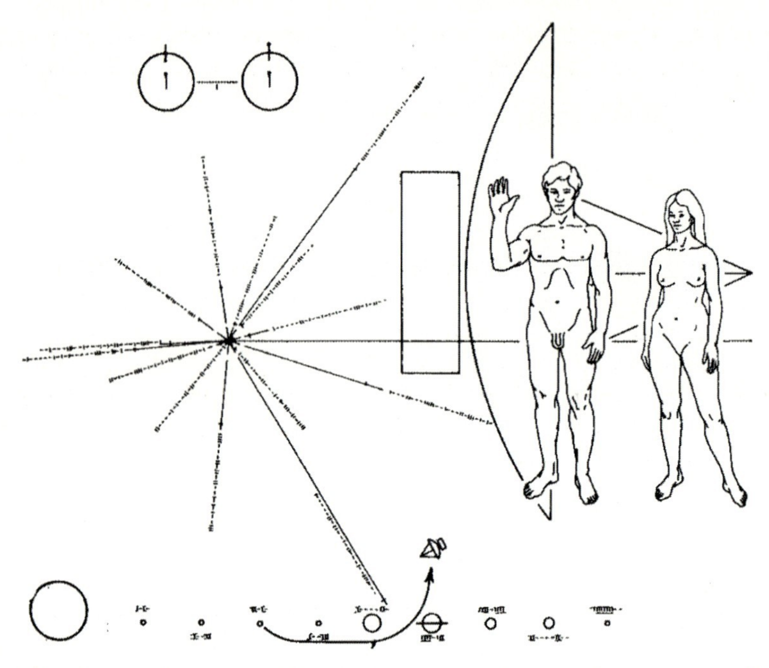

Fig. 13.4. Plate on the Pioneer 10 space probe, aimed at some possible extraterrestrial civilization. The message on this plate uses the 21 cm line of hydrogen as a "Rosetta stone" in order to tell where and when the probe was launched

in *space* and *time* the people who sent the message (perhaps thousands of years before).

Further Reading

- A.R. Edmonds, *Angular Momentum in Quantum Mechanics*, Princeton University Press, Princeton (1950).
- For the fine structure of the hydrogen atom, one can read T.W. Hänsch, A.L. Schawlow and G.W. Series, "The spectrum of the hydrogen atom", Sci. Am., March 1979.
- For developments in atomic clocks, see, for example, W. Itano and N. Ramsey, "Accurate measurement of time", Sci. Am., July 1993, p. 46.

Exercises

13.1. Permutation operator. Show that the permutation operator defined in (13.14) can be written as $\hat{P}^{\mathrm{s}}_{12} = (1 + \hat{\boldsymbol{\sigma}}_1 \cdot \hat{\boldsymbol{\sigma}}_2)/2$.

13.2. The singlet state. Consider two spin-1/2 systems, and the eigenbasis $\{|\pm\rangle_{\boldsymbol{u}} \otimes |\pm\rangle_{\boldsymbol{u}}\}$ of the two operators $\boldsymbol{S}_1 \cdot \boldsymbol{u}$ and $\boldsymbol{S}_2 \cdot \boldsymbol{u}$, $\boldsymbol{u}$ being any unit vector of R^3. Show that the singlet state can be written in that basis as

$$\frac{1}{\sqrt{2}} \left(|+\rangle_{\boldsymbol{u}} \otimes |-\rangle_{\boldsymbol{u}} - |-\rangle_{\boldsymbol{u}} \otimes |+\rangle_{\boldsymbol{u}} \right) .$$

13.3. Spin and magnetic moment of the deuteron. We denote by $\hat{\boldsymbol{J}}$ the total-angular-momentum observable of the electron cloud of an atom, and denote by $\hat{\boldsymbol{I}}$ the angular momentum of the nucleus. The respective magnetic moment observables are $\hat{\boldsymbol{\mu}}_J = g_J \mu_B \hat{\boldsymbol{J}}/\hbar$ and $\hat{\boldsymbol{\mu}}_I = g_I \mu_N \hat{\boldsymbol{I}}/\hbar$, where g_J and g_I are dimensionless factors. The magnetic-interaction Hamiltonian of the electron cloud with the nucleus is of the form $\hat{W} = a \hat{\boldsymbol{\mu}}_J \cdot \hat{\boldsymbol{\mu}}_I$, where a is a constant which depends on the electron distribution around the nucleus.

a. Suppose that the state of the nucleus (energy E_I, square of the angular momentum $I(I+1)\hbar^2$) and the state of the electron cloud (energy E_J, square of the angular momentum $J(J+1)\hbar^2$) are both fixed. What are the possible values $K(K+1)\hbar^2$ of the total angular momentum $\hat{\boldsymbol{K}}$ of the atom?
b. Express $\hat{W}$ in terms of $\hat{\boldsymbol{I}}^2$, $\hat{\boldsymbol{J}}^2$ and $\hat{\boldsymbol{K}}^2$. Express the hyperfine energy levels of the atom in terms of I, J and K.
c. Calculate the splitting between two consecutive hyperfine levels.
d. When one applies a uniform weak magnetic field B to a deuterium atom, it is observed that the two hyperfine levels (E_K and $E_{K'}$) of the ground state are split as a function of B, as shown in Fig. 13.5. Given that the electron is in its orbital ground state $\ell = 0$, what is the value of the deuteron spin?
e. Assuming that the proton and neutron inside the deuteron have zero orbital angular momentum, what is their spin state?
f. It can be shown that $a = -8\mu_0/12\pi a_1^3$, where a_1 is the Bohr radius and $\varepsilon_0 \mu_0 c^2 = 1$. Given that $g_J = 2$ and $g_I = 0.86$, to what frequency must a radio telescope be tuned in order to detect deuterium in the interstellar medium?

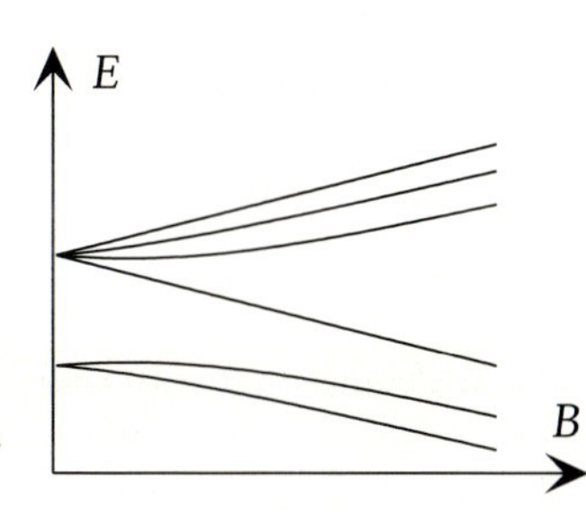

Fig. 13.5. Zeeman splitting of the ground state of deuterium

13.4. Determination of the Clebsch–Gordan coefficients.

a. Show that a Clebsch–Gordan coefficient $C^{j,m}_{j_1,m_1;j_2,m_2}$ is nonzero only if $m_1 + m_2 = m$.

b. Using $\hat{J}_+ = \hat{J}_{1+} + \hat{J}_{2+}$, prove the following recursion relation:

$$\begin{aligned} &\sqrt{j_1(j_1+1) - m_1(m_1-1)}\, C^{j,m}_{j_1,m_1-1;j_2,m_2} \\ &+ \sqrt{j_2(j_2+1) - m_2(m_2-1)}\, C^{j,m}_{j_1,m_1;j_2,m_2-1} \\ &= \sqrt{j(j+1) - m(m+1)}\, C^{j,m+1}_{j_1,m_1;j_2,m_2} \,. \end{aligned} \tag{13.34}$$

Deduce from this a relation between $C^{j,j}_{j_1,m_1-1;j_2,m_2}$ and $C^{j,j}_{j_1,m_1;j_2,m_2-1}$.

c. We impose the requirement that $C^{j,j}_{j_1,j_1;j_2,j-j_1}$ is real and positive. Show that the coefficients $C^{j,j}_{j_1,m_1;j_2,m_2}$ are defined unambiguously.

d. Prove the following recursion relation:

$$\begin{aligned} &\sqrt{j_1(j_1+1) - m_1(m_1+1)}\, C^{j,m}_{j_1,m_1+1;j_2,m_2} \\ &+ \sqrt{j_2(j_2+1) - m_2(m_2+1)}\, C^{j,m}_{j_1,m_1;j_2,m_2+1} \\ &= \sqrt{j(j+1) - m(m-1)}\, C^{j,m-1}_{j_1,m_1;j_2,m_2} \,. \end{aligned} \tag{13.35}$$

Deduce a method to calculate any given Clebsch–Gordan coefficient.

13.5. Scalar operators. An operator $\hat{\mathcal{O}}$ is said to be *scalar* if it commutes with the three components of the angular momentum $\hat{J}_i$, $i = x, y, z$. Consider the $(2j+1) \times (2j'+1)$ matrix elements

$$\langle \alpha, j, m | \hat{\mathcal{O}} | \beta, j', m' \rangle \,, \qquad m = -j, \ldots, j \,, \qquad m' = -j', \ldots, j' \,,$$

where α and β represent the set of quantum numbers which are necessary, in addition to the angular momentum quantum numbers j, m and j', m', in order to specify the state of the system. We want to show that all these matrix elements vanish if $j \neq j'$ or $m \neq m'$, and that the remaining matrix elements, i. e. those for which $j = j'$ and $m = m'$, are equal to each other.

a. Using $[\hat{J}_z, \hat{\mathcal{O}}] = 0$, show that $\langle \alpha, j, m | \hat{\mathcal{O}} | \beta, j', m' \rangle$ can be nonzero only if $m = m'$. We set $\mathcal{O}_m = \langle \alpha, j, m | \hat{\mathcal{O}} | \beta, j', m \rangle$.

b. Using $[\hat{J}_+, \hat{\mathcal{O}}] = 0$, show that

$$\sqrt{j(j+1) - m(m+1)}\, \mathcal{O}_m = \sqrt{j'(j'+1) - m(m+1)}\, \mathcal{O}_{m+1} \,.$$

Show also that

$$\sqrt{j(j+1) - m(m+1)}\, \mathcal{O}_{m+1} = \sqrt{j'(j'+1) - m(m+1)}\, \mathcal{O}_m \,.$$

c. Deduce from these results that $\mathcal{O}_m$ vanishes if $j \neq j'$, and that all these terms are equal if $j = j'$.

Examples: $\hat{O} = \hat{r}^2, \hat{p}^2, \hat{L}^2, \ldots$, where $\hat{\boldsymbol{r}}$, $\hat{\boldsymbol{p}}$, $\hat{\boldsymbol{L}}$ are the position, momentum and orbital-angular-momentum operators of a point-like particle.

13.6. Vector operators and the Wigner–Eckart theorem. A triplet of operators $(\hat{V}_x, \hat{V}_y, \hat{V}_z)$ is called a *vector operator* if the following commutation relations with the total-angular-momentum operator $\hat{\boldsymbol{J}}$ are satisfied:

$$[\hat{J}_j, \hat{V}_k] = \mathrm{i}\hbar\epsilon_{j,k,\ell}\,\hat{V}_\ell\,, \tag{13.36}$$

where $\epsilon_{j,k,\ell} = 1$ or -1 if (j,k,ℓ) is an even or odd permutation, respectively, of (x,y,z), and 0 otherwise.

a. Show that the position $\hat{\boldsymbol{r}}$, momentum $\hat{\boldsymbol{p}}$ and orbital-angular-momentum $\hat{\boldsymbol{L}}$ operators of a point-like particle are vector operators.
b. We set

$$\hat{V}_{+1} = \frac{-1}{\sqrt{2}}(\hat{V}_x + \mathrm{i}\hat{V}_y)\,, \qquad \hat{V}_0 = \hat{V}_z\,, \qquad \hat{V}_{-1} = \frac{1}{\sqrt{2}}(\hat{V}_x - \mathrm{i}\hat{V}_y)\,.$$

Show that the commutation relations (13.36) can also be written

$$[\hat{J}_z, \hat{V}_q] = \hbar q \hat{V}_q\,, \tag{13.37}$$

$$\left[\hat{J}_\pm, \hat{V}_q\right] = \hbar\sqrt{2 - q(q\pm 1)}\;\hat{V}_{q\pm 1}\,. \tag{13.38}$$

c. As in the previous exercise, we want to characterize the $3 \times (2j+1) \times (2j'+1)$ matrix elements

$$\langle \alpha, j, m|\hat{V}_q|\beta, j', m'\rangle\,, \qquad m = -j, \ldots, j\,, \qquad q = -1, 0, 1\,,$$
$$m' = -j', \ldots, j'\,.$$

(i) Using (13.37), show that the matrix elements can only be nonzero if $m = m' + q$.
(ii) Using (13.38), show that these matrix elements satisfy recursion relations whose structure is identical to (13.34) and (13.35).
(iii) Deduce that for $|j - j'| \le 1$, the matrix elements under consideration are proportional to the Clebsch–Gordan coefficients, i. e.

$$\langle \alpha, j, m|\hat{V}_q|\beta, j', m'\rangle \;\propto\; C^{j,m}_{j',m';1,q}\,, \tag{13.39}$$

where the proportionality coefficient for a given operator $\hat{V}$ depends only on α, β, j and j'.
(iv) Show that these matrix elements vanish if $|j - j'| > 1$.

This theorem, which proves the proportionality of all vector operators for a pair of subspaces characterized by (α, j) and (β, j'), is called the Wigner–Eckart theorem. It explains, in particular, the hypothesis $\hat{\mu} = \gamma\hat{J}$ that we made in Sect. 10.4, relating the magnetic moment and the angular momentum of a quantum system. It also allows to find in a simple way the *selection rules*, which tell us whether a given coupling (electric dipole, magnetic dipole, etc.) can induce a transition between two levels. The multiplicative factor in (13.39) is usually written $\langle\alpha, j||V||\beta, j'\rangle \,/\, (2j{+}1)$, i. e.

$$\langle\alpha, j, m|\hat{V}_q|\beta, j', m'\rangle = \frac{\langle\alpha, j||V||\beta, j'\rangle}{2j+1}\; C^{j,m}_{j',m';1,q}\,.$$

The quantity $\langle\alpha, j||V||\beta, j'\rangle$ is called the reduced matrix element of $\hat{V}$ between the subspaces (α, j) and (β, j').

14. Entangled States, EPR Paradox and Bell's Inequality

Written in collaboration with Philippe Grangier[1]

The way of paradoxes is the way of truth.
To test Reality we must see it on the tight-rope.
When the Verities become acrobats, we can judge them.
Oscar Wilde, The Picture of Dorian Gray

As soon as a quantum system has more than one degree of freedom, the corresponding Hilbert space $\mathcal{E}$ has a tensor structure, $\mathcal{E} = \mathcal{E}_\mathrm{a} \otimes \mathcal{E}_\mathrm{b} \otimes \dots$, where each of the $\mathcal{E}_i$ is associated with one of the degrees of freedom. Consider for instance a spin-1/2 particle; the Hilbert space is the product of the space $\mathcal{E}_\mathrm{external} = \mathcal{L}^2(R^3)$, in which we describe the *orbital* motion of the particle, and of the space $\mathcal{E}_\mathrm{spin}$ of dimension 2, in which we describe its *spin* state. This tensor structure of the Hilbert space generates specific quantum mechanical properties of systems, such that the various degrees of freedom are correlated or *entangled*.

Einstein, Podolsky and Rosen were the first to point out, in a celebrated paper,[2] the subtle and paradoxical character of quantum entanglement. They used this notion to demonstrate the conflict between quantum mechanics and a realistic, local theory of the physical world. In the last 10 to 15 years, this quantum property has been used in very clever and original setups. For instance, arrangements have been devised to perform the coding and treatment of information, with the aim of realizing quantum cryptography. Similar ideas have led to the concept of a quantum computer.

Consider a physical system with two degrees of freedom A and B, whose states are described in a space $\mathcal{E} = \mathcal{E}_A \otimes \mathcal{E}_B$. Some state vectors have a very simple factorized form:

$$|\Psi\rangle = |\alpha\rangle \otimes |\beta\rangle \,. \tag{14.1}$$

If the system is prepared in such a state, each subsystem is in a well-defined state $|\alpha\rangle$ for A, and $|\beta\rangle$ for B. However, an arbitrary state of $\mathcal{E}$ cannot be

[1] Philippe Grangier (CNRS), Institut d'Optique, bâtiment 503, B.P. 147, 91403 Orsay Cedex, France, e-mail: philippe.grangier@iota.u-psud.fr

[2] A. Einstein, B. Podolsky and N. Rosen, Phys. Rev. **47**, 777 (1935).

factorized. It consists of a (possibly infinite) sum of factorized states. Such a state is called an *entangled state.* Consider, for instance, the state

$$|\Psi\rangle = \frac{1}{\sqrt{2}} \left(|\alpha_1\rangle \otimes |\beta_1\rangle + |\alpha_2\rangle \otimes |\beta_2\rangle \right) . \tag{14.2}$$

In this state there are strong correlations between the degrees of freedom of A and B. If we measure separately the states of each of these degrees of freedom, we can find with probability 1/2 that A is in the state $|\alpha_1\rangle$ *and* B is in the state $|\beta_1\rangle$, or, also with probability 1/2, that A is in the state $|\alpha_2\rangle$ *and* B is in the state $|\beta_2\rangle$. However, we can never find that A is in the state $|\alpha_1\rangle$ *and* B is in the state $|\beta_2\rangle$, or that A is in the state $|\alpha_2\rangle$ *and* B in the state $|\beta_1\rangle$.

The aim of this chapter is to study the consequences of such correlations. We shall consider the EPR argument, and we shall show how Bell was able to put the notion of a realistic theory, which Einstein had in mind, into a quantitative form. Bell proved that the predictions for observable correlations which can be calculated within this class of theories are constrained by a specific inequality, which can be violated by quantum mechanics. We shall describe experiments showing that this Bell inequality is indeed violated under some specific experimental conditions. Next we shall describe quantum cryptography, which allows one to transmit a message and make sure that it has not been intercepted. Finally, in Sect. 14.3, we shall discuss the quantum computer. Up to now, this has been mainly a theoretical concept, whose practical design is only in its first stages. Such a device, which makes use of a generalization of the states (14.2), can in principle perform certain types of calculations with algorithms much more efficient than those used by conventional computers. This is at present the subject of very strong experimental and theoretical activity.

14.1 The EPR Paradox and Bell's Inequality

We now examine the celebrated argument of Einstein, Podolsky and Rosen, and the remarkable analysis done by Bell in 1964.

14.1.1 "God Does not Play Dice"

In their article published in 1935, Einstein, Podolsky and Rosen (EPR) showed an amazing characteristic of quantum mechanics which proved, in the opinions of the authors, that this theory could not constitute the ultimate description of the physical world. The EPR argument relies on the fundamental indeterminism of quantum mechanics. Let us be precise about what we mean by indeterminism in this context. Consider for instance a spin-1/2 particle prepared in the state[3]

$$|\psi\rangle = \frac{1}{\sqrt{2}}\left(|+z\rangle + |-z\rangle\right) . \tag{14.3}$$

If we measure the component S_z of the particle's spin along z, we know that there is a probability 1/2 of finding $+\hbar/2$ and a probability 1/2 of finding $-\hbar/2$. The result of a measurement is not certain, although we know perfectly well (in the quantum sense) the initial state of the system.

At first sight, this kind of indeterminism resembles the classical situation of the toss of a coin. However such an analogy does not stand up to a more advanced analysis. In a classical toss, we know that a sufficiently precise knowledge of the initial state (position, velocity, the state of the surface on which the coin falls, etc.) can in principle determine the result of the toss. The fact that we attribute a probability 1/2 to each possible outcome is a convenient and simple way to say we do not want, in practice, to worry about all that information. On the contrary, in the quantum case, if the initial spin state of the particle is (14.3), we cannot find any further information which would allow us to predict in advance the result $\pm\hbar/2$ of the measurement of S_z.

Quantum indeterminism is in complete opposition to the principles of classical theories. This provoked many debates and criticisms, such as the famous phrase of Einstein, "God does not play dice". What Einstein was hoping for was a super-theory, which would reproduce the predictions of quantum mechanics (no one would think of denying the practical successes of the theory), but which would be deterministic. Can such a super-theory exist? In fact, it cannot be ruled out by experiments if one considers only one-particle systems. However, if one considers entangled states involving two or more particles, such as those considered by Einstein, Podolsky and Rosen, it is possible to put constraints on the predictions of any possible super-theory. Such constraints, which were discovered by John Bell, are in some circumstances in contradiction to the predictions of quantum mechanics. This has provided the possibility of experimental tests, whose results have all clearly confirmed the validity of quantum mechanics. The deterministic, local[4] super-theory that Einstein dreamed of cannot exist.

14.1.2 The EPR Argument

We present the EPR argument in the form given by David Bohm in 1952. This presentation is more convenient to explain and to treat mathematically

[3] The state $|+z\rangle$ is identical to the state that we denoted by $|+\rangle$ in Chap. 12. Here, we mention explicitly the quantization axis, in order to avoid any ambiguity.

[4] In the present context, "locality" means that an action at a point in space cannot immediately have a *detectable* effect at some other point in space at a distance r. One must wait for a time of at least r/c in order for such an effect to be observable.

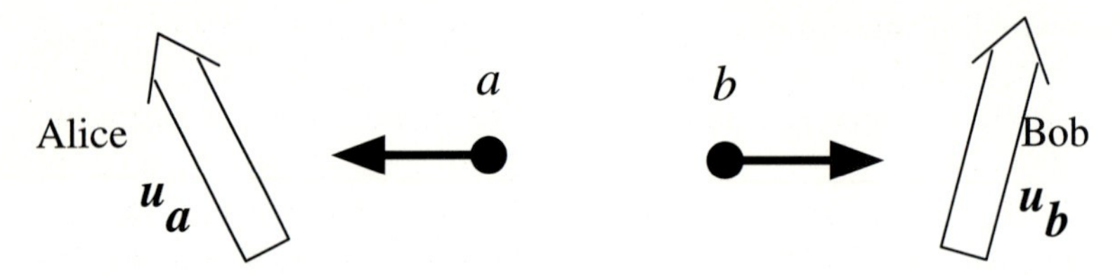

Fig. 14.1. Gedankenexperiment corresponding to the EPR argument. Two spin-1/2 particles a and b are prepared in the singlet state. Alice measures the component of the spin of particle a along an axis $\boldsymbol{u}_{\mathrm{a}}$. Bob measures the component of the spin of particle b along an axis $\boldsymbol{u}_{\mathrm{b}}$

than the initial version, although it is basically equivalent from the conceptual point of view. Suppose we prepare two spin-1/2 particles a and b in the singlet spin state

$$|\Psi_{\mathrm{s}}\rangle = \frac{1}{\sqrt{2}}\left(|a:+z\,;\,b:-z\rangle - |a:-z\,;\,b:+z\rangle\right) . \tag{14.4}$$

Particle a is detected by Alice, who measures the component of its spin along an axis of unit vector $\boldsymbol{u}_{\mathrm{a}}$ (Fig. 14.1); similarly, particle b is detected by Bob, who measures its spin component along an axis of unit vector $\boldsymbol{u}_{\mathrm{b}}$.

The results of the measurements of Alice and Bob are strongly correlated. Let us assume first that Alice and Bob choose the same axis z to do their measurements: $\boldsymbol{u}_{\mathrm{a}} = \boldsymbol{u}_{\mathrm{b}} = \boldsymbol{e}_z$. The argument presented in the introduction applies: with a probability 1/2, Alice will find $+\hbar/2$ and Bob will find $-\hbar/2$, and with the same probability 1/2, Alice will find $-\hbar/2$ and Bob will find $+\hbar/2$. Alice and Bob can *never* obtain the same result. There is a perfect correlation, or rather anticorrelation, of the two results.

This is easily generalized to all situations where Alice and Bob measure along the same axis. Suppose, for example, that they choose $\boldsymbol{u}_{\mathrm{a}} = \boldsymbol{u}_{\mathrm{b}} = \boldsymbol{e}_x$. The singlet state can be rewritten in the eigenbasis of the observables $\hat{S}_{ax}$ and $\hat{S}_{bx}$. Using

$$|\pm z\rangle = \frac{1}{\sqrt{2}}\left(|+x\rangle \pm |-x\rangle\right) \tag{14.5}$$

for both particles a and b, we obtain

$$|\Psi_{\mathrm{s}}\rangle = \frac{1}{\sqrt{2}}\left(|a:+x\,;\,b:-x\rangle - |a:-x\,;\,b:+x\rangle\right) . \tag{14.6}$$

The form of $|\Psi_{\mathrm{s}}\rangle$ in the basis corresponding to the x axis is the same as the form obtained using the z axis. This generalizes to any axis.[5] As a consequence, the perfect anticorrelation of the results of Alice and Bob remains

[5] The invariance of the structure of $|\Psi_{\mathrm{s}}\rangle$ under a change of the quantization axis is a consequence of the rotation invariance of the singlet state, which has zero angular momentum (Sect. 13.1.3).

true whatever spin component they choose, provided it is the same for both of them, i. e. $\boldsymbol{u}_\text{a} = \boldsymbol{u}_\text{b}$.

Such correlations appear frequently in daily life. Suppose we have two cards, one red and the other one yellow. We place each of them in a sealed envelope, we mix the envelopes in a closed box, and we give one of them to Alice and the other one to Bob. When Alice opens her envelope, she sees the color of her card (red with a probability $1/2$, yellow with a probability $1/2$). There is obviously a perfect anticorrelation with Bob's subsequent result: if Alice's card is red, Bob's card is yellow, and vice versa. There is no paradox in these anticorrelations: the fact that Alice looks at the color of her card does not affect the color of Bob's card. This is the central point in a key statement of the EPR article:

> *If, without in any way disturbing a system, we can predict with certainty (i. e. with a probability equal to unity) the value of a physical quantity, then there exists an element of physical reality corresponding to this physical quantity.*

This argument means that there is an *element of physical reality* associated with the color of Bob's card, since, without perturbing it in any manner, one can determine the color of this card by simply asking Alice what her result is. Similarly, there is an element of physical reality associated with the component S_{bz}, since, without perturbing particle b in any manner, one can determine the value of S_{bz} that one would measure in an experiment: it is sufficient to ask Alice to measure the component S_{az} and to tell Bob the result. If Alice finds $+\hbar/2$, Bob is sure to find $-\hbar/2$ by measuring S_{bz}, and vice versa.

Actually, the EPR argument goes one step further. Since we can transpose the argument about the z axis to the x axis, there must also exist an element of physical reality associated with the component S_{bx} of particle b. If one prepares a two-particle system in the singlet state, Bob can determine the component S_{bx} without "touching" particle b. It is sufficient to ask Alice to measure the component S_{ax} and to tell him her result. Although the term "element of physical reality" is quite vague at this point, we feel that we are reaching dangerous ground. The observables $\hat{S}_{bx}$ and $\hat{S}_{bz}$ do not commute. How can they possess simultaneously this element of physical reality?

In fact, all of the above argument is contrary to the basic principles of quantum mechanics. When the particles a and b are in an entangled state, such as the singlet state, it is risky to claim that one doesn't "act" on particle b when performing a measurement on a. Taken separately, particles a and b are *not* in well-defined states; only the global system $a+b$ is in a well-defined quantum mechanical state. It is only for factorized states, of the type (14.1), that the EPR argument can be applied safely. However, in that case, there is no paradox: a measurement on a gives no information about a measurement which might be performed on b later on.

At this stage, we can take either of the following attitudes. We can stick to the quantum description, which has this paradoxical nonlocal character: the two particles a and b, as far as they may be from one another (for example a on earth, b on the moon), do not have individual realities when their spin state is an entangled state. It is only after Alice (on earth) has measured S_{az} that the quantity S_{bz} (for the particle on the moon) acquires a well-defined value.[6] Alternatively, we can adopt the point of view of Einstein, and hope that some day someone will find a more "complete" theory than quantum mechanics. In that theory, the notion of locality would have the same meaning as it has in classical physics, and so would the notion of reality.

14.1.3 Bell's Inequality

In 1964, John Bell, an Irish physicist working at CERN, made a decisive theoretical breakthrough. This allowed scientists to carry this debate between two radically antagonistic conceptions of the physical world into the realm of experiment. Bell's formulation is the following: if the super-theory that Einstein was hoping for exists, it will involve, for any pair (a, b) of the EPR problem described above, a parameter λ which determines completely the results of the measurements of Alice and Bob. For the moment, we know nothing about the parameter λ, which is absent in any orthodox quantum description.

We denote by Λ the manifold in which the parameter λ evolves. In the super-theory framework, there must exist a function $A(\lambda, \boldsymbol{u}_\mathrm{a}) = \pm\hbar/2$ for Alice and a function $B(\lambda, \boldsymbol{u}_\mathrm{b}) = \pm\hbar/2$ for Bob which give the results of their measurements. These results therefore depend on the value of λ: for instance, if λ pertains to some subset $\Lambda_+(\boldsymbol{u}_\mathrm{a})$, then $A(\lambda, \boldsymbol{u}_\mathrm{a}) = \hbar/2$; if λ is in the complementary subset $\Lambda - \Lambda_+(\boldsymbol{u}_\mathrm{a})$, then $A(\lambda, \boldsymbol{u}_\mathrm{a}) = -\hbar/2$. Locality plays a crucial role in the above assumptions: we have assumed that the function A depends on the value of λ and on the direction of analysis $\boldsymbol{u}_\mathrm{a}$ chosen by Alice, but not on the direction of analysis $\boldsymbol{u}_\mathrm{b}$ chosen by Bob.

The parameter λ of the super-theory varies from one pair (a, b) to another, whereas, in quantum mechanics, all the pairs are prepared in the same state $|\Psi_\mathrm{s}\rangle$ and there is nothing that allows one to identify any difference between them. This parameter is therefore not accessible to a physicist who uses quantum mechanics: it is a *hidden variable*. The beauty of Bell's argument was that it proved that there exist strong constraints on theories with local hidden variables, and that these constraints can be established without any assumptions other than the ones given above. Notice that all correlations encountered in daily life can be described in terms of hidden-variable theories.

[6] One can check that this formulation does not allow the instantaneous transmission of information. In order to see the correlations with Alice's result, Bob must ask Alice what her result is and the corresponding information travels (at most) at the velocity of light (see Appendix Appendix D for more details).

In the previous example of cards of different colors, the hidden variable comes from the shuffling of the cards. If a careful observer memorizes the motion of the cards in this shuffling, he/she can predict with probability 1 the result of Alice (red or yellow) and that of Bob (yellow or red).

In order to obtain Bell's result, we introduce the correlation function $E(\boldsymbol{u}_\mathrm{a}, \boldsymbol{u}_\mathrm{b})$. This function is equal to the expectation value of the product of the results of Alice and Bob, for given directions of analysis $\boldsymbol{u}_\mathrm{a}$ and $\boldsymbol{u}_\mathrm{b}$, divided by $\hbar^2/4$ in order to obtain a dimensionless quantity. Whatever the underlying theory, we note the following property:

$$|E(\boldsymbol{u}_\mathrm{a}, \boldsymbol{u}_\mathrm{b})| \leq 1\,. \tag{14.7}$$

Indeed, for each pair, the product of Alice's and Bob's results is $\pm\hbar^2/4$.

For a hidden-variable theory, the function $E(\boldsymbol{u}_\mathrm{a}, \boldsymbol{u}_\mathrm{b})$ can be written as

$$E(\boldsymbol{u}_\mathrm{a}, \boldsymbol{u}_\mathrm{b}) = \frac{4}{\hbar^2}\int \mathcal{P}(\lambda)\, A(\lambda, \boldsymbol{u}_\mathrm{a})\, B(\lambda, \boldsymbol{u}_\mathrm{b})\, \mathrm{d}\lambda\,, \tag{14.8}$$

where the function $\mathcal{P}(\lambda)$ describes the (unknown) distribution law of the variable λ. The only constraints on $\mathcal{P}$ are

$$\text{for any } \lambda\,, \quad \mathcal{P}(\lambda) \geq 0 \qquad \text{and} \qquad \int \mathcal{P}(\lambda)\, \mathrm{d}\lambda = 1\,. \tag{14.9}$$

Here we assume that the function $\mathcal{P}(\lambda)$ does not depend on the directions of analysis $\boldsymbol{u}_\mathrm{a}$ and $\boldsymbol{u}_\mathrm{b}$. Indeed, these directions can be chosen by Alice and by Bob *after* the pair with the hidden parameter λ has been prepared.

In the framework of quantum mechanics, one can show that the value of the function $E(\boldsymbol{u}_\mathrm{a}, \boldsymbol{u}_\mathrm{b})$ is

$$E(\boldsymbol{u}_\mathrm{a}, \boldsymbol{u}_\mathrm{b}) = \frac{4}{\hbar^2}\langle \Psi_\mathrm{s}|\hat{\boldsymbol{S}}_\mathrm{a}.\boldsymbol{u}_\mathrm{a} \otimes \hat{\boldsymbol{S}}_\mathrm{b}.\boldsymbol{u}_\mathrm{b}|\Psi_\mathrm{s}\rangle = -\boldsymbol{u}_\mathrm{a}.\boldsymbol{u}_\mathrm{b}\,. \tag{14.10}$$

Bell's theorem can be stated in the following way:

1. For a local hidden-variable theory, the quantity

$$S = E(\boldsymbol{u}_\mathrm{a}, \boldsymbol{u}_\mathrm{b}) + E(\boldsymbol{u}_\mathrm{a}, \boldsymbol{u}'_\mathrm{b}) + E(\boldsymbol{u}'_\mathrm{a}, \boldsymbol{u}'_\mathrm{b}) - E(\boldsymbol{u}'_\mathrm{a}, \boldsymbol{u}_\mathrm{b}) \tag{14.11}$$

always satisfies the inequality

$$|S| \leq 2\,. \tag{14.12}$$

2. This inequality can be violated by the predictions of quantum mechanics.

We first prove the inequality satisfied by hidden variable theories. We introduce the quantity

$$\begin{aligned}\mathcal{S}(\lambda) &= A(\lambda, \boldsymbol{u}_\mathrm{a})\, B(\lambda, \boldsymbol{u}_\mathrm{b}) + A(\lambda, \boldsymbol{u}_\mathrm{a})\, B(\lambda, \boldsymbol{u}'_\mathrm{b}) \\ &\quad + A(\lambda, \boldsymbol{u}'_\mathrm{a})\, B(\lambda, \boldsymbol{u}'_\mathrm{b}) - A(\lambda, \boldsymbol{u}'_\mathrm{a}) B(\lambda, \boldsymbol{u}_\mathrm{b})\,,\end{aligned}$$

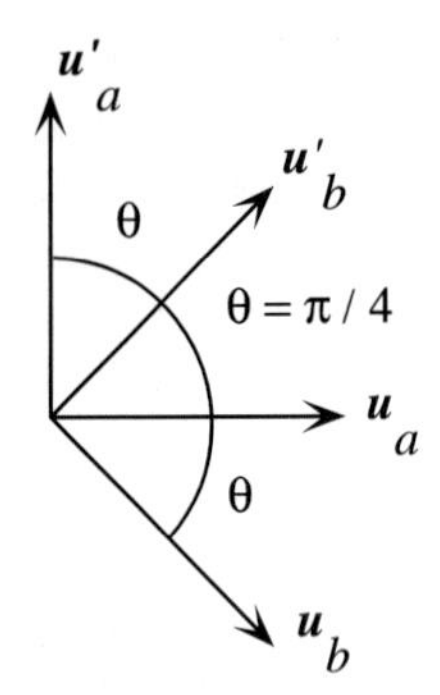

Fig. 14.2. A choice of the directions of the measurements of Alice and Bob which leads to a violation of Bell's inequality

which enters into the definition of S as follows:

$$S = \frac{4}{\hbar^2} \int \mathcal{P}(\lambda)\, \mathcal{S}(\lambda)\, \mathrm{d}\lambda\,.$$

This quantity $\mathcal{S}(\lambda)$ can be rewritten as

$$\begin{aligned} \mathcal{S}(\lambda) &= A(\lambda, \boldsymbol{u}_{\mathrm{a}})\,[B(\lambda, \boldsymbol{u}_{\mathrm{b}}) + B(\lambda, \boldsymbol{u}'_{\mathrm{b}})] \\ &\quad + A(\lambda, \boldsymbol{u}'_{\mathrm{a}})\,[B(\lambda, \boldsymbol{u}'_{\mathrm{b}}) - B(\lambda, \boldsymbol{u}_{\mathrm{b}})]\ , \end{aligned} \tag{14.13}$$

which is always equal to $\pm\hbar^2/2$. Indeed, the quantities $B(\lambda, \boldsymbol{u}_{\mathrm{b}})$ and $B(\lambda, \boldsymbol{u}'_{\mathrm{b}})$ can only take the two values $\pm\hbar/2$. Therefore they are either equal or opposite. In the first case, the second line of (14.13) vanishes, and the first line is equal to $\pm\hbar^2/2$. In the second case, the first line of (14.13) vanishes, and the second line is $\pm\hbar^2/2$. We then multiply $\mathcal{S}(\lambda)$ by $\mathcal{P}(\lambda)$ and integrate over λ in order to obtain the inequality we are looking for.

Concerning the second point of Bell's theorem, it suffices to find an example for which the inequality (14.12) is explicitly violated. Consider the vectors $\boldsymbol{u}_{\mathrm{a}}$, $\boldsymbol{u}'_{\mathrm{a}}$, $\boldsymbol{u}_{\mathrm{b}}$, and $\boldsymbol{u}'_{\mathrm{b}}$ represented in Fig. 14.2, where

$$\boldsymbol{u}_{\mathrm{b}}.\boldsymbol{u}_{\mathrm{a}} = \boldsymbol{u}_{\mathrm{a}}.\boldsymbol{u}'_{\mathrm{b}} = \boldsymbol{u}'_{\mathrm{b}}.\boldsymbol{u}'_{\mathrm{a}} = -\boldsymbol{u}_{\mathrm{b}}.\boldsymbol{u}'_{\mathrm{a}} = \frac{1}{\sqrt{2}}\,. \tag{14.14}$$

Using (14.10), we find

$$S = -2\sqrt{2}\,, \tag{14.15}$$

which obviously violates the inequality (14.12).

After this remarkable step forward due to Bell, which transformed a philosophical discussion into an experimental problem, experimentalists had to find the answer. Is quantum mechanics always right, even for a choice of angles such as in Fig. 14.2, which would eliminate any realistic, local super-theory, or, on the contrary, are there experimental situations where quantum mechanics can be falsified, which would allow for a more complete theory, as Einstein advocated?

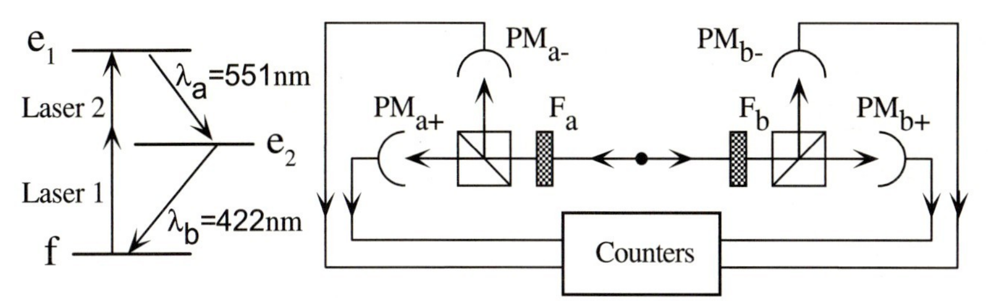

Fig. 14.3. *Left:* levels of atomic calcium used to produce photon pairs with correlated polarizations. *Right:* the photons a and b are first selected by frequency: $\mathrm{F_a}$ transmits photons a and stops photons b, and vice versa for $\mathrm{F_b}$. They are then detected by the photomultipliers PM_{a+}, PM_{a-}, PM_{b+} and PM_{b-}. The polarizing cubes are analogs of Stern–Gerlach devices. They transmit photons with the polarization $|\uparrow\rangle$ towards the detectors PM_{a+}, PM_{b+}, and they deflect photons with polarization $|\rightarrow\rangle$ towards the detectors PM_{a-}, PM_{b-}. Since the photons are emitted nearly isotropically, only a small fraction ($\sim 10^{-5}$) of the emitted pairs is actually used

14.1.4 Experimental Tests

The first experimental attempts to find a violation of Bell's inequality started at the beginning of the 1970s. These experiments were performed on photon pairs rather than on spin-1/2 particles, because it is experimentally simpler to produce a two-photon entangled state of the type (14.2).

The previous argument can be transposed with no difficulty to photon pairs. The spin states $|+z\rangle$ and $|-z\rangle$ are replaced by the polarization states of the photon $|\uparrow\rangle$ and $|\rightarrow\rangle$, corresponding to vertical and horizontal polarizations. The states $|+x\rangle$ and $|-x\rangle$, which are symmetric and antisymmetric combinations, respectively, of $|\pm : z\rangle$, are replaced by photon states linearly polarized at ± 45 degrees from the vertical direction:

$$|\nearrow\rangle = \frac{1}{\sqrt{2}}\left(|\uparrow\rangle + |\rightarrow\rangle\right) , \qquad |\nwarrow\rangle = \frac{1}{\sqrt{2}}\left(-|\uparrow\rangle + |\rightarrow\rangle\right) . \tag{14.16}$$

The first experimental tests, performed between 1970 and 1975 in the USA and in Italy, led to contradictory results concerning the violation of Bell's inequality. The experiments of Fry and Thomson in Texas in 1976 and, particularly those of Aspect and his group in Orsay between 1980 and 1982 led to the first undeniable violation of Bell's inequality in a situation close to the gedankenexperiment presented above.[7] The experiments of Aspect used pairs of photons emitted in an *atomic cascade* of calcium atoms (Fig. 14.3). These calcium atoms are prepared by lasers in an excited state $\mathrm{e_1}$. This excited state has a lifetime of 15 ns, and decays to an excited state $\mathrm{e_2}$ by emitting a photon a, of wavelength $\lambda_\mathrm{a} = 551\,\mathrm{nm}$. This latter level $\mathrm{e_2}$ has a lifetime

[7] A. Aspect, P. Grangier and G. Roger, Phys. Rev. Lett. **49**, 91 (1982); A. Aspect, J. Dalibard and G. Roger, Phys. Rev. Lett. **49**, 1804 (1982).

of 5 ns, and itself decays to the ground state f by emitting a second photon b, of wavelength $\lambda_\mathrm{b} = 422\,\mathrm{nm}$. The initial level e_1 and the final level f have zero angular momentum, whereas the intermediate level e_2 is has an angular momentum of 1. Under these conditions, one can show that the polarization state of the emitted photon pair is

$$|\Psi_\mathrm{p}\rangle = \frac{1}{\sqrt{2}} \left(|a:\uparrow\,;\, b:\uparrow\rangle + |a:\rightarrow\,;\, b:\rightarrow\rangle \right) . \tag{14.17}$$

This entangled state leads to the same type of correlations as does the singlet spin state considered above. The transposition of Bell's argument shows that some quantity S', involving correlation functions between the polarizations of detected photons, must satisfy $|S'| \le 2$ for any local hidden-variable theory. The Orsay result, $S' = 2.697 \pm 0.015$, violates this inequality, while it is in perfect agreement with the quantum mechanical prediction $S' = 2.70$. Therefore the universal, realistic, local super-theory which was supposed to replace quantum mechanics, as Einstein believed, cannot exist, at least for this system. Physicists must learn to live with the genuine indeterminism of quantum mechanics.

14.2 Quantum Cryptography

The aim of cryptography is to transmit a message from an sender (Alice) to a receiver (Bob) and to minimize the risk that a spy might intercept and decipher the message. In order to do that, classical cryptography uses sophisticated methods which cannot be "broken" in a reasonable amount of time, with the present capacities of computers. Quantum cryptography is based on a somewhat different principle. It allows Alice and Bob to *make sure* no spy has intercepted the message *before* it has actually been sent!

14.2.1 The Communication Between Alice and Bob

A message can always be coded in binary language, i. e. by a succession of numbers 0 and 1. Each number, 0 or 1, represents an elementary piece of information, or *bit*. In order to transmit her message, we assume that Alice sends Bob a beam of spin-1/2 particles in a well-controlled sequence, and that Bob detects these particles one after the other in a Stern–Gerlach type of apparatus. Each particle carries a bit, coded through the spin state of the particle.

Suppose first that Alice sends each particle in the state $|+z\rangle$ or $|-z\rangle$. By assumption, $|+z\rangle$ represents the value 1 and $|-z\rangle$ the value 0. Bob orients his Stern–Gerlach apparatus along the z axis also, he measures the spin states of the particles which arrive and he reconstructs Alice's message. Such a procedure has no quantum feature and is simple to spy upon. The spy just sits between Alice and Bob and places his own Stern–Gerlach apparatus in the z direction. He measures the spin state of each particle and reemits it towards Bob in the same spin state. He therefore reads the message and neither Alice nor Bob can detect his presence.

The situation changes radically if Alice chooses at random, for each of the particles she is sending, one of the four states $|+z\rangle$, $|-z\rangle$, $|+x\rangle$ or $|-x\rangle$, without telling anyone which axis she has chosen (x or z) for a given particle. Suppose Alice sends Bob a series of particles without trying, for the moment, to give it any intelligible form. There are 16 particles in the examples shown on Tables 14.1 and 14.2, but in practice one would work with much larger numbers. It is only at the end of the procedure, as we shall see, that Alice will decide which particles should be taken into account in order to construct the message she wants to transmit.

What can Bob do in this situation? He can orient the axis of his Stern–Gerlach apparatus in the direction x or z arbitrarily. On average, for half of the particles, his choice is the same as Alice's, in which case the bit he detects is significant. Indeed, if Alice sends a particle in the state $|+x\rangle$ and if Bob chooses the x axis, he measures + with probability 1. For the other half of the particles, Alice and Bob choose different axes and Bob's results are useless: if Alice sends $|+x\rangle$ and if Bob chooses the z axis, he will detect + with probability 1/2 and − with probability 1/2.

In order to make sure that no spy has intercepted the transmission, Bob announces openly the set of axes he has chosen, x or z, for all events. He also announces the results he has obtained, i. e. + or −, for a fraction of the particles. For instance, in the case of the 16 particles shown in Tables 14.1 and 14.2, Bob announces publicly his 16 choices of axes, and his first 8 results. Alice examines the results, and she can detect whether or not a spy has operated. Her argument is the following. The spy does not know the directions x or z she has chosen for each particle. Suppose that the spy ori-

Table 14.1. Detection of a possible spy: among Bob's results obtained along the same axis (particles 1, 3, 4 and 7), Alice looks for a possible difference which would mean a spy had operated. No anomaly appears here. In practice, in order to have a sufficient confidence level, one must use numbers of events much larger than 8

Number of particle	1	2	3	4	5	6	7	8
Axis chosen by Alice (kept secret)	z	z	x	z	z	x	x	z
State chosen by Alice (kept secret)	+	−	+	−	−	−	+	−
Axis chosen by Bob (broadcast openly)	z	x	x	z	x	z	x	x
State measured by Bob (broadcast openly)	+	−	+	−	−	+	+	+
Useful measurement?	Yes	No	Yes	Yes	No	No	Yes	No

Table 14.2. After making sure that there is no spy, Alice chooses, out of the useful measurements, those which allow her to communicate the message. For instance, to communicate the message "1, 1", i. e. "+, +", she openly asks Bob to look at the results of his measurements 11 and 15

Number of particle	9	10	11	12	13	14	15	16
Axis chosen by Alice (kept secret)	x	z	x	z	z	x	z	z
State chosen by Alice (kept secret)	+	−	+	+	−	−	+	−
Axis chosen by Bob (broadcast openly)	z	z	x	x	z	z	z	x
State measured by Bob (kept secret)	−	−	+	+	−	+	+	+
Useful measurement ?	No	Yes	Yes	No	Yes	No	Yes	No

ents his Stern–Gerlach apparatus in a random way along x or z, and that he reemits a particle whose spin state is the same as what he has measured: if he chooses the x axis and obtains the result +, he sends to Bob a particle in the $|+x\rangle$ state. This operation is detectable, because it induces errors in Bob's observations.

Consider for instance the case where Alice has sent a particle in the state $|+z\rangle$, and Bob has also oriented his detector along the z axis, but where the spy has oriented his own Stern–Gerlach apparatus along the x axis. The spy will measure + with a probability 1/2 and − with a probability 1/2. According to his result, he reemits to Bob a particle in the state $|+x\rangle$ or $|-x\rangle$. In both cases, since Bob's detector is oriented along z, Bob will measure + with a probability 1/2 and − with a probability 1/2. If the spy had not been present, Bob would have found + with probability 1.

Therefore, out of all the results announced by Bob, Alice looks at those where her own choice of axes is the same as Bob's (Table 14.1). If no spy is active, Bob's results must be identical to hers. If a spy is present, there must be differences in 25% of the cases. Therefore, if Bob announces publicly 1000 of his results, on average 500 will be useful for Alice (same axes), and the spy will have induced an error in 125 of them on average. The probability that a spy is effectively present but remains undetected by such a procedure is $(3/4)^{500} \sim 3 \times 10^{-63}$, which is completely negligible.

Once Alice has made sure that no spy has intercepted the communication, she tells Bob openly which measurements he must read in order to reconstruct the message she wants to send him. She simply chooses them out of the sequence of bits for which Bob and she have made the same choice of axes, and for which Bob did not announce his result openly (see Table 14.2).

14.2.2 The Quantum Noncloning Theorem

In the previous paragraph, we have assumed that the spy chooses at random the axis of his detector for each particle, and that he sends to Bob a particle in the state corresponding to his measurement. One may wonder whether this is his best strategy to remain unseen. In particular, if the spy could clone each incident particle sent by Alice into two particles in the same state, it would be possible to send one of them to Bob and to measure the other one. The spy would then become undetectable!

Fortunately for Alice and Bob, this cloning of an unknown state is impossible in quantum mechanics.[8] One cannot generate in a reliable way one or more copies of a quantum state unless some features of this state are known in advance. In order to prove this result, let us denote by $|\alpha_1\rangle$ an initial quantum state which we want to copy. The system on which the copy will be "printed" is initially in a known state, which we denote by $|\phi\rangle$ (the equivalent of a blank sheet of paper in a copying machine). The evolution of the total system *original* + *copy* during the cloning operation must therefore be

$$\text{cloning:} \quad |\text{original:}\alpha_1\rangle \otimes |\text{copy:}\phi\rangle \longrightarrow |\text{original:}\alpha_1\rangle \otimes |\text{copy:}\alpha_1\rangle\,. \qquad (14.18)$$

The evolution is governed by some Hamiltonian, which we need not specify, but which cannot depend on $|\alpha_1\rangle$, since this state is unknown by assumption. For another state $|\alpha_2\rangle$ of the original, orthogonal to $|\alpha_1\rangle$, we must also have

$$\text{cloning:} \quad |\text{original:}\alpha_2\rangle \otimes |\text{copy:}\phi\rangle \longrightarrow |\text{original:}\alpha_2\rangle \otimes |\text{copy:}\alpha_2\rangle\,. \qquad (14.19)$$

The impossibility of cloning is then obvious if we consider the initial state

$$|\alpha_3\rangle = \frac{1}{\sqrt{2}}\left(|\alpha_1\rangle + |\alpha_2\rangle\right)\,. \qquad (14.20)$$

If the cloning were successful for this state, we would find

$$\text{cloning:} \quad |\text{original:}\alpha_3\rangle \otimes |\text{copy:}\phi\rangle \longrightarrow |\text{original:}\alpha_3\rangle \otimes |\text{copy:}\alpha_3\rangle\,. \qquad (14.21)$$

However, the linearity of the Schrödinger equation imposes, by linear superposition of (14.18) and (14.19), the result

$$\begin{aligned} &|\text{original:}\alpha_3\rangle \otimes |\text{copy:}\phi\rangle \\ &\quad \longrightarrow \frac{1}{\sqrt{2}}\left(|\text{original:}\alpha_1\rangle \otimes |\text{copy:}\alpha_1\rangle + |\text{original:}\alpha_2\rangle \otimes |\text{copy:}\alpha_2\rangle\right)\,. \end{aligned}$$

This final state is an *entangled state.* It is therefore different from the desired state (14.21).

An inspection of this proof allows us to understand the way in which quantum mechanics can contribute to cryptography. If we limit ourselves to

---

[8] W.K. Wooters and W.H. Zurek, Nature **299**, 802 (1982).

a two-state transmission, where $|\alpha_1\rangle = |+z\rangle$ and $|\alpha_2\rangle = |-z\rangle$, then the spy can remain invisible as we have explained in the previous section. The two operations (14.18) and (14.19) are possible; we simply need to measure the spin state of the incident particle along the z axis and to reemit one or more particles in the same state. It is the fact that we can use simultaneously the states $|\alpha_1\rangle$, $|\alpha_2\rangle$ and linear combinations of these states $|\alpha_{3,4}\rangle = |\pm : x\rangle$ which makes quantum cryptography original and forbids any reliable duplication of a message intercepted by a spy.

14.2.3 Present Experimental Setups

As for experimental tests of Bell's inequality, current physical setups use photons rather than spin-1/2 particles. Various methods can be used to code information with photons. We shall consider only coding by means of polarization, which is the method effectively used in practice. Alice uses four states which define two nonorthogonal bases, each of which can code the bits 0 and 1, for instance in the form

$$|\uparrow\rangle : 1; \quad |\rightarrow\rangle : 0; \quad |\nearrow\rangle : 1; \quad |\nwarrow\rangle : 0. \tag{14.22}$$

In the present quantum cryptography devices, the challenge is to obtain sufficiently large distances of transmission. Distances of the order of 10 km are currently being reached, by using optical telecommunication techniques, in particular photons in optical fibers.

An important point is the light source. The noncloning theorem, which is crucial for the security of the procedure, applies only to individual photons. In contrast, the light pulses normally used in telecommunications contain very large numbers of photons, typically more than 10^6. If one uses such pulses with polarization coding, the noncloning theorem no longer applies. Indeed, it is sufficient for the spy to remove a small part of the light in each pulse and to let the remaining part propagate to Bob. The spy can measure in this way the polarization of the photons of the pulse without modifying the signal noticeably.

In order to guarantee the security of the procedure, each pulse must contain a single photon. This is a difficult condition to satisfy in practice, and one uses the following alternative as a compromise. Alice strongly attenuates the pulses so that the probability p of finding one photon in any one pulse is much smaller than one. The probability of finding two photons will be $p^2 \ll p$, which means that there will be very few pulses with two (or more) photons. Obviously, most of the pulses will contain no photons, which is a serious drawback of the method since Alice must code the information redundantly. In practice, a value of p between 0.01 and 0.1 is considered to be an acceptable compromise.

Once this basic question is solved, the essential part of the system uses optical telecommunication technologies. The source is a strongly attenuated pulsed laser, and the coding by means of polarization is performed directly in the optical fiber using integrated modulators. The attenuated pulses are detected with avalanche photodiodes, which transform a single photon into a macroscopic electrical signal by means of an electron multiplication process. In order to identify unambiguously the photons emitted by Alice and detected by Bob, electric pulses synchronized with the laser pulses are sent to Bob by conventional techniques, and they play the role of a clock. Finally, a computerized treatment of the data, involving a large number of pulses, realizes the various stages of the procedure described above, in particular those related to testing for the absence of a spy on the line.

At present, most systems which have been built are more demonstration prototypes than operational systems. Several relevant parameters have been tested, such as the transmission distance, the transmission rate and the error rate. Actually, developing these systems has for the moment a prospective character, since conventional (nonquantum) cryptographic systems are considered to be very reliable by civilian and military users. This confidence was shaken a little in 1994, as we shall see in the next section.

14.3 The Quantum Computer

We briefly sketch a fascinating concept which is currently under very active investigation.

14.3.1 The Quantum Bits, or "Q-Bits"

In the previous section, we have seen that one can code a bit of information (0 or 1) with two orthogonal states of a spin-1/2 particle or with a polarized photon. At this stage of a quantum mechanics course, a question arises naturally: in terms of information theory, what is the significance of a linear superposition of these two states? In order to account for this possibility, the notion of a "q-bit" is introduced, which, contrary to a classical bit, allows the existence of such intermediate states. The notion of a q-bit in itself is not very rich; however, it has interesting implications if one considers a quantum computer based on the manipulation of a large number of q-bits.

We use the very simplified definition of a computer as a system which is capable of performing operations on sets of N bits called "registers". The content of a register is a binary word, which represents a number memorized by the computer. For $N = 3$, we therefore have 8 possible words:

$$(+,+,+)\,(+,+,-)\,(+,-,+)\,(+,-,-)$$
$$(-,+,+)\,(-,+,-)\,(-,-,+)\,(-,-,-)$$

Consider now a q-register, made up of a set of N q-bits. The 2^N possible states of the corresponding classical register will define a basis of the space of states of the q-register, which can itself be in a linear superposition of all the basis states, of the form

$$|\Psi\rangle = \sum_{\sigma_1=\pm} \sum_{\sigma_2=\pm} \sum_{\sigma_3=\pm} C_{\sigma_1,\sigma_2,\sigma_3} |\sigma_1, \sigma_2, \sigma_3\rangle \qquad \text{for } N = 3\,.$$

Suppose that the computer calculates, i. e. it performs an operation on the state of the q-register. Since this operation is performed on a linear superposition of states, we can consider that it is done "in parallel" on 2^N classical numbers. This notion of quantum parallelism is the basis of the gain in efficiency of the computer. The gain may be exponential if the 2^N calculations corresponding to N q-bits are indeed performed simultaneously.

Naturally, many questions arise. From the fundamental point of view, what kind of calculations can one perform and what kind of algorithms can one use with such a device? In practice, how can one construct it?

14.3.2 The Algorithm of Peter Shor

In the previous section, we referred to nonquantum cryptography. The corresponding systems are often called algorithmic protocols. One of these protocols is based on the fact that some arithmetic operations are very easy to perform in one direction, but very difficult in the reverse direction. For instance, it is simple to calculate the product of two numbers, but it takes much more time to factorize a number into its prime divisors. If one considers the product P of two large prime numbers, one must perform approximately $\sqrt{P}$ divisions in order to identify the factors. The computing time increases exponentially with the number of digits (or of bits) of P: the factorization operation becomes impossible in practice for numbers of more than 300 digits, while the product operation leading to P can still be performed easily with a small computer. This "nonreversibility" is the origin of a cryptographic method due to Rivest, Shamir and Adleman (RSA), which is commonly used (in credit cards, electronic transactions, etc.), and which is considered to be extremely reliable.

This is why the 1994 paper of Peter Shor[9] created a shock in the community. Shor showed that a quantum computer could factorize the product of two prime numbers with a number of operations that was reduced exponentially as compared with known algorithms running on classical computers! The turmoil has now calmed down and the present situation is the following. The algorithm proposed by Shor is correct in principle, and it does provide the expected gain in efficiency. However, the practical development of a quantum computer seems outside the range of present technology, although no physical law forbids it.

[9] P.W. Shor, Proceeding of the 35th Annual Symposium on Foundations of Computer Science, Santa Fe, NM, USA, Nov. 20–22, 1994, IEEE Computer Society Press, Los Alamitos, CA, USA, pp. 124–134.

14.3.3 Principle of a Quantum Computer

We shall not attempt to explain Shor's algorithm here; we shall only give some intuitive ideas about how a quantum computer could perform a calculation. The basic principle is that the calculation must reduce to the evolution of a system with a Hamiltonian. This evolution starts from some initial state and ends with a "measurement" which determines the state of the q-register, and which interrupts the evolution. According to the principles of quantum mechanics, the value found in the measurement is one of the eigenvalues corresponding to the eigenstates of the measured observable. In this context, this value corresponds to the state of a classical register, i. e. a binary word. In order to perform the successive operations, the Hamiltonian of the system is made time-dependent, and it evolves under the action of a clock which determines the "rhythm" of the calculation. At first sight, the determination of this Hamiltonian for a given practical calculation seems to be a formidable task. Actually, however, one can show that the construction of the Hamiltonian can be done relatively easily. An actual calculation can be decomposed into a succession of simple operations which affect only one or two bits. These simple operations are performed by *logical gates*, such as the well-known classical NOT, AND and OR.

The quantum gates required for the Shor algorithm must have some particular features:

- They must be *reversible*, since they follow from a Hamiltonian evolution of the initial bits.
- They must handle q-bits, on which one can perform certain logical operations which are classically inconceivable.

A simple example of a quantum gate is a $\sqrt{\text{NOT}}$ gate. This gate transforms the q-bits 0 and 1 into the symmetric and antisymmetric linear superpositions of 0 and 1 (it is a rotation by $\pi/2$ for a spin of $1/2$). If one applies this gate twice, one inverts 0 and 1, which corresponds to a NOT gate, hence the name of gate $\sqrt{\text{NOT}}$.

One might think that it would be sufficient to let the computer evolve towards a one-component state, which would be the desired value. Actually, very few algorithms give rise to such a simple manipulation. In general, the final state of the computer is still a linear superposition, and the result of the calculation is therefore probabilistic. For instance, in Shor's algorithm, the result is an indication of a possible result. It is easy to check by a conventional method whether the answer is correct, and if not, to continue the calculation. Peter Shor has proven that this trial and error procedure gives the correct answer with a probability arbitrarily close to 1, using a number of trials which increases linearly (not exponentially) with the number of digits of the number we want to factorize.

14.3.4 Decoherence

The principle of a quantum computer is compatible with the laws of physics, and it seems as though it may be possible to construct such a computer, at least if one considers only simple calculations with a small number of gates. If, however, one wants to perform large computations, the global state of the computer has to be a quantum superposition of a large number of states, whose evolution must be controlled in such a way that all the properties of a linear superposition are preserved. It is not clear at present whether such a system can be devised. Studies are under way in two main directions:

- First, the register which evolves must be extremely well protected from the outside environment. Any coupling with this environment will induce a *decoherence effect*, which may destroy the interferences between the various terms of the linear superposition.
- Second, in case perturbations occur, one must prepare *error correction codes*, in order to place the computer in the same state as where it was before the external perturbation occurred.

These two directions – the choice of the system and the correction codes – are under intensive investigation, and the questions raised have been stimulating both for the development of algorithms and for experimental quantum mechanics. At present it is very difficult to predict the outcome of these investigations. However, one spin-off is that it is quite possible that simple logical operations will, in the medium term, be applied in quantum cryptography systems.

Further Reading

- On the EPR problem: *Quantum theory and measurement*, edited by J.A. Wheeler and W.H. Zurek, Princeton University Press, Princeton (1983); "The conceptual implications of quantum mechanics", Workshop of the Hugot Foundation of the "Collège de France", J. Physique **42**, colloque C2 (1981).
- J.S. Bell, Physics **1**, 195 (1964); see also J. Bell, *Speakable and Unspeakable in Quantum Mechanics*, Cambridge University Press, Cambridge (1993). The inequality derived here is a modified version of Bell's initial result, due to J.F. Clauser, M.A. Horne, A. Shimony and R.A. Holt, Phys. Rev. Lett. **23**, 880 (1969). F. Laloë, *"Do we really understand quantum mechanics?"* Am. J. Phys. **69**, 655 (2001). A. Peres, *Quantum Theory: Concepts and Methods*, Kluwer Academic Publishers, Dordrecht (1993).
- C. Bennett, G. Brassard and A. Ekert, "Quantum cryptography", Sci. Am., October 1992; R. Hughes and J. Nordholt, "Quantum cryptography takes the air", Phys. World, May 1999, p. 31.

- S. Lyod, "Quantum mechanical computers", Sci. Am., November 1995, p. 44; S. Haroche and J.-M. Raimond, "Quantum computing: dream or nightmare", Phys. Today, August 1996, p. 51; "Quantum information", special issue of Phys. World, March 1998; N. Gershenfeld and I.L. Chuang, "Quantum computing with molecules", Sci. Am., June 1998, p. 50; J. Preskill, "Battling decoherence: the fault-tolerant quantum computer", Phys. Today, June 1999, p. 24.
- The Physics of Quantum Information, D. Bouwmeester, A. Ekert, A. Zeilinger (Eds.), Springer-Verlag, Heidelberg (2000).
- Quantum Computation and Quantum Information, M. Nielsen and I.L. Chuang, Cambridge University Press, Cambridge (2000).

Exercises

14.1. Bell measurement. Consider a system of two particles of spin 1/2, whose spin state is written as

$$\alpha|+;+\rangle + \beta|+;-\rangle + \gamma|-;+\rangle + \delta|-;-\rangle \,, \tag{14.23}$$

where $|\alpha|^2 + |\beta|^2 + |\gamma|^2 + |\delta|^2 = 1$.

a. The component of the spin of each particle along the z axis is measured. What are the possible results and the corresponding probabilities?
b. Rather than the preceding measurement, a detection which projects the spin state of the two particles onto one of the four states of the *Bell basis*,

$$|\Psi_+\rangle = \frac{1}{\sqrt{2}}\left(|+;+\rangle + |-;-\rangle\right) , \qquad |\Phi_+\rangle = \frac{1}{\sqrt{2}}\left(|+;-\rangle + |-;+\rangle\right) ,$$
$$|\Psi_-\rangle = \frac{1}{\sqrt{2}}\left(|+;+\rangle - |-;-\rangle\right) , \qquad |\Phi_-\rangle = \frac{1}{\sqrt{2}}\left(|+;-\rangle - |-;+\rangle\right) ,$$

is performed. What is the probability of each of the four possible results?

14.2. Quantum teleportation of a spin state. Alice has a spin-1/2 particle A in the spin state

$$\alpha|+\rangle + \beta|-\rangle \,, \qquad \text{where} \qquad |\alpha|^2 + |\beta|^2 = 1 \,,$$

that she wants to teleport to Bob. Alice and Bob also have a pair of spin-1/2 particles B and C, prepared in the singlet state

$$\frac{1}{\sqrt{2}}\left(|+;-\rangle - |-;+\rangle\right)$$

(see Fig. 14.4)

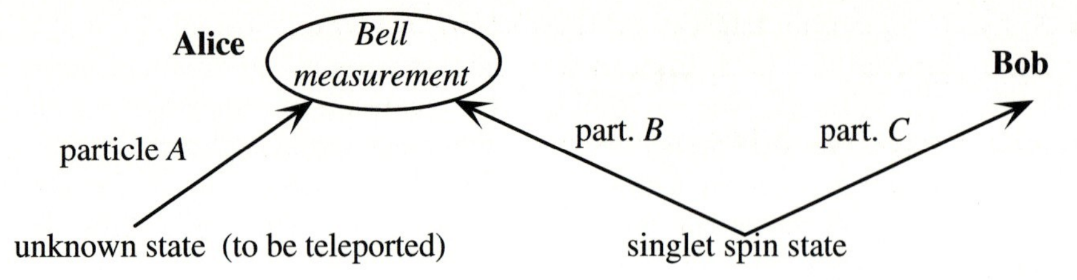

Fig. 14.4. Principle of the quantum teleportation of the quantum state of a particle

a. Alice performs a measurement of the spin state of AB, which projects this state onto one of the four vectors of the Bell basis of AB (see the preceding exercise). What are the probabilities of each of the four possible results?
b. Suppose that Alice finds the pair AB in the spin state $|\Phi_-\rangle$. What is the spin state of particle C after this measurement?
c. Deduce from the preceding questions the principle of quantum teleportation.
d. Can this principle be used to transmit information from Alice to Bob faster than with classical channels (and thus faster than the speed of light)?

15. The Lagrangian and Hamiltonian Formalisms, Lorentz Force in Quantum Mechanics

Nature always proceeds by the shortest means.
Pierre de Fermat

The Lorentz force $q\boldsymbol{v} \times \boldsymbol{B}$ acting on a particle of charge q moving with a velocity $\boldsymbol{v}$ in a magnetic field $\boldsymbol{B}$ cannot be derived from a potential. Therefore, the formulation of quantum mechanics we have used up to now does not apply. The aim of this chapter is to generalize the correspondence principle, in order to obtain the form of the Hamiltonian in such a problem.

Of course, it would suffice to simply introduce this Hamiltonian with no other justification than the fact that it accounts for observed phenomena. However, it is instructive to first develop some considerations related to classical mechanics. The development of analytic mechanics in the 18th and 19th centuries, which was due to the work of d'Alembert, Bernoulli, Euler, Lagrange, Hamilton and others, brought out an amazing geometric structure of the theory, based on a variational principle, the *principle of least action.* It was one of the first remarkable discoveries of Dirac, in 1925–1926, that the same basic structure underlies quantum mechanics. Starting from this observation, the correspondence principle can be stated in a much more profound manner, which enables one to treat complex problems which would be difficult to attack without this analysis.

In Sect. 15.1 we shall recall the basic elements of the Lagrangian formulation of mechanics, based on the principle of least action. In Sect. 15.2, we shall present the "canonical" formulation of Hamilton and Jacobi, which will allow us to exhibit, in Sect. 15.3, the parallelism between classical and quantum mechanics. The word *Hamiltonian*, which we have used so often, will then take on its full significance. In Sect. 15.4, we shall give the Lagrangian and Hamiltonian formulations of the problem of interest, i. e. the motion of a charged particle in a magnetic field. Finally, in Sect. 15.5, we shall transpose the result to quantum mechanics and we shall also take into account the possibility that the particle has spin.

15.1 Lagrangian Formalism and the Least-Action Principle

In his *Méchanique analytique*, published in 1787, one century after Newton's *Principia*, Lagrange proposed a new way to consider mechanical problems. Instead of determining the position $\boldsymbol{r}(t)$ and the velocity $\boldsymbol{v}(t)$ of a particle at time t knowing its initial state $\{\boldsymbol{r}(0), \boldsymbol{v}(0)\}$, Lagrange asked the following equivalent but different question: what is the *trajectory actually followed* by the particle if, leaving $\boldsymbol{r}_1$ at time t_1, it reaches $\boldsymbol{r}_2$ at t_2?

15.1.1 Least Action Principle

In order to simplify the discussion, consider first a one-dimensional problem. Among the infinite number of possible trajectories (see Fig. 15.1) such that

$$x(t_1) = x_1\,, \qquad x(t_2) = x_2\,, \tag{15.1}$$

what is the law that determines the correct one? Lagrange made use of the "principle of natural economy",[1] which is an expression due to Fermat that was adopted by Maupertuis and Leibniz (who called it the principle of "the best"). Lagrange's prescription is the following:

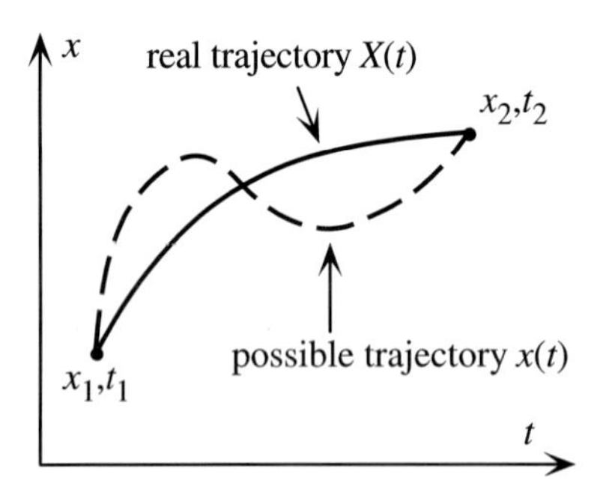

Fig. 15.1. Examples of trajectories starting from x_1 at time t_1 and arriving at x_2 at time t_2. Among all these possible trajectories, the trajectory actually followed by the particle is the one for which the action S is extremal

a. *Any mechanical system is characterized by a Lagrange function, or Lagrangian,* $\mathcal{L}(x, \dot{x}, t)$, *which depends on the coordinate* x, *on its derivative with respect to time* $\dot{x} = \mathrm{d}x/\mathrm{d}t$ *and possibly on time. The quantities* x *and* $\dot{x}$ *are called state variables. For instance, considering a particle in a one-dimensional potential, one has*

$$\mathcal{L} = \frac{1}{2}m\dot{x}^2 - V(x,t)\,. \tag{15.2}$$

[1] To be precise, the variational principle as we present it here was formulated by Hamilton in 1828. In order to simplify the presentation, we have omitted some of the history.

b. *For any trajectory $x(t)$ satisfying (15.1), the action S is defined as*

$$S = \int_{t_1}^{t_2} \mathcal{L}(x, \dot{x}, t)\, \mathrm{d}t\,. \tag{15.3}$$

The principle of least action states that the physical trajectory $X(t)$ is such that S is minimal, or, more generally, extremal.

15.1.2 Lagrange Equations

Let $X(t)$ be the physical trajectory. Consider a trajectory $x(t)$ infinitesimally close to $X(t)$, and also starting from x_1 at time t_1 and arriving at x_2 at time t_2:

$$x(t) = X(t) + \delta x(t)\,, \quad \dot{x}(t) = \dot{X}(t) + \delta\dot{x}(t)\,, \quad \delta\dot{x}(t) = \frac{\mathrm{d}}{\mathrm{d}t}\delta x(t)\,, \tag{15.4}$$

where, by assumption,

$$\delta x(t_1) = \delta x(t_2) = 0\,. \tag{15.5}$$

To first order in δx, the variation of S is

$$\delta S = \int_{t_1}^{t_2} \left(\frac{\partial \mathcal{L}}{\partial x}\,\delta x(t) + \frac{\partial \mathcal{L}}{\partial \dot{x}}\,\delta\dot{x}(t) \right) \mathrm{d}t\,.$$

Integrating the second term by parts and taking into account (15.5), we obtain

$$\delta S = \int_{t_1}^{t_2} \left[\frac{\partial \mathcal{L}}{\partial x} - \frac{\mathrm{d}}{\mathrm{d}t}\left(\frac{\partial \mathcal{L}}{\partial \dot{x}}\right) \right] \delta x(t)\, \mathrm{d}t\,. \tag{15.6}$$

It follows from the principle of least action that δS must vanish *whatever* the infinitesimal function $\delta x(t)$. Therefore, the equation which determines the physical trajectory is the *Lagrange equation*,

$$\frac{\partial \mathcal{L}}{\partial x} = \frac{\mathrm{d}}{\mathrm{d}t}\left(\frac{\partial \mathcal{L}}{\partial \dot{x}}\right)\,. \tag{15.7}$$

We can check readily that we recover the usual equation of motion $m\ddot{x} = -dV/\mathrm{d}x$ for a point particle placed in the potential $V(x, t)$ if we consider the Lagrangian (15.2).

The generalization to s degrees of freedom, i.e. $\{x_i, \dot{x}_i\}, i = 1, \ldots, s$ (where, for instance, $s = 3N$ for N particles in a three-dimensional space), is straightforward. One uses a Lagrangian $\mathcal{L}(\{x_i\}, \{\dot{x}_i\}, t)$ and obtains the set of Lagrange equations

$$\frac{\partial \mathcal{L}}{\partial x_i} = \frac{\mathrm{d}}{\mathrm{d}t}\left(\frac{\partial \mathcal{L}}{\partial \dot{x}_i}\right)\,, \qquad i = 1, \ldots, s\,. \tag{15.8}$$

Remarks

(1) The Lagrange equations keep the same form in any coordinate system. This is particularly useful when one makes changes of variables, e. g. to get from Cartesian coordinates (x, y, z) to spherical coordinates (r, θ, φ). The x_i are called *generalized coordinates.*

(2) It is remarkable that the laws of mechanics can be derived from a variational principle, which states that the physical trajectory minimizes a certain quantity, here the action. Almost all physical laws can be formulated in terms of variational principles like the Fermat principle in geometrical optics. The situations which are encountered physically appear to result from "optimizing" the effects of various "conflicting" contributions.

(3) In the absence of forces (i. e. a uniform potential in (15.2)), S is minimum for $\dot{x}$ = constant, corresponding to a linear, uniform motion. The presence of the potential can be visualized as a property of *space* which bends the trajectories. Forces and inertia appear to be in conflict. The particle follows a trajectory of minimal "length", this length being measured by the action S. Therefore one can say that a mechanical problem has been reduced to a geometric problem. The motion of a particle in a field of force derived from a potential in a *flat* Euclidean space can be transformed into the *free* motion of a particle in a *curved* space where it follows a geodesic. Einstein had this idea in mind in 1908 when he started to construct general relativity. It took him seven years to develop the mathematical structure of the theory.

The development of the fundamental concepts and principles of mechanics was performed during the 17th century. Copernicus had given us the notion of reference frames. Galileo had understood the principle of inertia: uniform linear motion is a *state* relative to the observer, and not a process. It is the modification of the velocity which constitutes a *process.* The final lines were written by Newton.

After the Newtonian synthesis and the publication in 1687 of the *Philosophiae Naturalis Principia Mathematica*, the 18th and the 19th centuries were marked by a fascinating endeavor. Through the impetus of d'Alembert, Maupertuis, the Bernoulli brothers (in particular Daniel), Euler and Lagrange, the basic structure of mechanics, i. e. its geometric structure, was discovered. A large class of problems could be reduced to problems of pure geometry. D'Alembert, who was the first to understand the importance of the abstract concepts of mass and of momentum, attacked the concept of force introduced by Newton. For d'Alembert, motion was the only observable phenomenon, whereas the "causality of motion" remained an abstraction. Hence the idea of studying not a particular trajectory of the theory, but the set of all motions that it predicts (to characterize a force by the set of all its effects is actually a very modern point of view.) In 1787, one century after the *Principia*, Lagrange published his *Méchanique analytique*, and gave a new formulation of mechanics where the geometric and global structure of the theory was emphasized.

The first formulation of a physical law in terms of the least-action principle originated from a dispute between Fermat and Descartes, around 1640, about the notion of proof (Descartes's "proof" of Snell's law of refraction was actually wrong). Fermat, who was a mathematician and who knew little, if any, physics, became interested in the laws of geometrical optics, in particular the equality between the angles of incidence and of reflection. He proved that the results were a geometrical property of the optical length of the light rays (in the case of reflection, this had been understood by Heron of Alexandria in 100 AD). The Snell–Descartes laws predicted which path would be followed by a light ray with given initial properties. In the more general point of view of Fermat, one determines the path effectively followed by a light ray which goes from A to B. At the end of his life, in 1661, Fermat stated his "principle of natural economy", which led to numerous fields of research, still very active today, related to variational principles.

15.1.3 Energy

We assume the system is isolated, i. e. $\partial\mathcal{L}/\partial t = 0$, and we calculate the time evolution of the quantity $\mathcal{L}(x, \dot{x})$ along the physical trajectory $x(t)$:

$$\frac{\mathrm{d}\mathcal{L}}{\mathrm{d}t}(x, \dot{x}) = \dot{x}(t)\,\frac{\partial\mathcal{L}}{\partial x} + \ddot{x}(t)\frac{\partial\mathcal{L}}{\partial\dot{x}} = \frac{\mathrm{d}}{\mathrm{d}t}\left(\dot{x}(t)\,\frac{\partial\mathcal{L}}{\partial\dot{x}}\right),$$

where we have transformed the first term using the Lagrange equation (15.7). Therefore

$$\frac{\mathrm{d}}{\mathrm{d}t}\left(\dot{x}(t)\,\frac{\partial\mathcal{L}}{\partial\dot{x}} - \mathcal{L}\right) = 0\,.$$

For an isolated system, the quantity:

$$E = \dot{x}(t)\,\frac{\partial\mathcal{L}}{\partial\dot{x}} - \mathcal{L}$$

$$\left(\text{and correspondingly } E = \sum_{i=1}^{s}\dot{x}_i(t)\frac{\partial\mathcal{L}}{\partial\dot{x}_i} - \mathcal{L}\right), \tag{15.9}$$

is conserved. It is a constant of the motion, called the energy of the system. In the simple case (15.2) we recover $E = m\dot{x}^2/2 + V(x)$.

15.2 Canonical Formalism of Hamilton

The so-called canonical formulation of Hamilton has led to an impressive number of applications, both in physics and mathematics. As we shall see, this formalism led Dirac to the modern formulation of quantum mechanics.

15.2.1 Conjugate Momenta

The quantity

$$p = \frac{\partial\mathcal{L}}{\partial\dot{x}} \qquad \left(\text{and correspondingly } p_i = \frac{\partial\mathcal{L}}{\partial\dot{x}_i}\right), \tag{15.10}$$

which appears in the definition of the energy (15.9), is called the *conjugate momentum*, or generalized momentum, of the variable x. In the simple case

(15.2), it reduces to the linear momentum $p = m\dot{x}$, but this is no longer true in non-Cartesian coordinates or, as we shall see, when the forces are velocity dependent. We notice that (15.7) implies

$$\dot{p} = \frac{\partial \mathcal{L}}{\partial x} \quad \left(\text{and correspondingly } \dot{p}_i = \frac{\partial \mathcal{L}}{\partial x_i}\right) . \tag{15.11}$$

15.2.2 Canonical Equations

The description of the state of a particle (or a system) by x and the conjugate momentum p, instead of x and the velocity $\dot{x}$, has some advantages. We assume that we can invert (15.10) and calculate $\dot{x}$ in terms of the *new* state variables x and p. The equations of motion are obtained by performing what is called a Legendre transformation. Let us introduce the Hamilton function, or *Hamiltonian*

$$H(x,p,t) = p\dot{x} - \mathcal{L}$$
$$\left(\text{and correspondingly } H(x_i,p_i,t) = \sum_i p_i\dot{x}_i - \mathcal{L}\right) . \tag{15.12}$$

We write down the total differential of H:

$$\mathrm{d}H = p\,\mathrm{d}\dot{x} + \dot{x}\,\mathrm{d}p - \frac{\partial \mathcal{L}}{\partial x}\,\mathrm{d}x - \frac{\partial \mathcal{L}}{\partial \dot{x}}\,\mathrm{d}\dot{x} - \frac{\partial \mathcal{L}}{\partial t}\,\mathrm{d}t .$$

If we take into account (15.10) and (15.11), the first and the fourth term cancel, and the third term is nothing but $-\dot{p}\,\mathrm{d}x$, and therefore

$$\mathrm{d}H = \dot{x}\,\mathrm{d}p - \dot{p}\,\mathrm{d}x - \frac{\partial \mathcal{L}}{\partial t}\,\mathrm{d}t . \tag{15.13}$$

This gives the equations of motion:

$$\dot{x} = \frac{\partial H}{\partial p} , \quad \dot{p} = -\frac{\partial H}{\partial x}$$
$$\left(\text{and correspondingly } \dot{x}_i = \frac{\partial H}{\partial p_i} , \ \dot{p}_i = -\frac{\partial H}{\partial x_i}\right) , \tag{15.14}$$

which are called the *canonical equations* of Hamilton and Jacobi. They are *first-order* differential equations in time, and are symmetric in x and p (up to a minus sign). They have the big technical advantage of representing the time evolution of the state variables directly in terms of these state variables. More generally, if we denote by $\boldsymbol{X} = (\boldsymbol{r},\boldsymbol{p})$ the coordinates of the system in phase space, these equations have the form $\dot{\boldsymbol{X}} = F(\boldsymbol{X})$. Such a problem, called a dynamical system, is of considerable interest in many fields, including mathematics.

Notice that in quantum mechanics, the Ehrenfest theorem, derived in Chap. 7, gives the time evolution of expectation values as follows

$$\frac{\mathrm{d}}{\mathrm{d}t}\langle x_i \rangle = \left\langle \frac{\partial \hat{H}}{\partial \hat{p}_i} \right\rangle , \qquad \frac{\mathrm{d}}{\mathrm{d}t}\langle p_i \rangle = \left\langle -\frac{\partial \hat{H}}{\partial \hat{x}_i} \right\rangle ,$$

where we can see some similarity to the canonical Hamilton–Jacobi equations.

15.2.3 Poisson Brackets

Consider two functions f and g of the state variables x, p and possibly of time, for instance two physical quantities. The *Poisson bracket* of f and g is defined as the quantity

$$\{f, g\} = \frac{\partial f}{\partial x}\frac{\partial g}{\partial p} - \frac{\partial f}{\partial p}\frac{\partial g}{\partial x}$$

$$\left(\text{or } \{f, g\} = \sum_{i=1}^{s} \frac{\partial f}{\partial x_i}\frac{\partial g}{\partial p_i} - \frac{\partial f}{\partial p_i}\frac{\partial g}{\partial x_i} \right) . \tag{15.15}$$

We find immediately

$$\{f, g\} = -\{g, f\} , \qquad \{x, p\} = 1 , \tag{15.16}$$

and, more generally,

$$\{x_i, x_j\} = 0 , \qquad \{p_i, p_j\} = 0 , \qquad \{x_i, p_j\} = \delta_{ij} , \tag{15.17}$$

and

$$\{x, f\} = \frac{\partial f}{\partial p} , \qquad \{p, f\} = -\frac{\partial f}{\partial x} . \tag{15.18}$$

Let us now calculate the time evolution of a quantity $f(x, \dot{x}, t)$:

$$\dot{f} = \frac{\mathrm{d}f}{\mathrm{d}t} = \frac{\partial f}{\partial x}\dot{x} + \frac{\partial f}{\partial p}\dot{p} + \frac{\partial f}{\partial t} . \tag{15.19}$$

Using Hamilton's equations (15.14), we obtain

$$\dot{f} = \{f, H\} + \frac{\partial f}{\partial t} . \tag{15.20}$$

In particular, the canonical equations (15.14) are written in the following completely symmetric way:

$$\dot{x} = \{x, H\} , \qquad \dot{p} = \{p, H\} . \tag{15.21}$$

In the canonical formalism, the Hamiltonian governs the time evolution of the system. If a physical quantity f does not depend explicitly on time, i.e. $\partial f/\partial t = 0$, then its time evolution is obtained through the Poisson bracket of f and the Hamiltonian: $\dot{f} = \{f, H\}$. If this Poisson bracket vanishes, f is a constant of the motion.

15.3 Analytical Mechanics and Quantum Mechanics

The results derived in the previous section reveal an amazing property. There is a strong analogy in the structures of analytical mechanics and quantum mechanics. Let us associate with any quantum observable $\hat{A}$ an observable $\dot{\hat{A}}$, such that, by definition,

$$\langle \dot{\hat{A}} \rangle \equiv \frac{d}{dt} \langle a \rangle$$

for any state of the system. The Ehrenfest theorem then implies

$$\dot{\hat{A}} = \frac{1}{i\hbar} [\hat{A}, \hat{H}] + \frac{\partial \hat{A}}{\partial t}, \tag{15.22}$$

to be compared with (15.20). Similarly, the canonical commutation relations

$$[\hat{x}_j, \hat{p}_k] = i\hbar \delta_{jk} \tag{15.23}$$

are highly reminiscent of (15.17).

This identity between the structures of the two kinds of mechanics was one of the first great discoveries of Dirac. Of course, the mathematical nature and the physical interpretation of the objects under consideration are different. But the equations which relate them are the same provided we apply the following correspondence, which was understood by Dirac during the summer of 1925:

Quantization rule. *The Poisson brackets of analytical mechanics are replaced by the commutators of the corresponding observables, divided by* $i\hbar$:

$$\textit{Analytical mechanics } \{f, g\} \longrightarrow \frac{1}{i\hbar} [\hat{f}, \hat{g}] \textit{ Quantum mechanics.} \tag{15.24}$$

This is the genuine form of the correspondence principle. In general, for complex systems (large number of degrees of freedom, constraints, etc.), the systematic method of obtaining the form and the commutation relations of the observables consists in referring to the Poisson brackets of the corresponding classical system. We shall see an example below when we treat the Lorentz force in quantum mechanics.

One can now understand why the name of Hamilton (1805–1865) appears so often in quantum mechanics, although Hamilton lived one century before its invention. Hamilton was one of the great geniuses of science. He made decisive contributions to analytical mechanics and invented vector analysis; he also invented, in the same year as Cayley and Grassmann (1843), noncommutative algebras and matrix calculus (the elements of Hamilton's quaternions are called ... Pauli matrices in quantum mechanics). He was the author of the synthesis of the geometrical and wave theories of light. He found in what limit the former theory is an approximation to the latter.

Hamilton was fascinated by variational principles, in particular by the similarity between Maupertuis's principle in mechanics and Fermat's principle in optics. In 1830 he made the remarkable statement that the formalisms of optics and of mechanics were basically the same and that Newtonian mechanics corresponded to the same limit as geometrical optics, which was only an approximation. This remark was ignored by his contemporaries, and the mathematician Felix Klein said in 1891 that it was a pity. It is true that, in 1830, no experimental fact could have revealed the existence of Planck's constant. However, in many ways, Hamilton can be considered as a precursor of quantum mechanics. Louis de Broglie refers to Hamilton's work in his thesis.

15.4 Classical Charged Particles in an Electromagnetic Field

Classically, a particle of charge q, placed in an electromagnetic field, is subject to the Lorentz force

$$\boldsymbol{f} = q\left(\boldsymbol{E} + \boldsymbol{v} \times \boldsymbol{B}\right) .$$

This force is velocity dependent and cannot be derived from a potential. Furthermore, the magnetic force $q\boldsymbol{v} \times \boldsymbol{B}$ does not do any work, and the energy of the particle is $E = m\boldsymbol{v}^2/2 + q\Phi$, where Φ is the scalar potential associated with the electric field $\boldsymbol{E}$.

The Hamiltonian is certainly different from $\boldsymbol{p}^2/2m + q\Phi$. Otherwise, the equations of motion would be strictly the same as in the absence of a magnetic field. We can use the previous considerations in order to determine the correct form of the Hamiltonian H.

Maxwell's equations, specifically the pair of equations

$$\boldsymbol{\nabla} \cdot \boldsymbol{B} = 0 , \qquad \boldsymbol{\nabla} \times \boldsymbol{E} = -\frac{\partial \boldsymbol{B}}{\partial t} , \tag{15.25}$$

allow us to express the fields $\boldsymbol{E}$ and $\boldsymbol{B}$ in terms of the scalar and vector potentials Φ and $\boldsymbol{A}$:

$$\boldsymbol{B} = \boldsymbol{\nabla} \times \boldsymbol{A} , \qquad \boldsymbol{E} = -\boldsymbol{\nabla}\Phi - \frac{\partial \boldsymbol{A}}{\partial t} . \tag{15.26}$$

Consider a particle of mass m and charge q placed in this electromagnetic field. We denote by $\boldsymbol{r}$ and $\dot{\boldsymbol{r}} = \boldsymbol{v}$ the position and the velocity of this particle. A possible Lagrangian for this particle can be written in terms of the potentials $\boldsymbol{A}$ and Φ:

$$\mathcal{L} = \frac{1}{2} m\dot{\boldsymbol{r}}^2 + q\,\dot{\boldsymbol{r}} \cdot \boldsymbol{A}(\boldsymbol{r}, t) - q\,\Phi(\boldsymbol{r}, t) . \tag{15.27}$$

Indeed, starting from the Lagrange equations and using

$$\frac{\mathrm{d}}{\mathrm{d}t}\boldsymbol{A}(\boldsymbol{r},t) = \frac{\partial \boldsymbol{A}}{\partial t} + \dot{x}\frac{\partial \boldsymbol{A}}{\partial x} + \dot{y}\frac{\partial \boldsymbol{A}}{\partial y} + \dot{z}\frac{\partial \boldsymbol{A}}{\partial z},$$

we can check that we obtain the desired equation of motion:

$$m\frac{\mathrm{d}\boldsymbol{v}}{\mathrm{d}t} = q(\boldsymbol{E} + \boldsymbol{v} \times \boldsymbol{B}).$$

Consider now the conjugate momentum $\boldsymbol{p}$. From the definition (15.10), we have

$$\boldsymbol{p} = m\dot{\boldsymbol{r}} + q\boldsymbol{A}(\boldsymbol{r},t). \tag{15.28}$$

In other words, the conjugate momentum $\boldsymbol{p}$ no longer coincides with the linear momentum, i. e. the product of the mass and the velocity $m\dot{\boldsymbol{r}}$!

Equation (15.28) is easily inverted: $\dot{\boldsymbol{r}} = [\boldsymbol{p} - q\boldsymbol{A}(\boldsymbol{r},t)]/m$, which gives the Hamiltonian we are looking for:

$$H = \frac{1}{2m}[\boldsymbol{p} - q\boldsymbol{A}(\boldsymbol{r},t)]^2 + q\Phi(\boldsymbol{r},t). \tag{15.29}$$

Like the Lagrangian, it is expressed in terms of the potentials $\boldsymbol{A}$ and Φ, and not the fields $\boldsymbol{E}$ and $\boldsymbol{B}$.

15.5 Lorentz Force in Quantum Mechanics

In order to treat the Lorentz force in quantum mechanics, we follow Dirac's quantization rules of Sect. 15.3.

15.5.1 Hamiltonian

The Hamiltonian of a charged particle in an electromagnetic field is

$$\hat{H} = \frac{1}{2m}[\hat{\boldsymbol{p}} - q\boldsymbol{A}(\hat{\boldsymbol{r}},t)]^2 + q\Phi(\hat{\boldsymbol{r}},t), \tag{15.30}$$

where the position and the conjugate momentum operators $\hat{\boldsymbol{r}}$ and $\hat{\boldsymbol{p}}$ satisfy the canonical commutation relations

$$[\hat{x}_j, \hat{x}_k] = 0, \qquad [\hat{p}_j, \hat{p}_k] = 0, \qquad [\hat{x}_j, \hat{p}_k] = \mathrm{i}\hbar\,\delta_{jk}.$$

In the wave function formalism, we can still choose $\hat{\boldsymbol{p}} = -\mathrm{i}\hbar\boldsymbol{\nabla}$. The *velocity* observable is no longer $\hat{\boldsymbol{p}}/m$, but

$$\hat{\boldsymbol{v}} = \frac{1}{m}[\hat{\boldsymbol{p}} - q\boldsymbol{A}(\hat{\boldsymbol{r}},t)]. \tag{15.31}$$

Note that two components of the velocity (e. g. $\hat{v}_x$ and $\hat{v}_y$) do not, in general, commute in the presence of a magnetic field. Using the Ehrenfest theorem, one can verify that (15.30) provides the appropriate structure of the equations of motion for the expectation values.

15.5.2 Gauge Invariance

One thing, though, seems surprising. The potentials Φ and $\boldsymbol{A}$ are not unique. Two sets $(\Phi, \boldsymbol{A})$ and $(\Phi', \boldsymbol{A}')$ related to each other by a *gauge transformation*

$$\boldsymbol{A}' = \boldsymbol{A} + \nabla\chi(\boldsymbol{r}, t)\,, \qquad \Phi' = \Phi - \frac{\partial \chi}{\partial t}\,, \tag{15.32}$$

where $\chi(\boldsymbol{r}, t)$ is an arbitrary function, correspond to the same electric and magnetic fields $\boldsymbol{E}$ and $\boldsymbol{B}$. Since the energy observable $\hat{H}$ is expressed in terms of $\boldsymbol{A}$ and Φ, the energy seems to depend on the gauge. However, we know that physical results should not depend on the gauge!

The answer to this problem is simple and remarkable. In a gauge transformation, the wave function also changes:

$$\psi(\boldsymbol{r}, t) \to \psi'(\boldsymbol{r}, t) = \mathrm{e}^{\mathrm{i}q\chi(\boldsymbol{r},t)/\hbar}\,\psi(\boldsymbol{r}, t)\,. \tag{15.33}$$

One can check that if ψ is a solution of the Schrödinger equation for the choice of potentials $(\boldsymbol{A}, \Phi)$, then ψ' is a solution for the choice $(\boldsymbol{A}', \Phi')$. For time-independent problems, this guarantees that the energy spectrum of the Hamiltonian for the choice $(\boldsymbol{A}, \Phi)$ coincides with the spectrum obtained with $(\boldsymbol{A}', \Phi')$.

The transformation (15.33) does not modify the probability density,

$$|\psi(\boldsymbol{r}, t)|^2 = |\psi'(\boldsymbol{r}, t)|^2\,,$$

which is of course crucial. The transformation simply affects the phase of the wave function by an amount which depends on the point in space.

One can verify, more generally, that the expectation values of all *measurable quantities* are gauge invariant. Consider, for instance, the velocity operator $\hat{\boldsymbol{v}} = \left(\hat{\boldsymbol{p}} - q\hat{\boldsymbol{A}}\right)/m$. We find

$$\begin{aligned}\left(\hat{\boldsymbol{p}} - q\hat{\boldsymbol{A}}'\right)\psi' &= \left(-\mathrm{i}\hbar\nabla - q\hat{\boldsymbol{A}} - q\boldsymbol{\nabla}\hat{\chi}\right)\,\mathrm{e}^{\mathrm{i}q\chi/\hbar}\,\psi \\ &= \mathrm{e}^{\mathrm{i}q\chi/\hbar}\left(-\mathrm{i}\hbar\nabla - q\hat{\boldsymbol{A}}\right)\psi = \mathrm{e}^{\mathrm{i}q\chi/\hbar}\left(\hat{\boldsymbol{p}} - q\hat{\boldsymbol{A}}\right)\psi\,,\end{aligned}$$

from which we deduce that

$$\psi'^*\left(\hat{\boldsymbol{p}} - q\hat{\boldsymbol{A}}'\right)\psi' = \psi^*\left(\hat{\boldsymbol{p}} - q\hat{\boldsymbol{A}}\right)\psi\,.$$

This proves that the probability current is the same in both gauges. If we integrate this relation over space, we find that the expectation value of the velocity is also gauge independent. In contrast, the momentum $\hat{\boldsymbol{p}}$ is not a gauge-invariant physical quantity.

If one *postulates* that the laws of physics are invariant under all gauge transformations (15.33), where $\chi(\boldsymbol{r}, t)$ is arbitrary, one can *derive* the result that the Hamiltonian has the structure (15.30). In quantum field theory, gauge invariance plays a crucial role in the physics of fundamental interactions and of the elementary constituents of matter.

The fact that the Hamiltonian (15.30) depends on the potentials and not on the fields can be verified experimentally, following a suggestion by Aharonov and Bohm in 1956. In a Young slit interference device, one places between the two slits a solenoid of small diameter, parallel to the slits (Fig. 15.2). When a current flows in the solenoid, one can observe a modification of the system of fringes. However, the magnetic field is zero everywhere outside the solenoid, in particular near the slits. Conversely, the vector potential is non zero outside the solenoid. This experiment has been performed and has confirmed the quantum mechanical predictions.[2]

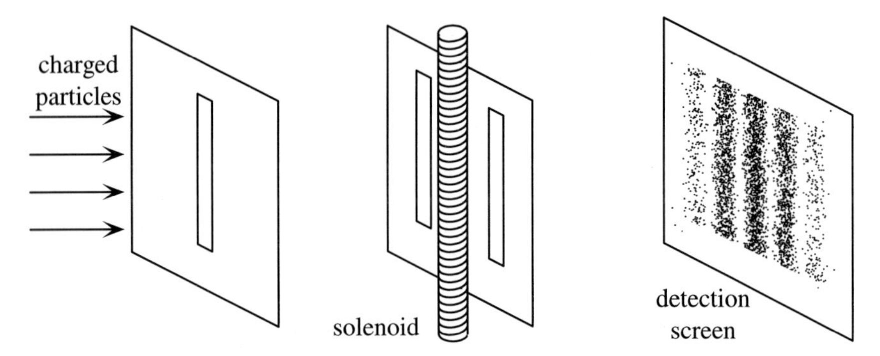

Fig. 15.2. The Aharonov–Bohm effect: in a Young slit experiment performed with charged particles, the interference pattern is shifted when a current is passed through the solenoid. However, the magnetic field created by this solenoid is zero everywhere except inside the solenoid itself. The corresponding phase shift has no classical counterpart (in terms of a force acting on the particles, for instance) and one refers to it as a topological phase shift

15.5.3 The Hydrogen Atom Without Spin in a Uniform Magnetic Field

We place a hydrogen atom in a constant uniform field $\boldsymbol{B}$, which is derived from the vector potential $\boldsymbol{A} = \boldsymbol{B} \times \boldsymbol{r}/2$, and we neglect spin effects for the moment. The Hamiltonian

$$\hat{H} = \frac{1}{2m_{\mathrm{e}}}(\hat{\boldsymbol{p}} + q\hat{\boldsymbol{A}})^2 + V(\hat{r}),$$

where $V(r) = -q^2/4\pi\varepsilon_0 r$ and $-q$ is the electron charge, can be expanded as

$$\hat{H} = \hat{H}_0 + \frac{q}{2m_{\mathrm{e}}}\left(\hat{\boldsymbol{p}} \cdot \hat{\boldsymbol{A}} + \hat{\boldsymbol{A}} \cdot \hat{\boldsymbol{p}}\right) + \frac{q^2}{2m_{\mathrm{e}}}\hat{\boldsymbol{A}}^2, \quad \text{where} \quad H_0 = \frac{\hat{\boldsymbol{p}}^2}{2m_{\mathrm{e}}} + \hat{V}(r).$$

[2] A. Tonomura et al., Phys. Rev. Lett. **56**, 792 (1986).

The first term $\hat{H}_0$ is simply the Hamiltonian studied in Chap. 11. The second term is called the *paramagnetic* term. We remark that, since $\hat{\boldsymbol{p}} \cdot \hat{\boldsymbol{A}} = \hat{\boldsymbol{A}} \cdot \hat{\boldsymbol{p}}$ in this gauge, we can rewrite this term as

$$\frac{q}{2m_\mathrm{e}}\,(\boldsymbol{B} \times \hat{\boldsymbol{r}}) \cdot \hat{\boldsymbol{p}} = \frac{q}{2m_\mathrm{e}}\,(\hat{\boldsymbol{r}} \times \hat{\boldsymbol{p}}) \cdot \boldsymbol{B} = -\gamma_0\,\hat{\boldsymbol{L}} \cdot \boldsymbol{B} = -\hat{\boldsymbol{\mu}}_L \cdot \boldsymbol{B}\,, \qquad (15.34)$$

where $\gamma_0 = -q/2m_\mathrm{e}$. We recover the magnetic dipole interaction term introduced in Chap. 10 (10.36).

The third term, $q^2\hat{\boldsymbol{A}}^2/2m_\mathrm{e}$, is called the *diamagnetic* term. One can check that for the lowest-lying levels E_n of the hydrogen atom, for magnetic fields below 1 T, the diamagnetic term is negligible: it is much smaller (by a factor of $\sim 10^{-4}$) than the paramagnetic term, which is itself small (by a factor of $\sim 10^{-4}$) compared with $|E_n|$.

15.5.4 Spin-1/2 Particle in an Electromagnetic Field

Consider a spin-1/2 particle, which may or may not be charged, with an intrinsic magnetic moment $\hat{\boldsymbol{\mu}}_S = \gamma_S \hat{\boldsymbol{S}}$, where $\hat{\boldsymbol{S}}$ is the spin observable. If we place this particle in an electromagnetic field, and possibly in another potential $V(\boldsymbol{r})$, its Hamiltonian is

$$\hat{H} = \frac{1}{2m}\left[\hat{\boldsymbol{p}} - q\boldsymbol{A}(\hat{\boldsymbol{r}},t)\right]^2 + q\Phi(\hat{\boldsymbol{r}},t) + V(\hat{\boldsymbol{r}}) - \hat{\boldsymbol{\mu}}_S \cdot \boldsymbol{B}(\hat{\boldsymbol{r}},t)\,, \qquad (15.35)$$

where q is the charge of the particle, and $\boldsymbol{A}$ and Φ are the electromagnetic potentials.

This Hamiltonian is called the Pauli Hamiltonian. It acts in the Hilbert space $\mathcal{E}_\mathrm{external} \otimes \mathcal{E}_\mathrm{spin}$ described in Chap. 12. For an electron, the form (15.35) can be directly obtained as the nonrelativistic limit of the Dirac equation, which predicts $\gamma_S = 2\gamma_0 = -q/m_\mathrm{e}$.

Further Reading

- Concerning the importance of variational principles in physics, see for instance R.P. Feynman, R.B. Leighton and M. Sands, *The Feynman Lectures on Physics*, Vol. II, Chap. 19, Addison-Wesley, Reading, MA (1964).
- Concerning variational principles in quantum mechanics, see R.P. Feynman and A.R. Hibbs, *Quantum Mechanics and Path Integrals*, McGraw-Hill, New York (1965).

Exercises

15.1. The Lorentz force in quantum mechanics. In this exercise, we want to check that with the prescription

$$\hat{H} = \frac{1}{2m}[\hat{\boldsymbol{p}} - q\boldsymbol{A}(\hat{\boldsymbol{r}})]^2$$

for the Hamiltonian of a charged particle in a magnetic field, one can recover, using the Ehrenfest theorem, the classical equations of motion.

a. We assume in all that follows that the field $\boldsymbol{B}$ is constant, uniform and directed along the z axis. We set $B = |\boldsymbol{B}|$. We introduce the vector potential $\boldsymbol{A} = \boldsymbol{B} \times \boldsymbol{r}/2$. Check that this choice gives the appropriate value of $\boldsymbol{B}$.
b. Write down the classical equation of motion of a particle of charge q and mass m in this field. Give the expression for the energy E of the particle. Describe the characteristics of the motion of the particle.
c. Consider the observable $\hat{\boldsymbol{u}} = \hat{\boldsymbol{p}} - q\hat{\boldsymbol{A}}$, where $\hat{\boldsymbol{A}} = \boldsymbol{A}(\hat{\boldsymbol{r}})$. Here $\hat{\boldsymbol{p}}$ is the usual momentum operator, i. e. $\hat{\boldsymbol{p}} = -\mathrm{i}\hbar\boldsymbol{\nabla}$ (which yields $[\hat{x}, \hat{p}_x] = \mathrm{i}\hbar$). Show that $\hat{\boldsymbol{p}} \cdot \hat{\boldsymbol{A}} = \hat{\boldsymbol{A}} \cdot \hat{\boldsymbol{p}}$. Write down the commutators $[\hat{u}_x, \hat{u}_y]$, $[\hat{u}_y, \hat{u}_z]$ and $[\hat{u}_z, \hat{u}_x]$ for the three components of the observable $\hat{\boldsymbol{u}}$.
d. We assume that the quantum Hamiltonian has the form $\hat{H} = \hat{\boldsymbol{u}}^2/2m$. Calculate $\mathrm{d}\langle \boldsymbol{r} \rangle / \mathrm{d}t$ and $\mathrm{d}\langle \boldsymbol{u} \rangle / \mathrm{d}t$. Compare the result with the classical equations of motion.
e. Deduce from this result the form of the velocity observable $\hat{\boldsymbol{v}}$.
f. Can the three components of the velocity be defined simultaneously in a magnetic field? Write down the corresponding uncertainty relations.

15.2. Landau levels. In this exercise we determine the energy levels of a spinless particle of charge q and mass m, which is placed in a constant, uniform magnetic field $\boldsymbol{B} = B\,\boldsymbol{u}_z$. We use here the Landau gauge $\boldsymbol{A}(\boldsymbol{r}) = Bx\,\boldsymbol{u}_y$.

a. Write down the eigenvalue equation for the Hamiltonian $\hat{H}$. The eigenfunction is denoted by $\Psi(\boldsymbol{r})$ and the corresponding eigenvalue by E_{tot}.
b. We look for particular solutions which are factorized in the following form

$$\Psi(x, y, z) = \mathrm{e}^{\mathrm{i}k_z z}\,\psi(x, y)\,.$$

Show that $\psi(x, y)$ is a solution of the eigenvalue equation

$$\frac{-\hbar^2}{2m}\left(\frac{\partial^2}{\partial x^2} + \left(\frac{\partial}{\partial y} - \mathrm{i}\frac{qB}{\hbar}x\right)^2\right)\psi(x, y) = E\,\psi(x, y)\,, \tag{15.36}$$

where $E = E_{\mathrm{tot}} - \hbar^2 k_z^2/2m$.
c. Equation (15.36) describes the motion of the charged particle in the xy plane. We look for particular solutions of this equation which are also factorized with respect to x and y:

$$\psi(x, y) = \mathrm{e}^{\mathrm{i}k_y y}\,\chi(x)\,.$$

(i) Write down the equation which determines $\chi(x)$. To which physical problem does it correspond? Introduce the cyclotron angular frequency $\omega_c = qB/m$.

(ii) Show that the possible eigenvalues for the energy E are

$$E = \left(n + \frac{1}{2}\right) \hbar\omega_c \,. \tag{15.37}$$

Do the eigenvalues depend on the wave vector k_y? The corresponding energy levels are called the *Landau levels*.

d. We now determine the degeneracy of a Landau level, assuming that the xy motion of the particle is confined in a rectangle $[0, X] \times [0, Y]$. We shall neglect any edge effect, assuming that $a_0 = (2\hbar/qB)^{1/2} \ll X, Y$, and we restrict ourselves to relatively low values of the quantum number n.

(i) We choose periodic boundary conditions for the motion along the y axis. Show that the wave vector k_y is quantized according to $k_y = 2\pi j/Y$, where j is an integer.

(ii) What are the relevant values of j, such that the wave function $\psi(x, y)$ is localized in the rectangle $X \times Y$ (and thus is physically acceptable)?

(iii) Express the degeneracy of a Landau level as a function of the flux $\Phi = BXY$ and of the *magnetic flux quantum* $\Phi_0 = h/q$.

15.3. The lowest Landau level (LLL). As in the previous exercise, we consider the quantum motion of a particle of charge q and mass m in a uniform magnetic field $\boldsymbol{B} = B\,\boldsymbol{u}_z$. Here we choose the symmetric gauge $\boldsymbol{A} = \boldsymbol{B} \times \boldsymbol{r}/2$. We restrict ourselves to the motion of the particle in the xy plane ($k_z = 0$) and we set, as above, $\omega_c = qB/m$.

a. Write down the eigenvalue equation for the energy. Introduce $\hat{L}_z = \hat{x}\hat{p}_y - \hat{y}\hat{p}_x$.

b. Consider the Landau level with the lowest energy $E_{\mathrm{LLL}} = \hbar\omega_c/2$ (see (15.37)). Show that the functions

$$\psi_\ell(x, y) = (x + \mathrm{i}y)^\ell \, \mathrm{e}^{-(x^2+y^2)/(2\,a_0^2)} \,,$$

where ℓ is an arbitrary integer and $a_0 = (2\hbar/qB)^{1/2}$ are all energy eigenstates for the eigenvalue E_{LLL}.

c. Recover for the LLL the degeneracy calculated in the previous exercise, assuming that the particle is confined in a disk centered on $x = y = 0$, with a radius $R \gg a_0$.

This eigenstate basis plays an important role in the study of the fractional quantum Hall effect, which was discovered in a two-dimensional electron gas placed in a magnetic field.

15.4. The Aharonov–Bohm effect. Consider the two-hole Young interference experiment shown in Fig. 15.3. A solenoid whose axis is perpendicular to the plane of the figure is placed between the two holes B and B'. A particle is emitted from the source point O at time t_1, and the impact of this particle on the detection screen is detected at a later time t_2. We shall assume that the probability amplitude $A(C)$ for detecting the particle at C is approximately given by[3]

$$A(C) = A_{OBC} + A_{OB'C} \propto \mathrm{e}^{\mathrm{i}S/\hbar} + \mathrm{e}^{\mathrm{i}S'\hbar} ,$$

where S and S' are the classical actions calculated along the paths OBC and $OB'C$, respectively.

a. In the absence of a current in the solenoid, recover the fringe spacing x_s found in Chap. 1 (Sect. 1.2.2).
b. Determine the change in the interference signal when a current flows in the solenoid. Express the result in terms of the total magnetic flux $\Phi = \pi r^2 B$ (where r is the radius of the solenoid and B the magnetic field inside the solenoid) and the *magnetic flux quantum* $\Phi_0 = h/q$.

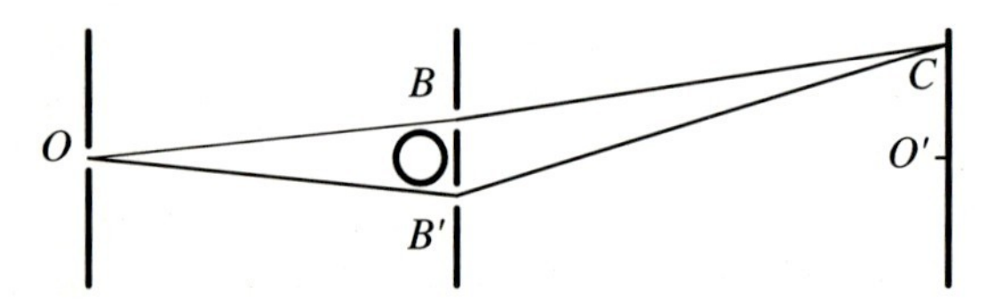

Fig. 15.3. Study of the Aharonov–Bohm effect: classical paths in a two-hole Young interference experiment

[3] This prescription can be deduced from the formulation of quantum mechanics based on path integrals; see R.P. Feynman and A.R. Hibbs, *Quantum Mechanics and Path Integrals*, McGraw-Hill, New York (1965).

16. Identical Particles and the Pauli Principle

If atoms existed ...,
some would be by nature indistinguishable ...,
which contradicts the most fundamental principles of reason.
Gottfried Wilhelm Leibniz

The origin of geometry (from Pythagoras to Euclid) lies in our environment, and in the observation that one can model the world in which we live by a space where each object is described by a point or a set of points. The concept of space itself came after the simpler concept of the "place" of an object. The idea of space arose from the question as to whether a place exists independently from the fact that some object occupies it or not. In this context, by definition, two objects cannot have the same position at the same time.

In this chapter, we address the quantum transposition of this problem. In the probabilistic quantum description, there is no reason a priori why the probability density for two particles to be at the same point in space should vanish, contrary to the classical observation. It is therefore legitimate to elevate the above question to state vectors (or wave functions) rather than positions. Can two particles be in the *same state* at the *same time*?

Naturally, two particles of different kinds, such as an electron and a proton, will never be in the same state: even if their wave functions coincide, their mass difference implies differences in the values of various physical quantities and one can always tell them apart from one another. However, there exist in nature *identical particles*: all electrons in the universe have the same mass, the same charge, etc. Can such particles, whose intrinsic properties are all the same, be in the same state? The answer lies in one of the simplest, but most profound principles of physics, whose consequences for the structure of matter are numerous and fundamental: the *Pauli principle.*

The principles of Chap. 5 do not suffice when one is dealing with systems containing identical particles. We show in Sect. 16.1 that there is a genuine physical problem: some predictions are ambiguous. A new fundamental principle must be added in order to get rid of this ambiguity. The basic property of two identical particles is that they can be interchanged in a physical system without modifying any property of this system. The mathematical tool which corresponds to the interchange of two particles is the *exchange oper-*

ator, which we introduce in Sect. 16.2. In Sect. 16.3 we express the Pauli principle as an additional axiom. Finally, in Sect. 16.4, we discuss some consequences of this principle.

16.1 Indistinguishability of Two Identical Particles

We first define what is meant by identical particles and the ensuing quantum mechanical problem.

16.1.1 Identical Particles in Classical Physics

By definition, two particles are identical if all their intrinsic properties are the same. In classical mechanics, for a two-particle system, it is always possible to measure at a given time the position of each particle. At that instant, we can define which particle we call 1 and which one we call 2. It is also possible to follow the trajectory of particle 1 and that of particle 2. We can keep on distinguishing unambiguously each particle at any later time. For instance, in a collision of two billiard balls of the same color, we can unambiguously tell the difference between the two processes shown in Fig. 16.1. Therefore, for any system which is described by classical physics, two particles are always distinguishable, whether or not they are identical (concerning macroscopic objects, the notion of identity is an idealization anyway).

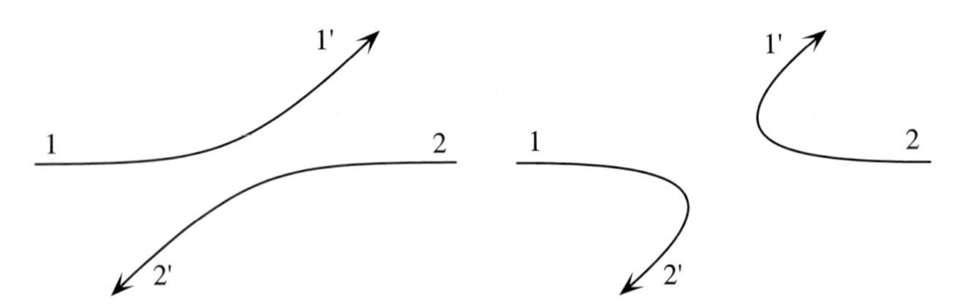

Fig. 16.1. Collision between two identical particles

16.1.2 The Quantum Problem

The situation is different in quantum mechanics. At a given time we can still measure the positions of the particles and label them with the indices 1 and 2. However, since the notion of a trajectory is not defined, it may be impossible to follow the two particles individually as time goes on. For instance, one cannot tell the difference between the two processes sketched in Fig. 16.1 if the two wave functions of particles 1 and 2 overlap. It is impossible to know whether particle 1 has become particle $1'$ or particle $2'$. In quantum mechanics two identical particles are indistinguishable.

Here, physics falsifies the famous "principle of the identity of indistinguishables", which was a basic principle of the philosophy of Leibniz, for whom two real objects were never exactly similar. We shall see that there exist cases where N particles can be in the same state (a Bose–Einstein condensate) although they are not a single entity. The *number* N of these particles is a measurable quantity, although they are indistinguishable from one another.

In the framework of the principles of Chap. 5, this indistinguishability leads to ambiguities in the predictions of physical measurements. Consider for instance two identical particles moving in a one-dimensional harmonic potential. We label the particles 1 and 2 and we assume that the Hamiltonian is

$$\hat{H} = \frac{\hat{p}_1^2}{2m} + \frac{1}{2}m\omega^2\hat{x}_1^2 + \frac{\hat{p}_2^2}{2m} + \frac{1}{2}m\omega^2\hat{x}_2^2 = \hat{h}^{(1)} + \hat{h}^{(2)} .$$

For simplicity, we suppose that the particles have no mutual interaction. Let $(n+1/2)\hbar\omega$ and $\phi_n(x)$ $(n = 0, 1, \ldots)$ be the eigenvalues and eigenfunctions of the one-particle Hamiltonian $\hat{h} = \hat{p}^2/2m + m\omega^2\hat{x}^2/2$.

No problem occurs in describing the physical situation where both particles are in the ground state of $\hat{h}$. The corresponding state is

$$\Phi_0(x_1, x_2) = \phi_0(x_1)\,\phi_0(x_2) ,$$

and its energy is $E_0 = \hbar\omega$.

In contrast, the description of the first excited state of the system is ambiguous. This corresponds to one of the particles being in the first excited state of $\hat{h}$ and the other in the ground state. The total energy is $2\hbar\omega$. One possible state is $\phi_1(x_1)\,\phi_0(x_2)$, another possible state is $\phi_0(x_1)\,\phi_1(x_2)$. Since these two states are possible candidates, then, according to the superposition principle, any linear combination

$$\Phi(x_1, x_2) = \lambda\,\phi_1(x_1)\,\phi_0(x_2) + \mu\,\phi_0(x_1)\,\phi_1(x_2)$$

also corresponds to an energy $2\hbar\omega$.

Therefore there are several different states which appear to describe the same physical situation. This might not be a problem, provided no measurement could tell the difference. Alas, this is not true! These various states lead to different predictions concerning physically measurable quantities. Consider for instance the product of the two positions, i. e. the observable $\hat{x}_1 \otimes \hat{x}_2$, for which the labeling of the two particles is irrelevant. Its expectation value is

$$\langle x_1 x_2 \rangle = \frac{\hbar}{m\omega}\,\mathrm{Re}(\lambda^* \mu) .$$

This prediction depends on λ and μ! However, nothing in the theory that we have presented tells us the values of these parameters. Therefore there is a basic ambiguity in the predictions of our principles, and we need a prescription in order to fix the values of λ and μ.

It is a remarkable fact of nature that the only allowed values are $\lambda = \pm\mu$, and that the sign depends only on the nature of the particles under consideration. The allowed states for a system of identical particles are therefore *restrictions* of the most general states that one could imagine if the particles were distinguishable.

16.2 Two-Particle Systems; the Exchange Operator

We consider a system of two identical particles. We shall elicit the constraint that this places on the two-particle state.

16.2.1 The Hilbert Space for the Two Particle System

Within the framework that we have used up to now, we describe a two-particle system (distinguishable or not) by labeling these particles. Here we choose to call them 1 and 2. The Hilbert space of the system is the tensor product of the Hilbert spaces of the two particles $\mathcal{E} = \mathcal{E}^{(1)} \otimes \mathcal{E}^{(2)}$. We denote by $\{|k\rangle\}$ and $\{|n\rangle\}$ a basis of $\mathcal{E}^{(1)}$ and of $\mathcal{E}^{(2)}$, respectively. A state of the system is therefore of the form

$$|\psi\rangle = \sum_{k,n} C_{k,n}\, |k\rangle \otimes |n\rangle \equiv \sum_{k,n} C_{k,n}\, |1:k\,;\,2:n\rangle\,. \tag{16.1}$$

16.2.2 The Exchange Operator Between Two Identical Particles

The labeling of the particles that we have used above has no absolute meaning if they are identical. Consequently, the predictions of experimental results must be independent of this labeling. In order to describe this property due to the exchange symmetry, we introduce the *exchange operator* $\hat{P}_{12}$ such that, for any pair (k, n),

$$\hat{P}_{12}\,|1:k\,;\,2:n\rangle = |1:n\,;\,2:k\rangle\,. \tag{16.2}$$

One can verify that this operator is Hermitian and that it satisfies

$$\hat{P}_{12}^2 = \hat{I}\,. \tag{16.3}$$

Examples.

a. For two spinless particles, we have

$$\hat{P}_{12} \equiv \hat{P}_{12}^{(\mathrm{external})}\,, \qquad \text{i.e.} \qquad \hat{P}_{12}\,\Psi(\boldsymbol{r}_1, \boldsymbol{r}_2) = \Psi(\boldsymbol{r}_2, \boldsymbol{r}_1)\,.$$

b. For two particles with spin, $\hat{P}_{12}$ exchanges both the orbital and the spin variables of the two particles:

$$\hat{P}_{12} = \hat{P}_{12}^{(\mathrm{external})} \otimes \hat{P}_{12}^{(\mathrm{spin})}\,.$$

c. *Permutation of two spin-1/2 particles.* In this case, one can write down explicitly the action of $\hat{P}_{12}$ by using the mixed representation introduced in Chap. 12:

$$\hat{P}_{12} \sum_{\sigma_1,\sigma_2} \Psi_{\sigma_1,\sigma_2}(\boldsymbol{r}_1,\boldsymbol{r}_2)\,|1:\sigma_1\,;\,2:\sigma_2\rangle$$
$$= \sum_{\sigma_1,\sigma_2} \Psi_{\sigma_1,\sigma_2}(\boldsymbol{r}_2,\boldsymbol{r}_1)\,|1:\sigma_2\,;\,2:\sigma_1\rangle\,,$$

where $\sigma_i = \pm$, with $i = 1, 2$. In order to discuss the properties of this permutation, it is convenient to shift from the decoupled basis $|1\!:\!\sigma_1;2\!:\!\sigma_2\rangle$ to the coupled basis $|S,m\rangle$, which is the eigenbasis of the square of the total spin $\hat{\boldsymbol{S}} = \hat{\boldsymbol{S}}_\mathbf{1} + \hat{\boldsymbol{S}}_\mathbf{2}$ and $\hat{S}_z$ (see Chap. 13):

$$|S=1,\,m=1\rangle = |1:+\,;\,2:+\rangle\,,$$
$$|S=1,\,m=0\rangle = \frac{1}{\sqrt{2}}\left(|1:+\,;\,2:-\rangle + |1:-\,;\,2:+\rangle\right)\,,$$
$$|S=1,\,m=-1\rangle = |1:-\,;\,2:-\rangle\,,$$
$$|S=0,\,m=0\rangle = \frac{1}{\sqrt{2}}\left(|1:+\,;\,2:-\rangle - |1:-\,;\,2:+\rangle\right)\,.$$

We have already noticed that:

- the triplet states ($S = 1$) are symmetric under the interchange of σ_1 and σ_2:

$$\hat{P}_{12}^{(\mathrm{spin})}|S=1,m\rangle \;=\; |S=1,m\rangle\,;$$

- the singlet state ($S = 0$) is antisymmetric under this interchange:

$$\hat{P}_{12}^{(\mathrm{spin})}|S=0,m=0\rangle \;=\; -\;|S=0,m=0\rangle\,.$$

16.2.3 Symmetry of the States

How can one fulfill the requirement that the experimental results must be unchanged as one goes from $|\Psi\rangle$ to $\hat{P}_{12}|\Psi\rangle$? These two vectors must represent the same physical state; therefore, they can differ only by a phase factor, i. e. $\hat{P}_{12}|\Psi\rangle = \mathrm{e}^{\mathrm{i}\delta}\,|\Psi\rangle$. Since $\hat{P}_{12}^2 = \hat{I}$, we have $\mathrm{e}^{2\mathrm{i}\delta} = 1$ and $\mathrm{e}^{\mathrm{i}\delta} = \pm 1$. Therefore

$$\hat{P}_{12}|\Psi\rangle = \pm|\Psi\rangle\,. \tag{16.4}$$

We then reach the following conclusion:

The only physically acceptable state vectors for a system of two identical particles are either symmetric or antisymmetric under the permutation of the two particles.

Referring to (16.1), this implies $C_{k,n} = \pm C_{n,k}$. The only allowed states are either symmetric under the exchange of 1 and 2, i. e.

$$|\Psi_S\rangle \propto \sum_{k,n} C_{k,n}\left(|1:k\,;\,2:n\rangle + |1:n\,;\,2:k\rangle\right), \quad \hat{P}_{12}|\Psi_S\rangle = |\Psi_S\rangle\,, \tag{16.5}$$

or antisymmetric, i. e.

$$|\Psi_A\rangle \propto \sum_{k,n\ (k\neq n)} C_{k,n}\left(|1:k\,;\,2:n\rangle - |1:n\,;\,2:k\rangle\right), \quad \hat{P}_{12}|\Psi_A\rangle = -|\Psi_A\rangle\,, \tag{16.6}$$

where the $C_{k,n}$'s are arbitrary.

This restriction to symmetric or antisymmetric state vectors is a considerable step forward in resolving the ambiguity pointed out in the previous section. For instance, the expectation value $\langle x_1 x_2\rangle$ considered in Sect. 16.1.2 can now take only two values $\pm\hbar/(2m\omega)$, corresponding to the two choices $\lambda = \pm\mu = 1/\sqrt{2}$. However, this is not yet sufficient, and two essential questions are still open:

a. Can a given species, e. g. electrons, behave in some experimental situations with the plus sign in (16.4) and in other situations with the minus sign?
b. Assuming that the answer to the first question is negative, what decides the sign which should be associated with a given species?

16.3 The Pauli Principle

The answers to the two questions above lead us to one of the simplest and most fundamental laws of physics. It is called the *Pauli principle*, although the general formulation was derived from Pauli's ideas by Fermi and Dirac.

16.3.1 The Case of Two Particles

Principle 4: the Pauli Principle

All particles in nature belong to one of the two following categories:

- *bosons*, for which the state vector of two identical particles is *always symmetric* under the operation $\hat{P}_{12}$;
- *fermions*, for which the state vector of two identical particles is *always antisymmetric* under the operation $\hat{P}_{12}$.

All particles of *integer spin* (including 0) are *bosons* (photon, π meson, α particle, gluons, etc.).
All particles of *half-integer spin* are *fermions* (electron, proton, neutron, neutrino, quarks, ^{3}He nucleus, etc.).

The state vectors of two bosons are of the form $|\Psi_\mathrm{S}\rangle$ (16.5), and those of fermions are of the form $|\Psi_\mathrm{A}\rangle$ (16.6). The Pauli principle therefore consists in restricting the set of accessible states for systems of two identical particles. The space of physical states is no longer the tensor product of the basis states, but only the subspace formed by the symmetric or antisymmetric combinations.

The Pauli principle also applies to composite particles such as nuclei or atoms, provided the experimental conditions are such that they are in the same internal state (be it the ground state or a metastable excited state). For instance, hydrogen atoms in their ground electronic and hyperfine state have a total spin $S = 0$ (Chap. 13) and behave as bosons.

This connection between the symmetry of states and the spin of particles is an experimental fact. However, it is both a triumph and a mystery of contemporary physics that this property, called the "spin–statistics connection", can be proven starting from general axioms of relativistic quantum field theory. It is a mystery because it is probably the only example of a simple physical law for which a proof exists but cannot be explained in an elementary way.

Examples

a. The wave function of two identical spin-zero particles must be symmetric: $\Psi(\boldsymbol{r}_1, \boldsymbol{r}_2) = \Psi(\boldsymbol{r}_2, \boldsymbol{r}_1)$.

b. The state of two spin-1/2 particles must be of the form

$$|\Psi\rangle = \Psi_{0,0}(\boldsymbol{r}_1, \boldsymbol{r}_2)\,|S=0, m=0\rangle + \sum_m \Psi_{1,m}(\boldsymbol{r}_1, \boldsymbol{r}_2)\,|S=1, m\rangle\,,$$

where $\Psi_{0,0}$ and $\Psi_{1,m}$ are symmetric and antisymmetric, respectively:

$$\Psi_{0,0}(\boldsymbol{r}_1, \boldsymbol{r}_2) = \Psi_{0,0}(\boldsymbol{r}_2, \boldsymbol{r}_1) \qquad \Psi_{1,m}(\boldsymbol{r}_1, \boldsymbol{r}_2) = -\Psi_{1,m}(\boldsymbol{r}_2, \boldsymbol{r}_1)\,.$$

Therefore the orbital state and the spin state of two identical fermions are correlated.

16.3.2 Independent Fermions and Exclusion Principle

Consider a situation where two fermions, for instance two electrons, are *independent*, i.e. they do not interact with each other. The total Hamiltonian then reads $\hat{H} = \hat{h}(1) + \hat{h}(2)$. In such conditions, the eigenstates of $\hat{H}$ are products of eigenstates $|n\rangle$ of $\hat{h}$: $|1:n\,;\,2:n'\rangle$. We remark that if $n = n'$, i.e. if the two particles are in the same quantum state, the state $|1:n\,;\,2:n\rangle$ is necessarily symmetric. This is forbidden by the Pauli principle, which results in the following (weaker) formulation:

Two independent fermions in the same system cannot be in the same state.

If $|n\rangle$ and $|n'\rangle$ are orthogonal, the only acceptable state is the antisymmetric combination

$$|\Psi_{\mathrm{A}}\rangle = \frac{1}{\sqrt{2}}\left(|1:n\,;\,2:n'\rangle - |1:n'\,;\,2:n\rangle\right) .$$

In this simplified form, the Pauli principle appears as an exclusion principle. This point of view is an approximation since two particles are never completely independent.

16.3.3 The Case of N Identical Particles

In the case of a system of N identical particles, we proceed in a similar manner. We introduce the exchange operator $\hat{P}_{ij}$ of the two particles i and j. The indistinguishability imposes the condition that $\hat{P}_{ij}|\Psi\rangle$ leads to the same physical results as $|\Psi\rangle$. The general form of the Pauli principle is as follows:

Principle 4: the Pauli Principle (General Form)

The state vector of a system of N identical bosons is completely symmetric under the interchange of any two of these particles.

The state vector of a system of N identical fermions is completely antisymmetric under the interchange of any two of these particles.

For instance, for $N = 3$, one has

$$\begin{aligned}\Psi_{\pm}(u_1,u_2,u_3) \propto & \left[f(u_1,u_2,u_3) + f(u_2,u_3,u_1) + f(u_3,u_1,u_2)\right] \\ & \pm \left[f(u_1,u_3,u_2) + f(u_2,u_1,u_3) + f(u_3,u_2,u_1)\right] ,\end{aligned}$$

where f is any function of the three sets of variables u_1, u_2, u_3. The plus sign corresponds to a function Ψ which is completely symmetric, and the minus sign to a function which is completely antisymmetric.

More generally, let us consider an orthonormal basis $\{|n\rangle\}$ of the one-particle states, and the $N!$ permutations P of a set of N elements. We want to describe the following physical situation: "*one particle in the state* $|n_1\rangle$, *one particle in the state* $|n_2\rangle$, ..., *one particle in the state* $|n_N\rangle$". In order to do this, we number the N particles in an arbitrary way from 1 to N and consider the following states.

- For bosons, we take

$$|\Psi\rangle = \frac{C}{\sqrt{N!}} \sum_P |1:n_{P(1)}\,;\,2:n_{P(2)}\,;\,\ldots\,;\,N:n_{P(N)}\rangle , \tag{16.7}$$

where $\sum_P$ denotes the sum over all permutations. Notice that two (or more) indices $n_i, n_j, \ldots$ labeling the occupied states may coincide. The normalization factor C is expressed in terms of the occupation numbers N_i of the states $|n_i\rangle$:

$$C = (N_1!\,N_2!\dots)^{-1/2} \ .$$

- In the case of fermions, the result is physically acceptable if and only if the N states $|n_i\rangle$ are pairwise orthogonal. The state $|\Psi\rangle$ is then

$$|\Psi\rangle = \frac{1}{\sqrt{N!}} \sum_P \varepsilon_P \, |1 : n_{P(1)} \, ; \, 2 : n_{P(2)} \, ; \, \dots \, ; \, N : n_{P(N)}\rangle \, , \tag{16.8}$$

  where ε_P is the signature of the permutation P: $\varepsilon_P = 1$ if P is an even permutation and $\varepsilon_P = -1$ if P is odd. This state vector is often written in the form of a determinant, called the Slater determinant:

$$|\Psi\rangle = \frac{1}{\sqrt{N!}} \begin{vmatrix} |1 : n_1\rangle & |1 : n_2\rangle & \dots & |1 : n_N\rangle \\ |2 : n_1\rangle & |2 : n_2\rangle & \dots & |2 : n_N\rangle \\ \vdots & \vdots & & \vdots \\ |N : n_1\rangle & |N : n_2\rangle & \dots & |N : n_N\rangle \end{vmatrix} \, . \tag{16.9}$$

  If two particles are in the same state, two columns are identical and this determinant vanishes.

The set of states which can be constructed using (16.7) or (16.8) forms a basis of the Hilbert space of an N-boson or N-fermion system, respectively.

16.3.4 Time Evolution

By definition, the Hamiltonian $\hat{H}$ of a system of N identical particles commutes with all exchange operators $\hat{P}_{ij}$. Otherwise this would mean that particle i and particle j do not have the same dynamics and are distinguishable. Therefore a state vector preserves its symmetry properties during its evolution. In the study of a N-boson or N-fermion system, the symmetrization or antisymmetrization is performed only once, and persists at all later times.

16.4 Physical Consequences of the Pauli Principle

We give below a few of the many physical consequences of the Pauli principle which concern both few-body systems and the macroscopic properties of a large number of bosons or fermions.

16.4.1 Exchange Force Between Two Fermions

Consider the helium atom and neglect magnetic effects, as we did for hydrogen in Chap. 11. We label the electrons 1 and 2, and the Hamiltonian is

$$\hat{H} = \frac{\hat{p}_1^2}{2m_e} + \frac{\hat{p}_2^2}{2m_e} - \frac{2e^2}{\hat{r}_1} - \frac{2e^2}{\hat{r}_2} + \frac{e^2}{\hat{r}_{12}}, \qquad \text{where} \qquad \hat{\boldsymbol{r}}_{12} = \hat{\boldsymbol{r}}_1 - \hat{\boldsymbol{r}}_2 .$$

The eigenvalue problem is technically complicated and can only be solved numerically, but the results of interest here are simple (Fig. 16.2). The ground state ($E_0 = -78.9\,\text{eV}$) corresponds to a symmetric spatial wave function, while the first two excited states $E_{1\text{A}} = -58.6\,\text{eV}$ and $E_{1\text{S}} = -57.8\,\text{eV}$ have antisymmetric and symmetric spatial wave functions, respectively. The symmetry of the wave function implies a specific symmetry of the spin state: E_0 and $E_{1\text{S}}$ are singlet spin states, and $E_{1\text{A}}$ is a triplet spin state. In the ground state, the two spins are antiparallel. In order to flip one of them to make them parallel, one must spend a considerable amount of energy ($\sim 20\,\text{eV}$).

This corresponds to a "force" which maintains the spins in the antiparallel state. It is *not* a magnetic coupling between the spins: this magnetic interaction can be calculated, and it corresponds to an energy of the order of 10^{-2} eV. The "force" that we are facing here has an electrostatic origin, i. e. the Coulomb interaction, and it is transformed into a constraint on the spins via the Pauli principle. Such an effect is called an *exchange interaction.* The same effect is the basic cause of ferromagnetism.

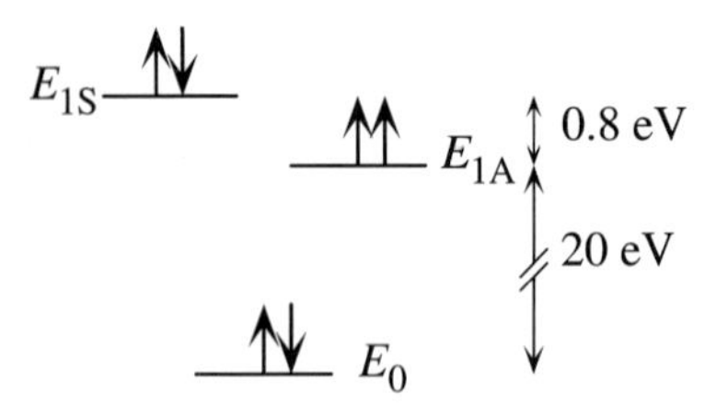

Fig. 16.2. The first three levels of the helium atom

16.4.2 The Ground State of N Identical Independent Particles

Consider N identical independent particles. The Hamiltonian is therefore the sum of N one-particle Hamiltonians:

$$\hat{H} = \sum_{i=1}^{N} \hat{h}^{(i)} . \tag{16.10}$$

Let $\{\phi_n, \varepsilon_n\}$ be the eigenfunctions and corresponding eigenvalues of $\hat{h}$: $\hat{h}\phi_n = \varepsilon_n\phi_n$, where we assume that the ε_n are ordered: $\varepsilon_1 \leq \varepsilon_2 \ldots \leq \varepsilon_n \ldots$.

From the previous considerations, we see that the ground state energy of a system of N *bosons* is:

$$E_0 = N\varepsilon_1 ,$$

whereas, for a system of *fermions*, we have

$$E_0 = \sum_{i=1}^{N} \varepsilon_i .$$

In this latter case, the highest occupied energy level is called the *Fermi energy* of the system; it is denoted by ϵ_{F}. The occupation of the states ϕ_n is represented in Fig. 16.3 for both a bosonic and a fermionic assembly.

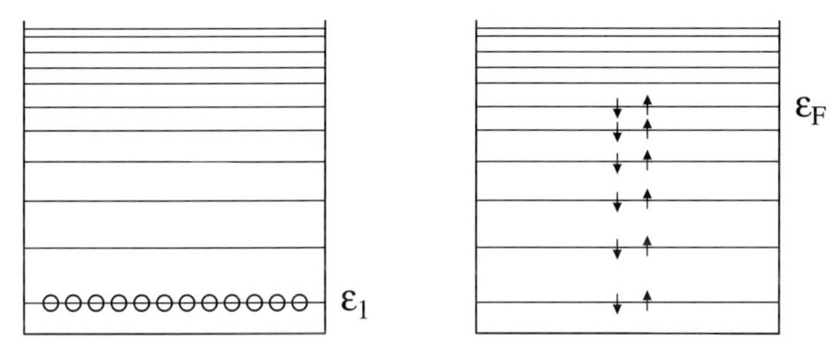

Fig. 16.3. Ground state of a system of N independent identical particles. *Left:* bosonic case, with all the particles in the ground state of the one-body Hamiltonian. *Right:* fermionic case, where the first $N/(2s+1)$ states of the one-body Hamiltonian are occupied (here $s = 1/2$)

Consider for instance N independent fermions of spin s confined in a cubic box of size L. We choose here a basis of states corresponding to periodic boundary conditions (see Sect. 4.4). Each eigenstate of the Hamiltonian $\hat{h}$ is a plane wave $\phi_{\boldsymbol{p}}(\boldsymbol{r}) = \mathrm{e}^{\mathrm{i}\boldsymbol{p}\cdot\boldsymbol{r}/\hbar}/\sqrt{L^3}$, associated with one of the $2s+1$ spin states corresponding to a well-defined component $m_s\hbar$ of the spin parallel to a given axis ($m_s = -s, -s+1, \dots, s$). The momentum $\boldsymbol{p}$ can be written as $\boldsymbol{p} = (2\pi\hbar/L)\boldsymbol{n}$, where the vector $\boldsymbol{n} = (n_1, n_2, n_3)$ stands for a triplet of positive or negative integers. The ground state of the N-fermion system is obtained by placing $2s+1$ fermions in each plane wave $\phi_{\boldsymbol{p}}$, as long as $|\boldsymbol{p}|$ is lower than the *Fermi momentum* p_{F}. This Fermi momentum p_{F} is determined using:

$$N = \sum_{\boldsymbol{p}\ (p<p_{\mathrm{F}})} (2s+1) .$$

For a large number of particles, we can replace this discrete sum by an integral (cf. (4.50)), which yields

$$N \simeq (2s+1)\frac{L^3}{(2\pi\hbar)^3}\int_{p<p_{\mathrm{F}}} \mathrm{d}^3p = \frac{2s+1}{6\pi^2}\,\frac{L^3 p_{\mathrm{F}}^3}{\hbar^3} . \tag{16.11}$$

The following equation relates the density $\rho = N/L^3$ of the gas to the Fermi momentum, independently of the size of the box:

$$\rho = \frac{2s+1}{6\pi^2}\left(p_\mathrm{F}/\hbar\right)^3 . \tag{16.12}$$

The average kinetic energy per particle can also be easily calculated:

$$\frac{\langle p^2 \rangle}{2m} = \frac{1}{N} \sum_{\boldsymbol{p}\ (p<p_\mathrm{F})} (2s+1)\frac{p^2}{2m} \simeq \frac{2s+1}{N}\frac{L^3}{(2\pi\hbar)^3}\int_{p<p_\mathrm{F}} \frac{p^2}{2m}\mathrm{d}^3p ,$$

which leads to

$$\frac{\langle p^2 \rangle}{2m} = \frac{3}{5}\frac{p_\mathrm{F}^2}{2m} . \tag{16.13}$$

16.4.3 Behavior of Fermion and Boson Systems at Low Temperature

The difference between the ground states of N-fermion and N-boson systems induces radically different behaviors of such systems at low temperature.

In a system of fermions at zero temperature and in the absence of interactions, we have just seen that all the energy levels of the one-body Hamiltonian are filled up to the Fermi energy ϵ_F. This simple model describes remarkably well the conduction electrons in a metal, and it accounts for many macroscopic properties of a solid, such as its thermal conductivity. Using the result (16.12), the Fermi energy $\epsilon_\mathrm{F} = p_\mathrm{F}^2/2m_\mathrm{e}$ can be written in terms of the number density ρ_e of conduction electrons as

$$\epsilon_\mathrm{F} = \frac{\hbar^2\left(3\rho_\mathrm{e}\pi^2\right)^{2/3}}{2m_\mathrm{e}} ,$$

where we have used $2s+1 = 2$ for electrons. This energy can reach large values ($\epsilon_\mathrm{F} = 3\,\mathrm{eV}$ for sodium). This is much larger than the thermal energy at room temperature ($k_\mathrm{B}T \simeq 0.025\,\mathrm{eV}$). This explains the success of the zero-temperature fermion gas model for conduction electrons. At room temperature, very few electrons participate in thermal exchanges.

The application of the Pauli principle to fermionic systems has many consequences, ranging from solid-state physics to the stability of stars such as white dwarfs or neutron stars. In nuclear physics, the Pauli principle explains why neutrons are stable with respect to β decay inside nuclei. An isolated neutron is unstable and decays through the process $\mathrm{n} \to \mathrm{p} + \mathrm{e} + \bar{\nu}_\mathrm{e}$ with a lifetime of the order of 15 minutes. Inside a nucleus, a neutron can be stabilized if all the final states allowed by energy conservation for the final proton are already occupied.

Concerning bosonic systems, a spectacular consequence of quantum statistics is the Bose–Einstein condensation. In the absence of interactions between the particles, if the number density $\rho = N/V$ is such that

$$\rho\Lambda_T^3 > 2.612\,, \qquad \text{where} \qquad \Lambda_T = \frac{h}{\sqrt{2\pi m k_\mathrm{B} T}}\,, \tag{16.14}$$

a macroscopic accumulation of particles occurs in a *single* quantum state, i. e. the ground state of the confining potential of the particles. This "gregarious" behavior of bosons contrasts with the "individualistic" character of fermions.

Until June 1995, the usual example of a Bose–Einstein condensation that was given in textbooks was the *normal liquid* $\rightarrow$ *superfluid liquid* transition of helium, which happens at a temperature of $T = 2.17\,\mathrm{K}$. However, the complicated interactions inside the liquid make the quantitative treatment of the superfluid transition quite involved, and different from the simple theory of the Bose–Einstein condensation of an ideal gas.

We now have at our disposal experiments performed on gases of alkali atoms (lithium, sodium, potassium, rubidium), which are initially cooled by lasers inside a vacuum chamber with a very low residual pressure (below 10^{-11} mbar). The atoms are then confined by an inhomogeneous magnetic field and are cooled further by evaporation until they reach the Bose–Einstein condensation, at a temperature below 1 microkelvin. The evaporative cooling technique consists in eliminating the more energetic atoms, in order to keep only the slower ones. Collisions between the trapped atoms maintain thermal equilibrium throughout the process. Starting from 10^9 atoms, one can obtain, after evaporation, a situation where the 10^6 remaining atoms are practically all in the ground state of the system. Figure 16.4 shows the time evolution of the momentum distribution of a gas of bosons (rubidium-87 atoms) confined in a magnetic trap and cooled down to the condensation point. These Bose–

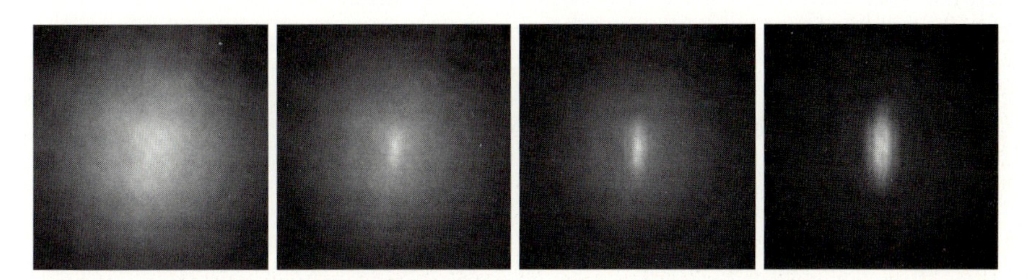

Fig. 16.4. Bose–Einstein condensation of a gas of ^{87}Rb atoms observed via the evolution of the momentum distribution of the particles in the xy plane. The atoms are confined in an anisotropic harmonic trap ($\omega_x < \omega_y$) and cooled by evaporation. The trapping potential is then turned off suddenly and the momentum distribution is measured using a time-of-flight technique (Sect. 4.6). *Left:* the temperature T is noticeably larger than the condensation temperature T_c. The momentum distribution is isotropic and close to a Maxwell–Boltzmann distribution ($m\langle v_i^2\rangle = k_\mathrm{B}T$ for $i = x, y$). *Middle images* ($T \leq T_\mathrm{c}$): a noticeable fraction of the atoms accumulate in the ground state of the magnetic trap. *Right image* ($T \ll T_\mathrm{c}$): a very large fraction of the atoms is in the ground state of the trap. The momentum distribution of the trap reflects the trap anisotropy ($m\langle v_i^2\rangle = \hbar\omega_i/2$). Strictly speaking, one needs to take into account the interactions between the atoms to explain quantitatively the details of this distribution (photographs by F. Chevy and K. Madison, ENS Paris)

Einstein condensates possess remarkable coherence and superfluid properties, and this has been a very active field of research in recent years.

16.4.4 Stimulated Emission and the Laser Effect

Consider a system of independent bosons (see Sect. 16.4.2), to which we apply, for a finite length of time, the one-body potential $\hat{V} = \sum_i \hat{v}^{(i)}$. Each potential $\hat{v}^{(i)}$ acts only on the particle i and can induce transitions between the various eigenstates of $\hat{h}^{(i)}$. We want to show that the probability for a particle to reach a given final state $|\phi_l\rangle$ is increased if this state is already occupied.

Consider first the case where we are dealing with a single particle initially in the state $|\phi_k\rangle$. If we assume that the effect of $\hat{v}$ is weak, the probability that the particle reaches the state $|\phi_l\rangle$ under the action of $\hat{v}$ is proportional to $|v_{kl}|^2 = |\langle\phi_k|\hat{v}|\phi_l\rangle|^2$. This result of time-dependent perturbation theory will be proven in Chap. 17 (see, e. g., (17.15)).

Suppose now that the state $|\phi_l\rangle$ is already occupied by N particles, but there is only one particle in the state $|\phi_k\rangle$. The properly symmetrized initial state of the $(N+1)$-boson system is

$$\begin{aligned}|\Psi_{\rm i}\rangle = \frac{1}{\sqrt{N+1}} \Big(& |1:\phi_k\,;\,2:\phi_l\,;\,\ldots\,;\,N:\phi_l\,;\,N+1:\phi_l\rangle \\ & + |1:\phi_l\,;\,2:\phi_k\,;\,\ldots\,;\,N:\phi_l\,;\,N+1:\phi_l\rangle + \ldots \\ & + |1:\phi_l\,;\,2:\phi_l\,;\,\ldots\,;\,N:\phi_l\,;\,N+1:\phi_k\rangle \Big) ,\end{aligned}$$

and we are interested in the probability of reaching the final state

$$|\Psi_{\rm f}\rangle = |1:\phi_l\,;\,2:\phi_l\,;\,\ldots\,;\,N:\phi_l\,;\,N+1:\phi_l\rangle .$$

The transition probability is now proportional to

$$|V_{\rm if}|^2 = |\langle\Psi_{\rm i}|\hat{V}|\Psi_{\rm f}\rangle|^2 = (N+1)\,|v_{kl}|^2 .$$

The presence of N particles in the state $|\phi_l\rangle$ increases, by a factor $N+1$, the probability that the particle initially in $|\phi_k\rangle$ reaches this state. The transition probability is the sum of the rate for a *spontaneous transition*, proportional to $|v_{kl}|^2$ and independent of N, and of the rate *stimulated* by the presence of the N bosons in the state $|\phi_l\rangle$ and proportional to $N\,|v_{kl}|^2$.

This gregarious behavior also manifests itself for photons, which are massless bosons. This explains the phenomenon of *stimulated emission of light*, which is the basis of the principle of the laser. An excited atom decays preferentially by emitting a photon in the quantum state occupied by the photons already present in the laser cavity. This leads to a chain reaction in the production of photons, which is the key point in the mechanism of lasers.

16.4.5 Uncertainty Relations for a System of N Fermions

Consider N independent fermions of spin s, each placed in a potential $V(r)$ centered at the origin. The Hamiltonian is therefore

$$\hat{H} = \sum_{i=1}^{N} \hat{h}^{(i)}\,, \qquad \text{where} \qquad \hat{h} = \frac{\hat{p}^2}{2m} + V(\hat{r})\,.$$

We denote by ε_n the energy levels of $\hat{h}$, and denote by g_n their degeneracies. The ground state E_0 of $\hat{H}$ is obtained by filling the lowest-lying levels ε_n up to the Fermi energy ϵ_{F}, each with $(2s+1)$ particles per state:

$$E_0 = (2s+1)\sum_{n=0}^{k} g_n \varepsilon_n\,,$$

where the number k is determined by the relation $N = (2s+1)\sum_{n=0}^{k} g_n$. For a harmonic potential $V(r) = m\omega^2 r^2/2$, we have

$$\varepsilon_n = (n+3/2)\hbar\omega\,, \qquad g_n = (n+2)(n+1)/2\,.$$

Therefore we find, for $N \gg 1$,

$$k \simeq \left(\frac{6N}{2s+1}\right)^{1/3}, \qquad E_0 \simeq \xi\, N^{4/3}\hbar\omega\,,$$

where $\xi = (3/4)6^{1/3}(2s+1)^{-1/3}$.

Consider now an arbitrary state $|\Psi\rangle$ of these N fermions. We define $\langle r^2\rangle = \langle r_i^2\rangle$ and $\langle p^2\rangle = \langle p_i^2\rangle$, where $i = 1,\ldots,N$. In this state, $\langle H\rangle \geq E_0$ and, consequently,

$$\langle H\rangle = N\frac{\langle p^2\rangle}{2m} + \frac{N}{2}m\omega^2\langle r^2\rangle \geq \xi N^{4/3}\hbar\omega\,,$$

which gives

$$\langle p^2\rangle + m^2\omega^2\langle r^2\rangle - 2\xi N^{1/3}\hbar m\omega \geq 0 \qquad \text{for} \quad N \gg 1\,.$$

This second-degree trinomial in $m\omega$ is positive for all values of $m\omega$. Therefore we obtain the result that in any state of these N fermions,

$$\langle r^2\rangle\,\langle p^2\rangle \geq \xi^2\, N^{2/3}\,\hbar^2 \qquad \text{for} \quad N \gg 1\,. \tag{16.15}$$

This relation is valid, in particular, in the center-of-mass frame, where $\langle \boldsymbol{p}\rangle = 0$. If we choose the origin of space at the center of the cloud, we obtain

$$\Delta x\,\Delta p_x \geq \frac{\xi}{3}\,N^{1/3}\,\hbar \qquad \text{for} \quad N \gg 1\,. \tag{16.16}$$

For spin-1/2 particles, $\xi/3 \sim 0.36$.

A similar calculation for fermions placed in a $1/r$ potential leads to

$$\langle p^2 \rangle \geq \gamma \hbar^2 \left\langle \frac{1}{r} \right\rangle^2 N^{2/3} \qquad \text{for} \qquad N \gg 1 \tag{16.17}$$

where $\gamma = 3^{-1/3}(2s+1)^{-2/3}$ (cf. exercise 16.4). The relation (16.17) plays an important role in the stability of self-gravitating systems, as we shall see in Chap. 19.

The Pauli principle therefore modifies the uncertainty relations. If we place N identical fermions in a volume $V \sim (\Delta x)^3$, each of these fermions must occupy a different quantum state. We can thus consider that the space accessible to each particle is a region of linear extension $\sim (V/N)^{1/3}$, so that the de Broglie wavelength of each particle is reduced by a factor $N^{1/3}$.

This brings a different, but equivalent, point of view to the physics of an N fermion system that we investigated in Sect. 16.4.3. In particular, consider again an ideal Fermi gas confined in a cubic box of size L at zero temperature. The position distribution in the box is uniform and the average momentum per particle can be deduced from (16.13) so that

$$\Delta x^2 = \frac{1}{L} \int_{-L/2}^{L/2} x^2 \, \mathrm{d}x = \frac{L^2}{12} ,$$

$$\Delta p_x^2 = \frac{\Delta p^2}{3} = \frac{p_\mathrm{F}^2}{5} = \frac{\hbar^2 N^{2/3}}{5L^2} \left(\frac{6\pi^2}{2s+1} \right)^{2/3} ,$$

from which one can check that (16.16) is well satisfied. Actually the product $\Delta x \; \Delta p_x$ calculated for an ideal Fermi gas confined in a square box exceeds by $\sim 10\%$ the rigorous lower bound (16.16).

16.4.6 Complex Atoms and Atomic Shells

We now indicate how one can describe in a simple, approximate way the structure of atoms with several electrons. The combination of a mean-field approximation to describe the electron–electron interaction with the Pauli principle will lead us to a qualitative understanding of the periodic classification of elements due to Mendeleev.

It was actually the Mendeleev classification, where series of elements appeared to be grouped into sets of twice the degeneracy of the hydrogen atom ($2n^2 = 2, 8, \ldots$), that led Pauli to the idea that electrons in atoms should be described by four quantum numbers, not three. The fourth quantum number was two-valued. In order to explain the classification, one had to apply the "Pauli verbot", i. e. that two electrons could not have the same values of the four quantum numbers.

The Hamiltonian of a Complex Atom. A complex neutral atom consists of a nucleus of charge Ze and Z electrons. Assuming the nucleus is infinitely heavy and neglecting magnetic effects, the Hamiltonian of the system is

$$\hat{H} = \sum_{i=1}^{Z} \frac{\hat{p}_i^2}{2m_e} - \sum_{i=1}^{Z} \frac{Ze^2}{\hat{r}_i} + \sum_{i=1}^{Z} \sum_{k=i+1}^{Z} \frac{e^2}{\hat{r}_{ik}} . \tag{16.18}$$

Owing to the Coulomb repulsion between the electrons, the diagonalization of this Hamiltonian is complicated. This repulsion term cannot be neglected or treated as a perturbation, since the distances r_{ik} between the electrons are of the same order of magnitude as the distances r_i between the electrons and the nucleus. There are $Z(Z-1)/2$ terms corresponding to the electron–electron repulsion e^2/r_{ik}, and their sum is of the same order of magnitude as the Z terms $-Ze^2/r_i$ corresponding to the attraction of the electrons by the nucleus. An ingenious method due to Hartree, however, gives an approximation method which can lead very far in the description of atoms.

Remark. Since we have neglected magnetic effects, the Hamiltonian (16.18) does not contain spin-dependent variables. Of course, although they are absent from the interaction, spin variables play a crucial role through the Pauli principle in this problem.

The Hartree Method and the Mean-Field Approximation. We rewrite the Hamiltonian (16.18) in the form

$$\hat{H} = \sum_i \frac{\hat{p}_i^2}{2m_e} + \underbrace{\sum_i -\frac{Ze^2}{\hat{r}_i} + V(\hat{r}_i)}_{\sum_i U(\hat{r}_i)} + \underbrace{\sum_i \left(\sum_{k>i} \frac{e^2}{\hat{r}_{ik}} - V(r_i) \right)}_{H_c} . \tag{16.19}$$

If the function $V(r)$ is chosen in an appropriate way, the term H_c can be sufficiently small to be neglected in a first approximation, although neither $\sum e^2/r_{ik}$ nor $\sum_i V(r_i)$ is separately small.

The potential $V(r)$ can be obtained by an iteration method called the self-consistent Hartree–Fock method. Its physical meaning is intuitive: it is a mean potential which represents as well as possible, for each electron, the repulsive Coulomb potential $\sum_{i \neq k} e^2/r_{ik}$ created by the set of the other electrons. We shall not describe here how one obtains $V(r)$.

Electronic Configurations. Suppose the function $V(r)$ has been determined sufficiently well that one can neglect the term $\hat{H}_c$. In this approximation, the Hamiltonian is a sum of *one-particle* terms:

$$\hat{H}_0 = \sum_{i=1}^{Z} \hat{h}^{(i)} , \qquad \text{where} \qquad \hat{h} = \frac{p^2}{2m_e} + U(r) . \tag{16.20}$$

The variables of the various electrons are separated and it is possible to determine either analytically or numerically the eigenfunctions $\psi_{n,\ell,m}(\boldsymbol{r})$ and the corresponding energies $E_{n,\ell}$ of the one-body Hamiltonian $\hat{h}$:

$$\left(\frac{p^2}{2m_\mathrm{e}} + U(r)\right) \psi_{n,\ell,m}(\boldsymbol{r}) = E_{n,\ell}\, \psi_{n,\ell,m}(\boldsymbol{r}) \,.$$

Using (16.8), one can then form a state vector for the Z-electron system with a total energy $E = E_1 + E_2 + \ldots + E_Z$.

An *electronic configuration* is a description of the state of the Z electrons in which we specify the number of electrons in a given energy level $E_{n,\ell}$. Consider, for instance, the ground states of some atoms:

- The ground state of the simplest atom, hydrogen, is the 1s state ($n = 1, \ell = 0$).
- The following atom, helium, is obtained by adding one electron, also in the 1s state. This is possible provided the spin state of the two electrons is antisymmetric, i. e. it is the singlet state. The configuration of the helium atom in its ground state is therefore $1s^2$. This set of two electrons with the quantum number $n = 1$ forms a complete shell, called the K shell. It is complete in the sense that the Pauli principle forbids a third electron to be in the same orbital state.
- The sodium atom in its ground state has the configuration $1s^2 2s^2 2p^6 3s^1$. The L shell, corresponding to $n = 2$, is also complete since it can accommodate eight electrons. The eleventh and last electron is in the M shell, corresponding to $n = 3$.
- For atoms with a larger number of electrons, the ℓ dependence of the energies produces modifications of this simple filling scheme. For instance, for potassium ($Z = 19$), the filling of the N shell ($n = 4$) starts before the M shell is complete, and the configuration of the ground state is $1s^2 2s^2 2p^6 3s^2 3p^6 4s^1$. The single valence electron in an s state is characteristic of alkali atoms (first column of the periodic table).

Therefore the simplest form of the Pauli principle, i. e. the exclusion principle, explains how the successive elements are grouped in the Mendeleev table. We insist, however, on the fact that this is only an *approximate* model. Several coincidences, such as the smallness of the fine structure constant, result in the fact that the mean-field approximation is excellent. This can be considered in some sense as a "miracle" of nature.

Further Reading

- The importance, profoundness, revolutionary aspects and philosophical implications of the Pauli principle are analyzed by H. Margenau, *The Nature of Physical Reality*, Chap. 20, McGraw-Hill, New York (1950). This author points out that this principle did not provoke the same interest as, for instance, relativity among philosophers, because it explained so many experimental facts (actually many more than relativity) that it had been immediately incorporated into the general theory of quantum mechanics.

- I. Duck and E.C.G. Sudarshan, *Pauli and the Spin–Statistics Theorem*, World Scientific, Singapore (1997).
- An elementary analysis of the collision process of two identical particles can be found in R.P. Feynman, R.B. Leighton and M. Sands, *The Feynman Lectures on Physics*, Vol. III, Chap. 3 (Sect. 4) and 4 (Sect. 1), Addison-Wesley, Reading, MA (1965). This shows the difference between fermions and bosons in an elementary interaction process. See also R.P. Feynman, *Elementary Particles and the Laws of Physics*, 1986 Dirac Memorial Lecture, Cambridge University Press, Cambridge (1987).
- "The discovery of superfluidity", Phys. Today, July 1995, p. 30; J. de Nobel, "The discovery of superconductivity", Phys. Today, September 1996, p. 40; R. Hallock, "The magic of helium 3 in two, or nearly two, dimensions", Phys. Today, June 1998, p. 30.
- E. Cornell and C. Wieman, "Bose–Einstein condensation", Sci. Am., March 1998, p. 26; W. Ketterle, "Experimental studies of Bose-Einstein condensation", Phys. Today, December 1999, p. 30; K. Helmerson et al., "Atom Lasers", Phys. World, August 1999, p. 31; Y. Castin et al., "Bose–Einstein condensates make quantum leaps and bounds", Phys. World, August 1999, p. 37.
- C.H. Townes, *How the Laser Happened: Adventures of a Scientist*, Oxford University Press, Oxford (1999); A. Siegman, *Lasers* (University Science Books, Mill Valley, 1986); O. Svelto, *Principles of Lasers*, Plenum, New York (1998).
- E.R. Scerri, "The evolution of the periodic system", Sci. Am., September 1999, p. 56.

Exercises

16.1. Identical particles incident on a beam splitter. Consider a particle prepared at an initial time t_i in a wave packet $\psi(\boldsymbol{r}, t_\mathrm{i}) = \phi_1(\boldsymbol{r})$, incident on a 50%–50% beam splitter (Fig. 16.5). At a later time t_f, the wave packet has crossed the beam splitter and the state of the particle can be written $\psi(\boldsymbol{r}, t_\mathrm{f}) = [\phi_3(\boldsymbol{r}) + \phi_4(\boldsymbol{r})]/\sqrt{2}$, where ϕ_3 and ϕ_4 correspond to normalized wave packets propagating in each of the output ports. We assume $\langle\phi_3|\phi_4\rangle \simeq 0$.

a. We prepare the particle in the state $\psi(\boldsymbol{r}, t_\mathrm{i}) = \phi_2(\boldsymbol{r})$, which is obtained from $\phi_1(\boldsymbol{r})$ by symmetry with respect to the beam splitter plane. The state of the particle at time t_f can then be written

$$\psi(\boldsymbol{r}, t_\mathrm{f}) = \alpha\phi_3(\boldsymbol{r}) + \beta\phi_4(\boldsymbol{r}) .$$

Determine (within a global phase factor) the coefficients α and β. Make use of the fact that the interaction of the particle with the beam splitter is described by a Hamiltonian (which need not be written explicitly). We take $\langle\phi_2|\phi_1\rangle = 0$.

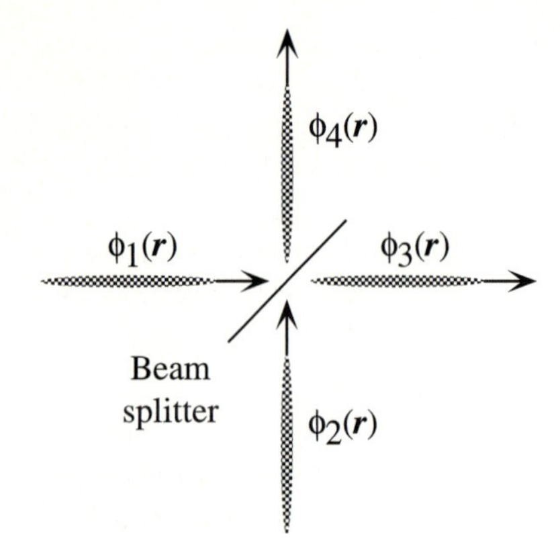

Fig. 16.5. An incident wave packet $\phi_1(\boldsymbol{r})$ or $\phi_2(\boldsymbol{r})$ crosses a 50%–50% beam splitter. After the crossing, this provides a coherent superposition of two outgoing wave packets $\phi_3(\boldsymbol{r})$ and $\phi_4(\boldsymbol{r})$

b. We prepare at time $t_{\rm i}$ two fermions in the same spin state, one in the external state $\phi_1(\boldsymbol{r})$, the other one in the state $\phi_2(\boldsymbol{r})$. What is the final state of the system? Can one detect both fermions in the same output port?
c. Consider the same problem for two bosons, also prepared in the same spin state, one in the external state $\phi_1(\boldsymbol{r})$, the other one in the state $\phi_2(\boldsymbol{r})$. Show that the two bosons always exit in the same port. This experiment has been performed with photons by C.K. Hong et al., Phys. Rev. Lett. **59**, 2044 (1987).

16.2. Bose–Einstein condensation in a harmonic trap. Consider N spin-0 bosons in an isotropic harmonic trap of angular frequency ω. We neglect interactions between the particles and we recall that the average number of particles in a given state of energy E is given by the Bose–Einstein law,

$$n_E = \left(\mathrm{e}^{(E-\mu)/k_{\rm B}T} - 1\right)^{-1} ,$$

where μ is the chemical potential and T the temperature.

a. Show that the chemical potential obeys the inequality $\mu < 3\hbar\omega/2$.
b. Show that the number of particles N' occupying a state different from the ground state of the trap cannot exceed $F(\xi)$, where

$$\xi = \frac{\hbar\omega}{k_{\rm B}T} , \qquad F(\xi) = \sum_{n=1}^{\infty} \frac{(n+1)(n+2)}{2(\mathrm{e}^{n\xi} - 1)} .$$

Recall that the degeneracy of an energy level $E_n = (n + 3/2)\,\hbar\omega$ is $g_n = (n+1)(n+2)/2$.
c. We consider the limiting case $k_{\rm B}T \gg \hbar\omega$, which allows us to replace the sum defining $F(\xi)$ by an integral. Show that the upper bound on the

total population of the excited states is

$$N'_{\max} = \zeta(3)\left(\frac{k_B T}{\hbar\omega}\right)^3 , \qquad \zeta(x) = \sum_{n=1}^{\infty} n^{-x} .$$

Here $\zeta(x)$ is the Riemann function, and $\zeta(3) \simeq 1.2$.

d. What happens if one places more than $N'_{\max}$ atoms in the trap? At what temperature can this phenomenon be observed with a trap containing 10^6 atoms, with $\omega/(2\pi) = 100\,\text{Hz}$?

16.3. Fermions in a square well. Consider two spin-1/2 particles of mass m confined in a one-dimensional infinite square well of size L (Sect. 4.3.3).

a. We neglect the interactions between the particles. Determine the four lowest energy levels.
b. We suppose that the particles interact via a contact potential $V(x_1 - x_2) = g\,\delta(x_1 - x_2)$, where the coupling constant g does not depend on spin. Determine to first order in g the four lowest energy levels of this two-particle system.

16.4. Heisenberg–Pauli inequality (continuation). Here we use the same type of arguments as in Sect. 16.4.5 for a Coulomb potential.

a. Consider the ground state of an N-fermion system placed in the external potential $V(r) = -e^2/r$. We assume that the fermions do not interact with each other and that $N \gg 1$. We denote by E_0 the energy of the ground state. Recall that the energy levels of the one-body Hamiltonian are $-E_I/n^2$, where n is a positive integer and $E_I = me^4/2\hbar^2$; the degeneracy of each level is $(2s+1)n^2$ (see Chap. 11). Show that there exists an integer k such that

$$N \simeq (2s+1)\frac{k^3}{3} , \qquad E_0 \simeq -(2s+1)E_I k .$$

b. Show that E_0 and N are related by

$$E_0 \simeq -\frac{me^4}{2\hbar^2}\,(2s+1)^{2/3}\,(3N)^{1/3} .$$

c. Consider an arbitrary state $|\Psi\rangle$ of this N-fermion system. Prove the result (16.17).

17. The Evolution of Systems

Written in collaboration with Gilbert Grynberg[1]

Stop the world, I want to get off!
Anonymous, June 1968, on a bus stop in Paris
(collected by Christian Nugue)

Any experimental observation or any practical use of quantum phenomena relies on *processes*, where one observes the evolution of a system at some time, knowing in what state it was initially. It is therefore essential to understand the various types of evolution a system can have, whether it is subject to external forces or not.

In this chapter, we present two characteristic processes: the oscillatory behavior of a two-state system under the influence of an external constant or oscillating field, and the irreversible evolution of a system coupled to a continuum. In Sect. 17.1, we introduce the notion of a transition probability and the basic method of calculation: time-dependent perturbation theory. In Sect. 17.2, we consider the atomic transitions induced by an external electromagnetic field, i. e. absorption and induced emission. We also present the physical processes which underlie the control of atomic motion by laser light. In Sect. 17.3, we consider the problem of the decay of a system, such as an excited atom or an excited nucleus. We show how the exponential decay law emerges, and how one can calculate the lifetime of a system. We also introduce the notion of the *width* of an unstable system. Finally, in Sect. 17.4, we discuss a few aspects of the time–energy uncertainty relation, $\Delta E\,\Delta t \geq \hbar/2$, which differs quite radically from the uncertainty relations we established in Chap. 7, and which illustrates the special role played by time in nonrelativistic quantum theory.

---

[1] Gilbert Grynberg, Laboratoire Kastler Brossel, Département de Physique, Ecole Normale Supérieure, 24, rue Lhomond, 75231 Paris Cedex 05, France, e-mail: gilbert.grynberg@lkb.ens.fr

17.1 Time-Dependent Perturbation Theory

Time-dependent perturbation theory consists of starting with a solvable problem and perturbing it with an additional Hamiltonian, which allows for transitions between the original eigenstates.

17.1.1 Transition Probabilities

Consider a system whose evolution can be derived from the Hamiltonian

$$\hat{H} = \hat{H}_0 + \hat{H}_1(t)\,, \tag{17.1}$$

where $\hat{H}_0$ is time independent. The eigenvectors $|n\rangle$ and eigenvalues E_n of $\hat{H}_0$ are assumed to be known:

$$\hat{H}_0|n\rangle = E_n|n\rangle\,. \tag{17.2}$$

The operator $\hat{H}_1(t)$ is an interaction term which may depend explicitly on time, and which, a priori, does not commute with $\hat{H}_0$. This term can induce transitions between two eigenstates $|n\rangle$ and $|m\rangle$ of $\hat{H}_0$.

Our aim is the following. Assuming the system is prepared at time t_0 in a given state $|\psi(t_0)\rangle = |n\rangle$, we want to calculate the probability

$$\mathcal{P}_{n\to m}(t) = |\langle m|\psi(t)\rangle|^2 \tag{17.3}$$

of finding the system in the eigenstate $|m\rangle$ of $\hat{H}_0$ at a later time t.

Example: a Collision Process. In addition to the examples that we shall meet in this chapter, a collision is a typical situation where this problem appears. Consider two particles a and b, each of which is prepared in a wave packet sharply defined in momentum p. At the initial time the centers of these two wave packets propagate towards each other. In the absence of interaction the particles propagate freely: $\langle p_a\rangle$ and $\langle p_b\rangle$ are constants of the motion. However, if we take into account the interaction potential $\hat{H}_1$ between the particles, a scattering process takes place. A measurement of the final momenta of the particles will give a result which generally differs from the initial values $\langle p_a\rangle$ and $\langle p_b\rangle$. The question is how to calculate the probability of measuring given final momenta, knowing the interaction potential.

17.1.2 Evolution Equations

At any time, the state of the system $|\psi(t)\rangle$ can be expanded in the basis $\{|n\rangle\}$ of eigenstates of $\hat{H}_0$:

$$|\psi(t)\rangle = \sum \gamma_n(t)\mathrm{e}^{-\mathrm{i}E_n t/\hbar}|n\rangle\,. \tag{17.4}$$

In this expression, we explicitly write the time evolution factors $\mathrm{e}^{-\mathrm{i}E_n t/\hbar}$ that would be present for $\hat{H}_1 = 0$. This simplifies the evolution equations of the coefficients $\gamma_n(t)$. Using the Schrödinger equation, we obtain

$$i\hbar \sum_n [\dot{\gamma}_n(t) - \frac{i}{\hbar} E_n \gamma_n(t)]\, \mathrm{e}^{-\mathrm{i}E_n t/\hbar} |n\rangle = \sum_n \gamma_n(t) \mathrm{e}^{-\mathrm{i}E_n t/\hbar} (\hat{H}_0 + \hat{H}_1)|n\rangle \,,$$

and therefore

$$\mathrm{i}\hbar \sum_n \dot{\gamma}_n(t) \mathrm{e}^{-\mathrm{i}E_n t/\hbar} |n\rangle = \sum_n \gamma_n(t) \mathrm{e}^{-\mathrm{i}E_n t/\hbar} \hat{H}_1 |n\rangle \,. \tag{17.5}$$

Multiplying by $\langle k|$, we obtain

$$\mathrm{i}\hbar \dot{\gamma}_k(t) = \sum_n \gamma_n(t)\, \mathrm{e}^{-\mathrm{i}(E_n - E_k)t/\hbar}\, \langle k|\hat{H}_1|n\rangle \,. \tag{17.6}$$

The problem is completely determined by this set of coupled differential equations and by the initial condition which specifies $|\psi(t_0)\rangle$.

17.1.3 Perturbative Solution

In general, this set of equations does not have an analytic solution. In order to make progress, we assume, as in Chap. 9, that $\hat{H}_1$ is "small" compared with $\hat{H}_0$. More precisely, we consider the Hamiltonian $\hat{H}_\lambda = \hat{H}_0 + \lambda \hat{H}_1$ and we assume that the corresponding coefficients $\gamma_k(t)$ are analytic functions of λ around the origin, including the case of $\lambda = 1$:

$$\gamma_k(t) = \gamma_k^{(0)}(t) + \lambda\, \gamma_k^{(1)}(t) + \ldots + \lambda^p\, \gamma_k^{(p)}(t) + \ldots \,. \tag{17.7}$$

Inserting this expansion in (17.6), we identify the coefficients of each power of, λ and we obtain the following

$$\text{to order 0: } \mathrm{i}\hbar \dot{\gamma}_k^{(0)}(t) = 0 \,, \tag{17.8}$$

$$\text{to order 1: } \mathrm{i}\hbar \dot{\gamma}_k^{(1)}(t) = \sum_n \gamma_n^{(0)}(t)\, \mathrm{e}^{-\mathrm{i}(E_n - E_k)t/\hbar}\, \langle k|\hat{H}_1|n\rangle \,; \tag{17.9}$$

$$\text{to order } r\text{: } \mathrm{i}\hbar \dot{\gamma}_k^{(r)}(t) = \sum_n \gamma_n^{(r-1)}(t)\, \mathrm{e}^{-\mathrm{i}(E_n - E_k)t/\hbar}\, \langle k|\hat{H}_1|n\rangle \,. \tag{17.10}$$

This system can be solved by iteration. The terms $\gamma_k^{(0)}(t)$ are determined by the knowledge of the initial state of the system. Inserting these into (17.9), we can calculate the terms of order 1, $\gamma_k^{(1)}(t)$, which in turn give the terms of order 2 through (17.10), and so on. One can therefore determine successively all the terms in the expansion.

17.1.4 First-Order Solution: the Born Approximation

The zeroth-order equation can be solved immediately. We find that $\gamma_k^{(0)}(t)$ is a constant. If we choose the initial condition $|\psi(t_0)\rangle = |i\rangle$, we obtain

$$\gamma_k^{(0)}(t) = \delta_{k,i} \,. \tag{17.11}$$

Inserting this in (17.9), we obtain, for $f \neq i$,

$$\mathrm{i}\hbar\dot{\gamma}_f^{(1)}(t) = \mathrm{e}^{\mathrm{i}(E_f - E_i)t/\hbar} \langle f|\hat{H}_1|i\rangle \,. \tag{17.12}$$

Taking into account the assumption $\gamma_f^{(1)}(t_0) = 0$, this gives

$$\gamma_f^{(1)}(t) = \frac{1}{\mathrm{i}\hbar} \int_{t_0}^{t} \mathrm{e}^{\mathrm{i}(E_f - E_i)t/\hbar} \langle f|\hat{H}_1|i\rangle \,\mathrm{d}t \,. \tag{17.13}$$

In this approximation, the transition probability from an initial state $|i\rangle$ to a final state $|f\rangle$ (with $f \neq i$) is given by

$$\mathcal{P}_{i\to f}(t) = |\gamma_f^{(1)}(t)|^2 \,.$$

This approximation is acceptable if $\mathcal{P}_{i\to f} \ll 1$ (a necessary condition).

17.1.5 Particular Cases

Constant Perturbation. We suppose the perturbation $\hat{H}_1$ is "switched on" at $t_0 = 0$ and "switched off" at a later time T. We suppose also that it does not depend on time between 0 and T. Setting $\hbar\omega_0 = E_f - E_i$, we obtain

$$\gamma_f^{(1)}(t \geq T) = \frac{1}{\mathrm{i}\hbar} \langle f|\hat{H}_1|i\rangle \,\frac{\mathrm{e}^{\mathrm{i}\omega_0 T} - 1}{\mathrm{i}\omega_0} \,, \tag{17.14}$$

and consequently

$$\mathcal{P}_{i\to f}(t \geq T) = \frac{1}{\hbar^2} |\langle f|\hat{H}_1|i\rangle|^2 \, y(\omega_0, T) \,. \tag{17.15}$$

We shall frequently make use of the above function $y(\omega, T)$, defined as

$$y(\omega, T) = \frac{\sin^2(\omega T/2)}{(\omega/2)^2} \,, \qquad \text{where} \qquad \int_{-\infty}^{+\infty} y(\omega, T)\,\mathrm{d}\omega = 2\pi T \,. \tag{17.16}$$

Its graph is given in Fig. 17.1.

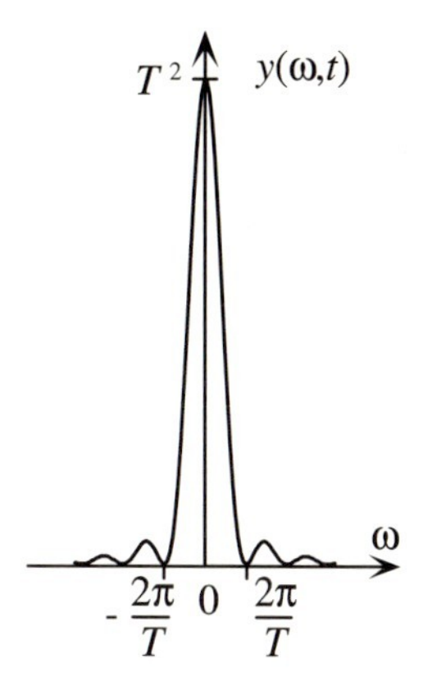

Fig. 17.1. Graph of the function $y(\omega, T)$

Sinusoidal Perturbation. Consider a coupling $\hat{H}_1(t)$ such that $\hat{H}_1(t) = \tilde{H}_1 e^{-i\omega t}$ for $0 < t < T$ and $\hat{H}_1(t) = 0$ otherwise. A simple calculation gives

$$\mathcal{P}_{i\to f}(t \geq T) = \frac{1}{\hbar^2} |\langle f|\tilde{H}_1|i\rangle|^2 \, y(\omega - \omega_0, T) . \qquad (17.17)$$

There is a *resonance* if the angular frequency ω of the perturbation is equal to the Bohr frequency $\omega_0 = (E_f - E_i)/\hbar$ of the system. The resonance curve giving the variation of $\mathcal{P}_{i\to f}$ as a function of ω has a full width at half maximum $\Delta\omega \sim 2\pi/T$. The resonance is sharper as the interaction time T increases.

17.1.6 Perturbative and Exact Solutions

We have already met the case of a two-state system in Chap. 12 when we studied the magnetic resonance of a spin-1/2 particle placed in a rotating magnetic field. For this particular problem, we know the exact solution of the evolution equations and it is instructive to compare it with the approximate result derived above.

Consider the specific example of the Rabi experiment. We denote by T the time spent by the molecular beam inside the cavity where the magnetic field rotating at frequency $\omega/2\pi$ is applied. The time-dependent coupling $\hat{H}_1$ is (cf. (12.28))

$$\langle +|\hat{H}_1|-\rangle = \frac{\hbar\omega_1}{2} e^{-i\omega t} \qquad (\omega_1 = -\gamma B_1) .$$

The exact formula obtained by Rabi (12.37) is

$$\mathcal{P}_{+\to-}(T) = \frac{\omega_1^2}{\Omega^2} \sin^2(\Omega T/2) , \qquad \text{where} \qquad \Omega = \left((\omega - \omega_0)^2 + \omega_1^2\right)^{1/2} ,$$

and the approximation (17.17) gives

$$\mathcal{P}_{+\to-}(T) = \frac{\omega_1^2}{(\omega-\omega_0)^2}\sin^2[(\omega-\omega_0)T/2]\,.$$

We notice that the two formulas nearly coincide in two cases:

- If the excitation frequency is sufficiently far from resonance, i. e. $|\omega-\omega_0| \gg \omega_1$. In this case $\Omega \simeq |\omega-\omega_0|$ and the two results coincide for all times.
- If the excitation is close to resonance ($|\omega-\omega_0| \ll \omega_1$) and if the interaction time is short enough, i. e. $\omega_1 T/2 \ll 1$.

17.2 Interaction of an Atom with an Electromagnetic Wave

Electromagnetic transitions in atomic and molecular systems play a central role in physics. Three basic processes are involved. Under the influence of an electromagnetic wave, an atom or a molecule can *absorb* energy. If it is in an excited state, it can also decay into a lower-energy state either by *spontaneous* emission of radiation or through the emission of radiation *induced* by an external electromagnetic wave. These three processes were introduced in 1917 by Einstein, who understood, in a remarkable piece of intuition, how a collection of atoms and photons could reach thermal equilibrium.

Here we want to study the behavior of an atom in a monochromatic wave whose electric field is

$$\boldsymbol{E}(\boldsymbol{r},t) = \mathcal{E}_0\,\boldsymbol{\epsilon}\,\cos(\omega t - \boldsymbol{k}\cdot\boldsymbol{r})\,. \tag{17.18}$$

This plane traveling wave has an amplitude $\mathcal{E}_0$, a wave vector $\boldsymbol{k}$ and a polarization $\boldsymbol{\epsilon}$ orthogonal to $\boldsymbol{k}$. We want to calculate the probabilities for the two processes of absorption and induced emission of light by the atom. Spontaneous emission cannot be treated quantitatively in this book, because a correct approach requires the quantization of the electromagnetic field. We shall simply make a few qualitative remarks concerning this latter process (see also Sect. 17.3).

17.2.1 The Electric-Dipole Approximation

We assume that we know the energy levels of the atomic system. We denote the ground state of energy E_1 and an excited state of energy E_2 by $|1\rangle$ and $|2\rangle$, respectively. We want to study here the absorption of light which results in a transition of the atom from the initial state $|1\rangle$ to the final state $|2\rangle$. The induced emission of light can be calculated in the same way, assuming the initial state is $|2\rangle$ and the final state is $|1\rangle$.

In order to describe this phenomenon, we consider the simple case of a one-electron atom. We denote by $\hat{\boldsymbol{D}} = q\hat{\boldsymbol{r}}$ the electric-dipole-moment operator, which is proportional to the position of the external electron with respect to the core of the atom. We treat the atom as infinitely heavy, and denote by $\boldsymbol{R}_0$ the position of the core. The coupling between the atom and the electric field (17.18) is given by

$$\hat{H}_1(t) = -\hat{\boldsymbol{D}} \cdot \boldsymbol{E}(\boldsymbol{R}_0, t)\,. \tag{17.19}$$

This coupling is called the *electric-dipole* interaction Hamiltonian.

17.2.2 Justification of the Electric Dipole Interaction

The complete interaction of an atom with an external electromagnetic field $(\boldsymbol{E}, \boldsymbol{B})$ derived from the potentials $(\boldsymbol{A}, \Phi)$ can be obtained using the considerations developed in Chap. 15. Let $\hat{\boldsymbol{r}}_i$ and $\hat{\boldsymbol{p}}_i$ be the position and momentum operators of the electrons $(i = 1, \dots, Z)$. Assuming the nucleus of charge Z is fixed, and omitting the spin magnetic interactions, the Hamiltonian of the system in the presence of the external fields is

$$\hat{H} = \sum_{i=1}^{N} \frac{1}{2m} \left[\hat{\boldsymbol{p}}_i - q_{\mathrm{e}} \boldsymbol{A}(\hat{\boldsymbol{r}}_i, t)\right]^2 + q_{\mathrm{e}} \Phi(\hat{\boldsymbol{r}}_i, t) - \frac{Z q_{\mathrm{e}}^2}{4\pi\varepsilon_0 \hat{r}_i}$$
$$+ \frac{1}{2} \sum_i \sum_{j \neq i} \frac{q_{\mathrm{e}}^2}{4\pi\varepsilon_0 |\hat{\boldsymbol{r}}_i - \hat{\boldsymbol{r}}_j|}\,. \tag{17.20}$$

As such, this expression is much too complicated. In practice, one must expand (17.20) and make approximations.

In a systematic expansion of the Hamiltonian (17.20), there exist terms due to the electric field of the incident wave, and others due to the magnetic field. We neglect this second type of interaction. In fact, we have $|\boldsymbol{B}| = |\boldsymbol{E}|/c$ for a plane wave in vacuum. Since the typical velocity of an external electron in an atom is of the order of $\alpha c \sim c/137$, i. e. much smaller than the velocity of light, the Lorentz force and the ensuing magnetic effects are very small compared with the electric part. If we were considering X rays and internal electrons, these magnetic effects would be comparable to the electric effects.

Even if we limit ourselves to the electric-dipole interaction of a one-electron atom, we should, in full rigor, keep the dependence of the incident field on $\boldsymbol{r}$. However, the typical spatial extension of the electron orbit is the atomic scale ($\langle r \rangle \sim 1$ Å). This is much smaller than the wavelength of radiation corresponding to the infrared, visible or ultraviolet part of the spectrum ($\lambda = 2\pi/k \geq 10^3$ Å). Consequently the variation of $\boldsymbol{E}$ with $\boldsymbol{r}$ is negligible and it is legitimate to replace $\boldsymbol{E}(\boldsymbol{R}_0 + \boldsymbol{r}, t)$ by $\boldsymbol{E}(\boldsymbol{R}_0, t)$.

To summarize, the simple expression that we choose for $\hat{H}_1$ in the case of a one-electron atom is the dominant term of the interaction between the electromagnetic field $(\boldsymbol{E}, \boldsymbol{B})$, and the charge and current densities inside the atom. It is the first term of a *multipole* expansion which also contains smaller effects, of magnetic and/or relativistic origin.

17.2.3 Absorption of Energy by an Atom

In order to simplify the notation, we assume that the center of mass of the atom is at $\boldsymbol{R}_0 = \boldsymbol{0}$. At time t, the atomic state is

$$|\psi(t)\rangle = \gamma_1(t)\,\mathrm{e}^{-\mathrm{i}E_1 t/\hbar}\,|1\rangle + \gamma_2(t)\,\mathrm{e}^{-\mathrm{i}E_2 t/\hbar}\,|2\rangle + \sum_{n\neq 1,2} \gamma_n(t)\,\mathrm{e}^{-\mathrm{i}E_n t/\hbar}\,|n\rangle\,,$$

with the initial conditions $\gamma_1(0) = 1$ and $\gamma_2(0) = \ldots = \gamma_n(0) = 0$. Inserting the expression (17.19) into the general result (17.13), we find

$$\gamma_2(t) = \frac{q_\mathrm{e}\mathcal{E}_0}{2\hbar}\,\langle 2|\hat{\boldsymbol{r}}\cdot\boldsymbol{\epsilon}|1\rangle \left(\frac{\mathrm{e}^{\mathrm{i}(\omega_0+\omega)t} - 1}{\omega_0+\omega} + \frac{\mathrm{e}^{\mathrm{i}(\omega_0-\omega)t} - 1}{\omega_0-\omega} \right), \tag{17.21}$$

where $\hbar\omega_0 = E_2 - E_1$.

A *resonance* phenomenon appears for $\omega \sim \omega_0$. In the above expression, the first term is of the order of $1/\omega = T_0/2\pi$, where T_0 is the period of the exciting field ($T_0 \sim 10^{-15}\,$s in the optical domain). For $\omega = \omega_0$, the second term increases linearly with the interaction time t. If $t \gg T_0$, we can neglect the first term compared with the second, and we obtain

$$\mathcal{P}_{1\to 2}(t) = \frac{q_\mathrm{e}^2\mathcal{E}_0^2}{4\hbar^2}\,|\langle 2|\hat{\boldsymbol{r}}\cdot\boldsymbol{\epsilon}|1\rangle|^2\, y(\omega-\omega_0,t)\,. \tag{17.22}$$

In this expression, the presence of the square of the matrix element $|\langle 2|\hat{\boldsymbol{r}}|1\rangle|^2$ is of great importance in determining which transitions are allowed, as we shall see. We also note the presence of the function $y(\omega-\omega_0,t)$ (17.16). This transition probability has a resonant behavior in the vicinity of $\omega = \omega_0$, and the width of the resonance is of the order of $1/t$.

Contribution of Spontaneous Emission. At resonance, the time t must be sufficiently small that $|\gamma_2(t)| \ll 1$, which is a necessary condition for the perturbative approach to be valid. Also, the time t has to be much smaller than the lifetime τ of the level $|2\rangle$ due to spontaneous emission. Otherwise, this process has to be taken into account in the above calculation, and it gives a finite width to the resonance line (see Sect. 17.3). We shall see in Sect. 17.2.5 that $\tau \gg T_0$, so that it is possible to fulfill simultaneously $t \gg T_0$ so that (17.22) holds, and $t \ll \tau$ so that spontaneous emission can be neglected.

The Concept of the Photon. We comment here on the result (17.22) that the transitions are important only when the frequency of the light wave is close to a Bohr frequency of the atom, $\hbar\omega = E_2 - E_1$. This phenomenon is analogous to the photoelectric effect: an electron jumps from one state to another provided the incoming frequency is tuned to a Bohr frequency. In the case of the photoelectric effect, an electron is emitted and the final state belongs to the continuum of ionized states. Contrary to a common prejudice due to the chronology of the discoveries, we obtain an explanation of the photoelectric effect although we have not quantized the electromagnetic field and we have not introduced the concept of the photon. This concept only becomes necessary when we wish to explain the properties of radiation itself or to account for the spontaneous emission of radiation.

Validity of the Perturbative Treatment. The electrostatic Coulomb field seen by an electron in an atom is of the order of 10^{11} V/m, which is enormous compared with the electric field of a "standard" light wave. In order to compete with the Coulomb field, one must use laser beams with an intensity of $\sim 10^{15}$ W/cm^2, which is considerable. We can see that the use of perturbation theory is justified, i. e. the external field appears as a very small fluctuation compared with the Coulomb field.

17.2.4 Selection Rules

We now derive from (17.22) the selection rules for electric-dipole absorption and induced emission. Consider the matrix element

$$\langle 2|\hat{\boldsymbol{r}}|1\rangle \equiv \langle n_2, \ell_2, m_2|\hat{\boldsymbol{r}}|n_1, \ell_1, m_1\rangle \,.$$

In spherical coordinates, we have $z = r\cos\theta$, $x \pm \mathrm{i}y = r\sin\theta\,\mathrm{e}^{\pm\mathrm{i}\varphi}$, i. e. the coordinates of $\boldsymbol{r}$ are expressed linearly in terms of $r\,Y_{1,m}(\theta,\varphi)$. In the above matrix element, the contribution of interest is the angular integral

$$\int [Y_{\ell_2,m_2}(\Omega)]^*\; Y_{1,m}(\Omega)\; Y_{\ell_1,m_1}(\Omega)\; \mathrm{d}^2\Omega \,.$$

Owing to the properties of spherical harmonics described in Chap. 10, this integral is nonzero if and only if

$$\ell_2 = \ell_1 \pm 1 \qquad \text{and} \qquad m_2 - m_1 = 1,\, 0,\, -1 \,. \tag{17.23}$$

This is the case, for instance, for the Lyman α line of hydrogen, 2p $\to$ 1s, and for the resonance line of sodium, 3p $\to$ 3s; in both cases, $\ell_1 = 1$ and $\ell_2 = 0$. For a pair of levels which does not fulfill (17.23), the transition is forbidden. An example is the transition corresponding to the 21 cm line of hydrogen, for which both levels have zero orbital angular momentum ($\ell_1 = \ell_2 = 0$). The dominant coupling between these two levels is a magnetic-dipole interaction, whose matrix element is much smaller than for an electric-dipole coupling.

17.2.5 Spontaneous Emission

A complete calculation of spontaneous emission requires the quantization of the electromagnetic field, and we shall not treat this subject here. However, it is interesting to give the main results and to discuss them.

Consider an excited atomic state $|i\rangle$ which is coupled by an electric-dipole transition to a state $|f\rangle$ with a lower energy. An atom prepared in the state $|i\rangle$ may decay to the state $|f\rangle$ by emitting spontaneously a photon with an energy $\hbar\omega_{if} = E_i - E_f$. One can show that the probability $\mathrm{d}\mathcal{P}_{i\to f}$ that the decay takes place during an arbitrarily short time interval $\mathrm{d}t$ is proportional to $\mathrm{d}t$. Therefore one defines a *probability per unit time* $\mathrm{d}\mathcal{P}_{i\to f}/\mathrm{d}t$, which is independent of $\mathrm{d}t$, and which is given by the formula

$$\frac{\mathrm{d}\mathcal{P}_{i\to f}}{\mathrm{d}t} = \frac{\omega_{if}^3}{3\pi\varepsilon_0\hbar c^3}\,|\langle i|\hat{\boldsymbol{D}}|f\rangle|^2\,, \tag{17.24}$$

where $\hat{\boldsymbol{D}}$ is the electric dipole moment introduced above. Since each photon carries an energy $\hbar\omega_{if}$, the energy radiated per unit time $\mathrm{d}I/\mathrm{d}t$ is therefore

$$\frac{\mathrm{d}I}{\mathrm{d}t} = \frac{\omega_{if}^4}{3\pi\varepsilon_0 c^3}|\langle i|\hat{\boldsymbol{D}}|f\rangle|^2\,.$$

We notice that these transitions follow the same selection rules as found in Sect. 17.2.4, since the same matrix element is concerned.

We can compare this result with the classical formula giving the total intensity radiated per unit time by an electric dipole of moment $\boldsymbol{p}(t) = \boldsymbol{P}\cos\omega t$:

$$\frac{\mathrm{d}I}{\mathrm{d}t} = \frac{1}{6\pi\varepsilon_0 c^3}\,|\ddot{\boldsymbol{p}}(t)|^2\,.$$

After performing a time average over a period $2\pi/\omega$, we obtain

$$\frac{\mathrm{d}I}{\mathrm{d}t} = \frac{\omega^4}{12\pi\varepsilon_0 c^3}\,\boldsymbol{P}^2\,.$$

We notice the analogy between the classical and quantum formulas, with the correspondence set out in Table 17.1. This substitution was made by Heisenberg in 1925. It was a basic ingredient of his matrix mechanics.

Table 17.1. Correspondence between classical and quantum formulas for spontaneous emission of radiation

	Classical		Quantum
Frequency	ω	$\to$	$\omega_{if} = (E_i - E_f)/\hbar$
Amplitude	P	$\to$	$2\,|\langle i|\hat{\boldsymbol{D}}|f\rangle|$

Lifetime of an Atomic Level; Orders of Magnitude. Consider an assembly of N_0 atoms all in state $|i\rangle$ at time 0. Since the probability that a given atom decays in a time step $\mathrm{d}t$ is proportional to $\mathrm{d}t$, the number of atoms $N(t)$ still in the state $|i\rangle$ at time t follows an exponential decay law:

$$N(t) = N_0\,\mathrm{e}^{-t/\tau}\,, \qquad \text{where} \qquad \frac{1}{\tau} = \frac{\mathrm{d}\mathcal{P}_{i\to f}}{\mathrm{d}t}\,.$$

The quantity τ is called the lifetime of the level $|i\rangle$.

For a monovalent atom, we know that the size a of an outer electron orbit is of the order of $\hbar^2/me^2 \simeq e^2/\hbar\omega$, which gives

$$\frac{1}{\tau} \sim \frac{\omega_{if}^3}{3\pi\varepsilon_0\hbar c^3} q_e^2 a^2 \sim \omega_{if}\,\alpha^3 , \qquad \text{where} \qquad \alpha = \frac{e^2}{\hbar c} \simeq \frac{1}{137} . \tag{17.25}$$

For optical radiation, the order of magnitude of the lifetime τ of atomic levels is 10^{-7} to 10^{-9}s. This is much longer than a typical Bohr period $2\pi/\omega_{if}$, owing to the smallness of the coefficient α^3 entering into (17.25).

17.2.6 Control of Atomic Motion by Light

In the above considerations, we have assumed that the atom is fixed at a point $\boldsymbol{R}_0$. If the atom has a velocity $\boldsymbol{v}$, our analysis must be modified, owing to the Doppler effect. The apparent angular frequency of the light wave becomes $\omega - \boldsymbol{k}\cdot\boldsymbol{v}$, which shifts the resonance position. For a typical atomic velocity at room temperature (500 m/s) and for a visible wavelength (0.5 μm), this shift is of the order of 1 GHz.

The recoil of the atom in an absorption or emission process should also be taken into account. Even if the atom is initially at rest, the fact that it absorbs or emits photons with momentum $\hbar\boldsymbol{k}$ will set it in motion. This effect is called *radiation pressure*, and we can estimate the corresponding force. In each absorption event, the atom gains a momentum $\hbar\boldsymbol{k}$. When it falls back to the ground state by induced emission, it "gives back" this momentum to the wave. In contrast, if it falls back to the ground state by a spontaneous emission process, the change of the atom momentum vanishes on average since the probability for spontaneous emission is the same in two opposite directions. Each *absorption–spontaneous-emission* cycle therefore results in a variation of the atomic velocity of $\hbar\,\mathrm{k/m}$. This *recoil velocity* is $\sim 3\,\mathrm{cm/s}$ for a sodium atom irradiated by a light wave tuned to its resonance line ($\lambda = 0.59\,\mu\mathrm{m}$).

If the light wave is of sufficient intensity, it is possible to repeat these cycles at a rate equal to $1/(2\tau)$, i.e. one cycle every two lifetimes τ of the excited state. The corresponding acceleration of the sodium atom ($\tau = 16\,\mathrm{ns}$) is then $\hbar k/(2m\tau) \sim 10^6\,\mathrm{m\,s^{-2}}$, i.e. 10^5 times the acceleration of gravity. This radiation pressure force allows one to deflect an atomic beam with an initial velocity $\sim 500\,\mathrm{m\,s^{-1}}$, and even to stop it in a distance of the order of one meter.

If the incident light wave has a more complex structure than a plane traveling wave, other forces will act on the atom. Suppose for instance that the atom is placed in a light wave with a gradient of intensity. The frequency of the light $\omega/2\pi$ is assumed to be close (but not equal) to the Bohr frequency $\omega_0/2\pi$ of the atom for its resonance transition, which connects the ground state and the first excited state. One can show that the light wave creates a *dipole force* on the atom which, for $aq\mathcal{E}_0 \ll \hbar\,|\omega - \omega_0|$ (a being a typical

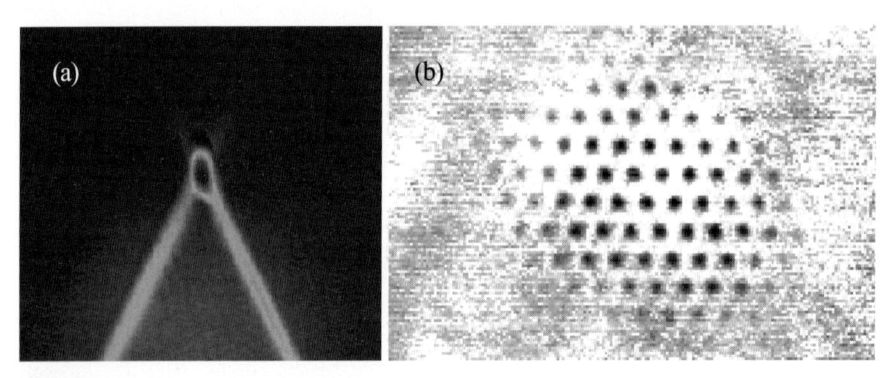

Fig. 17.2. Cesium atoms confined (**a**) at the intersection of two focused laser beams (dimensions of the focal spot ~80 μm), (**b**) in a hexagonal lattice formed by the interference pattern of three laser beams (distance between two adjacent sites ~28 μm). The temperature of the trapped atoms is 30 microkelvins. (Photograph (a): Hélène Perrin and Christophe Salomon. Photograph (b): D. Boiron et al., Phys. Rev. A **57**, R4106 (1998))

atomic size), derives from the potential

$$U(\boldsymbol{R}) = \frac{q^2 \mathcal{E}_0^2(\boldsymbol{R})}{4\hbar(\omega - \omega_0)} \, |\langle 2|\boldsymbol{r} \cdot \boldsymbol{\epsilon}|1\rangle|^2 \, .$$

This expression can be compared with the result found in Chap. 6, concerning ammonia molecules placed in a *static* electric-field gradient. According to the sign of the difference between the frequency of the light wave and the atomic Bohr frequency, the atom is attracted towards regions of high luminous intensities ($\omega < \omega_0$) or is repelled by these regions ($\omega > \omega_0$). It is therefore possible to confine atoms in a vacuum, in the vicinity of the focal point of an intense laser beam (see Fig. 17.2a). Using a standing light wave, one can also construct a regular lattice of atoms in space by trapping them at the nodes or the antinodes of the wave (see Fig. 17.2b).

The velocity dependence of the radiation forces opens up the possibility of cooling atomic gases. Here one uses particular electromagnetic field configurations, where the force created by the light is always in the direction opposite to the atomic velocity. This results in a damping of the motion of the atoms. This *optical molasses* allows investigators to reach temperatures in the microkelvin range. The manipulation and cooling of atoms by light have had many applications in recent years, in metrology (improvement of atomic clocks), atom interferometry, collisions and statistical physics.

17.3 Decay of a System

After studying the resonant or quasi-resonant coupling between two levels, we turn to another class of problems, where an initial state is coupled to a continuum of final states, i. e. a collection of states whose energies are very close and can be considered as a continuum. This is a central problem in collision physics and in the description of the decay of a system.

17.3.1 The Radioactivity of ^{57}Fe

In order to present in a concrete way the issues addressed in this section, we consider the specific case of a radioactive nucleus. We start with cobalt-57. This has the peculiarity that the isolated ^{57}Co nucleus is stable, but the atom is not. A proton of the nucleus can absorb an electron of the internal K shell. This gives rise to what is called an electron capture, or, equivalently, an inverse β reaction:

$$^{57}\mathrm{Co} + \mathrm{e}^- \to {}^{57}\mathrm{Fe}^{**} + \nu\,.$$

The ^{57}Co atom has a lifetime of 270 days.

The excited Fe nucleus ^{57}Fe** produced in this reaction emits a first photon γ_1, of energy 123 keV, with a very short lifetime $\tau_1 \simeq 10^{-10}$ s. This leaves the nucleus in another excited state, ^{57}Fe*. The ^{57}Fe* nucleus then emits a second photon γ_2, of energy 14 keV, with a lifetime $\tau_2 \simeq 1.4 \times 10^{-7}$ s, leaving the ^{57}Fe nucleus in its ground state:

$$\begin{aligned}
&^{57}\mathrm{Fe}^{**} \to {}^{57}\mathrm{Fe}^{*} + \gamma_1\,, \qquad \hbar\omega_1 = 123\,\mathrm{keV}\,;\\
&^{57}\mathrm{Fe}^{*} \;\;\to {}^{57}\mathrm{Fe} + \gamma_2\,, \qquad \hbar\omega_2 = 14\,\mathrm{keV}\,.
\end{aligned}$$

It is possible to measure, for each decay, the time interval between the emissions of the two photons γ_1 and γ_2.

By convention, we denote by $t_0 = 0$ the time when the photon γ_1 is detected. We want to calculate the probability $P(t)$ that the nucleus decays and falls back to its ground state between the times 0 and t. Experimentally, the answer to this question is well known: the decay $^{57}\mathrm{Fe}^{*} \to {}^{57}\mathrm{Fe} + \gamma$ obeys the exponential law

$$P(t) = 1 - \mathrm{e}^{-t/\tau}\,,$$

where in the present case, $\tau \simeq 1.4 \times 10^{-7}$ s. For $t \ll \tau$, the probability that the system decays between 0 and t is proportional to t:

$$t \ll \tau \qquad \to \qquad P(t) \simeq \frac{t}{\tau}\,. \tag{17.26}$$

The Hilbert Space of the Problem. For this example, we must consider the Hilbert space which describes the state of an iron nucleus, accompanied by a certain number of photons. This is a situation different from those we have met up to now. Strictly speaking, it requires the formalism of quantum field theory. Here we simply assume that some matrix elements between the relevant states exist, but we shall not attempt to calculate them explicitly.

There are two types of states to be considered:

- The initial state $|i\rangle \equiv |\mathrm{Fe}^*\rangle$ prepared at time $t = 0$, in the absence of photons (the photon γ_1, which is the signature that indicates that the Fe^* has been prepared, is absorbed by the detector at $t = 0$).
- The possible final states $|f\rangle \equiv |\mathrm{Fe} + \gamma, E_f\rangle$, representing the Fe nucleus in its ground state accompanied by one photon. E_f represents the sum of the energies of the γ photon and of the nucleus in its ground state. In full rigor, we must specify also the direction of propagation of the outgoing photon and its polarization, in order to define $|f\rangle$ completely.

The states $|i\rangle$ and $|f\rangle$ are eigenstates of the Hamiltonian $\hat{H}_0$, which describes nuclear forces on one hand and freely propagating photons on the other hand. These states are *not* eigenstates of the coupling $\hat{H}_1$ between the nucleus and the quantized electromagnetic field. In particular, a nucleus prepared in the state $|i\rangle$ will not remain in this state indefinitely. We want to calculate the evolution of the system assuming we know the matrix elements $\langle f|\hat{H}_1|i\rangle$.

Density of Final States. For simplicity, we neglect the recoil of the nucleus. The energy of the $^{57}\mathrm{Fe}$ nucleus is therefore fixed. The emitted photon can be in a whole series of energy states, which form a discrete set if we assume that the system is contained in a finite volume. Consider some energy band $\mathrm{d}E$. Inside this band, there is a number $\mathrm{d}N$ of photon states. As in Sect. 4.4.2, we define the *density of states* $\rho(E)$:

$$\rho(E) = \frac{\mathrm{d}N}{\mathrm{d}E} \, .$$

This allows us to replace a discrete sum over the possible final states by an integral over the final-state energy E_f, which is much easier to handle:

$$\sum_f \longrightarrow \int \rho(E_f) \, \mathrm{d}E_f \, .$$

Although we have not defined precisely the relation between photons and the electromagnetic field, all we need to know here is that photons are massless particles. The corresponding density of states can then be easily calculated. It is sufficient to make use of the fact that photons of momentum $\boldsymbol{p}$ have an energy $E = c\,|\boldsymbol{p}|$. As for the nonrelativistic particles that we studied in Chap. 4, the momenta of the photon are quantized if we suppose that the experiment takes place in an (arbitrarily large) cubic box of size L, with periodic boundary conditions:

$$\boldsymbol{p} = \frac{2\pi\hbar}{L}\boldsymbol{n}\,, \qquad \boldsymbol{n} = (n_1, n_2, n_3)\,, \qquad n_1, n_2, n_3 \text{ integers}\,. \tag{17.27}$$

Combining $E = cp$ and (17.27), we obtain

$$\rho(E) = \frac{L^3}{2\pi^2\hbar^3}\,\frac{E^2}{c^3}\,,$$

to be compared with (4.52) obtained in Chap. 4 for nonrelativistic massive particles. It remains to be checked at the end of the calculation that the predictions for any measurable quantity do not depend on the volume L^3 of the fictitious box that we have introduced. In the present case this is ensured by the expression for the matrix element $\langle f|\hat{H}_1|i\rangle$, which scales as $L^{-3/2}$.

17.3.2 The Fermi Golden Rule

We now come back to our decay problem. The nucleus $^{57}\mathrm{Fe}^*$ can decay into a continuous set of Fe + γ states. We are not interested in the probability that it decays to a specific state, but in the probability that it decays to some domain $\mathcal{D}_f$ of final states, characterized by their direction $\boldsymbol{\Omega}$ (within a small solid angle $\mathrm{d}^2\Omega$). We must therefore sum the formula which gives $\mathcal{P}_{i\to f}$ (17.15) over all *possible final states* of the domain $\mathcal{D}_f$:

$$\begin{aligned}\mathrm{d}^2\mathcal{P}_{i\to\mathcal{D}_f}(t) &= \frac{1}{\hbar^2}\sum_{f\in\mathcal{D}_f}|\langle f|\hat{H}_1|i\rangle|^2\,y(\omega_{if},t)\\ &= \frac{1}{\hbar^2}\int_{\mathcal{D}_f}|\langle f|\hat{H}_1|i\rangle|^2\,y(\omega_{fi},t)\,\rho(E_f)\,\mathrm{d}E_f\,\frac{\mathrm{d}^2\Omega}{4\pi}\,,\end{aligned} \tag{17.28}$$

where $\omega_{fi} = (E_f - E_i)/\hbar$. We now use the fact that, as the time t increases, the quantity y, considered as a function of E_f, becomes more and more peaked in the vicinity of $E_f = E_i$. Using (17.16), we obtain

$$\frac{1}{2\pi t}\,y(\omega_{fi},t) \approx \delta(\omega_{fi}) = \hbar\,\delta(E_f - E_i)\,. \tag{17.29}$$

We can therefore neglect the variations of $\rho(E)$ and of the matrix element $\langle f|\hat{H}_1|i\rangle$ in the integral over E_f. In other words, we extract the matrix element $\langle f|\hat{H}_1|i\rangle$ and the density of states ρ from the integral and evaluate them at the central point $E_f = E_i$. Using (17.16), this leads to

$$\mathrm{d}^2\mathcal{P}_{i\to\mathcal{D}_f}(t) = \frac{2\pi}{\hbar}\,|\langle f, E_f = E_i|\hat{H}_1|i\rangle|^2\,\rho(E_i)\,\frac{\mathrm{d}^2\Omega}{4\pi}\,t\,. \tag{17.30}$$

We recover the linear dependence on time observed experimentally for short times (see (17.26)). Let us assume for simplicity that the matrix element $\langle f, E_f = E_i|\hat{H}_1|i\rangle$ does not depend on the direction $\boldsymbol{\Omega}$ of the emitted photon. Summing over all solid angles, we deduce the lifetime of state $|i\rangle$:

$$\frac{1}{\tau} = \frac{2\pi}{\hbar} \, |\langle f, E_f = E_i|\hat{H}_1|i\rangle|^2 \, \rho(E_i) \, . \tag{17.31}$$

The fundamental relation (17.30) is called the *Fermi golden rule*. The range of times t for which it can be applied is limited by two constraints:

- The time t must be short enough that $\mathcal{P}_{i\to \text{all } f}(t) \ll 1$:

$$t \ll \tau \, . \tag{17.32}$$

  This is a necessary condition for the validity of first-order perturbation theory.
- The time t must be long enough that the frequency width $\sim 1/t$ of the function $y(\omega_{if}, t)$ in (17.28) is much smaller than the typical scale of variation of the two other terms, $\langle f|\hat{H}_1|i\rangle$ and ρ. Denoting by κ this scale of variation in the frequency domain, the second constraint reads

$$t^{-1} \ll \kappa \, . \tag{17.33}$$

In any problem where the Fermi golden rule is used, one must check that there exists a time interval during which these two constraints are simultaneously satisfied.

17.3.3 Orders of Magnitude

We have already given in (17.25) the scaling laws for the lifetime of an excited atomic level, which can decay by spontaneous emission by means of an electric-dipole transition. Aside from geometric factors, this decay rate reads

$$\frac{1}{\tau} \sim \alpha \, \frac{a_1^2 \omega_{if}^3}{c^2} = \frac{\hbar^3 \omega_{if}^3}{m_e^2 e^2 c^3} \, , \tag{17.34}$$

where m_e is the electron mass, $a_1 = \hbar^2/m_e e^2$ is the Bohr radius and $\hbar\omega_{if}$ is the energy of the emitted photon.

We can discuss the consistency of the Fermi golden rule in this example. The frequency scale κ for the variations of $\langle f|\hat{H}_1|i\rangle$ and ρ is typically $\kappa \sim \omega_{if}$. Therefore (17.32) and (17.33) can be simultaneously satisfied if:

$$\omega_{if}^{-1} \ll \tau \quad \Rightarrow \quad \frac{\hbar^3 \omega_{if}^2}{m_e^2 e^2 c^3} \ll 1 \, .$$

A typical Bohr frequency is $E_\mathrm{I}/\hbar$, where $E_\mathrm{I} = m_e e^4/2\hbar^2$ is the ionization energy of the hydrogen atom. The requirement for the consistency of our approach then reads

$$\alpha^3 \ll 1 \,.$$

Since $\alpha \simeq 1/137 \ll 1$, this inequality is well satisfied. The smallness of the fine structure constant guarantees that the perturbative treatment of the effect of electromagnetic interactions on atomic levels is a good approximation.

In going from atomic systems to nuclear systems, considering the expression $1/\tau \sim \alpha a_1^2 \omega_{if}^3/c^2$, we expect that the electric-dipole decay rates should be (i) reduced by a factor of order 10^{-10} owing to the change of size (10^{-15} m instead of 10^{-10} m), and (ii) enhanced by a factor of order 10^{18} owing to the change of energy scale (1 MeV instead of 1 eV).

We can therefore transpose (17.34) to the nuclear scale using $R \sim r_0 A^{1/3}$ for the radius of a nucleus, where $r_0 \sim 1.2$ fm and A is the number of nucleons. We obtain

$$\frac{1}{\tau} \sim \alpha \, \frac{r_0^2 \, A^{2/3} \, \omega^3}{c^2} \,. \tag{17.35}$$

One can check that the energies and lifetimes of the excited states of ^{57}Fe agree acceptably with this estimate. In particular, we can verify immediately that $\tau_2/\tau_1 \sim (\omega_1/\omega_2)^3 \sim 10^3$. In the case of nitrogen-13, there exists an excited state with $\hbar\omega = 2.38\,\mathrm{MeV}$ and a lifetime $\tau \sim 10^{-15}\,\mathrm{s}$. Using these parameters and (17.35), we obtain $\tau \sim 2 \times 10^{-15}\,\mathrm{s}$, which is a perfectly acceptable order of magnitude.

17.3.4 Behavior for Long Times

We have just found how the notion of a lifetime of an excited atomic or nuclear level emerges using the short-time approximation to the decay law. For longer times, first order perturbation theory no longer applies since we no longer have $\mathcal{P}_{i \to \mathrm{all}\, f} \ll 1$. In this case, one can recover the measured exponential decay law using another approximation due to Wigner and Weisskopf.

In order to illustrate this, we consider the following simple model. We assume that the only nonvanishing matrix elements are $\langle i|\hat{H}_1|f\rangle$ and $\langle f|\hat{H}_1|i\rangle$:

$$\langle i|\hat{H}_1|i\rangle = \langle f|\hat{H}_1|f\rangle = 0 \,.$$

The initial state is $|\psi(0)\rangle = |i\rangle$. Using the above form of the coupling, we can write the state of the system at a later time t as

$$|\psi(t)\rangle = \gamma_i(t)\, \mathrm{e}^{-\mathrm{i}E_i t/\hbar}\, |i\rangle + \int \gamma(f,t)\, \mathrm{e}^{-\mathrm{i}E_f t/\hbar}\, |f\rangle\, \rho(E_f)\, \mathrm{d}E_f \,. \tag{17.36}$$

Here we assume for simplicity that the energy is the only quantum number which characterizes the final states. We set $H(f) \equiv \langle i|\hat{H}_1|f\rangle$ ($H(f)$ is simply a function of E_f), and the Schrödinger equation gives

$$\mathrm{i}\hbar\dot{\gamma}_i(t) = \int \mathrm{e}^{\mathrm{i}(E_i-E_f)t/\hbar}\, H(f)\,\gamma(f,t)\,\rho(E_f)\,\mathrm{d}E_f\,, \tag{17.37}$$

$$\mathrm{i}\hbar\dot{\gamma}(f,t) = \mathrm{e}^{\mathrm{i}(E_f-E_i)t/\hbar}\, H^*(f)\,\gamma_i(t)\,, \tag{17.38}$$

with the initial conditions $\gamma_i(0) = 1$, $\gamma(f,0) = 0$.

We integrate (17.38) formally:

$$\gamma(f,t) = \frac{H^*(f)}{\mathrm{i}\hbar} \int_0^t \mathrm{e}^{\mathrm{i}(E_f-E_i)t'/\hbar}\, \gamma_i(t')\,\mathrm{d}t'\,, \tag{17.39}$$

and we insert this result into (17.37). We then obtain the integro-differential equation

$$\dot{\gamma}_i(t) = -\frac{1}{\hbar^2}\int \mathrm{d}E_f\,\rho(E_f)\int_0^t \mathrm{e}^{\mathrm{i}(E_i-E_f)(t-t')/\hbar}\,|H(f)|^2\,\gamma_i(t')\,\mathrm{d}t'\,, \tag{17.40}$$

which can be rewritten as

$$\dot{\gamma}_i(t) = -\int_0^t \mathcal{N}(t'')\,\gamma_i(t-t'')\,\mathrm{d}t''\,,$$

where

$$\mathcal{N}(t'') = \frac{1}{\hbar^2}\int \mathrm{e}^{\mathrm{i}(E_i-E_f)t''/\hbar}\,|H(f)|^2\,\rho(E_f)\,\mathrm{d}E_f\,.$$

The function $\mathcal{N}(t'')$ is proportional to the Fourier transform of the function of the final energy $G(E_f) = |H(f)|^2\,\rho(E_f)$. By the definition of a continuum, the width of the function $G(E_f)$ is large. Therefore $\mathcal{N}(t'')$ has a narrow width and it is nonvanishing only if t'' is close enough to 0. We denote by $t'' = \tau_\mathrm{c}$ the characteristic time above which the integrand oscillates so rapidly that $\mathcal{N}(t'')$ is negligible. We make the approximation (to be justified a posteriori in each case) that $\gamma_i(t-t'')$ varies slowly in the time interval $0 < t'' < \tau_\mathrm{c}$. We can then replace $\gamma_i(t-t'')$ by $\gamma_i(t)$ in the right hand side of the integro-differential equation, and we obtain

$$\dot{\gamma}_i(t) = -\gamma_i(t)\int_0^t \mathcal{N}(t'')\,\mathrm{d}t''\,.$$

For times t large compared with τ_c, the upper bound of the integral can be extended to infinity. Finally, we use the relation

$$\int_0^{+\infty} \mathrm{e}^{\mathrm{i}(\omega-\omega_0)t''}\,\mathrm{d}t'' = \pi\delta(\omega-\omega_0) + i\mathcal{PP}\left(\frac{1}{\omega-\omega_0}\right),$$

where $\mathcal{PP}$ is the principal-value integral, and we obtain

$$\dot{\gamma}_i(t) = -\left(\frac{1}{2\tau} + \mathrm{i}\delta\omega_i\right)\gamma_i(t)\,,$$

where

$$\frac{1}{\tau} = \frac{2\pi}{\hbar}\,|H(f)|^2 \rho(E_i) \quad \text{and} \quad \delta\omega_i = \mathcal{PP}\left(\int \frac{|H(f)|^2}{E_i - E_f}\rho(E_f)\mathrm{d}E_f\right). \quad (17.41)$$

The differential equation which gives the evolution of $\gamma_i(t)$ can be integrated immediately. This gives the probability that the system has decayed at time t:

$$P(t) = 1 - |\gamma_i(t)|^2 = 1 - \mathrm{e}^{-t/\tau}\,, \quad (17.42)$$

i. e. the exponential law. One can check that the value for τ derived in (17.41) coincides with the value (17.31) calculated previously using perturbation theory.

The quantity $\hbar\,\delta\omega_i$ corresponds to an energy shift of the excited state due to the coupling of the nucleus and the electromagnetic field. This shift is exactly the same as what one obtains in second-order time-independent perturbation theory (see (9.21)). Note that to first order, the energy shift vanishes because of our assumptions concerning the diagonal elements of $\hat{H}_1$. In the case of atomic levels, this second-order shift is called the *Lamb shift* (see Sect. 13.2).

We now insert the result for $\gamma_i(t)$ in (17.39), the equation giving $\gamma(f,t)$. We obtain the energy distribution of the final states as follows:

$$p(E_f) = |\gamma(f, t=\infty)|^2 = |H(f)|^2 \frac{1}{(E_f - \bar{E}_i)^2 + \Gamma^2/4}\,, \quad (17.43)$$

where $\bar{E}_i = E_i + \hbar\,\delta\omega_i$ and $\Gamma = \hbar/\tau$. If we assume that $|H(f)|^2$ varies slowly, this probability law is a Lorentz function, centered at $\bar{E}_i$, with a full width at half maximum $\Gamma = \hbar/\tau$ (see Fig. 17.3). In other words the energy of the final states is, on average, $\bar{E}_i$ with a dispersion ΔE, where

$$\Delta E = \Gamma/2 = \hbar/2\tau\,. \quad (17.44)$$

This dispersion in energy of the final state is characteristic of any unstable system: it occurs in beta decay of nuclei, radiative decay of atomic states,

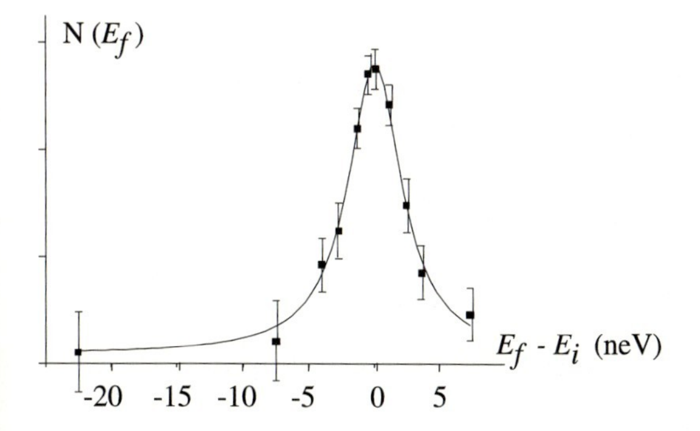

Fig. 17.3. Energy distribution of the photon γ_2 emitted in the decay of $^{57}Fe^*$

etc. It originates from the fact that the initial state $|i\rangle$ is an eigenstate of the Hamiltonian in the absence of the interaction but is not an eigenstate of the full Hamiltonian. Therefore, this initial state does not have a well-defined energy.

17.4 The Time-Energy Uncertainty Relation

One of the great controversial questions in the 1930s concerned the time–energy uncertainty relation

$$\Delta E\,\Delta t \geq \hbar/2\,. \tag{17.45}$$

Although this relation is commonly accepted, its interpretation varies considerably from one author to the other. It is indeed quite different from the uncertainty relations that we derived in Chap. 7. The relation $\Delta x\,\Delta p \geq \hbar/2$, for instance, follows directly from the principles of quantum mechanics and the commutation relation of the operators $\hat{x}$ and $\hat{p}$. It is therefore an *intrinsic* property of any system. On the other hand, in the Schrödinger equation, time is not an operator, but a parameter which has a well-defined value in the equations. Although we can measure it physically, time is not an *observable*.

We shall not give an exhaustive review of all points of view, neither will we adopt one attitude rather than another.[2] We simply wish to make a few observations which can be used as a starting point for further reflection.

17.4.1 Isolated Systems and Intrinsic Interpretations

We recall some results presented in the first few chapters for systems whose Hamiltonians do not depend on time.

Stationary States. These are eigenstates of the energy, whose time evolution reduces to a multiplicative global phase factor: $|\psi(t)\rangle = \mathrm{e}^{-\mathrm{i}Et/\hbar}|\psi(0)\rangle$. If the system is prepared in such a state, the expectation value $\langle a\rangle$ of any observable $\hat{A}$ does not change with time. This agrees with the relation (17.45):

An isolated system whose energy is well defined ($\Delta E = 0$) does not evolve from $t = -\infty$ to $t = +\infty$.

Evolution of a System. The state of a system $|\psi(t)\rangle$ can be a superposition of two or more energy eigenstates. For instance, in Chap. 2, we constructed a wave packet as an infinite sum of stationary states. Such a system does not have a well defined value of the energy, and the expectation values of observables evolve with time, unless they correspond to a conserved quantity.

[2] See for instance the article by Y. Aharonov and D. Bohm, Phys. Rev. **122**, 1649 (1961).

Consider a physical quantity A associated with the system, such as the position of a hand on a wristwatch. We denote by $\langle a \rangle_t$ and Δa the mean position and the mean square deviation of this quantity at time t. Let $v = \mathrm{d}\langle a \rangle_t / \mathrm{d}t$ be the velocity associated with $\langle a \rangle_t$. The characteristic time τ it takes the "wave packet" to cross a certain point, for instance $a = 0$, is $\tau = \Delta a / |v|$. In Chap. 7 we proved the following properties:

- Δa is related to the energy dispersion ΔE by

  $$\Delta a \, \Delta E \geq \frac{1}{2} |\langle \psi | [\hat{A}, \hat{H}] | \psi \rangle| \,,$$

  where $|\psi\rangle$ is the state of the system at time t.
- v is given by

  $$v = \frac{d\langle a \rangle_t}{\mathrm{d}t} = \frac{1}{\mathrm{i}\hbar} \langle \psi | [\hat{A}, \hat{H}] | \psi \rangle \,.$$

Combining these two relations, we obtain

$$\tau \, \Delta E \; = \; \frac{\Delta a}{|v|} \Delta E \; \geq \; \frac{\hbar}{2} \,.$$

We recover a relation similar to (17.45), and it appears here as an *intrinsic* property of the quantum system: the larger the dispersion ΔE of the energy, the shorter the *characteristic time of evolution* of any quantity. This formulation is due to Mandelstamm and Tamm.

Decay of an Unstable System. We have seen in Sect. 17.3.4 above that when a system is unstable and decays, its energy is not well defined. The energy distribution of the final products is peaked around some value with a dispersion related to the lifetime τ by $\Delta E = \hbar/(2\tau)$. This is also of the form (17.45), and it is again an intrinsic property of the system.

17.4.2 Interpretation of Landau and Peierls

This interpretation[3] comes from the analysis of the measurement of the energy E of a system. In order to perform such a measurement, we must couple the system of Hamiltonian $\hat{H}_\mathrm{s}$, to a detector of Hamiltonian $\hat{H}_\mathrm{d}$. The detector is initially in a state of known energy ϵ_d, and the coupling takes place for an interval of time T. When the coupling is switched off, the state of the set *system + detector* is a superposition of eigenstates of $\hat{H}_\mathrm{s} + \hat{H}_\mathrm{d}$, with an average energy $E' + \epsilon'_\mathrm{d}$ close to $E + \epsilon_\mathrm{d}$, up to an uncertainty $\hbar/T$:

$$|E' + \epsilon'_\mathrm{d} - E - \epsilon_\mathrm{d}| \sim \frac{\hbar}{T} \,.$$

[3] See, for instance, L. Landau and E. Lifshitz, *Quantum Mechanics*, Pergamon, Oxford (1965).

This results from the shape of the function $y(\omega = E/\hbar, t)$ introduced in Sect. 17.1.5. Suppose that we know precisely the initial and final energies of the detector ϵ_{d} and ϵ'_{d}. We can therefore deduce the uncertainty in $E' - E$: $\Delta(E' - E) \simeq \hbar/T$. In other words, even if the system is in a well-defined energy state before the measurement, an observer has access to this value only up to an uncertainty $\hbar/T$.

17.4.3 The Einstein–Bohr Controversy

In 1930, Einstein presented the following argument. A clock is placed in a box, hanging on a spring. It is set to open a shutter at time t_1 and to close it at time $t_2 = t_1 + T$, the interval T being determined with great accuracy. Some radiation escapes from the box when the shutter is open, and we measure the corresponding energy E by weighing the box before and after the experiment ($E = \delta m\, c^2$). Since we have as much time as we want to do this weighing, it can be very precise. Therefore this procedure seems to provide a counterexample to the relation $\Delta E\ T \geq \hbar/2$. Bohr disproved the argument in the following way:

(1) The position of the box which contains the clock is defined up to some quantum uncertainty Δz. Since the clock is placed in a gravitational field, its rate depends on the gravitational potential and, owing to *general relativity*, there is an uncertainty, given by

$$\frac{\Delta T}{T} = \frac{g\,\Delta z}{c^2}\,, \qquad (17.46)$$

in how long the shutter stays open.

(2) At time t_2, the determination of the decrease of the weight $\delta m\, g = Eg/c^2$ of the box is performed by measuring the momentum the box acquires during this time interval T,

$$p_z = \delta m\, gT = \frac{Eg}{c^2}T\,.$$

Here we assume that T is shorter than the oscillation period of the spring; a similar argument could be developed in the reverse case. Owing to the quantum uncertainty Δp_z in the initial momentum of the box, the accuracy in the measurement of the energy is

$$\Delta E = \frac{c^2}{gT}\,\Delta p_z\,. \qquad (17.47)$$

Combining the two equations (17.46) and (17.47) and using Heisenberg's inequality $\Delta z\, \Delta p_z \geq \hbar/2$ for the position and the initial momentum of the box, we recover the desired inequality. The story goes that Bohr, who had spent an entire night to find this counterargument, was quite proud of using Einstein's general relativity to solve the problem.

Further Reading

- For a discussion of the coupling between atoms and the electromagnetic field, see, for example, C. Cohen-Tannoudji, J. Dupont-Roc and G. Grynberg, *Atom–Photon Interactions, Basic Processes and Applications*, Wiley, New York (1992).
- For laser cooling of atoms and its applications to atomic clocks, see, for example, the 1997 Nobel lectures by S. Chu, C. Cohen-Tannoudji and W. Phillips, Rev. Mod. Phys. **70**, 685, 707 and 721 (1998); C. Cohen-Tannoudji and W.D. Phillips, "New mechanisms for laser cooling", Phys. Today, October 1990, p. 33; S. Chu, "Trapping neutral particles", Sci. Am., March 1992.
- The debate between Bohr and Einstein is presented in detail in: A. Pais, *Subtle is the Lord: the Science and the Life of Albert Einstein*, Oxford University Press, Oxford (1982); *N. Bohr, a Centenary Volume*, edited by A.P. French and P.J. Kennedy, Harvard University Press, Cambridge, MA (1985).

Exercises

17.1. Excitation of an atom with broadband light. Consider n two-level atoms driven by an electric field $\boldsymbol{E}(t) = E_0\boldsymbol{e}_z f(t)\cos\omega t$, where $f(t)$ is a function which is zero outside the interval $[-\tau,\tau]$. We consider an electric-dipole coupling between the atoms and the field. The ground and excited atomic states are denoted by a and b, respectively, and we set by convention $E_a = 0$ and $E_b = \hbar\omega_0$. We suppose that (i) the typical scale of variation of $f(t)$ is very large compared with the period $2\pi/\omega$ and (ii) the excitation frequency $\omega/2\pi$ is close to the Bohr frequency $\omega_0/2\pi$. We neglect the contribution of nonresonant terms.

a. We define $\hbar\Omega_1 = -dE_0$, where $d = \langle b|\boldsymbol{D}\cdot\boldsymbol{e}_z|a\rangle$ (Ω_1 is real) and we denote by $g(\Omega)$ the Fourier transform of $f(t)$. Using perturbation theory, calculate the average number of excited atoms at time τ. This number is denoted by $n_b(\tau)$.
b. The electric field now consists of a succession of wave packets:
$$\boldsymbol{E}(t) = E_0\,\boldsymbol{e}_z\sum_{p=1}^{\infty} f(t-t_p)\,\cos[\omega(t-t_p)]\,,\qquad \text{where}\quad t_1 < t_2 < \dots\,.$$
Consider T such that $t_\ell + \tau < T < t_{\ell+1} - \tau$. Calculate $n_b(T)$.
c. We suppose that the successive wave packets arrive in a random way, with γ wave packets per unit time on average. We denote by $\bar{n}_b(T)$ the statistical mean of $n_b(T)$. Calculate $\bar{n}_b(T)$ for $\gamma T \gg 1$. Show that one can define a transition probability per unit time from level a to level b. This quantity will be denoted $\Gamma_{a\to b}$.

d. We put $w(\omega+\Omega)=(\epsilon_0 c/2)E_0^2\gamma|g(\Omega)|^2$ and we denote by Φ the incident flux of energy. Relate w and Φ. Express $\Gamma_{a\to b}$ in terms of $w(\omega_0)$.
e. We suppose now that all atoms are initially in the state b. How can we transpose the previous reasoning?
f. Write down the evolution equations for the mean populations $n_a(t)$ and $n_b(t)$. What is the steady state of the system?

17.2. Atoms in equilibrium with black-body radiation. We consider the model of the previous exercise and suppose that the atomic assembly is irradiated by radiation from a black body of temperature T. We recall that we have in this case

$$w(\omega)=\mu\frac{\omega^3}{\mathrm{e}^{\hbar\omega/k_\mathrm{B}T}-1},$$

where μ depends only on fundamental constants. What must one add to the previous model in order to ensure the consistency of statistical physics? (This question was considered by Einstein in 1917.)

17.3. Ramsey fringes. A neutron (which is a spin-1/2 particle) propagates along the z axis. We denote by $|\pm\rangle$ the eigenstates of the operator $\hat{S}_z$, the projection of the neutron spin on the z axis. The neutron is initially prepared in the state $|+\rangle$, and it crosses two radio-frequency cavities of length L, separated by a distance $D \gg L$ (Fig. 17.4). In each cavity, a rotating magnetic field

$$\boldsymbol{B}_1=B_1(\cos\omega t\;\boldsymbol{u}_x+\sin\omega t\;\boldsymbol{u}_y)$$

is applied. The whole experimental setup is placed in a constant, uniform magnetic field $\boldsymbol{B}_0$ parallel to the z axis. The motion of the neutron is treated classically as a uniform linear motion with velocity v. We are interested only in the quantum evolution of the spin state of the neutron. The magnetic moment operator of the neutron is denoted by $\hat{\boldsymbol{\mu}}=\gamma\hat{\boldsymbol{S}}$, and we set $\omega_0=-\gamma B_0$ and $\omega_1=-\gamma B_1$.

Calculate to first order in B_1 the probability amplitude for finding the neutron in the spin state $|-\rangle$ at the output of the device. Show that the spin

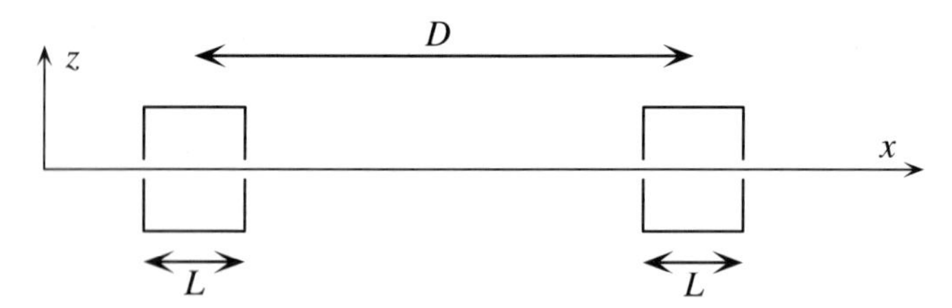

Fig. 17.4. Experimental setup for the observation of Ramsey fringes

flip probability varies rapidly with the detuning $\omega - \omega_0$, and that one can determine the resonance frequency ω_0 with a much better precision than one can if only one of the two cavities is used.

17.4. Damping of a quantum oscillator. Consider a harmonic oscillator of angular frequency ω (the "system") characterized by the creation and annihilation operators $\hat{a}^\dagger$ and $\hat{a}$. This oscillator is coupled to a "bath" of oscillators labeled by an index λ. An oscillator of the bath has an angular frequency ω_λ and is characterized by the creation and annihilation operators $\hat{b}_\lambda^\dagger$ and $\hat{b}_\lambda$. The total Hamiltonian reads

$$\hat{H} = \hbar\omega\hat{a}^\dagger\hat{a} + \sum_\lambda \hbar\omega_\lambda \hat{b}_\lambda^\dagger \hat{b}_\lambda + \sum_\lambda g_\lambda \left(\hat{a}^\dagger \hat{b}_\lambda + \hat{a}\hat{b}_\lambda^\dagger\right) ,$$

where the g_λ are real coefficients. We recall that $[\hat{a}, \hat{a}^\dagger] = [\hat{b}_\lambda, \hat{b}_\lambda^\dagger] = 1$. We use the Heisenberg representation (see exercise 5.3).

a. Write down the evolution equation of $\hat{a}(t)$ and of each $\hat{b}_\lambda(t)$.
b. Integrate the evolution equation of $\hat{b}_\lambda(t)$. Write down an evolution equation for $\hat{a}$ as a function of the operators $\hat{b}_\lambda(0)$ and of

$$\mathcal{N}(t'') = \frac{1}{\hbar^2} \sum_\lambda g_\lambda^2 \, \mathrm{e}^{\mathrm{i}(\omega - \omega_\lambda)t''} .$$

c. The bath of oscillators is assumed to be in its ground state. Using an approximation for the function $\mathcal{N}(t'')$ similar to that of Sect. 17.3.4, derive the evolution equation for the average values of $\hat{a}(t)$ and $\hat{a}^\dagger(t)$, for an arbitrary initial state of the system oscillator.

18. Scattering Processes

It is when you get too close to a lady...
that she says you're going too far.
Alphonse Allais

Scattering processes play a key role in the investigation of the properties of matter. Rutherford's experiment, which demonstrated the existence of a nucleus with a positive charge and a size 10^5 times smaller than the atomic radius, was based on the scattering of α particles by gold atoms. Modern laser spectroscopy, which provides information on the structure of atoms and molecules, can be viewed as a photon scattering process. The electric conductivity of a metal can only be understood quantitatively if one takes into account the scattering of conduction electrons by the impurities in the crystal. Last but not least, particle physics is entirely based on the analysis of scattering processes. The very large energies reached in modern accelerators (center-of-mass energies of approximately 100 GeV to a few TeV) allow us to probe matter at very short distances ($\leq 10^{-18}$ m).

This chapter gives an elementary approach to scattering processes. We first present the concept of the cross section, which plays a central role in the description of a collision, both in classical and in quantum physics. We shall then calculate cross sections to the lowest order in the scattering potential, and obtain what is known as the *Born approximation.* We shall use this expression to explain why and how scattering can be used to gain information on the structure of composite systems such as atoms and nuclei. In the last section, we explain how one can go beyond the Born approximation and obtain an exact (implicit) expression for the scattering cross section.

In all of this chapter we shall consider the scattering of a particle by a potential $V(\boldsymbol{r})$. This formalism also allows us to study a collision between two particles. Indeed, as we have already shown in Sect. 11.1, the solution of the two-body problem, whose Hamiltonian reads

$$\hat{H}_{\text{2body}} = \frac{\hat{\boldsymbol{p}}_1^2}{2M_1} + \frac{\hat{\boldsymbol{p}}_2^2}{2M_2} + V(\hat{\boldsymbol{r}}_1 - \hat{\boldsymbol{r}}_2)\,,$$

can be reduced to solving the one-body problem in the fixed potential $V(\boldsymbol{r})$. The motion of the *relative particle*, of coordinate $\boldsymbol{r} = \boldsymbol{r}_1 - \boldsymbol{r}_2$ and of mass

$m = M_1 M_2/(M_1 + M_2)$, is given by the Hamiltonian

$$\hat{H} = \frac{\hat{\boldsymbol{p}}^2}{2m} + V(\boldsymbol{r}) . \tag{18.1}$$

The scattering of a particle of mass m by the potential $V(\boldsymbol{r})$ therefore determines completely the result of a collision between the two particles of masses M_1 and M_2.

18.1 Concept of Cross Section

Classically, the exploration of a field of force consists in studying the trajectories of a particle placed in it. The smallness of atomic distances (≈1 nm) or nuclear distances (≈1 fm) and the fact that the notion of a trajectory loses its meaning in quantum mechanics mean that we must use another approach on such scales.

This analysis is based on the statistical concept of a *cross section.* This concept is used in classical physics and it remains valid in quantum mechanics. Its principle is based on sending a beam of particles with a well defined velocity to a target and measuring the angular and energy distribution of the final particles (Fig. 18.1).

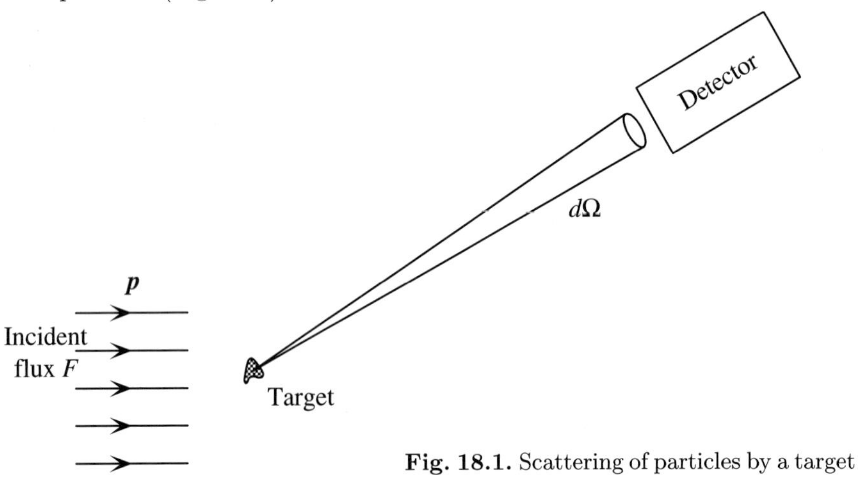

Fig. 18.1. Scattering of particles by a target

18.1.1 Definition of Cross Section

The notion of the cross section is basically experimental. Incident particles of momentum $\boldsymbol{p}$, whose flux is F, arrive at a target and are scattered. A detector counts the number of outgoing particles in a given solid angle[1] $\mathrm{d}\Omega$

[1] In this chapter, $\mathrm{d}\Omega$ denotes either a first-order $(2\pi \sin\theta \,\mathrm{d}\theta)$ or a second-order $(\sin\theta \,\mathrm{d}\theta \,\mathrm{d}\varphi)$ element according to whether or not we have integrated over the azimuthal angle φ.

in the vicinity of a direction $\mathbf{\Omega}$. We assume that the particles of the target are sufficiently far apart from one another that an incident particle interacts with only one target particle, and we neglect processes involving multiple scattering events.

Let N be the number of particles in the target. The average number dn of particles which are detected per unit time in the solid angle $d\Omega$ is proportional to N, to F and to $d\Omega$. We denote by $\mathrm{d}\sigma/\mathrm{d}\Omega$ the coefficient of proportionality:

$$dn = NF\,\frac{\mathrm{d}\sigma}{\mathrm{d}\Omega}\,d\Omega\,. \tag{18.2}$$

The quantity $\mathrm{d}\sigma/\mathrm{d}\Omega$, which has the dimensions of an area, is the *differential cross section* in the direction $\mathbf{\Omega}$. It is independent of the incident flux F and of the number N of scatterers in the target. The total number n of particles scattered per unit time is $n = NF\sigma$, where σ is the *total cross section*, given by

$$\sigma = \int \frac{\mathrm{d}\sigma}{\mathrm{d}\Omega}\,d\Omega\,. \tag{18.3}$$

18.1.2 Classical Calculation

In classical physics, a knowledge of the trajectories yields directly the differential scattering cross section if we use the following simple physical interpretation of $\mathrm{d}\sigma/\mathrm{d}\Omega$. The quantity $(\mathrm{d}\sigma/\mathrm{d}\Omega)\,\delta\Omega$ represents the area of the opaque surface that one would need to put perpendicular to the incident beam in order to block all particles which are scattered into the solid angle $\delta\Omega$ around the direction $\mathbf{\Omega}$ (Fig. 18.2).

Consider a single, fixed target particle and a flux F of incident particles of momentum $\boldsymbol{p}$. For each of these particles, one can calculate the trajectory in terms of the impact parameter b. We assume that the interaction potential

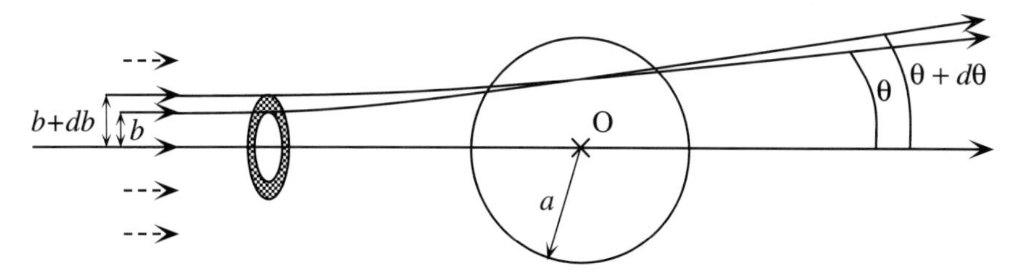

Fig. 18.2. Classical scattering of particles of momentum $\boldsymbol{p}$ by a spherically symmetric potential centered at the origin 0 and of range a. The *shaded area* represents the opaque surface that one would need to place perpendicular to the incident beam in order to block the particles scattered by the angle θ into the solid angle $\mathrm{d}\Omega = 2\pi\sin\theta\,\mathrm{d}\theta$. By definition, this area of surface is equal to $(\mathrm{d}\sigma/\mathrm{d}\Omega)\,\mathrm{d}\Omega$, where $\mathrm{d}\sigma/\mathrm{d}\Omega$ is the differential cross section

$V(r)$ is of finite range and that it has spherical symmetry. We denote the scattering angle by θ. Integration of the equations of motion gives the relation $b(\theta)$. The particles scattered into the interval between θ and $\theta + \mathrm{d}\theta$ are those whose impact parameter is between b and $b + \mathrm{d}b$ (see Fig. 18.2). The number $\mathrm{d}n$ of outgoing particles per unit time in the solid angle $\mathrm{d}\Omega = 2\pi \sin\theta \,\mathrm{d}\theta$ is equal to the number of incident particles whose impact parameter is between b and $b + \mathrm{d}b$. By the definition of the flux F, this number is $\mathrm{d}n = 2\pi\, b\, db\, F$, hence the differential cross section is

$$\frac{\mathrm{d}\sigma}{\mathrm{d}\Omega}(\theta) = \frac{b(\theta)}{\sin\theta} \left| \frac{\mathrm{d}b}{\mathrm{d}\theta} \right| . \tag{18.4}$$

A measurement of $\mathrm{d}\sigma/\mathrm{d}\Omega$ gives information about the relation $b(\theta)$, and therefore about the potential V.

18.1.3 Examples

Hard Spheres. Consider a particle incident on a hard sphere of radius R. Assuming the incident particle is point-like, one can show with no difficulty that $\mathrm{d}b/\mathrm{d}\theta = -(R/2)\sin(\theta/2)$. Since $b/R = \cos(\theta/2)$, the differential cross section is isotropic:

$$\frac{\mathrm{d}\sigma}{\mathrm{d}\Omega} = \frac{R^2}{4} .$$

The total cross section is $\sigma = \pi R^2$, which is the geometric cross section of the target sphere. In the case of two identical hard spheres, the scattering is isotropic in the center-of-mass frame and one obtains $\sigma = 4\pi R^2$.

Coulomb Scattering. A particle of charge $Z_1 e$, of kinetic energy $E_i = mv^2/2$, is scattered by a fixed center of charge $Z_2 e$ located at the origin (the potential is $V(r) = Z_1 Z_2 e^2/r$, where $e^2 = q^2/(4\pi\epsilon_0)$). In this hyperbolic Kepler motion, the relation between the impact parameter b and the scattering angle θ is

$$b = \frac{Z_1 Z_2 e^2}{2E_i} \cot\frac{\theta}{2} ,$$

and hence the differential cross section is

$$\frac{\mathrm{d}\sigma}{\mathrm{d}\Omega} = \left(\frac{Z_1 Z_2 e^2}{4E_i} \right)^2 \frac{1}{\sin^4(\theta/2)} . \tag{18.5}$$

This formula, called the Rutherford cross section, played a crucial role in the discovery of the nucleus inside the atom.

The Total Cross Section in Classical Mechanics. In classical mechanics, the total cross section is finite only if the potential $V(\boldsymbol{r})$ is a constant (i. e. the force vanishes) for distances r larger than some finite radius. Otherwise the combination of (18.3) and (18.4) leads to

$$\sigma \geq 2\pi \int_0^\infty b \,\mathrm{d}b = \infty \,.$$

The physical meaning of this infinite total cross section is simple. If the force $\boldsymbol{F} = -\boldsymbol{\nabla} V$ is not strictly zero outside some given volume, all incident particles are deflected by some finite amount, even when the impact parameter b is very large. This is very different from the situation in quantum mechanics, where one finds a finite total cross section even for potentials decreasing as r^{-n} at infinity (provided n is larger than 3). In quantum mechanics, there is a finite probability that an incident particle with a well defined momentum $\boldsymbol{p}_i$, i. e. in the state $\mathrm{e}^{\mathrm{i}\boldsymbol{p}_i\cdot\boldsymbol{r}/\hbar}$, remains in this state even when the force is different from zero everywhere in space.

18.2 Quantum Calculation in the Born Approximation

A exact calculation of the scattering cross section requires the determination of the eigenstates of the Hamiltonian (18.1). This is a difficult numerical task, except in some specific cases where the analytic form of these eigenstates is known. We shall address this problem in Sect. 18.4. Here we consider the case where the probability of scattering is small, and we treat the effect of the scattering potential as a small perturbation of the free motion of the particle. We derive an approximate value of the scattering cross section, valid to second order in V. This is the *Born approximation*, which is very useful in practice.

18.2.1 Asymptotic States

A scattering experiment is characterized by the following assumptions:

a. The particle has, initially, a well-defined momentum and is far from the scattering center, and therefore outside the range of the interaction potential $V(\boldsymbol{r})$. This preparation takes place at some initial time t_i in the "distant past": $t_\mathrm{i} \to -\infty$.
b. We are interested in the final momentum distribution of the particle when it is again far from the interaction region. This occurs at a time t_f in the "distant future": $t_\mathrm{f} \to +\infty$.

The interaction region has dimensions of the order of a nanometer for atomic collisions and of a femtometer for nuclear collisions. The measuring devices have macroscopic dimensions. Under these conditions, the states of the initial and final particles are free-particle states. These states are called *asymptotic states.*

In the asymptotic states, the particle has a well-defined momentum. Strictly speaking, we know that such states, which correspond to plane waves, are not physical. A real initial state is described by a wave packet whose dispersion in momentum Δp is small: $\Delta p/p \ll 1$. It is mathematically possible to consider such a wave packet at time t_i. One can calculate its evolution as it overlaps with the region of space where $V(\boldsymbol{r})$ is not negligible, and determine the final wave packet at time t_f. In this analysis, one extracts the scattering cross section by taking the limit $\Delta p/p \to 0$. However, this formalism is technically quite heavy, and does not bring out any new physical information compared with the much simpler procedure that we shall follow, and which is also based on a limiting procedure.

We assume that the initial and final states are plane waves, normalized in a cubic box of volume L^3 and defined with periodic boundary conditions as in Sect. 4.4:

$$|\boldsymbol{p}\rangle \to \phi_{\boldsymbol{k}}(\boldsymbol{r}) = \frac{\mathrm{e}^{\mathrm{i}\boldsymbol{k}\cdot\boldsymbol{r}}}{\sqrt{L^3}}, \qquad \text{where} \quad \boldsymbol{p} = \hbar\boldsymbol{k} = \frac{2\pi\hbar}{L}\boldsymbol{n}. \tag{18.6}$$

Here $\boldsymbol{n} = (n_1, n_2, n_3)$ is a triplet of positive or negative integers. We use the formalism developed in Chap. 17 to derive, to first order in V, the transition probability from the initial state $|\boldsymbol{p}\rangle$ to a set of final states whose momentum $\boldsymbol{p}'$ lies within a solid angle $d\Omega$ around the direction $\boldsymbol{\Omega}$.

18.2.2 Transition Probability

Consider the scattering of a spinless particle of mass m by a potential $V(\boldsymbol{r})$ centered at the origin. In the language of time-dependent perturbation theory, the dominant term in the Hamiltonian

$$\hat{H} = \frac{\hat{p}^2}{2m} + V(\boldsymbol{r}) \tag{18.7}$$

is the kinetic energy $\hat{H}_0 = \hat{p}^2/2m$ of the particle. We treat the potential $V(\boldsymbol{r})$ as a perturbation. The asymptotic states (18.6) are, by definition, eigenstates of the dominant term $\hat{H}_0$.

We have already derived in Chap. 17 the transition probability from an initial state $|i\rangle$ to a domain of final states, under the action of a small perturbation. Here we have $|i\rangle = |\boldsymbol{p}\rangle$, with energy $E_\mathrm{i} = p^2/2m$, and the domain of final states $\mathcal{D}_\mathrm{f}$ consists of states $|\boldsymbol{p}'\rangle$, with energy $E' = {p'}^2/2m$. The direction $\boldsymbol{p}'/p'$ is assumed to lie in a neighborhood $d\Omega$ of $\boldsymbol{\Omega}$. The transposition of the result (17.30) to our problem yields

$$\mathrm{d}\mathcal{P}_{i\to\mathcal{D}_\mathrm{f}} = \frac{2\pi}{\hbar}|\langle\boldsymbol{p}'|\hat{V}|\boldsymbol{p}\rangle|^2\,\rho(E_\mathrm{i})\,\frac{\mathrm{d}\Omega}{4\pi}\,t\,.$$

In this case, the energy conservation condition (17.29) leads to $|\boldsymbol{p}'| = |\boldsymbol{p}|$, and the density of states $\rho(E)$ for a nonrelativistic particle is given by (4.52): $\rho(E) = mL^3\sqrt{2mE}/(2\pi^2\hbar^3)$.

We recover here the linear variation of the transition probability with the interaction time which is characteristic of a continuum of final states. We can therefore define a transition rate $\mathrm{d}\Lambda_{i\to\mathcal{D}_\mathrm{f}}$, which is independent of t:

$$\mathrm{d}\Lambda_{i\to\mathcal{D}_\mathrm{f}} = \frac{\mathrm{d}\mathcal{P}_{i\to\mathcal{D}_\mathrm{f}}}{t} = \frac{2\pi}{\hbar}|\langle\boldsymbol{p}'|\hat{V}|\boldsymbol{p}\rangle|^2\ \rho(E_\mathrm{i})\ \frac{\mathrm{d}\Omega}{4\pi}\,. \tag{18.8}$$

Here the potential $\hat{V}$ plays the role of the transition Hamiltonian $\hat{H}_1$ of Chap. 17. Its matrix element between states of well defined momentum $|\boldsymbol{p}\rangle$ and $|\boldsymbol{p}'\rangle$ is

$$\langle\boldsymbol{p}'|V|\boldsymbol{p}\rangle = \frac{1}{L^3}\int \mathrm{e}^{\mathrm{i}(\boldsymbol{p}-\boldsymbol{p}')\cdot\boldsymbol{r}/\hbar}\ V(\boldsymbol{r})\,\mathrm{d}^3r = \frac{1}{L^3}\,\tilde{V}(\boldsymbol{p}-\boldsymbol{p}')\,, \tag{18.9}$$

where $\tilde{V}$ is the Fourier transform of the potential up to a constant factor of $(2\pi\hbar)^{3/2}$:

$$\tilde{V}(\boldsymbol{q}) = \int \mathrm{e}^{\mathrm{i}\boldsymbol{q}\cdot\boldsymbol{r}/\hbar}\ V(\boldsymbol{r})\,\mathrm{d}^3r\,. \tag{18.10}$$

18.2.3 Scattering Cross Section

By definition, the cross section $\mathrm{d}\sigma$ for scattering into the solid angle $\mathrm{d}\Omega$ around the value $\boldsymbol{\Omega}$ is

$$\mathrm{d}\sigma = \frac{\mathrm{d}\Lambda_{i\to\mathcal{D}_\mathrm{f}}}{F}\,, \tag{18.11}$$

where the flux F associated with the state $|\boldsymbol{p}\rangle$ is related to the velocity $v = p/m$ of the incident particle by

$$F = \frac{v}{L^3} = \frac{p}{mL^3}\,. \tag{18.12}$$

In fact, the particle is in the volume L^3 with probability 1 and the probability distribution associated with the plane wave $|\boldsymbol{p}\rangle$ is uniform (for particles of velocity v and density ρ, the flux is $F = \rho v$).

The exact expression for the transition rate (18.8), i.e. to all orders of perturbation theory, is of the form

$$\mathrm{d}\Lambda_{i\to\mathcal{D}_\mathrm{f}} = \frac{\mathrm{d}\mathcal{P}_{i\to\mathcal{D}_\mathrm{f}}}{t} = \frac{2\pi}{\hbar}|\langle\boldsymbol{p}'|\hat{T}|\boldsymbol{p}\rangle|^2\ \rho(E_\mathrm{i})\ \frac{\mathrm{d}\Omega}{4\pi}\,.$$

The quantity $L^3\langle\boldsymbol{p}'|\hat{T}|\boldsymbol{p}\rangle$ is called the *scattering amplitude*. The operator $\hat{T}$ is related to the potential $\hat{V}$ by an integral equation, called the Lippman–Schwinger equation, which reduces to $\hat{T} = \hat{V}$ in the Born approximation.

Using (18.8), (18.10) and (18.12), we find the following result:

$$\frac{d\sigma^{\text{Born}}}{d\Omega} = \left(\frac{m|\tilde{V}(\boldsymbol{p}-\boldsymbol{p}')|}{2\pi\hbar^2} \right)^2 . \tag{18.13}$$

The normalization volume L^3 cancels out identically, as expected. With this result, we see that in quantum mechanics, in the Born approximation, a measurement of the differential cross section gives us direct access to the Fourier transform of the potential, and therefore to the forces themselves. This result can be compared to diffraction, where the diffracted amplitude is the Fourier transform of the diffracting system. Notice that the modulus of the Fourier transform $|\tilde{V}(\boldsymbol{q})|$ is invariant under the translation $V(\boldsymbol{r}) \to V(\boldsymbol{r}-\boldsymbol{r}_0)$. Therefore the target need not be localized. This is one of the many reasons why the concept of the cross section is relevant in quantum mechanics.

18.2.4 Validity of the Born Approximation

The Born approximation is valid when the expansion of the eigenstates $\psi(\boldsymbol{r})$ of $\hat{H}$ in powers of the scattering potential V is possible and rapidly convergent. To zeroth order in V, the relevant eigenstates of $\hat{H}$ are simply the incident plane waves

$$\psi^{(0)}(\boldsymbol{r}) = \frac{e^{i\boldsymbol{k}\cdot\boldsymbol{r}}}{\sqrt{L^3}} . \tag{18.14}$$

Anticipating the exact results derived in Sect. 18.4 (see (18.46)), we use the following expression for the term of order 1 in V in the expansion of $\psi(\boldsymbol{r})$:

$$\psi^{(1)}(\boldsymbol{r}) = -\frac{m}{2\pi\hbar^2} \int \frac{e^{ik|\boldsymbol{r}-\boldsymbol{r}'|}}{|\boldsymbol{r}-\boldsymbol{r}'|} V(\boldsymbol{r}')\, \psi^{(0)}(\boldsymbol{r}')\, d^3r' . \tag{18.15}$$

Suppose that the maximum value of $|V(\boldsymbol{r})|$ is $V_0 > 0$, and that the region where $V(\boldsymbol{r})$ has a significant value is centered at $\boldsymbol{r} = 0$, with an extension a. For slow particles such that $ka \ll 1$, the value of $|\psi^{(1)}(\boldsymbol{r})|$ is small compared with $|\psi^{(0)}(\boldsymbol{r})|$ at any point $\boldsymbol{r}$ (in particular at $\boldsymbol{r} = 0$) if

$$\frac{m}{2\pi\hbar^2}(4\pi V_0 a^2) \ll 1 \qquad \Rightarrow \qquad V_0 \ll \frac{\hbar^2}{ma^2} . \tag{18.16}$$

For fast particles ($ka \gg 1$) the above condition (18.16) can be weakened, since the oscillating factor $e^{ik|\boldsymbol{r}-\boldsymbol{r}'|}$ reduces the value of the integral (18.15). In this case[2] one obtains the following validity criterion:

$$V_0 \ll \frac{\hbar^2}{ma^2} ka \qquad \text{if} \qquad ka \gg 1 . \tag{18.17}$$

[2] See, e. g., the books by Messiah and by Landau and Lifshitz, whose references are given at the end of this chapter.

18.2.5 Example: the Yukawa Potential

A very large class of elementary interactions can be described by superpositions of Yukawa potentials, whose basic form is

$$V(r) = g\,\frac{\hbar c}{r}\,\mathrm{e}^{-r/a}\,, \tag{18.18}$$

where a is the range of the potential and g is a dimensionless coupling constant. We denote the angle between $\boldsymbol{p}$ and $\boldsymbol{p}'$ by θ so that

$$|\boldsymbol{p} - \boldsymbol{p}'| = 2\hbar k \sin(\theta/2)\,.$$

We obtain

$$\tilde{V}(\boldsymbol{p} - \boldsymbol{p}') = \frac{4\pi g \hbar c a^2}{1 + 4a^2k^2\sin^2(\theta/2)}\,, \tag{18.19}$$

so that

$$\frac{\mathrm{d}\sigma^{\mathrm{Born}}}{\mathrm{d}\Omega} = \left(\frac{2mgca^2}{\hbar}\right)^2 \frac{1}{\left(1 + 4a^2k^2\sin^2(\theta/2)\right)^2}\,. \tag{18.20}$$

The total cross section is

$$\sigma^{\mathrm{Born}}(k) = \left(\frac{2mgca}{\hbar}\right)^2 \frac{4\pi a^2}{1 + 4k^2a^2}\,. \tag{18.21}$$

The validity condition for the Born approximation in this case is obtained by taking $V_0 \sim V(a) \sim g\hbar c/a$ in (18.16) or (18.17). For low-energy particles we obtain $gm \ll \hbar/(ac)$.

The Yukawa potential describes the interaction between two particles through the exchange of a massive particle of mass $M = \hbar/ac$. The validity condition given above can then be written as $gm \ll M$: the mass of the scattered particle m, multiplied by the dimensionless coupling constant g, must be much smaller than the mass M of the exchanged particle.

In the limit where the range a of the Yukawa potential tends to infinity, we recover the Coulomb potential. We take $g = Z_1Z_2\alpha$, which corresponds to the Coulomb scattering of a particle of charge Z_1e, mass m and momentum p (electron, muon, α particle) by a fixed charge Z_2e (proton, nucleus). If we use (18.19) in this case, we recover exactly the classical Rutherford formula (18.5). This result would not normally be expected the exact answer, since it has been obtained in the Born approximation. Fortunately, it is possible to calculate[3] an exact expression for the eigenstates of positive energy for the Coulomb problem (as we did in Chap. 11 for the negative-energy eigenstates) and to obtain the exact value of the cross section. Quite remarkably, one recovers the classical result (18.5). For the Coulomb problem, the exact cross section and the cross section calculated in the Born approximation with asymptotic states such as (18.6) coincide, and are both equal to the classical cross section.

[3] See, e. g., the book by Messiah (reference at the end of the chapter).

One reason for this coincidence is that the nonrelativistic Coulomb scattering problem for a particle of mass m with an incident energy E_i involves a natural length scale $\mathcal{L} = e^2/E_i$ where both $\hbar$ and c are absent. There is, of course, a quantum mechanical length scale $\mathcal{L}_Q = \hbar^2/(me^2)$, with which we can form a dimensionless constant $\mathcal{L}_Q/\mathcal{L} = \hbar^2 E_i/(me^4)$. From dimensional arguments, we therefore expect the quantum cross section to be of the form

$$\frac{d\sigma}{d\Omega} = \left(\frac{Z_1 Z_2 \mathcal{L}}{4}\right)^2 \frac{1}{\sin^4(\theta/2)} F[\theta, \hbar^2 E/(me^4)],$$

where F is a dimensionless function with $F(\theta, 0) = 1$. The surprising fact is that $F(\theta, x) = 1$ for all x, whereas it would be expected to contain quantum effects, i. e. a dependence on $\hbar$.

18.2.6 Range of a Potential in Quantum Mechanics

In all this chapter, the notion of the range of the scattering potential plays an essential role. Assume that the scatterer is centered at $\boldsymbol{r} = 0$. The range a is defined as the characteristic distance from the origin beyond which the incident particle does not feel appreciably the field of force created by the scatterer. For hard spheres or Yukawa potentials, this notion is quite obvious. For potentials which decrease as power laws at infinity, i. e.

$$V(\boldsymbol{r}) \underset{|\boldsymbol{r}|\to\infty}{\sim} \frac{C_n}{r^n}, \tag{18.22}$$

this notion is less obvious, but still relevant as we now show.

We first notice that, classically, a potential which behaves as (18.22) has an infinite range. A particle prepared at any distance from the origin with a kinetic energy $E_k \sim C_n/r^n$ will be accelerated significantly. Consider now a quantum particle prepared in a wave packet centered at $\boldsymbol{r}$ with an extension Δr. The hypothesis of a particle localized at $\boldsymbol{r}$ requires that $\Delta r \ll r$, so that the probability density of the wave packet is negligible at 0. From the Heisenberg inequality, we know that the dispersion Δp of the momentum distribution of the wave packet is such that $\Delta r \Delta p \geq \hbar/2$. Therefore the kinetic energy E_k of the particle is such that

$$E_k \geq \frac{\Delta p^2}{2m} \geq \frac{\hbar^2}{8m\,\Delta r^2} \gg E_k^{\min} = \frac{\hbar^2}{mr^2}.$$

This provides a lower bound on the kinetic energy of the particle localized at $\boldsymbol{r}$.

Suppose now that the exponent n of the potential (18.22) is strictly larger than 2. In this case, there is a characteristic distance beyond which the kinetic energy of the particle is necessarily much larger than its potential energy. This characteristic distance, which we denote by a, is such that the lower bound on the kinetic energy $E_k^{\min}$ is equal to $|V(\boldsymbol{r})|$:

$$a = \left(m\,|C_n|/\hbar^2\right)^{1/(n-2)}. \tag{18.23}$$

Therefore, in quantum mechanics, potentials decreasing at infinity faster than r^{-2} can be considered as finite-range interactions. The Coulomb potential has an infinite range, as it does in classical mechanics.

18.3 Exploration of Composite Systems

We now investigate the information which can be obtained about a composite system, such as an atom or a nucleus, when it is used as a target in a scattering experiment. Taking the example of Coulomb scattering, we show that the cross section gives access to the charge distribution inside the target. For simplicity, we perform this analysis in the Born approximation and neglect multiple scattering within the composite target.

18.3.1 Scattering Off a Bound State and the Form Factor

Consider the scattering of a particle a of mass m and position $\boldsymbol{r}$ by a particle b of mass m_1 and position $\boldsymbol{r}_1$. The particle b is supposed to be bound to a center of force by a potential $U(\boldsymbol{r}_1)$ and we denote by $\{\psi_n(\boldsymbol{r}_1)\}$ the corresponding eigenfunctions:

$$\left(\frac{\hat{p}_1^2}{2m_1} + U(\boldsymbol{r}_1)\right) \psi_n(\boldsymbol{r}_1) = E_n \, \psi_n(\boldsymbol{r}_1) \,. \tag{18.24}$$

We denote by $V(\boldsymbol{r} - \boldsymbol{r}_1)$ the interaction potential between a and b which is responsible for the scattering.

We assume that b is initially in the ground state $\psi_0(\boldsymbol{r}_1)$. The wave function of the initial asymptotic state is therefore

$$|i\rangle \quad \rightarrow \quad \Psi_{\mathrm{i}}(\boldsymbol{r}, \boldsymbol{r}_1) = \frac{\mathrm{e}^{\mathrm{i}\boldsymbol{p}\cdot\boldsymbol{r}/\hbar}}{\sqrt{L^3}} \, \psi_0(\boldsymbol{r}_1) \,. \tag{18.25}$$

In the final state, b can remain in the ground state $\psi_0(\boldsymbol{r}_1)$ of (18.24) or it can be excited into one of the states $\psi_n(\boldsymbol{r}_1)$. Here, we are interested in the case of an elastic collision, where b remains in the ground state after the collision.[4] The final state, such that the momentum of particle a has changed from $\boldsymbol{p}$ to $\boldsymbol{p}'$, is then

$$|f\rangle \quad \rightarrow \quad \Psi_{\mathrm{f}}(\boldsymbol{r}, \boldsymbol{r}_1) = \frac{\mathrm{e}^{\mathrm{i}\boldsymbol{p}'\cdot\boldsymbol{r}/\hbar}}{\sqrt{L^3}} \, \psi_0(\boldsymbol{r}_1) \,. \tag{18.26}$$

[4] The generalization to an inelastic collision, where b is left in an excited state $\psi_n(\boldsymbol{r}_1)$ and a loses the energy $E_n - E_0$, can be treated with no difficulty. Such a generalization can be used to provide a quantitative treatment of the Franck and Hertz experiment presented in Chap. 1.

In the Born approximation, the transition probability from $|i\rangle$ to the continuum of final states $|f\rangle$ (within a solid angle $\mathrm{d}\Omega$ around the direction of $\boldsymbol{p}'$) involves only the matrix element of the interaction potential $\hat{V}$:

$$\begin{aligned}\langle f|\hat{V}|i\rangle &= \int \mathrm{e}^{\mathrm{i}(\boldsymbol{p}-\boldsymbol{p}')\cdot\boldsymbol{r}/\hbar}\, |\psi_0(\boldsymbol{r}_1)|^2\, V(\boldsymbol{r}-\boldsymbol{r}_1)\, \mathrm{d}^3r\, \mathrm{d}^3r_1 \\ &= \tilde{V}(\boldsymbol{p}-\boldsymbol{p}')\, F(\boldsymbol{p}-\boldsymbol{p}'),\end{aligned}$$

where $\tilde{V}(\boldsymbol{q})$ is the same as in (18.10), and we have defined the *form factor* $F(\boldsymbol{q})$ as

$$F(\boldsymbol{q}) = \int \mathrm{e}^{\mathrm{i}\boldsymbol{q}\cdot\boldsymbol{r}_1/\hbar}\, |\psi_0(\boldsymbol{r}_1)|^2\, \mathrm{d}^3r_1\,. \tag{18.27}$$

The form factor $F(\boldsymbol{q})$ is the Fourier transform of the probability density of particle b in the ground state $\psi_0(\boldsymbol{r}_1)$.

Consequently, the cross section for the scattering of a by the bound particle b factorizes as the product of the *elementary* cross section σ_0 of a and b, i.e. the cross section we would observe if b were not bound, and of the modulus squared of the form factor:

$$\frac{\mathrm{d}\sigma}{\mathrm{d}\Omega} = \frac{\mathrm{d}\sigma_0}{\mathrm{d}\Omega}\, |F(\boldsymbol{p}-\boldsymbol{p}')|^2\,. \tag{18.28}$$

In other words, if we know the elementary cross section σ_0, a measurement of the variation of the cross section as a function of the momentum transfer $\boldsymbol{q} = \boldsymbol{p}-\boldsymbol{p}'$ provides a measurement of the wave function of the bound state.[5] We note that for $\boldsymbol{q} = 0$, $F(0) = 1$: at low momentum transfer, the target appears to be point-like. The larger the momentum transfer $|\boldsymbol{q}|$, the more accurately one detects the structure of the bound state. The form factor of a hydrogen-like wave function $\psi_0 \sim \mathrm{e}^{-r/2a}$ is $|F(\boldsymbol{q})|^2 = 1/(1+q^2a^2/\hbar^2)^2$. For a Gaussian wave function $\psi_0 \sim \mathrm{e}^{-r^2/2\sigma^2}$, one finds $|F(\boldsymbol{q})|^2 = \mathrm{e}^{-q^2\sigma^2/\hbar^2}$.

18.3.2 Scattering by a Charge Distribution

In the previous subsection we studied the scattering of a particle a by a single target particle b bound in an external potential $U(\boldsymbol{r}_1)$. In most situations, the target consists of n particles $b_1, \ldots, b_n$ forming a bound state of wave function $\psi_0(\boldsymbol{r}_1, \ldots, \boldsymbol{r}_n)$. In atomic physics, the target consists of a point-like nucleus surrounded by bound electrons. In nuclear physics, the n particles are the nucleons (protons and neutrons), bound together by nuclear forces.

[5] Some caution must be exercised in the above manipulations, in order to render all operations legitimate. For more details, see for instance the books by Messiah and by Mott and Massey (references at the end of this chapter).

Scattering Cross Section of a Composite System. As above, we shall consider only elastic scattering, for which the target remains in the state ψ_0 after the scattering. Therefore the asymptotic states read

$$|i\rangle \;\rightarrow\; \Psi_{\mathrm{i}}(\boldsymbol{r},\boldsymbol{r}_1,\ldots,\boldsymbol{r}_n) = \frac{\mathrm{e}^{\mathrm{i}\boldsymbol{p}\cdot\boldsymbol{r}/\hbar}}{\sqrt{L^3}}\;\psi_0(\boldsymbol{r}_1,\ldots,\boldsymbol{r}_n)\,,$$

$$|f\rangle \;\rightarrow\; \Psi_{\mathrm{f}}(\boldsymbol{r},\boldsymbol{r}_1,\ldots,\boldsymbol{r}_n) = \frac{\mathrm{e}^{\mathrm{i}\boldsymbol{p}'\cdot\boldsymbol{r}/\hbar}}{\sqrt{L^3}}\;\psi_0(\boldsymbol{r}_1,\ldots,\boldsymbol{r}_n)\,.$$

We work in the Born approximation. We calculate the cross section for the scattering of particle a to lowest order in the interaction potential

$$\hat{V} = \sum_{j=1}^{n} V_j(\hat{\boldsymbol{r}} - \hat{\boldsymbol{r}}_j)$$

and we neglect the contribution of multiple scattering. Note that the particle a may interact differently with the various particles b_j bound in the target, so that not all V_j's may be equal.

A calculation similar to that of the previous subsection leads to the scattering cross section

$$\frac{\mathrm{d}\sigma}{\mathrm{d}\Omega} = \left(\frac{m}{2\pi\hbar^2}\right)^2 |\langle f|\hat{V}|i\rangle|^2\,, \quad \text{where} \quad \langle f|\hat{V}|i\rangle = \sum_j \tilde{V}_j(\boldsymbol{q})\;F_j(\boldsymbol{q})\,. \tag{18.29}$$

Here we have set $\boldsymbol{q} = \boldsymbol{p} - \boldsymbol{p}'$ and have defined the form factors

$$F_j(\boldsymbol{q}) = \int \mathrm{e}^{\mathrm{i}\boldsymbol{q}\cdot\boldsymbol{r}_j/\hbar}\;|\psi_0(\boldsymbol{r}_1,...,\boldsymbol{r}_j,...,\boldsymbol{r}_n)|^2\;\mathrm{d}^3r_1\ldots\mathrm{d}^3r_j\ldots\mathrm{d}^3r_n\,.$$

Notice that the cross section $\mathrm{d}\sigma/\mathrm{d}\Omega$ given in (18.29) is *not* the sum of the individual cross sections for scattering by particles $b_1,\ldots,b_n$. The total scattering amplitude is the *sum* of the individual scattering *amplitudes* of $b_1,\ldots,b_n$ in the Born approximation. This leads to interference phenomena in scattering processes.

Coulomb Scattering by a Charge Distribution. To be more specific, we consider the Coulomb scattering of a particle a with a charge Ze by a bound system formed by the particles b_j, whose charges are denoted Z_je (as usual, we set $e^2 = q^2/(4\pi\epsilon_0)$, where q is the elementary charge). We introduce the charge density operator

$$\hat{\rho}(\boldsymbol{R}) = \sum_{j=1}^{n} Z_j e\;\delta(\boldsymbol{R} - \hat{\boldsymbol{r}}_j)\,,$$

whose expectation value $\rho(\boldsymbol{R}) = \langle\psi_0|\hat{\rho}(\boldsymbol{R})|\psi_0\rangle$ is the charge distribution of the composite system formed by the n particles. The result (18.29) can be written in this case as

$$\frac{\mathrm{d}\sigma}{\mathrm{d}\Omega} = \left(\frac{Ze}{4E_{\mathrm{i}}}\right)^2 \frac{|\tilde{\rho}(\boldsymbol{q})|^2}{\sin^4(\theta/2)}, \tag{18.30}$$

where $\boldsymbol{q} = \boldsymbol{p} - \boldsymbol{p}'$ and $\tilde{\rho}(\boldsymbol{q})$ is the Fourier transform of $\rho(\boldsymbol{R})$, with the same normalization as in (18.10).

We denote by a the characteristic extension of the bound state formed by the n particles and we consider the case of small momentum transfers, such that $|\boldsymbol{q}|a/\hbar \ll 1$. We find, in this case,

$$\tilde{\rho}(0) = \int \rho(\boldsymbol{r})\, \mathrm{d}^3 r = eZ_{\mathrm{tot}}\,, \qquad \frac{\mathrm{d}\sigma}{\mathrm{d}\Omega} = \left(\frac{ZZ_{\mathrm{tot}}e^2}{4E_{\mathrm{i}}}\right)^2 \frac{1}{\sin^4(\theta/2)}, \tag{18.31}$$

where eZ_{tot} is the total charge of the bound system. Consequently, at low momentum transfers, (18.30) reduces to the Rutherford cross section for a particle of charge Ze incident on a *point-like* particle of charge $Z_{\mathrm{tot}}e$. This is called *coherent* scattering by the bound state; the phases of the scattering amplitudes of each constituent are all equal.

Scattering of a Charged Particle by a Neutral Atom. We now suppose that the target is a neutral atom, from which we scatter a particle of charge Ze and incident energy E_{i}. The neutral atom is a composite system which consists of a point-like nucleus of charge $+Z_1 e$ localized at $\boldsymbol{R} = 0$, and an electron cloud with a charge density $\rho_{\mathrm{e}}(R)$ and a total charge $-Z_1 e$:

$$\rho(\boldsymbol{R}) = Z_1 e\, \delta(\boldsymbol{R}) + \rho_{\mathrm{e}}(\boldsymbol{R})\,, \qquad \tilde{\rho}(\boldsymbol{q}) = Z_1 e + \tilde{\rho}_{\mathrm{e}}(\boldsymbol{q})\,.$$

We set $\rho_{\mathrm{e}}(\boldsymbol{R}) = -Z_1 e F(\boldsymbol{R})$, where $F(\boldsymbol{R})$ gives the probability distribution of the electron cloud. The cross section for the scattering process is therefore:

$$\frac{\mathrm{d}\sigma}{d\Omega} = \left(\frac{ZZ_1 e^2}{4E_{\mathrm{i}}}\right)^2 \frac{[1 - \tilde{F}(\boldsymbol{q})]^2}{\sin^4(\theta/2)}\,. \tag{18.32}$$

Consider the case of momentum transfers such that $|\boldsymbol{q}|a/\hbar \gg 1$. In this case, $F(\boldsymbol{q})$ is much smaller than 1. For instance, for the electron distribution of the 1s state of the hydrogen atom $F(\boldsymbol{r}) = \mathrm{e}^{-2r/a_1}/(\pi a_1^3)$, we find

$$F(\boldsymbol{q}) = \frac{1}{[1 + (|\boldsymbol{q}|a_1/2\hbar)^2]^2} \qquad \rightarrow \qquad F(\boldsymbol{q}) \propto |\boldsymbol{q}|^{-4} \quad \text{for large } |\boldsymbol{q}|\,.$$

In the Rutherford experiment, α particles of energies $E_{\mathrm{i}} \sim$ 4–8 MeV and of mass $m_\alpha \simeq 4m_{\mathrm{p}}$ are scattered by neutral atoms. For scattering angles larger than 1 degree, we obtain $qa/\hbar > 10^3$, taking $a \sim 1$ Å as a typical extension of the electron density of an atom. In this case $F(\boldsymbol{q})$ plays a negligible role in (18.32): the incident α particles simply do not see the electrons and are scattered only by the nucleus.

Charge Distribution in Nuclei. The nucleus itself is a composite structure made up of protons and neutrons, whose extension is a few femtometers. A very efficient means to probe its internal structure, in particular the proton distribution, is to perform collisions of high-energy electrons ($E \sim$ 150–500 MeV) with nuclei; the results give direct information about the distribution of positive charges, i. e. protons. At small momentum transfers, or at small angles, one observes scattering off a point-like particle of charge $+Ze$, Z being the number of protons in the nucleus. At larger angles, the ratio of the measured cross section to the Rutherford cross section gives the square of the form factor, i. e. the proton distribution inside the nucleus.[6]

Electrons are very clean probes of the structure of matter. Unlike α particles, for instance, they have no nuclear interactions but only electromagnetic ones, which are well known and can be calculated in the Born approximation owing to the smallness of the fine structure constant. Figure 18.3 shows the angular dependence of the differential cross section for the scattering of 250 MeV electrons by calcium nuclei. The continuous line is a fit by a theoretical model assuming a "Saxon–Woods" charge density profile, $\rho(r) = \rho_0 \{1 + \exp[(r - r_0)/a]\}^{-1}$. These measurements show that to a good approximation, the charge density is constant inside a sphere of radius ≈ 6.2 fm. Similar experiments, due to Hofstadter, have measured the charge and magnetic moment distributions inside the proton and the neutron. These distributions are well accounted for by the quark model.

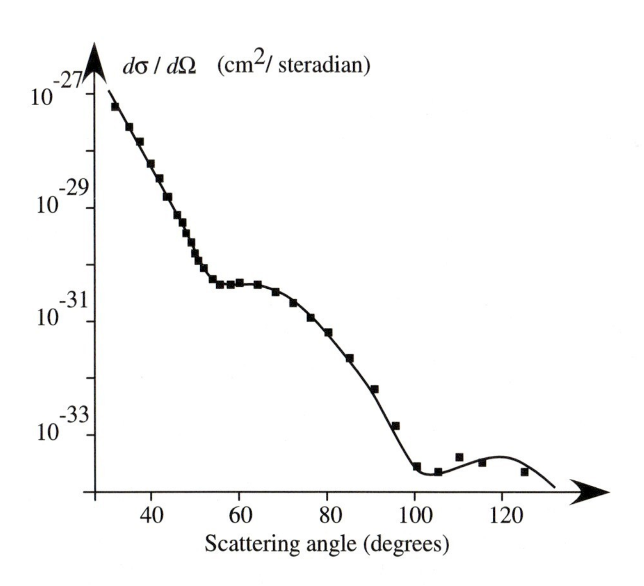

Fig. 18.3. Scattering of 250 MeV electrons by calcium nuclei (^{40}Ca)

[6] Further theoretical and experimental details can be found in R. Hofstadter, *Electron Scattering and Nuclear and Nucleon Structure*, W.A. Benjamin, New York (1963).

18.4 General Scattering Theory

In order to obtain the above results, we have used perturbation theory in the Born approximation. This approximation is very useful in many instances, but it is often necessary to go beyond it. We now present a general scattering theory, which is based on a formal derivation of the exact scattering states of the Schrödinger equation. Of course, we shall recover the Born approximation as a limiting case of this exact treatment, for weak enough scattering potentials.

18.4.1 Scattering States

In Sect. 3.5, we studied the scattering properties of potential barriers and wells in one dimension. We calculated the general form of the eigenstates of the total Hamiltonian, which is the sum of the kinetic energy and the potential energy associated with the barrier. We then considered eigenstates whose asymptotic behavior at $x = \pm\infty$ corresponded to a given physical situation. For instance, particles arriving from $-\infty$ can be either reflected or transmitted by a barrier located at $x = 0$. Finally, we derived the transmission and reflection coefficients of the barrier by calculating the ratio between the transmitted, reflected and incident probability currents.

We shall follow a similar procedure for the three-dimensional situation which is of interest here. Our goal in this subsection is to find an expression for the eigenstates of the Hamiltonian in the scattering problem which will allow a simple interpretation in terms of scattering cross sections.

Consider the scattering of a particle of mass m by a potential $V(r)$ centered at the origin. The Hamiltonian $\hat{H}$ is

$$\hat{H} = \hat{H}_0 + V(\hat{\boldsymbol{r}}) , \qquad \hat{H}_0 = -\frac{\hbar^2}{2m}\,\Delta . \tag{18.33}$$

We assume that the potential $V(\boldsymbol{r})$ tends to 0 faster than r^{-2} as $r \to \infty$. The stationary states of energy E are of the form $\Psi(\boldsymbol{r},t) = \psi(\boldsymbol{r})\,\mathrm{e}^{-\mathrm{i}Et/\hbar}$, where $\psi(\boldsymbol{r})$ satisfies the eigenvalue equation

$$\left(-\frac{\hbar^2}{2m}\,\Delta + V(\boldsymbol{r})\right)\,\psi(\boldsymbol{r}) = E\,\psi(\boldsymbol{r}) . \tag{18.34}$$

Since we are considering a scattering process, the energy E is positive and can be written $E = \hbar^2 k^2/(2m)$. We recall that, as in the one-dimensional case, the corresponding eigenstates $\psi(\boldsymbol{r})$ cannot be normalized (unlike bound states).

Consider the eigenvalue equation (18.34) at a point $\boldsymbol{r}$ such that $|\boldsymbol{r}| \gg a$, where a is the range of the potential $V(\boldsymbol{r})$. In this region, we have

$$\Delta\psi + k^2\psi \simeq 0 \qquad \text{for} \qquad |\boldsymbol{r}| \gg a .$$

Among all possible forms of states satisfying this equation, we shall cł the following one:

$$\psi(\boldsymbol{r}) \underset{|\boldsymbol{r}|\to\infty}{\sim} \mathrm{e}^{\mathrm{i}\boldsymbol{k}\cdot\boldsymbol{r}} + f(k,\boldsymbol{u},\boldsymbol{u}')\,\frac{\mathrm{e}^{\mathrm{i}kr}}{r}\,, \tag{18.35}$$

where the unit vectors $\boldsymbol{u}$ and $\boldsymbol{u}'$ are defined as $\boldsymbol{u} = \boldsymbol{k}/k$ and $\boldsymbol{u}' = \boldsymbol{r}/r$. An eigenstate of $\hat{H}$ with the asymptotic form (18.35) is called a *scattering state.* Such a form is intuitive in relation to our scattering problem. An incident plane wave $\mathrm{e}^{\mathrm{i}\boldsymbol{k}\cdot\boldsymbol{r}}$ interacts with a center located at the origin. The scattering process due to the potential $V(\boldsymbol{r})$ gives rise to a *divergent spherical wave* varying as $\mathrm{e}^{\mathrm{i}kr}/r$, whose amplitude depends on the energy $\hbar^2k^2/2m$, the incident direction $\boldsymbol{u}$ and the direction of observation $\boldsymbol{u}'$.

18.4.2 The Scattering Amplitude

We now relate the asymptotic form (18.35) to the physically observable quantities. This relation is based on the following property:

The differential cross section for the scattering of a particle of momentum $\boldsymbol{p} = \hbar\boldsymbol{k}$ *by the potential* $V(\boldsymbol{r})$ *is*

$$\frac{\mathrm{d}\sigma}{\mathrm{d}\Omega} = |f(k,\boldsymbol{u},\boldsymbol{u}')|^2\,. \tag{18.36}$$

$f(k,\boldsymbol{u},\boldsymbol{u}')$ is called the *scattering amplitude.*

As in the one-dimensional case, the relation between the asymptotic behavior of an eigenstate of the Hamiltonian and the physical properties of a scattering process is obtained using the probability current (see Sect. 3.5). The incident probability current corresponding to the incident plane wave $\mathrm{e}^{\mathrm{i}\boldsymbol{k}\cdot\boldsymbol{r}}$ is

$$\boldsymbol{J}_{\mathrm{inc}} = \frac{\hbar\boldsymbol{k}}{m}\,. \tag{18.37}$$

We can calculate the scattered current associated with the contribution of the spherical wave function $\mathrm{e}^{\mathrm{i}kr}/r$ to (18.35). Using spherical coordinates, we obtain the following for the radial component of the probability current:

$$J_{\mathrm{scatt},r}(\boldsymbol{r}) = \frac{\hbar k}{m}\,\frac{|f(k,\boldsymbol{u},\boldsymbol{u}')|^2}{r^2}\,. \tag{18.38}$$

At large distances, and away from the incident direction (i.e. for $\theta \neq 0$), the two other components of the probability current, $J_{\mathrm{scatt},\theta}$ and $J_{\mathrm{scatt},\varphi}$, decrease faster than $J_{\mathrm{scatt},r}$ (as r^{-3}). Therefore the scattered current is asymptotically radial.

The number $\mathrm{d}n$ of particles that are detected per unit time in a detector of area $\mathrm{d}\boldsymbol{S}$ at a large distance in the direction $\boldsymbol{u}'$ is therefore

$$\mathrm{d}n = \boldsymbol{J}_{\mathrm{scatt}}(\boldsymbol{r}) \cdot \mathbf{d}\boldsymbol{S} \underset{|r|\to\infty}{=} J_{\mathrm{scatt},r}(\boldsymbol{r})\, r^2 \,\mathrm{d}\Omega\,, \tag{18.39}$$

where $\mathrm{d}\Omega = \boldsymbol{u}' \cdot \mathbf{d}\boldsymbol{S}/r^2$ is the solid angle of the detector as seen from the origin. We therefore obtain

$$\mathrm{d}n = J_{\mathrm{inc}}\, |f(k, \boldsymbol{u}, \boldsymbol{u}')|^2 \,\mathrm{d}\Omega\,, \tag{18.40}$$

which proves (18.36), as a consequence of the definition of the cross section.

18.4.3 The Integral Equation for Scattering

The Schrödinger equation (18.34) can be cast into the form of an integral equation by introducing the *Green's functions* $G_{\mathrm{k}}(\boldsymbol{r})$, defined by

$$\frac{-\hbar^2}{2m}(\Delta + k^2)\, G_{\mathrm{k}}(\boldsymbol{r}) = \delta(\boldsymbol{r}) \qquad (k^2 > 0)\,, \tag{18.41}$$

two independent solutions of which are

$$G_{\mathrm{k}}^{\pm}(\boldsymbol{r}) = \frac{m}{2\pi\hbar^2}\, \frac{\mathrm{e}^{\pm ikr}}{r}\,. \tag{18.42}$$

These are called the *outgoing* (e^{+ikr}) and *incoming* (e^{-ikr}) Green's functions. Only the first one is of interest in the present discussion.

Let $\phi(\boldsymbol{r})$ be a solution of the Schrödinger equation for a free particle of energy $E = \hbar^2 k^2/2m$:

$$(\Delta + k^2)\, \phi(\boldsymbol{r}) = 0\,. \tag{18.43}$$

One can verify, using the definition (18.41), that any function $\psi(\boldsymbol{r})$ which satisfies the integral equation

$$\psi(\boldsymbol{r}) = \phi(\boldsymbol{r}) - \int G_{\mathrm{k}}^{\pm}(\boldsymbol{r} - \boldsymbol{r}')\, V(\boldsymbol{r}')\, \psi(\boldsymbol{r}')\, \mathrm{d}^3 r' \tag{18.44}$$

satisfies the Schrödinger equation (18.34) in the presence of the potential $V(\boldsymbol{r})$. For the scattering problem under consideration, we choose

$$\phi_{\boldsymbol{k}}(\boldsymbol{r}) = \mathrm{e}^{\mathrm{i}\boldsymbol{k}\cdot\boldsymbol{r}}\,, \tag{18.45}$$

and we obtain the *integral equation for scattering*:

$$\psi_{\boldsymbol{k}}(\boldsymbol{r}) = \mathrm{e}^{\mathrm{i}\boldsymbol{k}\cdot\boldsymbol{r}} - \int G_{\mathrm{k}}^{+}(\boldsymbol{r} - \boldsymbol{r}')\, V(\boldsymbol{r}')\, \psi_{\boldsymbol{k}}(\boldsymbol{r}')\, \mathrm{d}^3\boldsymbol{r}'\,. \tag{18.46}$$

We can check that the asymptotic form of this wave function corresponds to the requirement (18.35). The integration over r' in (18.46) runs over values of $|\boldsymbol{r}'|$ smaller than the range a of the potential. Consider a point $\boldsymbol{r}$ such that

$|\boldsymbol{r}| \gg a$. We find $|\boldsymbol{r}| \gg |\boldsymbol{r}'|$ and $|\boldsymbol{r}-\boldsymbol{r}'| \sim r - \boldsymbol{u}'\cdot\boldsymbol{r}'$, where $\boldsymbol{u}' = \boldsymbol{r}/r$. Inserting these results in the definition of the Green's function G_{k}^{+}, we obtain

$$G_{\mathrm{k}}^{+}(\boldsymbol{r}-\boldsymbol{r}') \sim \frac{m}{2\pi\hbar^2}\,\frac{\mathrm{e}^{\mathrm{i}kr}}{r}\,\mathrm{e}^{-\mathrm{i}\boldsymbol{k}'\cdot\boldsymbol{r}'}\,, \tag{18.47}$$

where we have set $\boldsymbol{k}' = k\boldsymbol{u}'$. Therefore the asymptotic variation of $\psi_{\boldsymbol{k}}$ has the expected form (18.35), where

$$f(k,\boldsymbol{u},\boldsymbol{u}') = -\frac{m}{2\pi\hbar^2}\int \mathrm{e}^{-\mathrm{i}\boldsymbol{k}'\cdot\boldsymbol{r}'}\,V(\boldsymbol{r}')\,\psi_{\boldsymbol{k}}(\boldsymbol{r}')\,\mathrm{d}^3r'\,, \tag{18.48}$$

and $\boldsymbol{k} = k\boldsymbol{u}$. This expression relates the scattering amplitude, i. e. the behavior of the scattering state at infinity, to the values of the scattering state inside the range of the potential.

The integral equation (18.46) is the basis of both analytical and numerical calculations concerning the scattering process.[7] The equation can be used to calculate $\psi_{\boldsymbol{k}}$ as a power series in the scattering potential in V. We have already given in (18.14) and (18.15) the first two terms of this expansion, which can be evaluated to arbitrary order in V. To order 1 in V, we can evaluate the scattering amplitude by taking $\psi_{\boldsymbol{k}}(\boldsymbol{r}') = \mathrm{e}^{\mathrm{i}\boldsymbol{k}\cdot\boldsymbol{r}'}$ in the integral of (18.48):

$$f(k,\boldsymbol{u},\boldsymbol{u}') \simeq -\frac{m}{2\pi\hbar^2}\int \mathrm{e}^{\mathrm{i}(\boldsymbol{k}-\boldsymbol{k}')\cdot\boldsymbol{r}'}\,V(\boldsymbol{r}')\,\mathrm{d}^3r'\,. \tag{18.49}$$

We can then derive the scattering cross section $\mathrm{d}\sigma/\mathrm{d}\Omega = |f|^2$ to first order in V. We can readily check that we recover the Born approximation (18.13). This shows the consistency between the two approaches based on time-dependent perturbation theory and on a formal solution of the Schrödinger equation for the scattering problem.

18.5 Scattering at Low Energy

Low-energy scattering is of considerable interest in nuclear, atomic and molecular physics. It is of great importance in the physics of cold quantum gases.

18.5.1 The Scattering Length

For a low enough incident kinetic energy ($ka \ll 1$), the scattering amplitude is independent of the incident and final directions $\boldsymbol{u}$ and $\boldsymbol{u}'$. This result can be seen from (18.48). If $ka \ll 1$, the term in the integral $\mathrm{e}^{-\mathrm{i}\boldsymbol{k}'\cdot\boldsymbol{r}'}$ can be

[7] This equation can be transformed into the Lippmann–Schwinger integral equation mentioned in Sect. 18.2.3, which is well suited for both formal and practical applications.

replaced by 1, since the points $\boldsymbol{r}'$ contributing to the integral are such that $r' \leq a$. This shows that $f(k, \boldsymbol{u}, \boldsymbol{u}')$ does not depend on the third variable $\boldsymbol{u}'$. Using the time-reversal invariance of the scattering process, one can prove the identity $f(k, \boldsymbol{u}, \boldsymbol{u}') = f(k, -\boldsymbol{u}', -\boldsymbol{u})$, which shows that f does not depend on $\boldsymbol{u}$ either, at low energies.

Consequently, the scattering process is isotropic for $ka \ll 1$ and the asymptotic behavior of the scattering state can be written as

$$\psi(\boldsymbol{r}) \underset{|\boldsymbol{r}|\to\infty}{\sim} \mathrm{e}^{\mathrm{i}\boldsymbol{k}\cdot\boldsymbol{r}} + f(k)\,\frac{\mathrm{e}^{\mathrm{i}kr}}{r}\,. \tag{18.50}$$

When the energy tends to zero, the function $f(k)$ usually tends to a finite *real* limit, and one can set

$$a_{\mathrm{s}} = -\lim_{k\to 0} f(k)\,. \tag{18.51}$$

The quantity a_{s} has the dimensions of a length and is called the *scattering length.* The total cross section in this case is $\sigma = 4\pi a_{\mathrm{s}}^2$.

In the Born approximation (18.49), the scattering length is given by

$$a_{\mathrm{s}} = \frac{m}{2\pi\hbar^2}\int V(\boldsymbol{r}')\,\mathrm{d}^3r'\,. \tag{18.52}$$

Because of the factor r'^2 appearing in the infinitesimal volume d^3r', this integral converges only if the potential decreases faster than r^{-3} at infinity. If this is not the case, there may remain an angular dependence of the scattering cross section at low energies, as is the case in Coulomb scattering.

18.5.2 Explicit Calculation of a Scattering Length

Assume for simplicity that the scattering potential is spherically symmetric: $V(\boldsymbol{r}) = V(r)$. We define $u(r) = r\int \psi(\boldsymbol{r})\,\mathrm{d}\Omega$ and integrate the Schrödinger equation (18.34) over a solid angle of 4π. Using the expression (10.22) for the Laplacian operator, we find that $u(r)$ is a solution of the equation

$$\frac{\mathrm{d}^2u}{\mathrm{d}r^2} - \frac{2mV(r)}{\hbar^2}u = 0\,, \qquad u(0) = 0\,, \tag{18.53}$$

where we explicitly take the limit $E \to 0$. For large r, $V(r)$ makes a negligible contribution and the asymptotic form of the solution is $u(r) = C(r-b)$, where C is an arbitrary multiplicative constant. The length b, which depends on the behavior of $V(r)$ at short distances, is nothing but the scattering length. In order to prove this, we simply integrate (18.50) over the solid angle in the limit $k \to 0$, and we obtain $u(r) \sim 4\pi(r - a_{\mathrm{s}})$. Consequently, the determination of the scattering length beyond the Born approximation (18.52) simply requires one to solve the one-dimensional differential equation (18.53) and to examine the asymptotic behavior of its solution.

Thermal Neutrons. Consider neutrons with momentum $p \sim \sqrt{m_n k_B T}$, where $T = 300\,\mathrm{K}$. This corresponds to a wave vector $k \sim 3 \times 10^{10}\,\mathrm{m}^{-1}$. When these neutrons propagate in matter, they interact with the nuclei through nuclear forces whose ranges are a few femtometers. Therefore ka is smaller than 10^{-4} and the scattering of the neutrons by the nuclei of the material is described very well by a scattering length a_s. As it is commonly done in optics, one can sum the spherical waves $a_s\,\mathrm{e}^{\mathrm{i}k|\boldsymbol{r}-\boldsymbol{r}_i|}/|\boldsymbol{r}-\boldsymbol{r}_i|$ scattered by the various nuclei of the material and define an index of refraction n for the propagation of the neutrons, obtaining

$$n = 1 - 2\pi \frac{\rho_s a_s}{k^2} \,,$$

where ρ_s is the density of scatterers in the medium.

Cold Atoms. Using laser cooling, one can obtain atoms with temperatures in the microkelvin range. Consider for instance rubidium atoms ($A = 85$ or 87) for which $k \sim 10^7\,\mathrm{m}^{-1}$. These atoms interact with one another at long distance through a van der Waals potential $-C_6/r^6$, where $C_6 \sim 3\,\mathrm{meV\,nm}^6$. Using (18.23) we find the characteristic range of interaction a to be of the order of 10 nm, so that $ka \sim 0.1$ in this case. A description of the interactions between cold atoms in terms of a scattering length is again well justified. For ^{85}Rb and ^{87}Rb, the value of a_s are -23 nm and 5.3 nm, respectively.

18.5.3 The Case of Identical Particles

Consider, finally, a collision between two identical particles, labeled 1 and 2. We assume for simplicity that the spins of the particles are polarized, so that the orbital wave function is symmetric for bosons, and antisymmetric for fermions. In the center-of-mass frame, the two-particle scattering state is therefore:

$$\text{bosons:}\quad \Psi(\boldsymbol{r}_1,\boldsymbol{r}_2) = \frac{1}{\sqrt{2}}\left[\psi_{\boldsymbol{k}}(\boldsymbol{r}) + \psi_{\boldsymbol{k}}(-\boldsymbol{r})\right] ;$$

$$\text{fermions:}\quad \Psi(\boldsymbol{r}_1,\boldsymbol{r}_2) = \frac{1}{\sqrt{2}}\left[\psi_{\boldsymbol{k}}(\boldsymbol{r}) - \psi_{\boldsymbol{k}}(-\boldsymbol{r})\right] .$$

Here we have set $\boldsymbol{r} = \boldsymbol{r}_1 - \boldsymbol{r}_2$ and we have used $\hat{P}_{12}\psi(\boldsymbol{r}) = \psi(-\boldsymbol{r})$. At low energies ($ka \ll 1$) the scattering state (18.50) leads to:

$$\text{bosons:}\quad \Psi(\boldsymbol{r}_1,\boldsymbol{r}_2) \underset{|\boldsymbol{r}|\to\infty}{\sim} \frac{\mathrm{e}^{\mathrm{i}\boldsymbol{k}\cdot\boldsymbol{r}} + \mathrm{e}^{\mathrm{i}\boldsymbol{k}\cdot\boldsymbol{r}}}{\sqrt{2}} + \sqrt{2}f(k)\,\frac{\mathrm{e}^{\mathrm{i}kr}}{r} ; \tag{18.54}$$

$$\text{fermions:}\quad \Psi(\boldsymbol{r}_1,\boldsymbol{r}_2) \underset{|\boldsymbol{r}|\to\infty}{\sim} \frac{\mathrm{e}^{\mathrm{i}\boldsymbol{k}\cdot\boldsymbol{r}} - \mathrm{e}^{\mathrm{i}\boldsymbol{k}\cdot\boldsymbol{r}}}{\sqrt{2}} . \tag{18.55}$$

The physical consequences of these two expressions are important. The scattering cross section of two bosons at low energy is increased by a factor 2 compared with the result one would find in the case of distinguishable particles interacting through the same potential $V(\boldsymbol{r})$. In particular, one obtains $\sigma = 8\pi a_s^2$ for the total cross section in the limit $k \to 0$.

For two polarized fermions, the result (18.55) shows that the scattering cross section is zero: two identical, polarized fermions "do not see each other" at very low energies. This leads to an important property of a gas of polarized fermions at low temperature. It always behave as a quasi-ideal gas in which the particles do not interact with each other. Using an expansion of the scattering state in spherical harmonics (which is called a *partial-wave expansion*), one can go one step further and show that the collision cross section varies as k^2 for small wave vectors k. On the other hand, if the Fermi gas is not completely polarized, it does not behave as an ideal gas, since the collision cross section between two fermions in different spin states does not vanish when k tends to 0.

Further Reading

- L. Landau and E. Lifshitz, *Quantum Mechanics*, Chap. 17, Pergamon, Oxford (1965).
- A. Messiah, *Quantum Mechanics*, Chap. 11 and 19, North-Holland, Amsterdam (1961).
- M.L. Goldberger and K.M. Watson, *Collision theory*, Wiley, New York (1964).
- N.F. Mott and H.S.W. Massey, *The Theory of Atomic Collisions*, Clarendon Press, Oxford (1965).
- C.J. Joachain, *Quantum Collision Theory*, North-Holland, Amsterdam (1983).

Exercises

18.1. Scattering length for a hard-sphere potential. Consider the low-energy scattering of particle of mass m by a potential $V(\boldsymbol{r})$ such that

$$V(r) = V_0 \quad \text{if} \quad r \le b\,, \qquad V(r) = 0 \quad \text{if} \quad r > b\,.$$

We assume that V_0 is positive.

a. Calculate the scattering length a in terms of V_0 and b. What is the sign of a? Show that one recovers the result of the Born approximation when V_0 is small enough.
b. How does the scattering length vary when V_0 tends to infinity, for a constant b (hard sphere)?

c. Consider the limit of a δ function potential, by letting V_0 tend to infinity and b to 0, the product $4\pi V_0 b^3/3 = \int V(\boldsymbol{r})\, \mathrm{d}^3r$ remaining constant. How does the scattering length vary?

18.2. Scattering length for a square well. Consider again the potential studied in the previous exercise, but assume now that V_0 is negative.

a. Show that the scattering length varies in a resonant manner when V_0 increases for a fixed b. Relate the position of these resonances to the number of bound states in the potential $V(\boldsymbol{r})$.
b. Recover the result of the Born approximation for a sufficiently small value of V_0. What is the sign of a in this case?

These "zero-energy resonances" play an important role in the physics of ultracold atoms. Using an external magnetic field, one can modify slightly the interaction potential between atoms, and change the intensity (the size of $|a|$) and the nature (attractive or repulsive depending on the sign of a) of the effective interactions between the particles. One can then shift continuously from the ideal-gas regime to the strongly interacting case.

18.3. The pseudopotential. There is no scattering in a potential $V(\boldsymbol{r}) = g\,\delta(\boldsymbol{r})$ (see exercise 18.1). In contrast, the pseudopotential[8] V defined by

$$\hat{V}\psi(\boldsymbol{r}) = g\delta(\boldsymbol{r})\,\frac{\partial}{\partial r}\left[r\,\psi(\boldsymbol{r})\right]$$

has scattering properties that can be calculated exactly.

a. Check that the wave function

$$\psi(\boldsymbol{r}) = \mathrm{e}^{\mathrm{i}\boldsymbol{k}\cdot\boldsymbol{r}} - \frac{a}{1+\mathrm{i}ka}\,\frac{\mathrm{e}^{\mathrm{i}kr}}{r}$$

is solution of the eigenvalue equation for the Hamiltonian $\hat{H} = \hat{\mathbf{p}}^2/2m + \hat{V}$, for a value of a that can be expressed in terms of g and m.
b. What is the scattering length associated with the pseudopotential?
c. Characterize the angular dependence of the scattering.
d. How does the total scattering cross section vary with the incident energy $\hbar^2k^2/2m$?

This pseudopotential also plays an important role in the theory of ultracold gases. It allows one to treat quantitatively and in a simple manner the interactions between particles, provided the scattering length associated with this potential is equal to the actual scattering length.

---

[8] See, e. g., K. Huang, *Statistical Physics*, Chap. 13, Wiley, New York (1963).

19. Qualitative Physics on a Macroscopic Scale

Written in collaboration with Alfred Vidal-Madjar[1]

We spend our time romanticizing our motivations
and simplifying facts.
Boris Vian

Quantum mechanics constitutes a complete theory of the structure of matter and of its interaction with the electromagnetic field. It allows us to calculate the size, geometry and binding energies of atomic and molecular structures. The accurate quantitative description of a given phenomenon requires one to solve equations which are often complicated. In this chapter, our purpose is to show how a good understanding of the fundamental laws of quantum mechanics enables us to estimate important physical effects qualitatively or semiquantitatively. Starting from fundamental constants and elementary processes, we shall obtain, with very good accuracy, the orders of magnitude of some macroscopic phenomena which belong to "everyday life" (or nearly).

The majority of phenomena that we see around us originate from electromagnetic interactions, mainly Coulomb forces between charged particles (electrons and nuclei), and gravitational interactions. Therefore we shall express the characteristic orders of magnitude of the behavior of matter as we see it, starting from the fundamental constants which are involved: the proton mass m_{p} (close to the neutron mass), the electron mass m_{e}, the elementary charge q (with $e^2 = q^2/4\pi\epsilon_0$), the velocity of light c, Newton's constant G, and Planck's constant $\hbar$.

We recall that the fine structure constant, which is the fundamental constant of electromagnetic interactions, is

$$\alpha = \frac{e^2}{\hbar c} \simeq \frac{1}{137} .$$

Consider Newton's universal law of gravitation $f = -Gmm'/r^2$. Since the mass of an atom is, to a good approximation, an integer multiple of the proton mass, we can define in a similar way a dimensionless constant α_G which characterizes the gravitational interactions:

[1] Alfred Vidal-Madjar, Institut d'Astrophysique de Paris, 98bis Boulevard Arago, 75014 Paris, France, e-mail: alfred@iap.fr.

$$\alpha_G = \frac{Gm_\mathrm{p}^2}{\hbar c} \simeq 5.90 \times 10^{-39} \qquad (G \simeq 6.67 \times 10^{-11}\,\mathrm{m^3\,kg^{-1}\,s^{-2}})\,.$$

We see that electromagnetic forces are considerably stronger than gravitational forces. However, gravitational forces are always additive, whereas electric neutrality screens electrostatic interactions at large distances.

One of our aims is to determine the amount of matter beyond which gravitational forces "beat" electric forces, or, in more sophisticated terms, to understand how we can get from a crystal, whose geometric structure is dominated by electric forces and which has an arbitrary shape, to a spherical planet.

We shall first recall the relevant orders of magnitude for microscopic systems. We shall extend and complete these orders of magnitude for N-fermion systems. We shall then obtain an estimate of the conditions under which gravitational and Coulomb forces are of the same order, which will lead us to an estimate of the maximum size of mountains on a planet. Finally, using a very simple model, we shall understand the characteristics of white dwarf stars and neutron stars, including the existence of a "gravitational catastrophe" for systems whose mass is too large.

19.1 Confined Particles and Ground State Energy

We first examine how the uncertainty relations lead to semiquantitative estimates for isolated atoms as well as macroscopic objects.

19.1.1 The Quantum Pressure

Consider a particle which is confined in a region of dimension ξ, i.e. its dispersion in position is $\Delta x = \xi$. If the particle is in the ground state of the confining potential, the uncertainty relation requires that its kinetic energy is of the order of

$$E_\mathrm{k} = \frac{\langle p^2 \rangle}{2m} \sim \frac{\hbar^2}{2m\xi^2}\,. \tag{19.1}$$

If the particle is in some excited state, the de Broglie wavelength is shorter than ξ and the kinetic energy is correspondingly larger.

The confinement energy (19.1) becomes larger as ξ becomes smaller. One can view this increase of energy as a quantum pressure exerted by the particle against this confinement. If we use the thermodynamic definition of this pressure as the derivative (up to a minus sign) of the kinetic energy with respect to the volume, this quantum pressure (also called the Schrödinger pressure) scales as

$$P_\mathrm{S} \sim \frac{\hbar^2}{m\xi^5}\,.$$

19.1.2 Hydrogen Atom

For the hydrogen atom, the mean value of the energy is

$$\langle H \rangle = \frac{\langle p^2 \rangle}{2m_e} - e^2 \left\langle \frac{1}{r} \right\rangle .$$

We denote by ξ_1 the size of the ground state of the hydrogen atom. More precisely, we set $\xi_1^{-1} = \langle 1/r \rangle$. As mentioned above, $\langle p^2 \rangle \simeq \hbar^2/\xi_1^2$, and

$$\langle H \rangle \simeq \frac{\hbar^2}{2m_e \xi_1^2} - \frac{e^2}{\xi_1} .$$

The minimum value of this expression as a function of ξ_1 gives the order of magnitude of the ground-state energy. This minimum corresponds to

$$\xi_1 = a_1 \qquad \text{and} \qquad E_1 = \langle H \rangle_{\min} = -E_{\mathrm{I}} ,$$

where a_1 is the Bohr radius and E_{I} the Rydberg energy, i. e. the ionization energy of the hydrogen atom:

$$a_1 = \frac{\hbar^2}{m_e e^2} , \qquad E_{\mathrm{I}} = \frac{m_e e^4}{2\hbar^2} .$$

Here we obtain the exact result because the general rigorous inequality $\langle p^2 \rangle \geq \hbar^2 \langle 1/r \rangle^2$ (see Sect. 9.2.3) is saturated if the average is taken over the hydrogen ground-state wave function.

One could perform this calculation for excited states also. The principle of this calculation can be also applied to the case of a particle confined in a square well or a harmonic potential.

19.1.3 *N*-Fermion Systems and Complex Atoms

For N identical fermions with spin 1/2, the ground state can be obtained by a similar argument. However, because of the Pauli principle, the Heisenberg relations must be replaced by:

$$\langle r^2 \rangle \, \langle p^2 \rangle \geq \xi^2 \, N^{2/3} \, \hbar^2 , \qquad \text{where} \qquad \xi = 3^{4/3}/4 \quad \text{and} \quad N \gg 1 , \tag{19.2}$$

or

$$\langle p^2 \rangle \geq \gamma \, N^{2/3} \, \hbar^2 \left\langle \frac{1}{r} \right\rangle^2 , \qquad \text{where} \qquad \gamma = (12)^{-1/3} \quad \text{and} \quad N \gg 1 , \tag{19.3}$$

as seen in Chap. 16, (16.15) and (16.17) (we assume that $\langle \boldsymbol{r} \rangle = 0$ and $\langle \boldsymbol{p} \rangle = 0$). Both inequalities (19.2) and (19.3) may be used, and we shall choose in the following the one which leads to the simplest calculations.

Consider for instance a neutral complex atom with Z electrons. The Hamiltonian is

$$\hat{H} = \sum_{i=1}^{Z} \frac{\hat{p}_i^2}{2m_e} - \sum_{i=1}^{Z} \frac{Ze^2}{\hat{r}_i} + \frac{1}{2}\sum_i \sum_{j \neq i} \frac{e^2}{\hat{r}_{ij}} .$$

We repeat the previous argument, setting $\langle p_i^2 \rangle = \langle p^2 \rangle$ and $\langle 1/r_i \rangle = 1/\xi$ for each electron. In addition, we make the approximation[2] $\langle 1/r_{ij} \rangle \simeq 1/\xi$ and we obtain the following for large Z:

$$\begin{aligned} \langle H \rangle &\simeq Z\frac{\langle p^2 \rangle}{2m_e} - \frac{Z^2e^2}{\xi} + \frac{Z(Z-1)e^2}{2\xi} \\ &\simeq Z\frac{\langle p^2 \rangle}{2m_e} - \frac{Z^2e^2}{2\xi} . \end{aligned} \tag{19.4}$$

Now, we assume that the minimum-energy state saturates the inequality (19.3) ($\langle p^2 \rangle \simeq \gamma Z^{2/3}\hbar^2/\xi^2$), and we minimize the resulting expression, choosing ξ as a variational parameter. We obtain

$$\xi \simeq 0.9\, \frac{a_1}{Z^{1/3}} \quad \text{and} \quad \langle H \rangle_{\min} \simeq -0.6\, Z^{7/3} E_{\mathrm{I}} \quad \text{for } Z \gg 1 . \tag{19.5}$$

If the orders of magnitude (Bohr radius and Rydberg energy) were expected, the powers of Z are not that obvious. They arise from the combination of the Heisenberg inequalities with the Pauli principle. The ground-state energy of a Z electron atom increases as $Z^{7/3}$, or, equivalently, as $Z^{4/3}$ per electron (to be compared with Z^2 for hydrogen-like atoms with a single electron), whereas the mean distance decreases as $Z^{-1/3}$ (to be compared with Z^{-1} for hydrogen-like atoms). These results can be recovered using more sophisticated methods such as Hartree's.

Notice that ξ is the average distance between an electron and the nucleus, and *not* the "size" d of the atom. This size is usually defined as the mean distance between the small number of external electrons and the nucleus, and it is more or less independent of Z. Typically,

$$a_1 \; < \; d \; < \; 6a_1 .$$

Correspondingly, the ionization energy, i. e. the energy necessary to remove the least bound electron, is in the range $0.2E_{\mathrm{I}}$ to E_{I}.

19.1.4 Molecules, Liquids and Solids

Consider now two neutral atoms separated by a distance D. When D is large compared with the atomic size d, the electrostatic interaction between the

[2] For particles uniformly distributed in a sphere of radius R, one finds $\langle 1/r_i \rangle = 3/(2R)$ and $\langle 1/r_{ij} \rangle = 5/(3R)$. Making this distinction between $\langle 1/r_i \rangle$ and $\langle 1/r_{ij} \rangle$ would affect the final results only marginally.

two atoms results in an attractive potential, which varies as D^{-6}. This is called the van der Waals interaction and can be understood within a simple classical picture. The instantaneous electric dipole $\boldsymbol{P}$ of one of the two atoms creates an electric field at the second atom, whose magnitude varies as D^{-3}. This field polarizes the second atom, and the induced dipole ($\propto D^{-3}$) creates an electric field $\boldsymbol{E}$ at the first atom, whose magnitude varies as $D^{-3} \times D^{-3}$. The van der Waals interaction corresponds to the interaction energy $-\boldsymbol{P} \cdot \boldsymbol{E}$ between the initial dipole and this electric field.[3]

For a distance D of the order of the atomic size d, this attractive interaction becomes even stronger owing to the tunnel effect, which allows the external electrons to jump from one atom to the other. Depending on the type of material, other types of bonding may also come into the game, such as ionic or hydrogen bonds. If one tries to reduce further the distance between the two atoms ($D < d$), the interaction becomes strongly repulsive: the two electron clouds start to overlap, and Pauli exclusion together with the quantum pressure term dominates. Consequently, the typical distance between the two atoms when they form a molecule is the atomic size d. The dissociation energy of this molecule is

$$D = \beta E_{\rm I}\,, \qquad \text{where} \qquad 0.2 < \beta < 0.5\,.$$

The electrostatic interaction between two atoms, due either to the van der Waals or the tunneling interaction, decreases very rapidly as the distance D increases: it is a *short-range* interaction. This allows us to define the binding energy B of an atom in a solid or a liquid as the energy necessary to extract the atom from the surface. This energy is typically $B \sim \left|\sum_{i=1}^{N} V(\boldsymbol{r}_{\rm A} - \boldsymbol{r}_i)\right|$, where $\boldsymbol{r}_{\rm A}$ is the position of the atom at the surface of the solid. The sum runs over the N atoms which constitute the solid. Owing to the fast decrease of V as the interatomic distance tends to infinity, this sum does not depend on the shape of the solid, nor on its size (this would not be true if V were to decrease more slowly than D^{-3}, since the three-dimensional integral over the various positions $\boldsymbol{r}_i$ would then be divergent for a solid of infinite size). The typical binding energy B of an atom or a molecule in a liquid or a solid is

$$B = \gamma E_{\rm I}\,, \qquad \text{where} \qquad \begin{cases} 0.05 < \gamma_{\rm liquid} < 0.1 \\ 0.1 < \gamma_{\rm solid} < 0.3 \end{cases}.$$

19.1.5 Hardness of a Solid

We can now understand an elementary property of matter: why is a solid ... solid? The hardness of a solid can be measured by its resistance to compres-

[3] Here we neglect retardation effects related to the propagation of the electromagnetic field, assuming that $D < \lambda$, where λ is the characteristic electromagnetic wavelength of the problem ($\lambda \sim hc/E_{\rm I} \sim 10^{-7}$ m). For $D > \lambda$, the interaction energy varies as D^{-7}.

sion, defined by the compression modulus

$$C = PV/\Delta V\,.$$

This is the ratio between the applied pressure P and the relative variation of the volume $\Delta V/V$.

We can evaluate the order of magnitude of C for a metal, assuming for simplicity that the atoms are monovalent. The N positive ions form a structure inside an N-electron gas. Each electron occupies freely a volume V/N, and therefore has a kinetic energy

$$E_{\mathrm{k}} \sim \frac{\hbar^2}{md^2}\,,$$

where $d = (V/N)^{1/3}$ represents the distance between two ions of the crystal. For a more accurate result, we can use the expressions (16.12) and (16.13), which give the energy of an N-fermion gas confined in a box of volume V: $NE_{\mathrm{k}} = \xi N\hbar^2/md^2$, where $\xi = 3(3\pi^2)^{2/3}/10 \simeq 2.9$. The corresponding Schrödinger pressure is

$$P_{\mathrm{S}} \sim \frac{\hbar^2}{m_{\mathrm{e}}d^5}\,. \tag{19.6}$$

For a typical value $d = 4a_1$, we find $P_{\mathrm{S}} \sim 5\times 10^{10}$ Pa. Such a pressure does not cause a piece of metal to blow up, simply because it is exactly compensated by the Coulomb attraction forces between the ions and the electrons. However, in order to compress a metal and change its volume by a significant amount, one must exert a pressure of the same order. Therefore, we expect C to be also of the order of 10^{10} Pa. This is indeed the typical value which is observed in metals and also in other crystalline solids (i. e. compression moduli of 10^{10} to 10^{11} Pa are observed).

19.2 Gravitational Versus Electrostatic Forces

Up to know, we have neglected gravitational interactions in our estimates of the internal energy of an atom and the binding energy of an atom in a solid. This is legitimate as long as the solid is small enough, but it cannot hold for arbitrarily large bodies, as we shall see.

19.2.1 Screening of Electrostatic Interactions

Consider a system of $N \gg 1$ neutral atoms or molecules bound, as described in the previous section, by electrostatic forces. The electrostatic binding energy of this system (a crystal or a piece of rock) is proportional to N and is of the order of

$$E_{\text{elec.}} \sim -N E_{\text{atom}} \qquad \text{for} \qquad N \gg 1\,, \tag{19.7}$$

where E_{atom} is the electrostatic binding energy of a single atom. In other words, the binding energy *per atom* (or molecule) is a constant, independent of the size of the sample.

This important property is a direct consequence of the screening of electrostatic forces in neutral matter. We have seen above that the van der Waals and tunneling interactions between two atoms are of short range. Consequently, the energy necessary to fragment a piece of rock or some other solid material is small compared with the total binding energy: it is a *surface* energy, not a volume energy. This conclusion is confirmed by observation in everyday life: breaking an object at a given place is easier if the object is thin at that place, independent of the size of the object itself.

19.2.2 Additivity of Gravitational Interactions

Consider a material body which contains a large number N of atoms of atomic mass A (mass of the nucleus $= Am_{\text{p}}$). The gravitational potential energy of this body is

$$E_{\text{grav.}} = -\frac{1}{2}\sum_{i=1}^{N}\sum_{j\neq i}\frac{GA^2m_{\text{p}}^2}{r_{ij}}\,. \tag{19.8}$$

Note that we neglect here the gravitational attraction of the electrons, since they are much lighter than the nuclei.

Suppose now that the distance between neighboring particles d is still determined by electrostatic interactions and that it remains constant as one increases the total number N of atoms. For a spherical object of total mass $M = NA\,m_{\text{p}}$ with a uniform spatial density of matter, the number of atoms is $N = (4\pi/3)\,(R/d)^3$ and the gravitational energy is

$$\langle E_{\text{grav.}}\rangle = -\frac{3}{5}\frac{GM^2}{R} \sim -N^{5/3}\frac{GA^2m_{\text{p}}^2}{d}\,. \tag{19.9}$$

This potential energy increases faster than the number of constituents in the system.

At this stage, a question naturally arises: what is the critical number of atoms N_{c} above which the gravitational energy becomes dominant compared with the electrostatic energy? For N smaller than N_{c}, one can still rely on the determination of the interatomic distance d from purely electrostatic considerations, which leads to values of d of the order of a few Bohr radii. For N larger than N_{c}, the gravitational interaction becomes dominant and one has to turn to a more elaborate determination of d.

The critical number N_c is obtained when the two energies (19.7) and (19.9) are equal:

$$N_c^{2/3} = \frac{E_{\text{atom}}\, d}{GA^2 m_p^2}\,.$$

In this context, the relevant energy E_{atom} is the energy of the k valence electrons ($k \sim 1$ to 4), whose average positions determine the atomic size d. Therefore we take as a typical value $E_{\text{atom}} \sim kE_I$ and we obtain

$$N_c \sim \frac{k^{3/2}}{A^3}\, N_0\,, \qquad \text{where} \qquad N_0 = \left(\frac{d}{2\,a_1}\right)^{3/2} \left(\frac{\alpha}{\alpha_G}\right)^{3/2}. \tag{19.10}$$

The fact that N_0 is a function of the ratio of dimensionless constants α/α_G was certainly expected, but the power 3/2, due to the Pauli principle, is less obvious. Inserting the values of α and α_G and taking $d \sim 4\,a_1$ as a typical value for the interatomic spacing, we obtain $N_0 \sim 4\times 10^{54}$. The corresponding critical radius and mass are

$$R_c = d\left(\frac{3N_c}{4\pi}\right)^{1/3} \sim 2.5\, a_1 N_c^{1/3}\,, \tag{19.11}$$

$$M_c = AN_c\, m_p = \frac{k^{3/2}}{A^2}\, M_0\,, \tag{19.12}$$

where

$$M_0 = m_p\, N_0 \simeq 6 \times 10^{27}\ \text{kg} \ \sim 3 \times 10^{-3}\, M_\odot$$

The quantity $M_\odot$ is the mass of the sun: $M_\odot = 2 \times 10^{30}$ kg. It is instructive to compare these orders of magnitude with objects in our close environment:

- The planet Jupiter is mainly composed of hydrogen ($A = 1, k = 1$) and its mass is 1.9×10^{27} kg, of the same order of magnitude as M_c.
- The Earth is composed of "ferrous" materials $A \sim 50$ (notice that the molecular mass of silica SiO_2 is very close to the atomic mass of iron) and its mass is 6.0×10^{24} kg. Taking as a typical value $k = 2$, we obtain $M_c \sim 7 \times 10^{24}$ kg, of the same order as the mass of our planet.

Consequently, gravitational forces play a dominant role in the cohesion of these planets. However, the gravitational interactions do not overwhelm the electrostatic interactions, which still play a role in the structure of matter both in the crust and in the core of these planets.

19.2.3 Ground State of a Gravity-Dominated Object

As we have just seen, the gravitational energy dominates over the electrostatic energy for sufficiently large objects. In particular, the average spacing between two neighboring atoms can be reduced with respect to what is found in a smaller object, since the gravitational forces induce a compression which can be significantly larger than the standard Schrödinger pressure. The purpose of this subsection is to address the theoretical description of objects of this type, and to derive their ground-state size and energy.

Consider N nuclei of atomic mass A, surrounded by kN valence electrons (as above, k is between 1 and 4). At normal densities of matter, only the valence electrons are delocalized. The $Z-k$ internal electrons of each atom are bound to a particular nucleus, which forms with them an effective nucleus of mass Am_p and charge k. We shall consider in Sect. 19.3 a situation (white dwarf) where all Z electrons are delocalized.

The following remarks are of interest for the determination of the ground state of the system:

a. Since the electrons are fermions, the spatial extent of their wave function is constrained by the Heisenberg–Pauli relations (19.2) and (19.3).
b. Local electric neutrality imposes the requirement that the spatial distribution of the nuclei is the same as that of the electrons, whatever the spin of the nuclei and their statistical nature (bosons or fermions).
c. Since the nuclei are confined in the same spatial volume and are much heavier than the electrons, their kinetic energy is much smaller. The kinetic energy of the system is, to a good approximation, the kinetic energy of the electrons.
d. By assumption, the potential energy is dominated by the Newtonian attraction between the nuclei.

Therefore, to a first approximation, the Hamiltonian is

$$\hat{H} \simeq \sum_{i=1}^{kN} \frac{\hat{p}_i^2}{2m_\mathrm{e}} - \frac{1}{2}\sum_{i=1}^{N}\sum_{j\neq i} \frac{GA^2m_\mathrm{p}^2}{\hat{r}_{ij}} .$$

Let us assume that we are dealing with a spherical object of radius R and a uniform spatial density. As in (19.9), the gravitational energy is given by $-(3/5)GM^2/R$, where $M = NAm_\mathrm{p}$ is the total mass, so that

$$\langle H \rangle \sim kN\frac{\langle p^2 \rangle}{2m_\mathrm{e}} - \frac{3}{5}\frac{GM^2}{R} . \tag{19.13}$$

We now use (19.2) to relate $\langle p^2 \rangle$ and R. We first notice that for a uniform distribution within a sphere of radius R, $\langle r^2 \rangle = 3R^2/5$. We then assume that the inequality (19.2) is saturated[4] and we minimize the total energy (19.13) with respect to R. This yields the approximate ground state radius

$$R \sim 3a_1 \,\frac{\alpha}{\alpha_G}\,\frac{k^{5/3}}{A^2N^{1/3}} \sim R_\mathrm{c}\left(\frac{N_\mathrm{c}}{N}\right)^{1/3} , \tag{19.14}$$

[4] For a uniform distribution of particles inside a sphere, one can easily check that inequality (19.3) is weaker (and therefore less useful) than (19.2).

where N_c and R_c are defined in (19.10) and (19.11) (as above, we choose $d = 4a_1$ and $k = 2$). The corresponding energy is

$$E_{\text{grav.}} = \langle H \rangle_{\min} \sim 0.2\, E_{\text{I}}\, \frac{A^4 N^{7/3}}{k^{5/3}} \left(\frac{\alpha_G}{\alpha} \right)^2 .$$

Let us emphasize that this result is valid only if the object is large enough, so that the electrostatic energy $E_{\text{elec.}}$ of the valence electrons can be neglected. The approximation $E_{\text{elec.}} \ll E_{\text{grav.}}$ is correct if the number of atoms N exceeds the critical number N_c.

To summarize, the variation of the size of a celestial body with the number N of atoms is quite complex. First, for N much smaller than N_c (given in (19.10)), the size R increases as $N^{1/3}$, the distance between two neighboring atoms staying constant, of the order of a few Bohr radii. Then, when $N \sim N_c$, the size reaches a maximum of the order of the critical radius R_c (see (19.11)). For a larger number of atoms, the size decreases as $N^{-1/3}$ (see (19.14)), since gravitational forces become dominant. We shall see in the next section how this picture must be refined for very massive objects, such as white dwarf stars and neutron stars.

19.2.4 Liquefaction of a Solid and the Height of Mountains

Even in a relatively small celestial body, in which the gravitational interactions are not strongly dominant, the gravitational force may have a very significant impact on the shape of this body. A mountain on this celestial body (a planet or an asteroid) cannot be too high, otherwise the base of the mountain starts to flow. Indeed it may be energetically favorable to have a reduction of the height of the mountain, accompanied by a conversion of potential energy into latent heat associated with the fusion of the base of the mountain.

In order to liquefy a given fraction of a solid, containing δN atoms, one has to supply an energy which is typically $10^{-2}\, E_{\text{I}} \times \delta N$. More precisely, the necessary energy per atom is a small fraction ($\sim 5\%$) of the binding energy of the atom to the solid, which itself is a small fraction ($\sim 20\%$) of the Rydberg energy.

Consider first a mountain, which we assume to be cylindrical for simplicity (Fig. 19.1), on earth. The potential energy gained by reducing the height of the mountain is $\delta E_{\text{p}} = \delta M\, gH$, where $\delta M = A m_{\text{p}}\, \delta N$ represents the mass which has been liquefied. This reduction of height is energetically favorable if δE_{p} is larger than the energy δE_ℓ necessary to liquefy the mass δM: $\delta E_\ell = 10^{-2}\, E_{\text{I}}\, \delta N$. This process will occur until the height of the mountain is reduced to a value H such that

$$A m_{\text{p}} g H \leq 10^{-2}\, E_{\text{I}} . \tag{19.15}$$

For the earth, $A \sim 50$, and the critical height is found to be $H \sim 27$ km, whereas the actual height of the highest mountains is ~ 10 km. The agreement

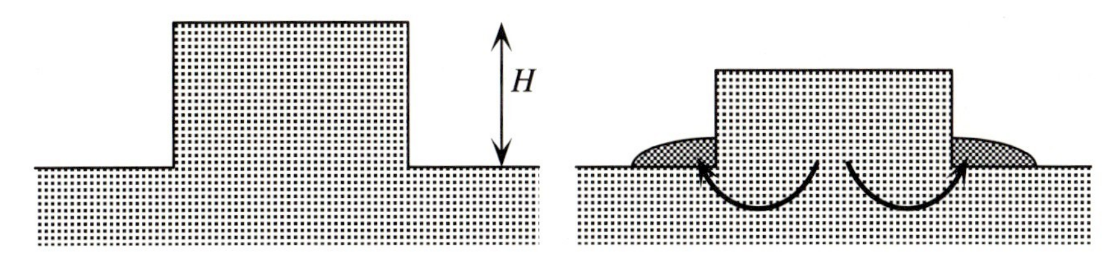

Fig. 19.1. Following an argument by Weisskopf, one can show that a mountain is stable only if it is not too high. Otherwise it is energetically favorable to convert a fraction of the potential energy of the mountain into liquefaction energy associated with the fusion of the base of the mountain

of these orders of magnitude is remarkable, considering the very simplified model that we have considered.

We can generalize these results to any planet, and relate the maximum height H to the radius R of the planet. The gravity at the surface of the planet is

$$g = \frac{GM}{R^2} ,$$

where $M = ANm_{\mathrm{p}}$ is the mass of the planet. The total number of atoms N is given by $4\pi R^3/3 = Nd^3$. Assuming $d \sim 4a_1$, we derive from (19.15) the following relation between R and the maximum mountain height H:

$$HR \sim 0.1\, a_1^2 \, \frac{1}{A^2} \, \frac{\alpha}{\alpha_G} . \tag{19.16}$$

The maximum height of mountains on a planet is proportional to the inverse of the *radius* of this planet. This is the case for Mars, where the mountains are twice as high as on earth. It is not, however, the case on the moon, where the tectonic activity has not been sufficient to create appreciable mountains. Otherwise, these mountains could reach ~60 km in height

Finally, from (19.16), one can calculate the size that a planet should have in order for it *not* to be spherical, i.e. for which $H \sim R$. This occurs for a critical radius given by

$$R \sim 0.3\, a_1 \, \frac{1}{A} \left(\frac{\alpha}{\alpha_G} \right)^{1/2} . \tag{19.17}$$

One finds $R \sim 350$ km for $A = 50$, in interesting agreement with observations, since the smallest spherical asteroids known have diameters of 400 to 700 km, whereas the satellite of Jupiter Amalthea, of diameter 250 km, is clearly nonspherical. This is another example where the competition between gravity and electromagnetism, together with the Pauli principle, governs the states of matter on all scales, including astronomical objects.

19.3 White Dwarfs, Neutron Stars and the Gravitational Catastrophe

The sun is quietly burning its hydrogen and transforming it into helium. In 5 billion years, the fuel will be exhausted and there will no longer exist a thermal pressure to balance the gravitational pressure. The system will implode until it reaches temperature and density conditions such that fusion reactions of helium into carbon and oxygen can start: 3 ^{4}He$\to$ ^{12}C and 4 ^{4}He$\to$ ^{16}O. After this second phase, which will be much shorter ($\sim 10^8$ years) than the previous one, a new effect will appear, which will prevent further thermonuclear reactions from starting, such as burning carbon and oxygen into ^{23}Na, ^{28}Si and ^{31}P, and these latter nuclei into ^{56}Fe, the most tightly bound nucleus. The density will become so high that the quantum pressure of the degenerate electron gas will stop the gravitational collapse. The system will then become a white dwarf star, whose only fate will be to lose its heat by radiating. For initial masses somewhat above $M_\odot$, a star can reach the ^{28}Si or ^{56}Fe stage before this happens

All stars do not end up as white dwarfs. Beyond some critical value of the final mass of the star, called the Chandrasekhar mass, the pressure of the degenerate electron gas cannot compete any longer with gravitation, and this leads to a gravitational catastrophe, the explosion of a supernova and the formation, in its center, of a neutron star.

White dwarfs have masses of the order of the solar mass $M_\odot$, sizes comparable to the radius of the earth (i. e. $0.01R_\odot \sim 10^4$ km) and densities of order 10^6 g cm^{-3} (i. e. $\sim 10^6$ larger than normal densities). Neutron stars are much more compact objects, with masses $\sim M_\odot$ and radii ~ 10 km. Their densities reach 10^{15} g cm^{-3}. We wish to understand these orders of magnitude.

19.3.1 White Dwarfs and the Chandrasekhar Mass

Consider N nuclei of mass Am_p and charge Z, surrounded by NZ electrons. As in Sect. 19.2.3, the dominant terms in the total energy of the system are assumed to be the following:

1. The potential energy E_p is dominated by the gravitational attraction between these nuclei. Assuming for simplicity a spherical object of radius R and with a uniform spatial density, we take

$$E_\mathrm{p} = -\frac{3}{5}\frac{GM^2}{R},$$

where $M = NA\,m_\mathrm{p}$ is the mass of the star.

2. The kinetic energy E_k is that of the electrons, which may become relativistic:[5]

$$E_\mathrm{k} = \sum_{i=1}^{NZ} \sqrt{p_i^2 c^2 + m_\mathrm{e}^2 c^4} \qquad \Rightarrow \qquad \langle E_\mathrm{k} \rangle \sim NZc\sqrt{\langle p^2 \rangle + m_\mathrm{e}^2 c^2}\,.$$

As seen before, the average squared momentum $\langle p^2 \rangle$ is bounded from below owing to the confinement of the electrons in the volume $V = 4\pi R^3/3$ of the star. This induces a competition between the gravitational forces, which tend to reduce R, and the quantum pressure term, which tends to increase it. Since we wish to explore a domain of energy where the electron motion may be relativistic, the inequality (19.2), which was derived in Chap. 16 using a nonrelativistic argument, is not an exact result anymore. However, we can still consider that each electron is confined in a volume $v = 2V/(NZ)$ (the factor 2 is due to the spin degeneracy), and we can take

$$\langle p^2 \rangle \simeq \frac{\pi^2 \hbar^2}{v^{2/3}}\,. \tag{19.18}$$

For an ideal Fermi gas, the total kinetic energy derived within this approximation coincides with the exact result to within 10%,[6] in the two limits of nonrelativistic and ultrarelativistic systems.

The radius R of the star and the average squared momentum $\langle p^2 \rangle$ are now obtained by minimizing the total energy $E_\mathrm{p} + E_\mathrm{k}$, taking into account the constraint (19.18) relating these two quantities. A calculation using $\langle p^2 \rangle$ as a variational parameter and very similar to that of Sect. 19.2.3 leads to

$$\frac{\langle p^2 \rangle}{\langle p^2 \rangle + m_\mathrm{e}^2 c^2} = \left(\frac{M}{M_\mathrm{Ch}}\right)^{4/3}, \tag{19.19}$$

where we have introduced the *Chandrasekhar mass* M_Ch:

$$M_\mathrm{Ch} \sim \frac{4}{\alpha_G^{3/2}} \frac{Z^2}{A^2} m_\mathrm{p}\,. \tag{19.20}$$

We notice that (19.19) has a solution only if the mass M is smaller than the Chandrasekhar mass. For $A = 2Z$ (which holds for carbon and oxygen), the Chandrasekhar mass is $M_\mathrm{Ch} \sim 3.8 \times 10^{30}$ kg, i.e. 1.9 times the mass of the sun. For masses larger than the Chandrasekhar mass, the Fermi pressure of the electron gas cannot compete with the gravitational pressure: the system is unstable and it undergoes a gravitational collapse.

[5] Concerning the use and properties of the operator $\sqrt{p^2c^2 + m^2c^4}$ in the Schrödinger equation, see J.L. Basdevant and S. Boukraa, Z. Phys. **C28**, 413 (1985); Z. Phys. **C30**, 103 (1986) and references therein.

[6] See, e.g., K. Huang, *Statistical Mechanics*, Chap. 11, Wiley, New York (1963).

The collapse arises from a simple property of degenerate Fermi gases. In the nonrelativistic regime, the pressure P of a Fermi gas is related to the density ρ by $P \propto \rho^{5/3}$. This can always balance the gravitational inward pressure $P_{\rm grav} \propto \rho^{4/3}$, provided the density is large enough. If the conditions are such that the Fermi gas is relativistic, then its pressure is related to the density by $P \propto \rho^{4/3}$, and there is a value of the mass such that the gravitational pressure prevails and the system collapses.

For a mass lower than the Chandrasekhar mass, the equilibrium radius is given in our model by

$$R = R_{\rm Ch} \left(\frac{M_{\rm Ch}}{M}\right)^{1/3} \left(1 - \left(\frac{M}{M_{\rm Ch}}\right)^{4/3}\right)^{1/2}, \tag{19.21}$$

where

$$R_{\rm Ch} \sim 2.5 \, \frac{Z}{A} \frac{1}{\sqrt{\alpha_G}} \frac{\hbar}{m_e c} \qquad (\sim 6300\,\text{km for } A = 2Z)\,.$$

This result is in good agreement with predictions obtained from more elaborate treatments. The value of $M_{\rm Ch}$ predicted by such treatments is $1.4\,M_\odot$. For a relatively low mass ($M \ll M_{\rm Ch}$), we recover the scaling law $R \propto N^{-1/3}$ of the nonrelativistic treatment of Sect. 19.2.3:

$$R \sim 7800\text{ km} \times \left(\frac{M_\odot}{M}\right)^{-1/3} \qquad \text{for} \quad A = 2Z. \tag{19.22}$$

Consider for instance von Maanen's star, which was one of the first white dwarfs to be discovered. Its radius is $\sim 8900\,$km (78 times smaller than the radius of the sun) and its mass is $0.68 M_\odot$, in good agreement with (19.22).

The equilibrium radius (19.21) is a decreasing function of the mass of the white dwarf, and, owing to relativistic effects, it shrinks to zero as the mass approaches the Chandrasekhar mass. Therefore, the most massive white dwarfs correspond to ultrarelativistic electrons and they all have the same mass $M_{\rm Ch}$.

19.3.2 Neutron Stars

At higher densities, it become energetically favorable for protons to capture electrons according to an inverse β process: $\mathrm{p} + \mathrm{e}^- \to \mathrm{n} + \nu$. The neutrinos escape from the star, which forms a neutron star.

Neutron stars were discovered in the mid 1960s as pulsars. They are gigantic nuclei, in the sense of nuclear physics. They are made of neutrons (electrically neutral) bound by the gravitational force, and packed together at nuclear distances $\sim 10^{-15}$ m. The size of such objects is of the order of 10 km, their mass is of the order of a few solar masses $M_\odot$ and their density is of the order of 10^{14} to 10^{15} g cm^{-3}.

Following a procedure very similar to that of Sect. 19.3.1, we shall assume that the dominant terms in the energy of a neutron star containing N neutrons are:

a. The potential energy E_{p}, dominated by the gravitational attraction. Assuming for simplicity a uniform distribution of the N neutrons, we take as before

$$E_{\mathrm{p}} = -\frac{3}{5}\,\frac{GN^2 m_{\mathrm{p}}^2}{R}\,.$$

b. The relativistic kinetic energy of the neutrons:

$$E_{\mathrm{k}} = \sum_{i=1}^{N} \sqrt{p_i^2 c^2 + m_{\mathrm{p}}^2 c^4} \qquad \Rightarrow \qquad \langle E_{\mathrm{k}} \rangle \sim Nc\sqrt{\langle p^2 \rangle + m_{\mathrm{p}}^2 c^2}\,.$$

We now minimize the total energy $E_{\mathrm{p}} + E_{\mathrm{k}}$. Since neutrons are fermions, (19.18), which relates $\langle p^2 \rangle$ and R, still holds. Taking $\langle p^2 \rangle$ as a variational parameter, we find that the minimum energy is obtained for

$$\frac{\langle p^2 \rangle}{\langle p^2 \rangle + m_{\mathrm{p}}^2 c^2} = \left(\frac{N}{N_1}\right)^{4/3}, \qquad \text{where} \quad N_1 \sim \frac{4}{\alpha_G^{3/2}} \sim 9 \times 10^{57}\,.$$

As for the case of a white dwarf star, this equation has a solution only if the number of neutrons is below a critical number N_1. In this case, the radius of the star is given by

$$R = R_1 \left(\frac{N_1}{N}\right)^{1/3} \left(1 - \left(\frac{N}{N_1}\right)^{4/3}\right)^{1/2},$$

where

$$R_1 \sim 2.5\,\frac{\hbar}{m_{\mathrm{p}} c}\,\frac{1}{\sqrt{\alpha_G}} \sim 7\ \mathrm{km}\,.$$

Because the neutrons have an average velocity close to c, the mass M_{s} of the star differs from the mass Nm_{p} of its constituents. The mass M_{s} is obtained using the total energy ($E_{\mathrm{p}} + E_{\mathrm{k}} = M_{\mathrm{s}} c^2$), which yields

$$M_{\mathrm{s}} = Nm_{\mathrm{p}} \left(1 - \left(\frac{N}{N_1}\right)^{4/3}\right)^{1/2}.$$

The mass is maximum when $N \sim 0.7 N_1$; the value of this maximum mass is

$$M_{\mathrm{s}}^{\max} \sim 6.5 \times 10^{30}\ \mathrm{kg} \sim 3 M_{\odot}\,, \tag{19.23}$$

corresponding to a radius $\sim 5\,\mathrm{km}$.

Beyond the critical number N_1, a *gravitational catastrophe* happens: the nonsaturation (or the additivity) of gravitational forces leads to a binding energy *larger* in absolute value than the mass energy of the particles. If, in its complicated evolution, which we shall not describe (neutronization of matter by absorption of the electrons by the protons), the system can lose energy by radiating neutrinos, it can fall into a "catastrophic" ground state where a new kind of physics applies, and it becomes a black hole.

One can refine this extremely simple model, in particular by taking into account the inhomogeneous density profile in the neutron star. The theory, due to Landau, Oppenheimer and Volkov, leads to a critical mass very similar to (19.23).

Further Reading

- V.F. Weisskopf, "Of atoms, mountains and stars: a study in qualitative physics", Science **187**, 605 (1975).
- E. Salpeter, "Dimensionless ratios and stellar structure", in *Perspectives in Modern Physics*, edited by R.E. Marshak, Wiley, New York (1966); "New views on neutron stars", Phys. Today, February 1999, p. 40.
- C. Kittel, *Introduction to Solid State Physics*, Wiley, New York (1966).
- S. Weinberg, *Gravitation and Cosmology*, Wiley, New York (1972).

20. Early History of Quantum Mechanics

Horresco referens
(I shiver when I recall it).
Virgil, The Aeneid

20.1 The Origin of Quantum Concepts

The first encounter of physics with the quantum world could have occurred in one of many problems, for instance, in attempting to understand atomic structure or the properties of solid-state matter. Historically, it just happens that it occurred in connection with a phenomenon which may seem less fundamental: the spectrum of radiation inside a hot oven, also called the black-body radiation spectrum.

20.1.1 Planck's Radiation Law

The problem was the following. Consider an isothermal enclosure at temperature T; how can one determine the frequency distribution $u(\nu, T)$ of the energy density of radiation inside this enclosure? It was one of the most remarkable results of Kirchhoff that, as a consequence of the second law of thermodynamics, $u(\nu, T)$ does not depend on the chemical nature or size of the enclosure, and that it is a universal function.

In 1900, Lord Rayleigh showed that in classical physics one necessarily obtains $u(\nu, T) = 8\pi\nu^2\, k_{\mathrm{B}}T/c^3$. This result demonstrates a severe inconsistency of classical physics, since the total radiation energy $\int u\, \mathrm{d}\nu$ would be infinite if it were correct. Experimentally, it is observed that Rayleigh's expression is well satisfied at low frequencies, but that at high frequencies the distribution obeys the law obtained by Paschen and Wien, $u = \alpha\nu^3\, \mathrm{e}^{-\gamma\,\nu/T}$, where α and γ are constants.

In 1900, Max Planck discovered empirically an interpolating formula between these two regimes. In Planck's formula,

$$u_\nu = \frac{8\pi\nu^2}{c^3}\frac{h\nu}{\mathrm{e}^{h\nu/k_{\mathrm{B}}T} - 1}, \tag{20.1}$$

there occurs a fundamental constant h. This formula is in excellent agreement with experimental observations. Planck succeeded in giving it an explanation within statistical thermodynamics, by assuming that matter behaves as a collection of perfect harmonic oscillators and that the energy absorbed or emitted by an oscillator of frequency ν is, at any time, an integer multiple of an "energy quantum" $h\nu$. This was the first occurrence of "quantization", although its meaning remained obscure.

20.1.2 Photons

In 1905, Einstein performed a critical analysis of Planck's argument, in particular concerning the absorption and emission of light. In order for the argument to be consistent, Einstein realized that the quantum aspects must be present in the radiation itself, and he introduced the concept of the photon. This also led to an explanation of the photoelectric effect. With Einstein's photons, one could find a much clearer *raison d'être* for Planck's law. Einstein also had a premonition of the double manifestations of the properties of light, which are both wave-like and particle-like.

20.2 The Atomic Spectrum

The next step in the development of quantum physics was an investigation of a much more fundamental nature. This concerned the interpretation of atomic spectra.

20.2.1 Empirical Regularities of Atomic Spectra

After a series of intense experimental analyses, spectroscopists had been led, between 1885 and 1908, to a set of empirical results concerning the distribution of spectral lines. These conclusions can be summarized as follows.

a. The emission and absorption eigenfrequencies of atoms are differences between "spectral terms":

$$\nu_{m,n} = A_m - A_n \tag{20.2}$$

(Rydberg and Ritz), where A_n is a function of an integer n, called a *quantum number* (the first use of this term).

b. In the case of hydrogen-like atoms, one has $A_n = K/n^2$ (Balmer, 1885). Other empirical formulas, similar but not so simple, had been found by Rydberg for alkali atoms.

20.2.2 The Structure of Atoms

In 1903, J.J. Thomson had constructed a so-called "plum-pudding" model of atomic structure. He assumed that, within an extended spherical, continuous distribution of positive charge, elastically bound electrons radiated or absorbed light at characteristic frequencies.

However, in 1908, Marsden and Geiger, by directing 4 MeV α particles from a radioactive source onto gold atoms, observed that some of the particles were scattered backward at large angles, up to 150 degrees. This observation was in opposition to Thomson's theory, which only allowed small scattering angles. Rutherford understood that these experimental observations could be explained by a model in complete opposition to J.J. Thomson's ideas. Rutherford proposed a "planetary" model consisting of a central positively charged nucleus, containing most of the atomic mass, surrounded by electrons orbiting under the Coulomb attraction.

20.2.3 The Bohr Atom

In 1913, a decisive step forward was made by Niels Bohr when he visited Rutherford in Manchester. Bohr postulated that there exist discrete energy levels where the electrons can remain without radiating. Radiation occurs suddenly, when an electron makes a transition between a stationary level of energy E_n to another stationary level of energy E_m by emitting a photon. Owing to energy conservation, the photon energy $h\nu$ is

$$h\nu = E_n - E_m \,. \tag{20.3}$$

In order to calculate the energies E_n, Bohr assumed there were "quantum restrictions" on the classical trajectories. These restrictions consisted of assuming that the action along a trajectory is an integer multiple of h, in other words $\oint \boldsymbol{p}\cdot \mathrm{d}\boldsymbol{r} = nh$. By applying this condition to circular trajectories in the case of the hydrogen atom, Bohr obtained the celebrated formula

$$E_n = -\frac{E_{\mathrm{I}}}{n^2}\,, \qquad \text{where} \qquad E_{\mathrm{I}} = \frac{2\pi^2 m_{\mathrm{e}} q^4}{(4\pi\varepsilon_0)^2 h^2}\,, \tag{20.4}$$

in remarkable agreement with experiment. This expression has the same form as Balmer's empirical formula, but, quite remarkably, the constant, known as Rydberg's constant, is expressed only in terms of fundamental physical constants (q and m_{e}, the charge and mass of the electron, and h).

In fact, both physics and Niels Bohr were quite lucky. We now know that his argument can only determine the values E_n for large values of n (in which case the trajectory becomes semiclassical). The miracle is that, in the case of hydrogen, the formula remains valid for small values of n. It is only the Schrödinger equation which, later on, accounted for that fact. One may wonder what the history of quantum mechanics would have been without this happy coincidence.

The validity of (20.4) was confirmed when Bohr identified some lines in the spectrum of the star ζ Puppis, which had first been attributed to hydrogen, as belonging to the spectrum of ionized helium. However, it was the experiments of Franck and Hertz (1914–1919) which confirmed the most revolutionary aspect of Bohr's theory, i. e. the existence of discrete stationary states.

20.2.4 The Old Theory of Quanta

Arnold Sommerfeld was one of the main theorists who developed Bohr's discovery into the form of what is now called the Bohr–Sommerfeld old theory of quanta. This was a very sophisticated theory, in the sense that it used all techniques of analytical mechanics, the basic assumption being that, for any pair of Lagrange conjugate variables p and q, one assumes the quantization of the integral $\oint p \, \mathrm{d}q = nh$. However, it turned out that it was impossible to repeat Bohr's original success. In 1925, the "quantum theory" was, from the methodological point of view, an inextricable list of hypotheses, principles, theorems and ad hoc recipes which reflected more a scholarly form of ignorance than a theory.

20.3 Spin

The rule for quantizing angular momentum had been discovered empirically by Ehrenfest in June 1913, *before* Bohr's model. Ehrenfest managed to improve considerably the theory of Einstein and Stern of the specific heats of diatomic gases, especially their temperature dependence, by assuming that the rotation energies of these molecules are quantized in a similar way to Planck's oscillators: $E_{\mathrm{rot}} = I(2\pi\nu_n)^2/2 = nh\nu_n/2$, where n is an integer and I is the moment of inertia of the molecule. This formula can be written in the equivalent forms $\nu_n = n\hbar/(2\pi I)$ (quantization of the rotation frequency) and $E_{\mathrm{rot}} = n^2\hbar^2/(2I)$, which displays the quantization of the square of the angular momentum. In July 1913, Bohr, in his celebrated paper on the hydrogen atom, was led to a similar result concerning the square of the angular momentum. Within the Bohr–Sommerfeld theory, it appeared that the projection of the angular momentum along a given axis z had values equal to $L_z = m\hbar$, where m is an integer between $-\ell$ and $+\ell$, the value of ℓ being a characteristic of the atomic level.

In order to test this peculiar prediction, Otto Stern engaged in a series of experiments which ended, in 1921, with the Stern–Gerlach experiment (Chap. 8). The observation that a beam of silver atoms could be split into two outgoing beams appeared to be the first experimental evidence in favor of the suggestion that quantization could apply to *trajectories* of particles. This appeared to be a triumph of the Bohr–Sommerfeld theory.

Needless to say, things were more subtle. The discovery of spin was one of the most breathtaking and fascinating steps of quantum theory. The first decisive step was made by Pauli. The problem he addressed was to understand why the electrons of a complex atom fill "shells" rather than all being in the same ground state, which would minimize the energy. In 1924, Pauli proposed an "exclusion principle". The electronic states in an atom are characterized not by three quantum numbers (n, ℓ, m) but by four (n, ℓ, m, m_s), where the fourth one, $m_s = \pm 1$, is "by essence a quantum quantity which has no classical analogue". The exclusion principle states that there can be at most one electron in a state of given quantum numbers.

The correct interpretation of this new quantum number was given in 1925 by Uhlenbeck and Goudsmit. They postulated that the electron possesses an intrinsic angular momentum, to which there corresponds an intrinsic magnetic moment. The only two possible values for the projection of this angular momentum on any axis are $\pm\hbar/2$. The only two possible values for the projection of the associated magnetic moment on any axis are $\pm q\hbar/2m_e$. This theory gave, in particular, an explanation for the "anomalous" Zeeman effect, which had been for 25 years a genuine challenge to the scientific community.

20.4 Heisenberg's Matrices

In the same year, 1925, the Bohr–Sommerfeld theory was replaced by quantum mechanics. The starting point of Heisenberg (in 1924) was to reject the classical notions of position and momentum for a particle, which no experiment could measure at that time, in favor of observable quantities, i. e. the positions and intensities of spectral lines. The problem Heisenberg faced was no longer to guess the quantum answer to a classical question, but to find the mathematical structure which governs quantum phenomena.

His first analysis concerned quantum kinematics, or, in other words, the mathematical nature of quantum quantities. In classical physics, the position $X(t)$ of a particle performing a periodic motion of frequency ν_0 can be expanded in a Fourier series as

$$X(t) = \sum X_n \, \mathrm{e}^{-2\pi \mathrm{i}\, n\nu_0 t} \,. \tag{20.5}$$

However, as Heisenberg noted, one observes quantum frequencies of the form

$$h\nu_{nm} = E_n - E_m \,, \tag{20.6}$$

which depend on *two* integers, and are more complicated than the Fourier frequencies $n\nu_0$. Heisenberg therefore postulated that, since two indices (n, m) are necessary in order to characterize quantum frequencies, the set of numbers which generalizes the Fourier coefficients of a physical quantity in the quantum case is a two-dimensional array $X_{nm}(t)$, whose time-dependence[1]

[1] Here we work in the *Heisenberg representation* (see exercise 5.3). In this representation, which is related to the Schrödinger picture used in this book by a unitary transformation, states are time independent whereas observables are time dependent.

is given by

$$X_{nm}(t) = X_{nm}\, e^{2\pi i\, \nu_{nm} t} . \tag{20.7}$$

In order to find the mathematical relations, i. e. the algebra, between these quantities, Heisenberg relied on (20.6). In fact, if we associate with the physical quantity X^2 the generalized Fourier coefficients $(X^2)_{nm}$, then, in order for the relation (20.7) to be satisfied, one must assume the "symbolic multiplication rule"

$$(X^2)_{nm} = \sum_q X_{nq} X_{qm} , \tag{20.8}$$

in which case one obtains

$$\begin{aligned} (X^2)_{nm}(t) &= \sum_q X_{nq}(t)\, X_{qm}(t) \\ &= \sum_q e^{it(E_n - E_q + E_q - E_m)/\hbar}\, X_{nq}\, X_{qn} \\ &= e^{2\pi i \nu_{nm} t}\, (X^2)_{nm} . \end{aligned}$$

Max Born remarked that this symbolic multiplication was nothing but matrix multiplication, and that the X_{nm} were matrices. Very rapidly, Born discovered the fundamental commutation relation between the matrices X and P of the elements X_{nm} and P_{nm}:

$$XP - PX = i\, \frac{h}{2\pi}\, I , \tag{20.9}$$

where I is the identity matrix. Born called this relation the fundamental equation of *Quanten Mechanik*, i. e. a mechanics pertaining to quantum phenomena.

Matrix Mechanics. The new quantum mechanics, or matrix mechanics, was presented in 1925 by Born, Heisenberg and Jordan. The theory consists in postulating that the fundamental physical quantities are matrices of the type $X = \{X_{nm}\}$. The position and momentum matrices satisfy, by definition, the fundamental commutation relation (20.9). The energy of a quantum system has the same formal expression as in classical mechanics, but it must be interpreted as a relation between matrices. Furthermore, the time derivative of any matrix A is given by the fundamental hypothesis of quantum dynamics:

$$\dot{A} = -\frac{i}{\hbar}(AH - HA) . \tag{20.10}$$

Consider a particle of mass M in a potential $V(X)$. Taking the time derivative of (20.7), we obtain

$$(\dot{X})_{nm}(t) = -\frac{2\pi i}{h}\, (E_n - E_m)\, X_{nm}\, e^{2\pi i \nu_{nm} t} .$$

Since the energy $H = P^2/2M + V$ is time independent, it must be represented by a diagonal matrix $(H)_{nm} = E_n\delta_{nm}$, from which one can deduce the relation between matrices

$$\frac{P}{M} \equiv \dot{X} = -\frac{\mathrm{i}}{\hbar}(XH - HX)$$

as a particular case of (20.10).

In October 1925, starting from these well-defined assumptions, Pauli succeeded in showing that this formalism leads to the observed energy levels of the hydrogen atom (i. e. that the corresponding matrix H has the correct eigenvalues E_n), including the case where an external static electric field is applied to the atom (the Stark effect). He was also able to give the solutions to problems which had remained unsolved in the context of the Bohr–Sommerfeld theory, such as the motion of a particle in crossed electric and magnetic fields.

Dirac's Genius. In July 1925, Heisenberg delivered a lecture on his work at Cambridge. Fowler asked his most brilliant student, the young Dirac, to read and study Heisenberg's papers. At first, Dirac was skeptical because of the awkward mathematical formulation used by Heisenberg. Two weeks later, however, he dropped into Fowler's office and said that the key point of quantum theory was contained in Heisenberg's work. Dirac knew of the existence of noncommutative algebras, and he recognized that noncommutativity was the essential point of Heisenberg's theory. Since no fundamental principle dictates that physical quantities should commute, Dirac worked to incorporate the fundamental property of noncommutativity into the classical equations of analytical mechanics. In a few months, he constructed his own version of quantum mechanics based on "q-numbers" (instead of classical "c-numbers"), independently of the Göttingen group. In particular, he understood the relationship between the quantum commutators and the classical Poisson brackets.

In 1927, Darwin and Pauli showed independently how the electron spin could be incorporated into the formalism. In 1928, Dirac published his relativistic equation, which gives a natural account of the electron spin.

20.5 Wave Mechanics

The de Broglie Hypothesis (1923). At the same time as the matrix approach, wave mechanics developed. In 1923, shortly after Compton proved that Einstein's photons of energy $E = h\nu$ also possessed a momentum $\boldsymbol{p} = \hbar\boldsymbol{k}$, Louis de Broglie proposed a systematic association of waves and particles. He wrote down the fundamental relation between the momentum and the wave vector for any material particle, $\boldsymbol{k} = \boldsymbol{p}/\hbar$. Louis de Broglie believed that the quantization of energies could emerge as the result of a stationary-wave problem.

The Schrödinger Equation (1926). The chemist Victor Henri gave the work of Louis de Broglie to Schrödinger in 1925. Schrödinger, who was skeptical at first, was encouraged by Einstein's enthusiasm. After some unsuccessful attempts with a relativistic wave equation (now called the Klein–Gordon equation), which he found more reasonable, Schrödinger found his celebrated equation in a nonrelativistic limit (which he said he didn't understand). In an impressive series of articles, published in 1926, he solved many problems, including the calculation of the energy levels of hydrogen as an eigenvalue problem, perturbation theory, the Stark effect, etc. He also proved the equivalence of matrix mechanics and wave mechanics by expanding the equations of wave mechanics in the eigenbasis of energy states.

The Experimental Confirmation. In 1927, Davisson and Germer, at the end of a long series of experimental investigations started by Davisson in 1919, demonstrated the diffraction of electrons by a crystal, in agreement with de Broglie's formula. The same year, G.P. Thomson, the son of J.J. Thomson, obtained a similar result with a different experimental technique.

20.6 The Mathematical Formalization

1926 In order to interpret electron collision experiments, Max Born suggests that $|\Psi(\boldsymbol{r})|^2\, \mathrm{d}^3r$ is a probability. He therefore understands the probabilistic interpretation of the coefficients c_n in the expansion of the wave function $\psi(\boldsymbol{r}) = \sum c_n\, f_n(\boldsymbol{r})$ in a complete basis of orthonormal functions.
Dirac deepens the structure of the theory and introduces the δ "function".

1927 Hilbert and Nordheim clarify the mathematical foundations of quantum theory. They introduce Hilbert space and give a general theory of operators.
Von Neumann gives a rigorous formulation of the continuous spectrum of an operator which avoids using the δ function, but which has the disadvantage of being cumbersome. He introduces the density operator.

With these latter developments, the axioms of quantum theory were established. Quantum mechanics had become a consistent and predictive theory.

20.7 Some Important Steps in More Recent Years

We restrict ourselves to discoveries which are directly related to the quantum phenomena studied in this book.

1927 Dirac quantizes the electromagnetic field and explains completely the emission and absorption of photons by atoms.

1928 Dirac proposes the relativistic theory of the electron and predicts the existence of the antielectron, or positron.
Bloch develops the quantum theory of electric conduction in metals.
Gamow, Gurney and Condon explain α decay by quantum tunneling.

1930 Invention of magnetic resonance by Rabi.

1931 Wigner applies group theory to quantum mechanics and clarifies the notions of spin, parity and time reversal.

1932 Anderson discovers the positron in cosmic rays.

1938 Kapitza discovers the superfluidity of helium. London relates this macroscopic quantum phenomenon to Bose–Einstein condensation.

1941 Landau theory of superfluidity.

1945 Bloch and Purcell independently discover nuclear magnetic resonance.

1946 Theory of semiconductors and invention of the transistor (Bardeen, Brattain and Shockley).
Tomonaga proposes a covariant relativistic mathematical formalism of quantum field theory.

1947 Measurement of the Lamb shift of the $2s_{1/2}$ and $2p_{1/2}$ levels of hydrogen, which was not predicted by Dirac's equation.

1948 Kusch measures the magnetic moment of the electron accurately and finds a value slightly different from the prediction of the Dirac equation.

1949 Theory of quantum electrodynamics which explains all electromagnetic phenomena (Feynman, Schwinger and Tomonaga). Accurate calculations of the effects found by Lamb and Kusch.

1950 Kastler and Brossel invent the optical pumping technique.

1956 Lee and Yang suggest that parity is not conserved in weak interactions.

1957 A fundamental problem of solid state physics, superconductivity, is solved by Bardeen, Cooper and Schrieffer.

1962 Townes invents the laser.
Discovery of the Josephson effect in superconductors.

1965 Time-reversal invariance is slightly violated in weak interactions (Fitch and Cronin).
Bell proves his celebrated inequalities.

1977–1987 Experimental proofs of the violation of Bell's inequalities and of the validity of quantum mechanics in that respect.

1985 Invention of the tunneling-effect microscope by Binnig and Rohrer.

1995 Observation of the Bose–Einstein condensation of atomic gases at temperatures below 1 microkelvin (Nobel prize 2001 awarded to Cornell, Ketterle and Wieman).

1997 The Nobel prize is awarded to Chu, Cohen-Tannoudji and Phillips for the development of methods for cooling and trapping atoms with light.

Further Reading

- M. Jammer, *The Conceptual Development of Quantum Mechanics*, McGraw-Hill, New York (1966).
- B.L. Van Der Waerden, *Sources of Quantum Mechanics*, North-Holland, Amsterdam (1967).
- J. Mehra and H. Rechenberg, *The Historical Development of Quantum Theory*, Springer, New York (1982).

Appendix A. Concepts of Probability Theory

One is always forced to let something to chance.
Napoleon Bonaparte

1 Fundamental Concepts

Consider a set of phenomena of the same nature on which we repeatedly make the same observation or measurement. For instance, we might play dice, deal shuffled cards, measure an outside temperature or an economic parameter, etc. Each observation belongs to some set Ω of possibilities. This set can be discrete (the numbers on a die), continuous (the set of observable temperatures) or a more complicated object such as a set of functions (such as curves of noise intensities between t_0 and t_1).

The set Ω is the set of a priori possible *outcomes* of the experiment. One also speaks of *events*: "the number obtained in a spin of the roulette wheel is even", "the observed temperature is between T_0 and T_1", etc. Each event is therefore defined by a set of possible outcomes (a part of Ω).

We can introduce the notion of frequency. Suppose we repeat an experiment a large number of times N, Ω being the set of possible outcomes. Consider a specific event α and suppose that N_α is the number of times, out of a total of N, that α happens. The observed number N_α obviously depends on the specific sequence of experiments. The ratio

$$f_\alpha(N) = N_\alpha/N$$

is called the empirical frequency of the event α in this sequence of experiments. A fundamental empirical observation is the following: when N becomes large, if the successive repetitions of the experiment are performed *independently* (the result of an experiment has no a priori influence on the conditions under which the other experiments are done), the frequencies $f_\alpha(N)$ tend, for each event α, to a well defined limit. This leads us to suppose that to each event α there corresponds a number $P(\alpha)$, called the *probability* of event α, which is related to the empirical frequency by the relation

$$P(\alpha) = \lim_{N\to\infty} f_\alpha(N)\,.$$

Clearly, $P(\alpha) \geq 0$, $P(\Omega) = 1$, $P(\emptyset) = 0$ and, if $(A_i)_{i\in I}$ is a finite family of disjoint events,

$$P\left(\bigcup_{i\in I} A_i\right) = \sum_{i\in I} P(A_i)\,.$$

As a mathematical theory, probability theory assumes a priori the existence of probabilities. A sequence of experiments constitutes a more complicated event. One can then prove a theorem of the following form:

The probability that the frequency $f_\alpha(N)$ differs from $P(\alpha)$ by more than ϵ tends to zero as N tends to infinity.

This type of convergence, which is of a quite particular form, is called *stochastic* (from the Greek word *stokhastikos*, meaning "conjectural").

2 Examples of Probability Laws

2.1 Discrete Laws

The Simple Alternative. In this example, there are only two possible outcomes, $\alpha = 1$ or 2 (for example, heads or tails). We denote the probability of outcome 1 by p and that of outcome 2 by q. We obviously have $p + q = 1$.

The Generalized Alternative. In this case there are n possible outcomes $\alpha = 1, 2 \ldots, n$. For instance, one can place m_1 balls marked $1, m_2$ balls marked 2, ... in an urn. If the process of drawing a ball does not distinguish between the balls, the probability law is specified by the set of numbers p_1, p_2, ..., p_n such that

$$p_\alpha = \frac{m_\alpha}{\sum_{\beta=1}^{n} m_\beta}\,, \qquad \text{where} \qquad \sum_{\alpha=1}^{n} p_\alpha = 1\,.$$

2.2 Continuous Probability Laws in One or Several Variables

A probability law P on one ($\boldsymbol{R}$) or n ($\boldsymbol{R}^n$) continuous real variables is said to be of density p, p being a positive integrable function such that $\int_{-\infty}^{+\infty} p(x)\,\mathrm{d}x = 1$ or $\int_{R^n} p(x)\,\mathrm{d}^n x = 1$, respectively, if, for any interval or any volume I,

$$P(I) = \int_I p(x)\,\mathrm{d}x\,.$$

Mathematically, it is useful to treat the discrete and continuous cases in the same formalism by working with the distribution function

$$F(t) = P(]-\infty, t])\,.$$

A probability law on $\boldsymbol{R}$ is entirely determined by the values it takes for the events $]-\infty, t]$ for all t.

Examples.

a. Exponential law:

$$p(x) = \begin{cases} \lambda \mathrm{e}^{-\lambda x} & \text{if } x \geq 0 \ (\lambda > 0) \\ 0 & \text{if } x < 0 \end{cases},$$

which yields

$$F(t) = \int_{-\infty}^{t} p(x)\, \mathrm{d}x = \begin{cases} 0 & \text{if } t < 0 \\ 1 - \mathrm{e}^{-\lambda t} & \text{if } t > 0 \end{cases}.$$

b. Gaussian law, with parameters μ, σ (Fig. A.1):

$$p(x) = \frac{1}{\sigma\sqrt{2\pi}} \exp -\frac{(x-\mu)^2}{2\sigma^2}, \qquad \text{where} \quad \mu \in \boldsymbol{R}, \sigma \in \boldsymbol{R}^* . \qquad \text{(A.1)}$$

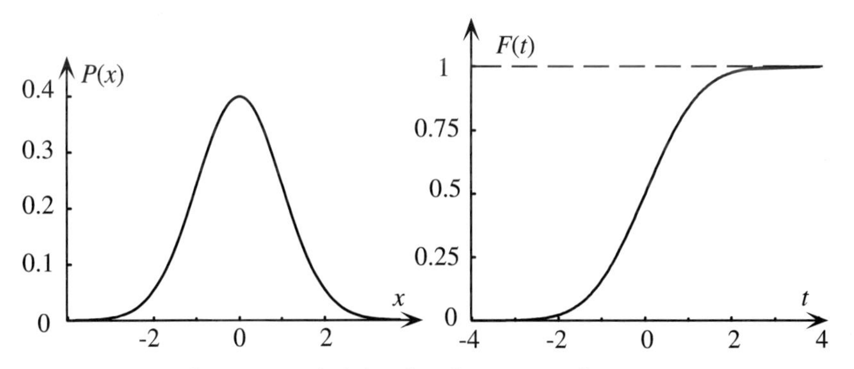

Fig. A.1. The Gaussian probability law for $\mu = 0$ and $\sigma = 1$

3 Random Variables

3.1 Definition

Consider the example of a game with n possible outcomes $\alpha_1, \ldots, \alpha_n$ of respective probabilities $p_1, \ldots, p_n$. If, in this game, we win some amount of money x_α when the outcome is α, the number x_α, which is a function of the (random) outcome of the experiment, is called a *random variable*.

In the above example, the set of the $\{x_\alpha\}$ is discrete. A random variable x is called a discrete random variable if there is a set of numbers x_α (which may be positive, negative or complex) each of which is associated with a possible outcome of a discrete random event. The pairs $\{x_\alpha, p_\alpha\}$ define the *probability law* of the random variable x.

In the same way, one can consider continuous random variables. Let x be a random variable which takes values in an interval $[a, b]$. The probability density $p(x)$ (positive or zero) defines the probability law of this random variable if the probability of finding, in an experiment, a value between x and $x + \mathrm{d}x$ is $p(x)\,\mathrm{d}x$. We obviously have $\int_a^b p(x)\,\mathrm{d}x = 1$. The generalization to $\boldsymbol{R}^n$ is straightforward.

3.2 Conditional Probabilities

Consider two types of events $[A]$ and $[B]$. We are led to defining the *conditional probability* of the event B if we know that the event A has happened, denoted by $P(B/A)$, as

$$P(B/A) = \frac{P(B \cap A)}{P(A)} \quad \text{as long as } P(A) > 0\,.$$

If X is a discrete random variable, we can define the *conditional probability* $P(B|X = x)$ *of the event* B *when* $X = x$, i.e. knowing that the event $\{X = x\}$ has happened.

Example: the Exponential Decay Law. If a radioactive particle exists at time t, its probability of decaying in the time interval $]t, t+\Delta t]$ is independent of its past history. Therefore the conditional probability that the time X at which the particle decays is between t and $t + \Delta t$, knowing that $\{X > t\}$, is independent of t and equal to $P\{0 < X \leq \Delta t\}$:

$$P\{0 < X \leq \Delta t\} = \frac{P\{t < X \leq t + \Delta t\}}{P\{X > t\}}\,.$$

If F is the distribution function of X, we obtain the functional relation

$$F(\Delta t) = \frac{F(t + \Delta t) - F(t)}{1 - F(t)}\,.$$

The function F therefore satisfies the differential equation

$$F'(t) = F'(0)\,[1 - F(t)]\,.$$

Therefore, setting $\lambda = p(0) = F'(0)$ (where λ is a decay rate), we obtain $F(t) = 1 - \mathrm{e}^{-\lambda t}$. The density of the probability law for X is then

$$p(x) = \begin{cases} \lambda \mathrm{e}^{-\lambda x} & \text{for } x \geq 0 \\ 0 & \text{for } x < 0 \end{cases}\,.$$

Since λ has the dimensions of the inverse of a time, we can denote it by $1/\tau$, where τ is the *lifetime* (or *mean life*). This exponential law is met with in many practical applications (physics, pharmacology, reliability, etc.).

3.3 Independent Random Variables

Consider two discrete random variables X and Y with values in E_1 and E_2, respectively. One says that X and Y are two independent variables if an observation of X does not give any information on Y, and vice versa. In other words, the conditional probability of finding x if one knows y is independent of y (and vice versa).

This can be expressed in a symmetric form in x and y as

$$P(\{X = x, Y = y\}) = P(\{X = x\})\, P(\{Y = y\}) .$$

The variables X and Y are independent if and only if the law for the pair (X, Y) is the product of the laws for X and Y.

This condition can easily be extended to real variables. If X and Y are real independent variables of respective densities p_1 and p_2, the law for the pair $\{X,Y\}$ is $p(x, y) = p_1(x)\, p_2(y)$.

3.4 Binomial Law and the Gaussian Approximation

Consider an experiment that consists of repeating an experiment with two possible outcomes (for instance heads and tails) N times, independently. The first possible outcome, denoted by 1, has a probability p of happening, and the second, denoted by 0, has a probability $q = 1 - p$ of happening. Such a sequence of experiments is called a *Bernoulli sequence.*

Since the successive partial experiments are assumed to be independent, the probability of a given sequence $(x_1, \dots, x_N)$ is given by

$$P(x_1, \dots, x_N) = P[X_1 = x_1] \dots P[X_N = x_N] = p^k\, q^{N-k} ,$$

where k is the number of 1's in the sequence $(x_1, \dots, x_N)$. We now consider the random variable $X = X_1 + \dots + X_N$ representing the number of times 1 appears in the N successive draws:

$$P[X = k] = \binom{N}{k}\, p^k q^{N-k} \equiv b(k; N, p) .$$

This law $b(k; N, p)$ is called the *binomial law*, with parameters N and p.

Normal Approximation to the Binomial Law. Using Stirling's formula, $n! \sim \sqrt{2\pi n}\, n^n\, \mathrm{e}^{-n}$, we obtain, for $n \gg 1$:

$$b(k; n, p) \sim \sqrt{\frac{1}{2\pi npq}} \exp\left(-\frac{(k - np)^2}{2npq}\right) ,$$

i. e. a Gaussian law for k, with $\mu = np$ and $\sigma = \sqrt{npq}$.

4 Moments of Probability Distributions

4.1 Mean Value or Expectation Value

Consider a function $\varphi(x)$ of a random variable x ($\varphi(x)$ is a new random variable). We define its *mean value*, or *expectation value*, $\langle\varphi\rangle$ as

$$\langle\varphi\rangle = \begin{cases} \sum_\alpha \varphi(x_\alpha)\, p_\alpha & \text{discrete case} \\ \int_a^b \varphi(x)\, p(x)\, \mathrm{d}x & \text{continuous case} \quad (a < x < b) \end{cases} .$$

We denote by $\langle x\rangle$ the mean value of the variable x itself:

$$\langle x\rangle = \int x\, p(x)\, \mathrm{d}x .$$

This quantity is also called the mathematical expectation or the *expectation value*: if we gain an amount x_α when the result is α, then we expect to gain $\langle x\rangle$ on average.

Expectation Values of Some Common Laws

a. The simple alternative: $\langle X\rangle = p$.
b. Binomial law $b(k;n,p)$: $\langle k\rangle = np$.
c. Geometric law $P\{X = k\} = (1-p)p^k$ $(k \geq 0)$: $\langle X\rangle = p/(1-p)$.
d. Poisson law $P\{X = k\} = \mathrm{e}^{-\lambda}\, \lambda^k/k!$ $(k \geq 0)$: $\langle X\rangle = \lambda$.

Example. In the exponential decay above, the mean time that the particle spends before it decays, or the expectation value of its lifetime, is

$$\langle t\rangle = \int_0^\infty \frac{t}{\tau}\, \mathrm{e}^{-t/\tau} \mathrm{d}t = \tau .$$

4.2 Variance and Mean Square Deviation

Consider a real random variable x whose expectation value is $\langle x\rangle = m$. The *mean square deviation* of x, denoted by σ or Δx, is defined by

$$(\Delta x)^2 = \sigma^2 = \langle (x - \langle x\rangle)^2\rangle ;$$

σ^2 is also called the *variance* of the probability law. One can readily check, by expanding the squared term, that

$$\sigma^2 = \langle x^2\rangle - 2\langle x\rangle\langle x\rangle + \langle x\rangle^2 = \langle x^2\rangle - \langle x\rangle^2 .$$

The smaller σ is, the more probable it is to find a value of x close to the mean value. The quantity σ measures the deviation from the mean value.

Variance of Some Common Laws.

a. The simple alternative: $\sigma^2 = p(1-p)$.
b. Binomial law: $\sigma^2 = np(1-p)$. Note that the relative dispersion $\sigma/\langle X \rangle$ tends to zero as $n^{-1/2}$ when $n \to \infty$.
c. Gaussian law: the variance coincides with the parameter σ^2 of (A.1).
d. Geometric law: $\sigma^2 = p/(1-p)^2$.
e. Poisson law of parameter λ: $\sigma^2 = \lambda$.

4.3 Bienaymé–Tchebycheff Inequality

We denote the mean value of the discrete real variable X by m and its variance by σ^2. It can be shown that

$$P(\{|X - m| \geq \tau\sigma\}) \leq 1/\tau^2 \,, \tag{A.2}$$

which proves that for a small variance, there is a small probability of finding X far from its expectation value.

Error Function. In the particular case of the Gaussian law (m, σ), the quantity $P(\{|X - m| \leq \tau\sigma\})$ is called the *error function* $\Phi(\tau)$. We have

$$\Phi(\tau) = \int_{\tau\sigma}^{+\tau\sigma} \frac{1}{\sigma\sqrt{2\pi}} \, \mathrm{e}^{-x^2/2\sigma^2} \, \mathrm{d}x \,.$$

Some values of $\Phi(\tau)$ are the following:

τ	1	2	3
$\Phi(\tau)$	0.68	0.95	0.99

Example. Suppose one tosses a coin 10^6 times ($n = 10^6$, $p = q = 1/2$). The expectation value is obvious: $m = 5 \times 10^5$. The mean square deviation is not so obvious: $\sigma = 500$. Using the Bienaymé–Tchebycheff theorem, we see that the probability for finding, in 10^6 draws, a number k outside the interval $m \pm 2\sigma$ is smaller than 25%. This probability is actually 5% (this can be obtained from the Gaussian approximation of the binomial law).

4.4 Experimental Verification of a Probability Law

Quantum mechanics predicts probability laws. How can one verify such predictions experimentally, to a given accuracy?

Consider a specific example. It is predicted that the decay law of an elementary particle is of the form $\mathrm{e}^{-t/\tau}\,\mathrm{d}t/\tau$. We study the decay of a large number n of particles, and we count the number of decays that occur in the time interval between t and $t + \Delta t$. Each decay is an independent event. We therefore want to know how many times the outcome 1 (a decay between t and $t+\Delta t$) happens. The theoretical probability is

$$p = \int_t^{t+\Delta t} \mathrm{e}^{-t/\tau} \mathrm{d}t/\tau = \mathrm{e}^{-t/\tau}(1 - \exp[-\Delta t/\tau]) \, .$$

We choose n sufficiently large and Δt not too small compared with τ so that np is large. We are then in the case where the binomial law reduces to a Gaussian. The probability law for observing a number k of decays during the time interval Δt is a Gaussian with an expectation value np and a mean square deviation $\sigma = \sqrt{np(1-p)}$. Given the values of the error function Φ, we expect a probability of 95% that k is inside the interval $np \pm 2\sqrt{np(1-p)}$, and 99% that it is in the interval $np \pm 3\sqrt{np(1-p)}$, if the predicted value of p is correct.

This determines the order of magnitude of the experimental effort required to achieve a given accuracy. We know that the observed frequency $f = k/n$ tends to p. In order to verify the probability law p up to an accuracy of δp with a 95% confidence level, we must perform a number of observations such that

$$\delta p \geq 2\sqrt{p(1-p)/n}, \qquad \text{i.e.} \qquad n \geq 4p(1-p)/(\delta p)^2 \, .$$

Conversely, if we do not have a theoretical prediction for p, then if we measure an experimental frequency $f = k/n$, we can claim with a confidence level $\Phi(\tau)$ that

$$p = f \pm \tau\sqrt{f(1-f)/n} \, .$$

Note that the "errors" which appear here have a probabilistic character. They are called *statistical* errors, as opposed to *systematic* errors, which originate from the fact that one is actually measuring a phenomenon slightly different from the desired one (because of the influence of the operator or of an external influence on the measured system).

Exercises

A.1. Distribution of impacts. We observe some impacts on a target in the xy plane. The observable is assumed to obey a probability law with a density $p(x,y) = (2\pi\sigma^2)^{-1}\exp[-\rho^2/(2\sigma^2)]$, where $\rho = (x^2+y^2)^{1/2}$ is the distance of the impact point from the origin. What is the probability law for ρ?

A.2. Is is a fair game? Suppose that someone offers you the following game: *Bet one euro and throw three dice. If the number 6 (or any number you choose in advance) does not show up, you lose your bet; you get paid 2 euros if this number shows up on one dice, 3 euros if it shows up on two dice,* *and 6 euros if it shows up on all three of them.* Calculate the expectation value of what you gain (which is negative if you lose) and find out if it is reasonable to play.

A.3. Spatial distribution of the molecules in a gas. Consider N molecules (6×10^{23}, for instance) in a volume V (22.4 liters, for instance). Suppose that a part of this volume, of volume v ($10^{-3}\,\mathrm{cm}^{-3}$), is enclosed. How many molecules are there on average in v? What are the fluctuations of this number?

Appendix B. Dirac Distribution, Fourier Transformation

Under King Louis XVIII,
there was, at the Academy of Sciences,
a famous Fourier that posterity has forgotten.
Victor Hugo, Les Misérables

1 Dirac Distribution, or δ "Function"

We often refer to point-like objects in physics. The mass density $\rho(\boldsymbol{r})$ (or the charge density) of such an object is not a function in the usual sense, since it is everywhere zero except at a point $\boldsymbol{r}_0$, but its "integral" is finite:

$$\int \rho(\boldsymbol{r})\, \mathrm{d}^3 r = m\,.$$

The δ "function", introduced by Paul Dirac, can describe such a density. Its mathematical definition was developed by the mathematician Laurent Schwartz in the framework of distribution theory, which we shall briefly describe in the next section.

1.1 Definition of $\delta(x)$

In this section we present the (mathematically improper) names and formalism used by physicists. For a real variable x, the "function" $\delta(x)$ has the following properties:

$$\delta(x) = 0 \quad \text{for} \quad x \neq 0 \qquad \text{and} \qquad \int_{-\infty}^{+\infty} \delta(x)\, \mathrm{d}x = 1. \tag{B.1}$$

For any function $F(x)$ that is regular at $x = 0$, we have, by definition,

$$\int F(x)\, \delta(x)\, \mathrm{d}x = F(0)\,. \tag{B.2}$$

By a change of variables, we can define the function $\delta(x - x_0)$, for which

$$\int F(x)\,\delta(x-x_0)\,\mathrm{d}x = F(x_0)\,. \tag{B.3}$$

The generalization to several dimensions is straightforward. Consider, for instance $\boldsymbol{r} = (x, y, z)$. We then have

$$\delta(\boldsymbol{r}-\boldsymbol{r}_0) = \delta(x-x_0)\,\delta(y-y_0)\,\delta(z-z_0)\,, \tag{B.4}$$

that is to say,

$$\int F(\boldsymbol{r})\,\delta(\boldsymbol{r}-\boldsymbol{r}_0)\,\mathrm{d}^3r = F(\boldsymbol{r}_0)\,.$$

1.2 Examples of Functions Which Tend to $\delta(x)$

One can construct distributions which are nearly point-like, using functions concentrated in the vicinity of a point x_0 (Fig. B.1). In order to do so, we consider sequences of functions depending on a parameter which determines their width ($y_\varepsilon(x)$ and $g_\sigma(x)$ in the first two of the following examples). Although these functions have no limit in the usual sense when their width goes to zero, the integral of their product with any function F that is regular at $x = x_0$ remains well defined and tends to the limit $F(x_0)$. The following are examples of sequences of functions of this kind.

a. Consider the sequence of functions $y_\varepsilon(x)$ (Fig. B.1a) defined by

$$y_\varepsilon = \begin{cases} 1/\varepsilon & \text{for } |x| \le \varepsilon/2 \\ 0 & \text{for } |x| > \varepsilon/2 \end{cases}\,. \tag{B.5}$$

Then

$$\int_{-\infty}^{+\infty} F(x)\,y_\varepsilon(x)\,\mathrm{d}x = \frac{1}{\varepsilon}\int_{-\varepsilon/2}^{\varepsilon/2} F(x)\,\mathrm{d}x = F(\theta\varepsilon/2)\,,$$

where $-1 \le \theta \le 1$. In the limit $\varepsilon \to 0$, $y_\varepsilon(x)$ "tends" to $\delta(x)$.

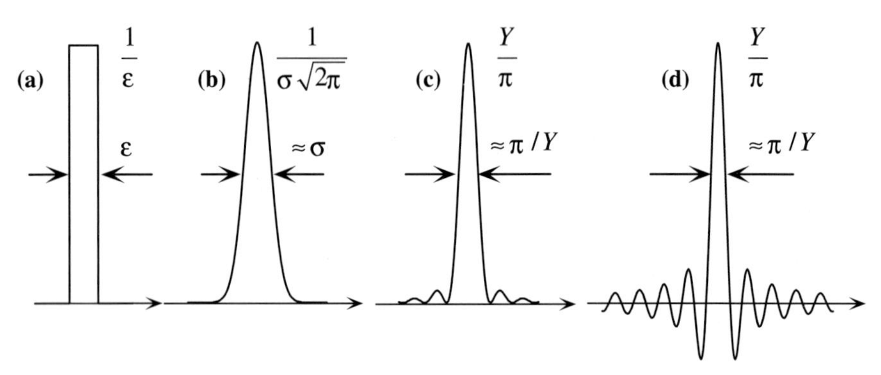

Fig. B.1a–d. Examples of functions concentrated in the vicinity of a point, whose limit, in the sense of distributions, is equal to $\delta(x)$

b. Gaussian function (Fig. B.1b):

$$g_\sigma(x) = \frac{1}{\sqrt{2\pi}\sigma}\,\exp(-x^2/2\sigma^2)\,.$$

By the change of variables $y = x/\sigma$, we obtain

$$\int_{-\infty}^{+\infty} F(x)\, g_\sigma(x)\, \mathrm{d}x = \frac{1}{\sqrt{2\pi}}\int_{-\infty}^{+\infty} \mathrm{e}^{-y^2/2}\, F(\sigma y)\, \mathrm{d}y\,.$$

In the limit $\sigma \to 0$, the above integral remains well defined and gives the result $F(0)$. As $\sigma \to 0$, $g_\sigma(x)$ "tends" to $\delta(x)$.

c. Square of a "sine cardinal" function (Fig. B.1c): $\sin^2(xY)/(\pi x^2 Y)$, where $Y \to \infty$. Since

$$\int_{-\infty}^{+\infty} \frac{\sin^2 x}{x^2}\,\mathrm{d}x = \pi\,,$$

we have

$$\int_{-\infty}^{+\infty} \frac{\sin^2 xY}{\pi x^2 Y}\,\mathrm{d}x = 1\,.$$

d. "Sine cardinal" function (Fig. B.1d): $\sin(xY)/(\pi x)$, where $Y \to \infty$. Since

$$\int_{-\infty}^{+\infty} \frac{\sin x}{x}\mathrm{d}x = \pi\,,$$

we have, for all Y,

$$\int_{-\infty}^{+\infty} \frac{\sin xY}{\pi x}\mathrm{d}x = 1\,.$$

The last case differs from the previous examples in the sense that for $x \neq 0$, the function $\sin(xY)/(\pi x)$ does not tend to zero in the sense of a function as $Y \to \infty$. Instead, it oscillates more and more rapidly. It is only "on the average" that it vanishes.

1.3 Properties of $\delta(x)$

a. $\delta(x)$ is an even function: $\delta(x - x_0) = \delta(x_0 - x)$ (just make a change of variables in (B.3)).

b. We have $\delta(ax) = \dfrac{1}{|a|}\delta(x)$ (a real). Indeed, we find, for $a > 0$,

$$\int_{-\infty}^{+\infty} F(x)\,\delta(ax)\,\mathrm{d}x = \int_{-\infty}^{+\infty} F(u/a)\,\delta(u)\,\frac{\mathrm{d}u}{a} = \frac{1}{a}F(0)\,.$$

For $a < 0$, we use the fact that δ is even.

2 Distributions

The ideas outlined above can be formalized rigorously using distribution theory. We sketch this theory in order to extract some useful results.

2.1 The Space S

We shall consider the vector space S whose elements are complex valued functions $\varphi(x)$ of one (or several) real variables x and which satisfy the following conditions: the functions $\varphi(x)$ are infinitely differentiable and, as x tends to infinity, they and all their derivatives tend to zero, more rapidly than any power of $1/|x|$. For example, the functions e^{-x^2} and $x^n\mathrm{e}^{-x^2}$ are elements of S.

2.2 Linear Functionals

A *continuous linear functional* f on the space S is a mapping of S onto the complex numbers ($f : S \to \boldsymbol{C}$), such that to each φ in S, there corresponds a complex number, denoted by (f, φ). This mapping has the following properties:

a. Linearity:

$$(f, \alpha_1\varphi_1 + \alpha_2\varphi_2) = \alpha_1(f, \varphi_1) + \alpha_2(f, \varphi_2)\,, \tag{B.6}$$

whatever the complex numbers α_1 and α_2 and the functions φ_1 and φ_2 belonging to S.

b. Continuity: if the sequence of functions $\varphi_1, \varphi_2, \ldots, \varphi_n$ in S tends to zero, the sequence of numbers $(f, \varphi_1), (f, \varphi_2), \ldots, (f, \varphi_n)$ tends to zero. N.B. We say that the sequence φ_n tends to zero if $x^k(\mathrm{d}/\mathrm{d}x)^{k'}\varphi_n$ tends to zero uniformly in x whatever the integers k and k', provided that they are positive or zero.

These functionals are called *tempered distributions* and their set is called S'.

Examples

a. Let $f(x)$ be a locally integrable function which remains bounded by a power of $|x|$ as $|x| \to \infty$. We can associate with it a functional, also denoted by f, using the formula

$$(f, \varphi) = \int f(x)\,\varphi(x)\,\mathrm{d}x \qquad \text{for any } \varphi \text{ in } S\,. \tag{B.7}$$

b. The Dirac δ distribution is the functional which associates the number $\varphi(0)$ with any function $\varphi(x)$ in S. This is written as

$$(\delta, \varphi) = \varphi(0)\,. \tag{B.8}$$

It is convenient for physicists (but improper) to write

$$\int \delta(x)\,\varphi(x)\,\mathrm{d}x = \varphi(0)\,.$$

In the examples in the previous section, the statement that y_ε or g_σ tends to δ is incorrect. However, the statement

$$(g_\sigma, \varphi) \underset{\sigma\to 0}{\to} (\delta, \varphi) \qquad \text{for any } \varphi \text{ in } S \tag{B.9}$$

is perfectly correct. We say that g_σ (or $y_\varepsilon, \ldots$) tends to δ *in the sense of distributions.*

2.3 Derivative of a Distribution

When the distribution is associated with a differentiable function $f(x)$, as in (B.7), we can write the following, all operations being legitimate:

$$(f', \varphi) = \int_{-\infty}^{+\infty} \frac{\mathrm{d}f(x)}{\mathrm{d}x}\,\varphi(x)\,\mathrm{d}x = -\int_{-\infty}^{+\infty} f(x)\,\frac{\mathrm{d}\varphi}{\mathrm{d}x}\,\mathrm{d}x = -(f, \varphi')\,.$$

We define the *derivative* $\mathrm{d}f/\mathrm{d}x$, or f' of an arbitrary linear functional f in the set of tempered distributions by the relation

$$\left(\frac{\mathrm{d}f}{\mathrm{d}x}, \varphi\right) = -\left(f, \frac{\mathrm{d}\varphi}{\mathrm{d}x}\right)\,. \tag{B.10}$$

Examples

a. Derivative δ' of δ:

$$(\delta', \varphi) = -(\delta, \varphi') = -\varphi'(0)\,, \tag{B.11}$$

which physicists write as $\int \delta'(x)\,\varphi(x)\,\mathrm{d}x = -\varphi'(0)$.

b. Consider the step function (Heaviside function) defined by

$$\Theta(x) = \begin{cases} 0 & x < 0 \\ 1 & x \geq 0 \end{cases}\,. \tag{B.12}$$

This function is locally integrable, and Θ belongs to the space of tempered distributions. We can calculate its derivative:

$$(\Theta', \varphi) = -(\Theta, \varphi') = -\int_{-\infty}^{+\infty} \Theta(x)\,\varphi'(x)\,\mathrm{d}x = -\int_{0}^{+\infty} \varphi'(x)\,\mathrm{d}x = \varphi(0)\,,$$

and hence the remarkable equality (in the sense of distributions)

$$\frac{\mathrm{d}\Theta(x)}{\mathrm{d}x} = \delta(x)\,. \tag{B.13}$$

c. In three dimensions, we have (see exercise B.1):

$$\Delta(1/r) = -4\pi\delta(\boldsymbol{r})\,. \tag{B.14}$$

2.4 Convolution Product

The *convolution product* of two integrable functions f and g is the function

$$h(x) = \int f(y)\, g(x-y)\, \mathrm{d}y\,; \tag{B.15}$$

this can be written $h = f * g$. By making the change of variables $u = x - y$, we obviously obtain $f * g = g * f$: the convolution product is commutative.

The natural framework for the convolution product is the space of distributions. A distribution h is associated with the function $h(x)$ above such that

$$(h, \varphi) = \int h(x)\, \varphi(x)\, \mathrm{d}x = \iint f(y)\, g(z)\, \varphi(y+z)\, \mathrm{d}y\, \mathrm{d}z\,.$$

We define the convolution product $f * g$ of two distributions f and g by

$$(f * g, \varphi) = (f(x)g(y), \varphi(x+y))\,. \tag{B.16}$$

With this definition, one can check that the Dirac distribution is the identity in the convolution algebra: $\delta * f = f$.

3 Fourier Transformation

We state here a few properties of the Fourier transformation, which is an essential tool both in physics and mathematics.

3.1 Definition

Consider a function $f(k)$ in the above space S. We call the function $g(x)$ defined by the integral

$$g(x) = \frac{1}{\sqrt{2\pi}} \int_{-\infty}^{+\infty} f(k)\, \mathrm{e}^{\mathrm{i}kx}\, \mathrm{d}k \tag{B.17}$$

the Fourier transform of the function $f(k)$. From the definition of S, this integral exists and is infinitely differentiable with respect to x. The generalization to several variables $\boldsymbol{x} = (x_1, \dots, x_n)$, $\boldsymbol{k} = (k_1, \dots, k_n)$ is obvious:

$$g(\boldsymbol{x}) = \frac{1}{\left(\sqrt{2\pi}\right)^n} \int f(\boldsymbol{k})\, \mathrm{e}^{\mathrm{i}\boldsymbol{k}\cdot\boldsymbol{x}}\, \mathrm{d}^n k\,. \tag{B.18}$$

3.2 Fourier Transform of a Gaussian

Consider the particular case of a Gaussian function:

$$f(k) = \frac{\mathrm{e}^{-k^2/(2\sigma^2)}}{\sigma\sqrt{2\pi}} \quad \Rightarrow \quad g(x) = \frac{1}{\sqrt{2\pi}} \int_{-\infty}^{+\infty} \frac{\mathrm{e}^{-k^2/(2\sigma^2)}}{\sigma\sqrt{2\pi}} \, \mathrm{e}^{\mathrm{i}kx} \, \mathrm{d}k \,. \tag{B.19}$$

This integral can be calculated in several ways. For instance, we can note that $g(x)$ satisfies the differential equation

$$g'(x) + x\sigma^2 g(x) = 0$$

and that $g(0) = 1/\sqrt{2\pi}$, and therefore

$$g(x) = \frac{\mathrm{e}^{-x^2\sigma^2/2}}{\sqrt{2\pi}} \,. \tag{B.20}$$

The Fourier transform of a Gaussian is also a Gaussian. We notice that

$$\frac{1}{\sqrt{2\pi}} \int \mathrm{e}^{-\mathrm{i}kx} \, g(x) \, \mathrm{d}x = \frac{1}{\sqrt{2\pi}} \int \mathrm{e}^{-\mathrm{i}kx} \, \frac{\mathrm{e}^{-x^2\sigma^2/2}}{\sqrt{2\pi}} \, \mathrm{d}x = \frac{\mathrm{e}^{-k^2/(2\sigma^2)}}{\sigma\sqrt{2\pi}} = f(k) \,. \tag{B.21}$$

3.3 Inversion of the Fourier Transformation

The Fourier transformation $f(k) \to g(x)$ is a mapping of $S(k)$ in $S(x)$. This transformation can be inverted as follows:

$$f(k) = \frac{1}{\sqrt{2\pi}} \int_{-\infty}^{+\infty} \mathrm{e}^{-\mathrm{i}kx} \, g(x) \, \mathrm{d}x \,. \tag{B.22}$$

In order to demonstrate this result, we consider the integral

$$\begin{aligned} h_\sigma(k) &= \frac{1}{\sqrt{2\pi}} \int_{-\infty}^{+\infty} \mathrm{e}^{-\mathrm{i}kx} \, \mathrm{e}^{-x^2\sigma^2/2} \, g(x) \, \mathrm{d}x \\ &= \frac{1}{\sqrt{2\pi}} \int_{-\infty}^{+\infty} \mathrm{e}^{-\mathrm{i}kx} \, \frac{\mathrm{e}^{-x^2\sigma^2/2}}{\sqrt{2\pi}} \left(\int_{-\infty}^{+\infty} \mathrm{e}^{\mathrm{i}xk'} \, f(k') \, \mathrm{d}k' \right) \mathrm{d}x \,. \end{aligned}$$

In the limit $\sigma \to 0$ we recover the right-hand side of (B.22); furthermore, in the double integral above, for $\sigma \neq 0$, all integrals are absolutely convergent and we can interchange the integrations. After integrating over x and using the result (B.21), we obtain

$$h_\sigma(k) = \frac{1}{\sigma\sqrt{2\pi}} \int_{-\infty}^{+\infty} \mathrm{e}^{-(k'-k)^2/2\sigma^2} \, f(k') \, \mathrm{d}k' \,.$$

By a change of variable $y = (k' - k)/\sigma$, we obtain

$$h_\sigma(k) = \frac{1}{\sqrt{2\pi}} \int_{-\infty}^{+\infty} \mathrm{e}^{-y^2/2} f(k+\sigma y)\,\mathrm{d}y\,.$$

In the limit $\sigma \to 0$, we obtain the desired result:

$$\lim_{\sigma\to 0} h_\sigma(k) = \frac{1}{\sqrt{2\pi}} \int_{-\infty}^{+\infty} \mathrm{e}^{-y^2/2}\, f(k)\,\mathrm{d}y = f(k)\,.$$

Therefore the Fourier transformation is a reciprocal transformation:

$$f(k) = \frac{1}{\sqrt{2\pi}} \int_{-\infty}^{+\infty} \mathrm{e}^{-\mathrm{i}kx}\, g(x)\,\mathrm{d}x \longleftrightarrow g(x) = \frac{1}{\sqrt{2\pi}} \int_{-\infty}^{+\infty} \mathrm{e}^{\mathrm{i}kx}\, f(k)\,\mathrm{d}k\,. \tag{B.23}$$

For this reason, we can say indiscriminately that $f(k)$ and $g(x)$ are Fourier transforms of one another.

3.4 Parseval–Plancherel Theorem

Consider two functions $f_1(k)$ and $f_2(k)$ in S, and their Fourier transforms $g_1(x)$ and $g_2(x)$. A fundamental property of the Fourier transformation is the Parseval–Plancherel theorem:

$$\int f_1^*(k)\, f_2(k)\,\mathrm{d}k = \int g_1^*(x)\, g_2(x)\,\mathrm{d}x\,. \tag{B.24}$$

Using the definition of $g_2(x)$, we obtain

$$\begin{aligned}\int g_1^*(x)\, g_2(x)\,\mathrm{d}x &= \frac{1}{\sqrt{2\pi}} \int g_1^*(x) \left(\int \mathrm{e}^{\mathrm{i}kx}\, f_2(k)\,\mathrm{d}k\right) \mathrm{d}x \\ &= \frac{1}{\sqrt{2\pi}} \iint \mathrm{e}^{\mathrm{i}kx}\, g_1^*(x)\, f_2(k)\,\mathrm{d}k\,\mathrm{d}x\,.\end{aligned}$$

On the other hand, from the definition of $f_1(k)$ (and therefore of f_1^*, which is its complex conjugate), we find

$$\begin{aligned}\int f_1^*(k)\, f_2(k)\,\mathrm{d}k &= \frac{1}{\sqrt{2\pi}} \int \left(\int \mathrm{e}^{\mathrm{i}kx}\, g_1^*(x)\,\mathrm{d}x\right) f_2(k)\,\mathrm{d}k \\ &= \frac{1}{\sqrt{2\pi}} \iint \mathrm{e}^{\mathrm{i}kx}\, g_1^*(x)\, f_2(k)\,\mathrm{d}k\,\mathrm{d}x\,,\end{aligned}$$

which proves the result.

We can introduce, in the space S, a scalar product defined by

$$\langle f_1, f_2\rangle = \int_{-\infty}^{+\infty} f_1^*(k)\, f_2(k)\,\mathrm{d}k\,, \tag{B.25}$$

with which we associate the norm $\|f\| = \sqrt{\langle f, f\rangle}$. This scalar product is invariant under a Fourier transformation (i.e. $\langle f_1, f_2\rangle = \langle g_1, g_2\rangle$), which is called an isometric transformation.

3.5 Fourier Transform of a Distribution

One can extend the definition of the Fourier transformation to the space S' of distributions. We note here a particularly useful formula, which allows us to recover quickly all the important results, such as (B.22) and (B.24). If, in (B.19), we let σ tend to zero, the function $f(k)$ tends, as we have seen, to the Dirac distribution $\delta(k)$. Therefore we find, in this limit,

$$\delta(k) = \frac{1}{2\pi} \int_{-\infty}^{+\infty} \mathrm{e}^{-ikx}\, \mathrm{d}x\,. \tag{B.26}$$

In the calculation of the inverse Fourier transform (B.22), we actually used this limit. The Dirac distribution $\delta(x)$ is the Fourier transform of the constant $(2\pi)^{-1/2}$. We can recover, using another limiting procedure, the result of Sect. 1.2, example 4:

$$\delta(x) = \lim_{Y\to\infty} \frac{1}{2\pi} \int_{-Y}^{+Y} \mathrm{e}^{+ikx}\, \mathrm{d}k = \lim_{Y\to\infty} \frac{\sin(xY)}{\pi x}\,.$$

Some one-dimensional Fourier transforms are given in Table B.1.

Table B.1. Fourier transforms of some commonly encountered functions

function or distribution $f(k)$	Fourier transform $g(x)$
$(\mathrm{d}/\mathrm{d}k)^n\, f(k)$	$(-\mathrm{i}x)^n g(x)$
$k^n f(k)$	$[-\mathrm{i}(\mathrm{d}/\mathrm{d}x)]^n\, g(x)$
$f(ak)$	$(1/\lvert a\rvert) g(x/a)$
$\mathrm{e}^{\mathrm{i}x_0 k} f(k)$	$g(x+x_0)$
$f(k+k_0)$	$\mathrm{e}^{-\mathrm{i}k_0 x} g(x)$
$\delta(k)$	$1/\sqrt{2\pi}$
1	$\sqrt{2\pi}\,\delta(x)$
$\mathrm{e}^{-k^2/(2\sigma^2)}$	$\sigma\,\mathrm{e}^{-x^2\sigma^2/2}$

We also see that the Fourier transform of a product of two functions $f_1(k) f_2(k)$ is proportional to the *convolution product* of the Fourier transforms $g_1(x)$ and $g_2(x)$:

$$f_1(k)\, f_2(k) \xleftrightarrow{\text{F.T.}} \frac{1}{\sqrt{2\pi}} \int g_1(x')\, g_2(x-x')\, \mathrm{d}x' = \frac{1}{\sqrt{2\pi}}\, g_1 * g_2\,.$$

This can be proved using (B.26).

3.6 Uncertainty Relation

Consider a function $f(k)$ and assume that $|f(k)|^2$ is the probability law for the random variable k. The function is therefore normalized as follows:

$$\int |f(k)|^2 \,\mathrm{d}k = 1 \,. \tag{B.27}$$

With this probability law, we can define the expectation value $\langle k \rangle$ of k. By a change of variable $k = \langle k \rangle + q$, we can shift to the centered variable q, with zero expectation value. We assume this has been done and refer to the centered variable as k in the following treatment. The mean square deviation Δk of k is therefore given by

$$(\Delta k)^2 = \int k^2 \,|f(k)|^2 \,\mathrm{d}k \,. \tag{B.28}$$

Because of the isometry, the Fourier transform $g(x)$ of $f(k)$ satisfies $\int |g(x)|^2 \,\mathrm{d}x = 1$. The function $|g(x)|^2$ can be considered as the probability law for the random variable x. This variable will be centered if $\langle x \rangle = \int x \,|g(x)|^2 \,\mathrm{d}x = 0$. If this is not the case, we can, by a change of variables, switch to a centered variable (a *translation* in x does not affect $|f(k)|^2$.) The mean square deviation Δx of x is given by

$$(\Delta x)^2 = \int x^2 \,|g(x)|^2 \,\mathrm{d}x \,. \tag{B.29}$$

We have the following theorem:

Whatever the function f, we have the inequality

$$\Delta x \, \Delta k \geq 1/2 \,. \tag{B.30}$$

The equality occurs only for Gaussian functions.

Proof. Consider the integral $I(\lambda) = \int |kf(k) + \lambda(\mathrm{d}f\mathrm{d}k)|^2 \,\mathrm{d}k$, where λ is a real number. We have

$$I(\lambda) = \int k^2 \,|f(k)|^2 \,\mathrm{d}k + \lambda \int k \left(f^* \frac{\mathrm{d}f}{\mathrm{d}k} + \frac{\mathrm{d}f^*}{\mathrm{d}k} f \right) \mathrm{d}k + \lambda^2 \int \left| \frac{\mathrm{d}f}{\mathrm{d}k} \right|^2 \mathrm{d}k \,.$$

The first term is equal to $(\Delta k)^2$. The second term, after integration by parts, gives

$$\int_{-\infty}^{+\infty} k \,\frac{\mathrm{d}|f|^2}{\mathrm{d}k} \,\mathrm{d}k = - \int_{-\infty}^{+\infty} |f|^2 \,\mathrm{d}k = -1 \,.$$

As for the third term, we notice that $\mathrm{d}f/\mathrm{d}k$ is the Fourier transform of

$-ixg(x)$. Owing to (B.24), we therefore have

$$\int \left| \frac{\mathrm{d}f}{\mathrm{d}k} \right|^2 \mathrm{d}k = \int x^2 |g(x)|^2 \, \mathrm{d}x = (\Delta x)^2 \, .$$

We obtain, finally, $I(\lambda) = (\Delta k)^2 - \lambda + \lambda^2 (\Delta x)^2$. But $I(\lambda)$ is positive for all λ, which is possible only if $1 - 4(\Delta x)^2 (\Delta k)^2 \leq 0$. This is nothing but the inequality (B.30).

The inequality (B.30), called an *uncertainty relation*, shows that when the "width" of a function is large, the "width" of its Fourier transform is narrow. It can be shown that in the case of a Gaussian, and only in that case, we obtain saturation of the inequality, i. e. $\Delta x \, \Delta k = 1/2$.

Exercises

B.1. Laplacian operator in three dimensions. By applying (B.10), show that, in three dimensions, $\Delta(1/r) = -4\pi\delta(\boldsymbol{r})$.

B.2. Fourier transform and complex conjugation. We choose a function $f(k)$ in the space S and denote its Fourier transform by $g(x)$. Determine the Fourier transform of $f^*(k)$. Characterize the properties of $g(x)$ when $f(k)$ is real and symmetric (i. e. $f(k) = f(-k)$).

Appendix C. Operators in Infinite-Dimensional Spaces

It is not wise for everyone to read the following pages.
Only a few will safely enjoy these bitter fruits.
Lautréamont

1 Matrix Elements of an Operator

Here, we extend to infinite-dimensional spaces the concept of "associativity", which we have already established for finite-dimensional spaces:

$$\langle\phi|\hat{A}|\psi\rangle = \langle\phi|\left(\hat{A}|\psi\rangle\right) = \left(\langle\phi|\hat{A}\right)|\psi\rangle\,.$$

We first define the action of $\hat{A}$ on a bra $\langle\phi|$. In order to do this, we remark that the mapping

$$|\psi\rangle \longrightarrow \langle\phi|\left(\hat{A}|\psi\rangle)\right)$$

is a linear form. Therefore there exists a ket $|\chi\rangle$ and its associated bra $\langle\chi|$ which correspond to this linear form:

$$\langle\chi|\psi\rangle = \langle\phi|\left(\hat{A}|\psi\rangle)\right) \qquad \text{for any } |\psi\rangle \text{ in } \mathcal{E}_{\mathrm{H}}\,. \tag{C.1}$$

This enables us to define the action of $\hat{A}$ on the bra $\langle\phi|$. We set, by definition,

$$\langle\chi| = (\langle\phi|\hat{A})\,.$$

We now obtain the desired associativity since (C.1) can be written as

$$\left(\langle\phi|\hat{A}\right)|\psi\rangle = \langle\phi|\left(\hat{A}|\psi\rangle\right)\,. \tag{C.2}$$

We therefore write this expression as $\langle\phi|\hat{A}|\psi\rangle$, as in the finite-dimensional case.

Examples. Consider the following operators in the space $\mathcal{L}^2(R)$:

- $\hat{A} = \hat{x}$: *Position operator along* x.
  Equation (C.2) can then be written very simply:

$$\int [x\phi(x)]^* \, \psi(x) \, \mathrm{d}x = \int [\phi(x)]^* [x\psi(x)] \, \mathrm{d}x \, ,$$

  which just means that the action of $\hat{x}$ on the bra $\langle\phi|$ is also a multiplication of the wave function by x.
- $\hat{A} = \hat{p}_x$: *Momentum operator along* x.
  Integrating by parts, we find

$$\int \left(\frac{\hbar}{\mathrm{i}}\frac{\mathrm{d}\phi}{\mathrm{d}x}\right)^* \psi(x) \, \mathrm{d}x = \int \phi^*(x) \left(\frac{\hbar}{\mathrm{i}}\frac{\mathrm{d}\psi}{\mathrm{d}x}\right) \, \mathrm{d}x \, ,$$

  which means that the action of $\hat{p}_x$ on the bra $\langle\phi|$ corresponds to $\hbar/\mathrm{i}$ times the operation of *differentiation with respect to* x.

These two examples show the limitations of Dirac's notation and of the associativity described above in infinite-dimensional spaces. For instance, in the second case, nothing guarantees that it is legitimate to integrate by parts and that the integrals converge. It may happen that the derivative of $\psi(x)$ exists but not that of $\phi(x)$. In that case, the right-hand side of (C.2) is meaningful, but not the left-hand side. These problems arise because, in infinite-dimensional spaces, even operators as trivial as x or $\partial/\partial x$ are not defined on *all* the space $\mathcal{E}_\mathrm{H}$.

2 Continuous Bases

In an infinite-dimensional Hilbert space, it happens that, for some operators $\hat{A}$, the eigenvalue problem $\hat{A}|\alpha\rangle = a_\alpha|\alpha\rangle$ has no solutions, because no vector of the space $\mathcal{E}_\mathrm{H}$ can satisfy this equation. For instance, in the space $\mathcal{L}^2$, this is the case for the following operators:

a. *The momentum operator* $p_x = -\mathrm{i}\hbar(\partial\,/\,\partial x)$. The solution of

$$\hat{p}_x \psi_{p_0}(x) = p_0 \psi_{p_0}(x) \, ,$$

where the eigenvalue p_0 is any real number, is a plane wave,

$$\psi_{p_0}(x) = A \, \exp(\mathrm{i}p_0 x/\hbar) \, , \tag{C.3}$$

which is not square integrable.

b. *The position operator.* This operator corresponds to multiplying the wave function $\psi(x)$ by x. The eigenvalue equation is

$$x\,\psi_{x_0}(x) = x_0\,\psi_{x_0}(x)\,,$$

where the eigenvalue x_0 is any real number. This equation cannot be satisfied by a square-integrable function, but it is satisfied by a Dirac distribution,

$$\psi_{x_0}(x) = B\,\delta(x - x_0)\,. \tag{C.4}$$

It is therefore necessary to *generalize* the eigenvalue problem to vectors which do not belong to the Hilbert space. Physicists retain the same language, although it is mathematically incorrect. A rigorous mathematical formulation would be very heavy. It is convenient, at our level, to continue using the improper language. This does not mean that mathematical rigor is useless, but simply that the cases we shall consider are not too pathological.

We shall call α an eigenvalue of $\hat{A}$ if it is such that $\hat{A}|v_\alpha\rangle = \alpha|v_\alpha\rangle$, even if $|v_\alpha\rangle$ does not belong to the space $\mathcal{E}_{\mathrm{H}}$, provided that

$$\int C(\alpha)|v_\alpha\rangle\,\mathrm{d}\alpha \text{ belongs to } \mathcal{E}_{\mathrm{H}} \quad \text{if} \quad \int |C(\alpha)|^2\,d\alpha < \infty\,.$$

In the two examples above, this condition is fulfilled. Indeed, any square-integrable function $\psi(x)$ can be written as

$$\psi(x) = \int \mathrm{e}^{\mathrm{i}p_0 x/\hbar}\,\varphi(p_0)\,\mathrm{d}p_0 \quad \text{(Fourier transformation)},$$

$$\psi(x) = \int \delta(x - x_0)\,\psi(x_0)\,\mathrm{d}x_0 \quad \text{(definition of Dirac distribution)}.$$

Consider the operator $\hat{Z} = \hat{p}\hat{x}^3 + \hat{x}^3\hat{p}$, which is apparently symmetric, and would be Hermitian if $\hat{x}$ and $\hat{p}$ were finite-dimensional matrices. Inserting the expression $\hat{p} = -\mathrm{i}\hbar(\mathrm{d}/\mathrm{d}x)$, one can check that the square-integrable function $\phi(x) = \lambda x^{-3/2}\exp(-a^2/4x^2)$, where a is real and λ is a normalization constant, satisfies the eigenvalue equation for $\hat{Z}$ with a pure imaginary eigenvalue $z_0 = -4\mathrm{i}a^2/\hbar$. The operators $\hat{x}$ and $\hat{p}$ are "good" self-adjoint operators, in the sense that their domain (i. e. the functions they may be applied to, the results belonging to $\mathcal{E}_{\mathrm{H}}$) is *dense* in $\mathcal{E}_{\mathrm{H}}$, which is not the case for the operator $\hat{Z}$ above.

The orthogonality relation of two such "eigenvectors" $|v_\alpha\rangle$ and $|v_{\alpha'}\rangle$ becomes (up to a normalization constant)

$$\langle v_\alpha|v_{\alpha'}\rangle = \delta(\alpha - \alpha')\,. \tag{C.5}$$

The vectors $|v_\alpha\rangle$ belong to spaces larger than $\mathcal{L}^2$; they are eigendistributions and not eigenfunctions.

Let us come back to the specific case of position and momentum operators. We denote by $|p\rangle$ the eigenvector of $\hat{p}_x$ corresponding to the eigenvalue p (p

real) and denote by $|x\rangle$ the eigenvector of $\hat{x}$ corresponding to the eigenvalue x (x real). From (C.5), we have

$$\langle p_0|p_1\rangle = \delta(p_0 - p_1)\,, \qquad \langle x_0|x_1\rangle = \delta(x_0 - x_1)\,.$$

We can check these results directly, by returning to the results (C.3) and (C.4). If we choose $A = 1/\sqrt{2\pi\hbar}$ and $B = 1$, we have

$$\langle x_0|x_1\rangle = \int \psi^*_{x_0}(x)\psi_{x_1}(x)\,\mathrm{d}x = \int \delta(x - x_0)\,\delta(x - x_1)\,\mathrm{d}x = \delta(x_0 - x_1)\,,$$
$$\langle p_0|p_1\rangle = \int \psi^*_{p_0}(x)\psi_{p_1}(x)\,\mathrm{d}x = \frac{1}{2\pi\hbar}\int \mathrm{e}^{\mathrm{i}(p_1-p_0)x/\hbar}\,\mathrm{d}x = \delta(p_0 - p_1)\,.$$

Consider now a vector $|\psi\rangle$ of $\mathcal{L}^2$. We calculate the scalar products $\langle x_0|\psi\rangle$ and $\langle p_0|\psi\rangle$,

$$\langle x_0|\psi\rangle = \int \psi^*_{x_0}(x)\,\psi(x)\,\mathrm{d}x = \int \delta(x - x_0)\,\psi(x)\,\mathrm{d}x\,,$$
$$\langle p_0|\psi\rangle = \int \psi^*_{p_0}(x)\,\psi(x)\,\mathrm{d}x = \int \frac{1}{\sqrt{2\pi\hbar}}\mathrm{e}^{-\mathrm{i}p_0x/\hbar}\,\psi(x)\,\mathrm{d}x\,,$$

and obtain the very simple result

$$\langle x_0|\psi\rangle = \psi(x_0)\,, \tag{C.6}$$
$$\langle p_0|\psi\rangle = \varphi(p_0)\,, \tag{C.7}$$

where $\varphi(p)$ is the Fourier transform of $\psi(x)$.

We take now the operator $\hat{O} = \int |x\rangle\langle x|\,\mathrm{d}x$. For two vectors $|\psi\rangle$ and $|\chi\rangle$ of $\mathcal{E}_{\mathrm{H}}$, we consider the quantity $\langle\psi|\hat{O}|\chi\rangle$. Using (C.6), we find

$$\langle\psi|\hat{O}|\chi\rangle = \int \langle\psi|x\rangle\langle x|\chi\rangle\,\mathrm{d}x = \int \psi^*(x)\chi(x)\,\mathrm{d}x = \langle\psi|\chi\rangle\,.$$

This relation, valid for any pair $|\psi\rangle, |\chi\rangle$, yields the result $\hat{O} = I_{\mathrm{H}}$. Therefore, the closure relation in the continuous basis $\{|x\rangle\}$ can be written as

$$\int |x\rangle\langle x|\,\mathrm{d}x = I_{\mathrm{H}}\,.$$

Similarly, for the continuous basis $\{|p\rangle\}$, we obtain

$$\int |p\rangle\langle p|\,\mathrm{d}p = I_{\mathrm{H}}\,.$$

These two identities replace the relation (5.25) that we had in the case of countable bases.

Similarly, the results

$$|\psi\rangle = \int |x\rangle\langle x|\psi\rangle\,\mathrm{d}x = \int \psi(x)\,|x\rangle\,\mathrm{d}x$$
$$= \int |p\rangle\langle p|\psi\rangle\,\mathrm{d}p = \int \varphi(p)\,|p\rangle\,\mathrm{d}p$$

generalize the decomposition

$$|\psi\rangle = \sum_n C_n |n\rangle \,, \qquad C_n = \langle n|\psi\rangle \,.$$

The numbers $\psi(x)$ and $\varphi(p)$ can be interpreted as the components of the vector $|\psi\rangle$ in the continuous bases $|x\rangle$ and $|p\rangle$.

In this book we call the quantity $\psi(x)$ the *wave function*. It is now clear that this is just one particular representation of the state vector $|\psi\rangle$, which has no more intrinsic value than any other (for instance $\varphi(p)$ or C_n). The function $\psi(x)$ is more convenient for describing the position in space of the particle; similarly, $\varphi(p)$ is more convenient for describing its momentum. It follows that one frequently calls any representation of the state vector $|\psi\rangle$ the wave function. However, in order to avoid any confusion, in this book we give the name "wave function" only to the representation $\psi(x)$ associated with the position basis.

Finally, the transformation matrix from the x basis to the p basis is given by

$$\langle x_0|p_0\rangle = \int \psi^*_{x_0}(x)\psi_{p_0}(x)\,\mathrm{d}x = \int \delta(x-x_0)\,\frac{1}{\sqrt{2\pi\hbar}}\mathrm{e}^{\mathrm{i}p_0x/\hbar}\,\mathrm{d}x \,;$$

in other words,

$$\langle x_0|p_0\rangle = \frac{1}{\sqrt{2\pi\hbar}}\,\mathrm{e}^{\mathrm{i}p_0x_0/\hbar} \,.$$

Remarks

a. In the above notation, the Fourier transformation appears as a change of basis:

$$\psi(x) = \langle x|\psi\rangle = \int \langle x|p\rangle\langle p|\psi\rangle\,\mathrm{d}p = \int \frac{1}{\sqrt{2\pi\hbar}}\mathrm{e}^{\mathrm{i}px/\hbar}\varphi(p)\,\mathrm{d}x \,,$$

$$\varphi(p) = \langle p|\psi\rangle = \int \mathrm{d}x\,\langle p|x\rangle\langle x|\psi\rangle = \int \frac{1}{\sqrt{2\pi\hbar}}\mathrm{e}^{-\mathrm{i}px/\hbar}\psi(x)\,\mathrm{d}x \,.$$

b. The *position* and *momentum* operators are written, in this language, as

$$\hat{x} = \int x\;|x\rangle\langle x|\;\mathrm{d}x \,, \qquad \hat{p} = \int p\;|p\rangle\langle p|\;\mathrm{d}p \,.$$

Appendix D. The Density Operator

Alas! I only found a dreadful mixture.
Jean Racine, Athalie

The formulation of quantum mechanics that we have used in this book relies implicitly on the principle that one can prepare a system in a well defined state vector $|\psi\rangle$ of the Hilbert space. Knowing $|\psi\rangle$ we can then calculate the probabilities for all possible results of a given measurement. Such states are called *pure states.* However, in many physical situations one cannot or does not have such a complete knowledge of the state of the system under consideration. The two following examples illustrate this point.

The first one deals with the Stern–Gerlach experiment. The silver atoms emerging from the oven are *unpolarized.* By this, we mean that *whatever* the orientation of the magnetic field gradient, we always observe two spots of equal intensities on the detecting plate. How can we describe the magnetic or spin state of a silver atom in this situation? Following the line of thought developed for pure states, we could attempt to use a state of the type

$$|\psi\rangle = \frac{1}{\sqrt{2}}(|+\rangle + e^{i\phi}|-\rangle)\,. \tag{D.1}$$

With such a state, one can indeed account for the probability 1/2 of finding $\pm\mu_0$ in a measurement of $\hat{\mu}_z$, whatever the value of the phase ϕ. However, (D.1) cannot account for the observations. In fact, a Stern–Gerlach apparatus oriented in the direction $\boldsymbol{u} = \cos\phi\, \boldsymbol{e}_x + \sin\phi\, \boldsymbol{e}_y$ will yield, with probability 1, the sole result $+\mu_0$. This contradicts the observation that *any* orientation of the magnetic gradient yields $\pm\mu_0$ with equal probabilities.

The second example concerns an incomplete measurement. Consider an isotropic harmonic oscillator in two dimensions x and y, with an angular frequency ω. Suppose that we perform a measurement of its energy and find the result $2\,\hbar\omega$. What is the state of the system after this measurement? There is no unique answer to this question, since $|n_x = 1,\, n_y = 0\rangle$, $|n_x = 0,$ $n_y = 1\rangle$ and any linear superposition of these two states are all acceptable choices.

Consequently, it is not always possible to define in an unambiguous manner the state vector of a quantum system. We have to replace this notion by a more general one which can deal with systems that are not well prepared, i. e. for which all the physical quantities associated with a CSCO have not been measured. In order to do so, we shall reformulate the principles of quantum mechanics presented in Chap. 5, in terms of what is called the *density operator.* We shall do so first in cases where the system can indeed be described by a state vector. In such cases, the system is said to be in a *pure state.* We shall then turn to the general situation, corresponding to the two examples presented above, for which the description in terms of a state vector fails. We shall show that the description in terms of a density operator is indeed suitable for such cases, which are referred to as *impure states* or *statistical mixtures.* We shall then give a few examples of applications of the density operator, such as the Wigner distribution and the description of measurements on systems consisting of entangled states of two subsystems.

1 Pure States

We first give the density operator formalism for pure states that we have dealt with throughout this book.

1.1 A Mathematical Tool: the Trace of an Operator

Consider a Hilbert space with an orthonormal basis $\{|n\rangle\}$ and an operator $\hat{A}$ acting in this space. The trace of $\hat{A}$ is defined as

$$\mathrm{Tr}(\hat{A}) = \sum_n \langle n|\hat{A}|n\rangle \,. \tag{D.2}$$

The quantity $\mathrm{Tr}(\hat{A})$ does not depend on the particular choice of the basis $\{|n\rangle\}$. If we consider another orthonormal basis $\{|\bar{n}\rangle\}$, we have

$$\begin{aligned}
\sum_n \langle n|\hat{A}|n\rangle &= \sum_{n,\bar{n},\bar{m}} \langle n|\bar{n}\rangle\, \langle \bar{n}|\hat{A}|\bar{m}\rangle\, \langle \bar{m}|n\rangle \\
&= \sum_{\bar{n},\bar{m}} \langle \bar{n}|\hat{A}|\bar{m}\rangle \left(\sum_n \langle n|\bar{n}\rangle\, \langle \bar{m}|n\rangle \right) \\
&= \sum_{\bar{n},\bar{m}} \langle \bar{n}|\hat{A}|\bar{m}\rangle\, \delta_{\bar{n},\bar{m}} = \sum_{\bar{n}} \langle \bar{n}|\hat{A}|\bar{n}\rangle \,,
\end{aligned}$$

where we have used the closure relations $\sum_n |n\rangle\langle n| = \sum_{\bar{n}} |\bar{n}\rangle\langle\bar{n}| = \hat{1}$.

An important property of the trace is that $\mathrm{Tr}(\hat{A}\hat{B}) = \mathrm{Tr}(\hat{B}\hat{A})$ for any pair of operators $\hat{A}$ and $\hat{B}$, whether or not they commute:

$$\mathrm{Tr}(\hat{A}\,\hat{B}) = \sum_n \langle n|\hat{A}\,\hat{B}|n\rangle = \sum_{n,m} \langle n|\hat{A}|m\rangle\,\langle m|\hat{B}|n\rangle = \sum_m \langle m|\hat{B}\,\hat{A}|m\rangle\,,$$
$$= \mathrm{Tr}(\hat{B}\,\hat{A})\,, \qquad \text{(D.3)}$$

where we have again used the closure relation.

In the following we shall consider particular operators such as $\hat{A} = |\psi\rangle\langle\phi|$, where $|\psi\rangle$ and $|\phi\rangle$ are two states of the Hilbert space. In such a case, we obtain the matrix element of $\hat{B}$:

$$\mathrm{Tr}(\hat{A}\,\hat{B}) = \sum_n \langle n|\;(|\psi\rangle\langle\phi|)\,\hat{B}\;|n\rangle = \sum_n \langle n|\psi\rangle\;\langle\phi|\hat{B}|n\rangle = \langle\phi|\hat{B}|\psi\rangle\,. \qquad \text{(D.4)}$$

In particular, if we take $\hat{B} = \hat{1}$ and $|\psi\rangle = |\phi\rangle$, where $|\psi\rangle$ is a vector of norm 1, we have

$$\mathrm{Tr}(|\psi\rangle\langle\psi|) = \langle\psi|\psi\rangle = 1\,.$$

1.2 The Density Operator of Pure States

Consider a system in the state $|\psi(t)\rangle$. We define the *density operator* $\hat{\rho}(t)$ as

$$\hat{\rho}(t) = |\psi(t)\rangle\langle\psi(t)|\,. \qquad \text{(D.5)}$$

The operator $\hat{\rho}$ is Hermitian and is the projection operator on the state $|\psi\rangle$. From the principles of quantum mechanics presented in Chap. 5, we deduce the following properties:

(1) If we perform a measurement of a physical quantity A, corresponding to the observable $\hat{A}$, the probability of finding the eigenvalue a_α is

$$\mathcal{P}(a_\alpha) = ||\hat{P}_\alpha|\psi\rangle||^2 = \langle\psi|\hat{P}_\alpha|\psi\rangle\,,$$

where $\hat{P}_\alpha$ is the projector on the eigensubspace of $\hat{A}$ corresponding to the eigenvalue a_α. This can also be written

$$\mathcal{P}(a_\alpha) = \mathrm{Tr}\left(\hat{P}_\alpha\hat{\rho}\right)\,. \qquad \text{(D.6)}$$

In particular, the expectation value $\langle a\rangle = \langle\psi|\hat{A}|\psi\rangle$ is equal to

$$\langle a\rangle = \mathrm{Tr}\left(\hat{A}\,\hat{\rho}\right)\,. \qquad \text{(D.7)}$$

(2) Immediately after a measurement yielding the value a_α, the state of the system is $|\psi'\rangle = \hat{P}_\alpha|\psi\rangle/||\hat{P}_\alpha|\psi\rangle||$. The corresponding density operator is therefore

$$\hat{\rho}' = \frac{\hat{P}_\alpha\,\hat{\rho}\,\hat{P}_\alpha}{\mathcal{P}(a_\alpha)}\,. \qquad \text{(D.8)}$$

(3) Let $\hat{H}(t)$ be the Hamiltonian of the system. As long as this system does not undergo any observation, the evolution of the system is given by the Schrödinger equation. Therefore the density operator evolves as

$$\begin{aligned} \mathrm{i}\hbar\frac{\mathrm{d}\hat{\rho}}{\mathrm{d}t} &= \mathrm{i}\hbar\frac{\mathrm{d}|\psi(t)\rangle}{\mathrm{d}t}\langle\psi(t)| + \mathrm{i}\hbar|\psi(t)\rangle\frac{\mathrm{d}\langle\psi(t)|}{\mathrm{d}t} \\ &= \hat{H}|\psi(t)\rangle\langle\psi(t)| - |\psi(t)\rangle\langle\psi(t)|\hat{H}\,, \end{aligned}$$

which is nothing but

$$\mathrm{i}\hbar\frac{\mathrm{d}\hat{\rho}}{\mathrm{d}t} = \left[\hat{H}(t), \hat{\rho}(t)\right]\,. \tag{D.9}$$

1.3 Alternative Formulation of Quantum Mechanics for Pure States

We can reformulate the principles of quantum mechanics state in Chap. 5 in terms of the density operator instead of the state vector:

Principle 1 (Alternative Formulation)

With each physical system one can associate an appropriate Hilbert space $\mathcal{E}_{\mathrm{H}}$. At each time t, the state of the system is completely determined by a density operator $\hat{\rho}(t)$. This operator is Hermitian and satisfies the normalization condition $\mathrm{Tr}(\hat{\rho}(t)) = 1$.

Particular Case

If the system is in a pure state, $\hat{\rho}(t)$ has one eigenvalue equal to 1 and all the other eigenvalues are zero.

Principles 2 and 3

The equations concerning the measurement of physical quantities and the time evolution (5.40), (5.41) and (5.48) are replaced by (D.6), (D.8) and (D.9).

Notice that the assumption concerning the eigenvalues of $\hat{\rho}(t)$ is equivalent to the formulation in terms of a state vector $|\psi\rangle$. Indeed $\hat{\rho}$ can be diagonalized since it is a Hermitian operator. The condition $\mathrm{Tr}(\hat{\rho}) = 1$ implies that

$$\sum_n \Pi_n = 1\,,$$

where the Π_n are the eigenvalues of $\hat{\rho}$. By assumption, all eigenvalues of $\hat{\rho}$ but one are zero. We denote by $\Pi_1 = 1$ the nonzero eigenvalue, and denote the corresponding eigenvector by $|\psi\rangle$. We have, by construction, $\hat{\rho} = |\psi\rangle\langle\psi|$.

At this stage, all we have done is to slightly complicate the principles of Chap. 5. However, we shall show in the next section that this allows us to describe more general situations corresponding to the incompletely prepared systems mentioned in the introduction.

2 Statistical Mixtures

We now turn to the general situation of statistical mixtures or impure states.

2.1 A Particular Case: an Unpolarized Spin-1/2 System

Consider the unpolarized atoms emerging from the oven in a Stern–Gerlach apparatus. As mentioned above, the probabilities for finding $\pm\hbar/2$ in the measurement of $\hat{S}_{\boldsymbol{u}} = \hat{\boldsymbol{S}} \cdot \boldsymbol{u}$ are equal to $1/2$ for any orientation $\boldsymbol{u}$ of the magnet. We have already explained that there is no state vector $|\psi\rangle$ which can account for that result. In contrast, it is easy to find the corresponding density operator, assuming that (D.7) still holds:

$$\hat{\rho}_{\text{unpol.}} = \frac{1}{2}\begin{pmatrix} 1 & 0 \\ 0 & 1 \end{pmatrix} = \frac{1}{2}\,\hat{1} \tag{D.10}$$

(which obviously keeps the same form in any basis). Indeed, one finds in this case $\text{Tr}(\hat{S}_{\boldsymbol{u}}\hat{\rho}_{\text{unpol.}}) = (1/2)\ \text{Tr}(\hat{S}_{\boldsymbol{u}}) = 0$, since the traces of all three Pauli matrices are zero.

The physical meaning of (D.10) is the following. In the eigenbasis $|\pm\rangle$ of $\hat{S}_z$, the density operator $\hat{\rho}$ can be written

$$\hat{\rho} = \frac{1}{2}|+\rangle\langle+| \;+\; \frac{1}{2}|-\rangle\langle-|\,.$$

We can describe the calculation of the expectation value of any operator $\hat{A}$ as follows.

a. Assume that the system is in the state $|+\rangle$ and calculate the corresponding expectation value:
$$\langle a\rangle^{(+)} = \langle+|\hat{A}|+\rangle = \text{Tr}(\hat{A}\ |+\rangle\langle+|)\,.$$
b. Similarly, assume that the system is in the state $|-\rangle$ and calculate $\langle a\rangle^{(-)}$.
c. Average the two results with equal weights:
$$\text{Tr}(\hat{A}\hat{\rho}_{\text{unpol}}) = \frac{1}{2}\langle a\rangle^{(+)} + \frac{1}{2}\langle a\rangle^{(-)}\,.$$

This procedure is very different from the one we would follow if the system was in the pure state $|+\rangle_x = (|+\rangle + |-\rangle)/\sqrt{2}$, for example. In the pure-state case, we work with probability amplitudes, and interference phenomena arise from the fact that we are dealing with a coherent superposition of the $|+\rangle$ and $|-\rangle$ states. In contrast, for an unpolarized spin-1/2 system, we simply average the results associated with the two possible states $|\pm\rangle$, as one would do in a classical probabilistic description. The situation described by (D.10) is often referred to as an incoherent mixture, in contrast to the coherent superposition $|+\rangle_x$, for which the density operator reads

$$\hat{\rho}_{\text{pol. along } x} = \frac{1}{2}\begin{pmatrix} 1 & 1 \\ 1 & 1 \end{pmatrix}\,. \tag{D.11}$$

2.2 The Density Operator for Statistical Mixtures

We can generalize to any quantum system what we have seen for an unpolarized spin-1/2. We assume that the most general density operator $\hat{\rho}$ for a quantum system has the following properties:

Principle 1 (Continued)

General Case

The density operator is Hermitian and satisfies the normalization condition $\mathrm{Tr}(\hat{\rho}(t)) = 1$. All eigenvalues Π_i of $\hat{\rho}$ satisfy

$$0 \leq \Pi_i \leq 1 . \tag{D.12}$$

Principles 2 and 3 are still described mathematically by (D.6)–(D.9).

The physical interpretation of this definition is as follows. Considering an eigenbasis $\{|\psi_i\rangle\}$ of $\hat{\rho}$, we have, by definition,

$$\hat{\rho} = \sum_i \Pi_i \, |\psi_i\rangle\langle\psi_i| .$$

The set of positive numbers Π_i can be interpreted as a probability distribution since $\mathrm{Tr}(\hat{\rho}) = \sum_i \Pi_i = 1$. The calculation of the expectation value of any physical quantity A can be written as

$$\langle a \rangle = \mathrm{Tr}(\hat{A}\hat{\rho}) = \sum_i \Pi_i \, \langle\psi_i|\hat{A}|\psi_i\rangle .$$

Therefore one can first calculate the various expectation values $\langle a \rangle^{(i)}$ assuming that the system is the pure state $|\psi_i\rangle$, and then average over the various $\langle a \rangle^{(i)}$ with the statistical weights Π_i.

Suppose we perform an experiment with one system (e. g. one atom in a Stern–Gerlach experiment) whose state corresponds to a statistical mixture $\hat{\rho} = \sum_i \Pi_i |\psi_i\rangle\langle\psi_i|$. Everything happens as if we were dealing with a state $|\psi_j\rangle$ chosen at random by some "external actor" out of the collection of available states $\{|\psi_i\rangle\}$. The probability that this actor chooses $|\psi_j\rangle$ is Π_j. The whole experimental sequence of measurements (e. g. a series of Stern–Gerlach magnets) on this particular system must be analyzed as if the initial state were the pure state $|\psi_j\rangle$, and no other. Next, the same experiment performed a second time (e. g. with another atom) will correspond to some other state $|\psi_k\rangle$, also picked at random out of the collection $\{|\psi_i\rangle\}$ of eigenstates of $\hat{\rho}$, and so on.

The statistical mixture is completely different from the pure-state situation, where the system is prepared in a coherent superposition, for instance $|\psi\rangle = \sum_i \sqrt{\Pi_i} |\psi_i\rangle$. In this latter case, if we perform the same sequence of measurement on many systems all prepared in the state $|\psi\rangle$, the experiment can reveal interferences between the various probability amplitudes $\sqrt{\Pi_i}$.

3 Examples of Density Operators

There are numerous applications of the density operator formalism. We have chosen a few interesting cases.

3.1 The Micro-Canonical and Canonical Ensembles

Consider a system for which the only available information is the following:

The system is with certainty (i. e. with probability 1) is a given subspace $\mathcal{F}$ of the total Hilbert space $\mathcal{E}$.

For instance, the state of the harmonic oscillator mentioned in the introduction is in the two-dimensional subspace spanned by $\{|n_x = 1, n_y = 0\rangle, |n_x = 0, n_y = 1\rangle\}$ if a measurement of its energy has yielded the result $2\,\hbar\omega$.

If the dimension d of $\mathcal{F}$ is strictly larger than 1, the system is not in a pure state, and it has to be described by a density operator. We postulate that this operator is

$$\hat{\rho} = \frac{1}{d}\hat{P}_{\mathcal{F}}\,, \tag{D.13}$$

where $\hat{P}_{\mathcal{F}}$ is the projector on this subspace. If we introduce a orthonormal basis set $|\psi_i\rangle$ $(i = 1, \ldots, d)$ of $\mathcal{F}$, the density operator reads

$$\hat{\rho} = \frac{1}{d}\sum_{i=1}^{d} |\psi_i\rangle\langle\psi_i|\,. \tag{D.14}$$

This choice, called the microcanonical density operator, is rather intuitive. Since we have no information whatsoever concerning the particular state which is occupied within $\mathcal{F}$, we attribute to all possible states of this subspace equal probabilities. It is a direct generalization of what we did in (D.10) for an unpolarized spin-1/2 particle.

This principle is the basis of quantum statistical physics. Using this principle, one can construct, for instance, the canonical ensemble, corresponding to the description of a system $\mathcal{S}$ interacting very weakly with a large energy reservoir. The density operator of $\mathcal{S}$ is then

$$\hat{\rho} = \frac{\mathrm{e}^{-\beta\hat{H}}}{\mathrm{Tr}(\mathrm{e}^{-\beta\hat{H}})}\,, \tag{D.15}$$

where β is related to the temperature T of the reservoir ($\beta = (k_{\mathrm{B}}T)^{-1}$) and where $\hat{H}$ is the Hamiltonian of $\mathcal{S}$, in absence of coupling to the reservoir.

From (D.15), we can deduce Boltzmann's law. Consider two energies E_n and E_m, which are nondegenerate eigenvalues of $\hat{H}$. In an energy measurement, the respective probabilities P_n and P_m of finding the results E_n and E_m are such that $P_n/P_m = \exp[-(E_n - E_m)/k_{\mathrm{B}}T]$.

3.2 The Wigner Distribution of a Spinless Point Particle

Consider a point particle of mass m and spin zero. The density operator $\hat{\rho}$ describing the state of this particle can be expanded in the continuous basis $|\boldsymbol{r}\rangle$ associated with the position operator. We set

$$\rho(\boldsymbol{r},\boldsymbol{r}') = \langle \boldsymbol{r}|\hat{\rho}|\boldsymbol{r}'\rangle ,$$

which is a complex function of $\boldsymbol{r}$ and $\boldsymbol{r}'$. From the principles outlined above, we deduce that the quantity $\rho(\boldsymbol{r},\boldsymbol{r})$ is positive, and that it gives the probability of finding the particle at $\boldsymbol{r}$ (within d^3r) as

$$\mathrm{d}^3P = \rho(\boldsymbol{r},\boldsymbol{r})\,\mathrm{d}^3r , \qquad \text{and} \qquad \int \rho(\boldsymbol{r},\boldsymbol{r})\,\mathrm{d}^3r = 1 .$$

The Wigner representation $w(\boldsymbol{r},\boldsymbol{p})$ of the density operator $\hat{\rho}$ is defined as

$$w(\boldsymbol{r},\boldsymbol{p}) = \frac{1}{(2\pi\hbar)^3}\int \rho\left(\boldsymbol{r}-\frac{\boldsymbol{u}}{2},\boldsymbol{r}+\frac{\boldsymbol{u}}{2}\right)\mathrm{e}^{\mathrm{i}\boldsymbol{u}\cdot\boldsymbol{p}/\hbar}\,\mathrm{d}^3u , \tag{D.16}$$

where $\boldsymbol{p}$ has the dimensions of a linear momentum. Since $\rho(\boldsymbol{r},\boldsymbol{r}') = \rho^*(\boldsymbol{r}',\boldsymbol{r})$, this quantity satisfies

$$w(\boldsymbol{r},\boldsymbol{p}) \ \ \text{real,} \qquad \iint w(\boldsymbol{r},\boldsymbol{p})\,\mathrm{d}^3r\,\mathrm{d}^3p = 1 . \tag{D.17}$$

The expectation value of a physical quantity depending only on position, $A(\boldsymbol{r})$, or on momentum, $B(\boldsymbol{p})$, has a very simple expression in terms of the Wigner distribution:

$$\langle a\rangle = \mathrm{Tr}\,(A(\hat{\boldsymbol{r}})\,\hat{\rho}) = \iint A(\boldsymbol{r})\,w(\boldsymbol{r},\boldsymbol{p})\,\mathrm{d}^3r\,\mathrm{d}^3p , \tag{D.18}$$

$$\langle b\rangle = \mathrm{Tr}\,(B(\hat{\boldsymbol{p}})\,\hat{\rho}) = \iint B(\boldsymbol{p})\,w(\boldsymbol{r},\boldsymbol{p})\,\mathrm{d}^3r\,\mathrm{d}^3p . \tag{D.19}$$

From (D.17)–(D.19), it would be tempting to infer that $w(\boldsymbol{r},\boldsymbol{p})$ is the phase space density for the particle, i. e. that $w(\boldsymbol{r},\boldsymbol{p})\,\mathrm{d}^3r\,\mathrm{d}^3p$ gives the probability of finding the particle at position $\boldsymbol{r}$ (within d^3r) and with momentum $\boldsymbol{p}$ (within d^3p). This is of course wrong, since such a statement is manifestly meaningless in quantum mechanics if $\mathrm{d}^3r\,\mathrm{d}^3p$ is smaller than $\hbar^3$. Besides, nothing guarantees that $w(\boldsymbol{r},\boldsymbol{p})$ is a positive quantity: actually, one can easily find situations where $w(\boldsymbol{r},\boldsymbol{p})$ is locally negative in some regions of phase space.

Suppose that the particle is moving in the potential $V(\boldsymbol{r})$, so that the Hamiltonian is $\hat{H} = \hat{p}^2/2m + V(\hat{\boldsymbol{r}})$. We leave it as a simple exercise to the reader to show that the equation of motion of $w(\boldsymbol{r},\boldsymbol{p},t)$ deduced from (D.9) can be cast into the form

$$\frac{\partial w}{\partial t} + \frac{\boldsymbol{p}}{m}\cdot\boldsymbol{\nabla}_{\boldsymbol{r}}w = K[w] , \tag{D.20}$$

where

$$K[w] = \int \mathcal{N}(\boldsymbol{r},\boldsymbol{q})\; w(\boldsymbol{r},\boldsymbol{p}-\boldsymbol{q},t)\; \mathrm{d}^3q \tag{D.21}$$

and

$$\mathcal{N}(\boldsymbol{r},\boldsymbol{q}) = \frac{1}{\mathrm{i}\hbar}\,\frac{1}{(2\pi\hbar)^3} \int \left[V\left(\boldsymbol{r}-\frac{\boldsymbol{u}}{2}\right) - V\left(\boldsymbol{r}+\frac{\boldsymbol{u}}{2}\right)\right] \mathrm{e}^{\mathrm{i}\boldsymbol{u}\cdot\boldsymbol{q}/\hbar}\; \mathrm{d}^3u\,. \tag{D.22}$$

The integro-differential equation of motion (D.20), (D.21) for w is more complicated than the classical equivalent for the evolution of the phase-space density. In the classical case, we have the Liouville equation, which has the same structure as (D.20) but where (D.21) is replaced by

$$K[w]_{\text{class.}} = \boldsymbol{\nabla}_{\boldsymbol{r}} V \cdot \boldsymbol{\nabla}_{\boldsymbol{p}} w\,. \tag{D.23}$$

The fact that the classical equation, which is local in $\boldsymbol{r}$ and $\boldsymbol{p}$, is replaced in the quantum world by a nonlocal integro-differential equation is a direct manifestation of the quantum properties that we have investigated throughout this book. For instance, in the double-well problem, because of the nonlocality of (D.22) with respect to $V(\boldsymbol{r})$, a particle initially located in the left-hand well can "feel" the existence of the right-hand well and eventually reach it after some finite amount of time.

There are two cases where the classical and quantum evolutions coincide, at least approximately:

(1) If the potential $V(\boldsymbol{r})$ varies linearly or quadratically with position, then (D.21) and (D.23) are identical. Take, for instance, an isotropic harmonic potential $V(\boldsymbol{r}) = m\omega^2 r^2/2$. The kernel $\mathcal{N}(\boldsymbol{r},\boldsymbol{q})$ is in this case

$$\mathcal{N}(\boldsymbol{r},\boldsymbol{q}) = m\omega^2 \boldsymbol{r}\cdot\boldsymbol{\nabla}_{\boldsymbol{q}}\delta(\boldsymbol{q})\,,$$

where we have used $\delta(\boldsymbol{q}) = (2\pi\hbar)^{-3}\int \mathrm{e}^{\mathrm{i}\boldsymbol{u}\cdot\boldsymbol{q}/\hbar}\,\mathrm{d}^3u$. An integration by parts in (D.21) then yields $K[w] = m\omega^2\boldsymbol{r}\cdot\boldsymbol{\nabla}_{\boldsymbol{p}}w = K[w]_{\text{class.}}$. The equivalence between the evolutions of the classical phase space density and of $w(\boldsymbol{r},\boldsymbol{p})$ for a harmonic oscillator generalizes the result of Sect. 7.3.2, where we demonstrated that the evolutions of $\langle x\rangle$ and $\langle p\rangle$ coincide with the classical prediction in this case.

(2) Suppose that $w(\boldsymbol{r},\boldsymbol{p}-\boldsymbol{q})$ varies smoothly with $\boldsymbol{q}$ in the integral (D.21). We can then set

$$w(\boldsymbol{r},\boldsymbol{p}-\boldsymbol{q}) \simeq w(\boldsymbol{r},\boldsymbol{p}) - (\boldsymbol{q}\cdot\boldsymbol{\nabla}_{\boldsymbol{p}})w\,.$$

Using $\int \mathcal{N}(\boldsymbol{r},\boldsymbol{q})\;\mathrm{d}^3q = 0$ and $\int \boldsymbol{q}\;\mathcal{N}(\boldsymbol{r},\boldsymbol{q})\;\mathrm{d}^3q = -\boldsymbol{\nabla}_{\boldsymbol{r}}V$, we obtain $K[w] \simeq K[w]_{\text{class.}}$. For this approximation to be valid, the width in $\boldsymbol{q}$ of $\mathcal{N}$ must be much smaller than the momentum scale Δp over which $w(\boldsymbol{r},\boldsymbol{p})$ varies appreciably. The $\boldsymbol{q}$ variation of $\mathcal{N}$ is directly related to the Fourier transform of the potential $V(\boldsymbol{r})$. Therefore, if we denote by r_0 the typical scale of variation of $V(\boldsymbol{r})$, this approximation is valid if $r_0 \gg \hbar\,/\,\Delta p$. This generalizes the discussion of the validity of the Ehrenfest theorem (Sect. 7.3.2).

4 Entangled Systems

Entanglement is at the heart of the debates concerning the interpretation of quantum mechanics. The density operator formalism allows for a quantitative description of entangled states, in particular as far as individual measurements are performed.

4.1 Reduced Density Operator

Consider a quantum system $\mathcal{S}$ formed from two subsystems $\mathcal{A}$ and $\mathcal{B}$. The Hilbert space associated with $\mathcal{S}$ is $\mathcal{E}_A \otimes \mathcal{E}_B$. We denote a basis of $\mathcal{E}_A$ by $\{|\psi_n\rangle\}$ and a basis of $\mathcal{E}_B$ by $\{|\phi_m\rangle\}$.

For a given density operator $\hat{\rho}$ of the whole system, we define the reduced density operators $\hat{\rho}_A$ and $\hat{\rho}_B$, acting in $\mathcal{E}_A$ and $\mathcal{E}_B$, respectively, as follows:

$$\langle \psi_n|\hat{\rho}_A|\psi_{n'}\rangle = \sum_m \langle \psi_n; \phi_m|\hat{\rho}|\psi_{n'}; \phi_m\rangle \tag{D.24}$$

$$\langle \phi_m|\hat{\rho}_B|\phi_{m'}\rangle = \sum_n \langle \psi_n; \phi_m|\hat{\rho}|\psi_n; \phi_{m'}\rangle \,. \tag{D.25}$$

These reduced density operators are also called "partial traces" of $\hat{\rho}$ over B and A and are denoted by $\hat{\rho}_A = \mathrm{Tr}_B(\hat{\rho})$ and $\hat{\rho}_B = \mathrm{Tr}_A(\hat{\rho})$.

It is straightforward to check that $\hat{\rho}_A$ and $\hat{\rho}_B$ satisfy the properties of density operators in $\mathcal{E}_A$ and $\mathcal{E}_B$, respectively. They are both Hermitian and their traces are equal to 1, since $\mathrm{Tr}(\hat{\rho}) = 1$. In addition, for any projector $\hat{P}_A$ on a subspace of $\mathcal{E}_A$, we have

$$\mathrm{Tr}\left(\hat{P}_A\, \hat{\rho}_A\right) = \mathrm{Tr}\left((\hat{P}_A \otimes \hat{1}_B)\, \hat{\rho}\right) \geq 0\,,$$

where the last inequality holds because all eigenvalues of $\hat{\rho}$ are positive. We deduce that all eigenvalues of $\hat{\rho}_A$ are positive (and similarly for $\hat{\rho}_B$).

The relevance of $\hat{\rho}_A$ and $\hat{\rho}_B$ appears if when we want to measure a physical quantity associated with only one of the two subsystems. Suppose for instance that the observable of interest is $\hat{A} \otimes \hat{1}_B$. The probability of finding the eigenvalue a_α of $\hat{A}$ is (cf. (D.6))

$$\mathcal{P}(a_\alpha) = \mathrm{Tr}\left((P_\alpha \otimes \hat{1}_B)\, \hat{\rho}\right) = \mathrm{Tr}\left(P_\alpha\, \hat{\rho}_A\right) \,.$$

In particular, the expectation value of A is $\langle a\rangle = \mathrm{Tr}(\hat{A}\, \hat{\rho}_A)$.

To summarize, when a measurement is performed on only the subsystem A, all predictions concerning this measurement can be made by using the reduced density operator $\hat{\rho}_A$.

4.2 Evolution of a Reduced Density Operator

Suppose that the system $\mathcal{S}$ is isolated and that the subsystems $\mathcal{A}$ and $\mathcal{B}$ do not interact with one another. The total Hamiltonian $\hat{H}$ therefore has the form $\hat{H} = \hat{H}_A + \hat{H}_B$, where $\hat{H}_A = \hat{H}_A \otimes \hat{1}_B$ and $\hat{H}_B = \hat{1}_A \otimes \hat{H}_B$ are the Hamiltonians of the systems A and B, respectively. Using the general formula (D.9) for the evolution of the density operator, we readily obtain

$$i\hbar \frac{\mathrm{d}\hat{\rho}_A}{\mathrm{d}t} = [\hat{H}_A, \hat{\rho}_A] \,, \qquad i\hbar \frac{\mathrm{d}\hat{\rho}_B}{\mathrm{d}t} = [\hat{H}_B, \hat{\rho}_B] \,. \tag{D.26}$$

If the two subsystems do not interact, each reduced density operator $\hat{\rho}_A$ or $\hat{\rho}_B$ evolves under the action of the Hamiltonian of the subsystem $\hat{H}_A$ or $\hat{H}_B$, respectively, only.

4.3 Entanglement and Measurement

In general, the density operator $\hat{\rho}$ of the combined system differs from the tensor product $\hat{\rho}_A \otimes \hat{\rho}_B$. Consider for instance a system $\mathcal{S}$ consisting of two spin-1/2 systems, prepared in the singlet state: $|\Psi\rangle = (|+\,;\,-\rangle - |-\,;\,+\rangle)/\sqrt{2}$. The reduced density operators are easily calculated: $\hat{\rho}_A = \hat{1}_A/2$ and $\hat{\rho}_B = \hat{1}_B/2$. This implies that in a measurement performed on $\mathcal{A}$ only (or $\mathcal{B}$ only), everything happens as if the spin $\mathcal{A}$ (or $\mathcal{B}$) were unpolarized. In this particular case, we obtain $\hat{\rho}_A \otimes \hat{\rho}_B = (1/4)\hat{1}_S \neq |\Psi\rangle\langle\Psi|$. Physically, the difference between $\hat{\rho}$ and $\hat{\rho}_A \otimes \hat{\rho}_B$ means that one can infer the initial state $|\Psi\rangle$ of the total system only if one *correlates* the results of measurements on $\mathcal{A}$ and on $\mathcal{B}$.

Finally, let us show that a measurement performed by Bob on the subsystem $\mathcal{B}$, and the ensuing "reduction of the wave packet", has no influence[1] on the expectation value of any possible measurement performed by Alice on the subsystem $\mathcal{A}$. First if Alice measures a physical quantity A on the subsystem $\mathcal{A}$ when Bob has not done anything yet, the expectation value of the result is $\langle a \rangle = \mathrm{Tr}(\hat{A}\,\hat{\rho})$. Suppose now that Bob performs a measurement of a physical quantity B on the subsystem $\mathcal{B}$ *before* Alice measures A. The question is the following: does Bob's measurement change the expectation value of Alice's measurement?

After the measurement of the quantity B on the system $\mathcal{B}$, which has yielded the result b_β, the new density operator of the whole system is (cf. (D.8)

$$\hat{\rho}'(b_\beta) = \frac{(\hat{1}_A \otimes \hat{P}_\beta)\,\hat{\rho}\,(\hat{1}_A \otimes \hat{P}_\beta)}{\mathcal{P}(b_\beta)} \,. \tag{D.27}$$

For a given outcome b_β of the measurement of B, the expectation value of the physical quantity A is $\langle a \rangle_{b_\beta} = \mathrm{Tr}(\hat{A}\,\hat{\rho}'(b_\beta))$. In general, this depends on the result b_β found by Bob, since there may exist correlations between the two subsystems $\mathcal{A}$ and $\mathcal{B}$.

[1] This guarantees that there is no instantaneous transmission of information in an EPR-type experiment, such as the one discussed in Chap. 14.

Suppose that Alice knows that Bob has performed a measurement of B, but she does not know the result b_β. In order to calculate the expectation value of her own measurement, she must average over the possible results of Bob:

$$\langle a \rangle = \sum_{b_\beta} \mathcal{P}(b_\beta)\, \langle a \rangle_{b_\beta} = \sum_{b_\beta} \mathrm{Tr}\left((\hat{A} \otimes P_\beta)\, \hat{\rho} \right) = \mathrm{Tr}\left((\hat{A} \otimes \hat{1}_B)\, \hat{\rho} \right) .$$

This coincides with the prediction for the expectation value of A when Bob has not made any measurement. Therefore, Alice, by measuring a quantity connected only with $\mathcal{A}$, cannot infer whether or not Bob has made a measurement on $\mathcal{B}$.

To summarize the results of this section, when a system $\mathcal{A}$ is isolated during a time interval (t_i, t_f), we may calculate the physical properties of this system (probabilities of the outcomes of a measurement on $\mathcal{A}$) using only the reduced density operator of this system at the initial time t_i and the corresponding evolution of $\hat{\rho}_A$ given by (D.26). This is valid even if $\mathcal{A}$ has interacted with another system $\mathcal{B}$ (or several others) before time t_i and if the total system $\mathcal{A} + \mathcal{B}$ is in an entangled state at time t_i. In contrast, it is only by using the total density operator $\hat{\rho}$ (and not $\hat{\rho}_A \otimes \hat{\rho}_B$) that the correlations between measurements on $\mathcal{A}$ and $\mathcal{B}$ can be accounted for correctly.

Further Reading

- General properties of $\hat{\rho}$: U. Fano, Rev. Mod. Phys. **29**, 74 (1957); D. Ter Haar, Rep. Prog. Phys. **24**, 304 (1961).
- Concerning the Wigner distribution: E.P. Wigner, Phys. Rev. **40**, 749 (1932); T. Takabayasi, Prog. Theor. Phys. **11**, 341 (1954); S.R. De Groot and L. G. Suttorp, *Foundations of Electrodynamics*, North-Holland, Amsterdam (1972).
- For recent developments concerning the experimental measurement of the Wigner distribution of material particles, see M. Freyberger et al., "The art of measuring quantum states", Phys. World, November 1997, p. 41; D. Leibfried, T. Pfau and C. Monroe, "Shadows and mirrors: reconstructing quantum states of atomic motion", Phys. Today, April 1998, p. 22.

Exercises

D.1. Trace of $\hat{\rho}^2$. Show that $\mathrm{Tr}(\hat{\rho}^2) \leq 1$, and that $\mathrm{Tr}(\hat{\rho}^2) = 1$ only for a pure state.

D.2. Evolution of a pure state. Using the result of the above exercise and the evolution equation (D.9), show that a pure state remains a pure state during its time evolution.

D.3. Inequalities related to $\boldsymbol{\rho}$. Show that $|\langle\psi|\hat{\rho}|\phi\rangle|^2 \leq \langle\psi|\hat{\rho}|\psi\rangle\langle\phi|\hat{\rho}|\phi\rangle$ for any pair of states $|\psi\rangle, |\phi\rangle$.

D.4. Density operator of a spin-1/2 system. Show that the density operator of a spin-1/2 system can always be written $\hat{\rho} = \left(a_0\hat{1} + \boldsymbol{a}\cdot\hat{\boldsymbol{\sigma}}\right)/2$, where the σ_i's $(i = x, y, z)$ are the Pauli matrices and the four numbers a_0, a_i $(i = x, y, z)$ are real. Give the constraints on these four numbers. Calculate the average spin $\langle\boldsymbol{S}\rangle$ for this density operator, and identify the two limiting cases of a completely polarized and a completely unpolarized spin.

Solutions to the Exercises

If there is no solution,
then there is no problem.
Fourth Shadok principle

Chapter 1

1.1. Photoelectric effect in metals. a. $h = 4.1 \times 10^{-15}$ eV s.
b. $E = 1.74\,\mathrm{eV}$. c. $\lambda = 0.71\,\mu\mathrm{m}$.

1.2. Photon fluxes. a. $N \sim 1.5 \times 10^{30}$. b. $N \sim 5000$ for a pupil of surface area $12\,\mathrm{mm}^2$.

1.3. Orders of magnitude for de Broglie wavelengths. (a) $\lambda = 0.124\,\mathrm{nm}$; (b) for $E = 0.025\,\mathrm{eV}$, $\lambda = 0.18\,\mathrm{nm}$. Both wavelengths are of the order of atomic sizes or spacings, and can be used in diffraction experiments on crystals or molecules.

1.4. De Broglie relation in the relativistic domain. Such electrons are ultrarelativistic ($E \simeq pc$) and we obtain $\lambda \simeq hc/E \sim 10^{-17}\,\mathrm{m}$. In order to probe matter at distances smaller than one fermi (10^{-15} m), wavelengths smaller than this value are necessary.

Chapter 2

2.1. Phase velocity and group velocity.

a. $\omega^2 = k^2c^2 + m^2c^4/\hbar^2$. The frequency ω must be larger than a cutoff frequency, $|\omega| > \omega_0 = mc^2/\hbar$. Below that frequency, free waves cannot propagate.
b. Owing to the relation $E = \sqrt{p^2c^2 + m^2c^4}$, one recovers the relation $E = \hbar\omega$, and the cutoff frequency simply corresponds to the condition $E \geq mc^2$.
c. $v_\mathrm{g} = \mathrm{d}\omega/\mathrm{d}k = kc^2/\omega \leq c$ and $v_\varphi = \omega/k \geq c$; therefore $v_\mathrm{g}v_\varphi = c^2$.

2.2. Spreading of a wave packet of a free particle.

a. Using a procedure similar to that of (2.26), we obtain from the Schrödinger equation

$$\frac{\mathrm{d}\langle x^2\rangle_t}{\mathrm{d}t} = \frac{\mathrm{i}\hbar}{m}\int x\left(\psi\,\frac{\partial\psi^*}{\partial x} - \psi^*\frac{\partial\psi}{\partial x}\right)\,\mathrm{d}x\,.$$

b. The Schrödinger equation gives

$$\frac{\mathrm{d}A}{\mathrm{d}t} = \frac{\hbar^2}{2m^2}\int x\left(\psi\,\frac{\partial^3\psi^*}{\partial x^3} - \frac{\partial^2\psi}{\partial x^2}\frac{\partial\psi^*}{\partial x}\right)\,\mathrm{d}x \,+\, \text{c.c.}\,.$$

Using

$$\psi\,\frac{\partial^3\psi^*}{\partial x^3} - \frac{\partial^2\psi}{\partial x^2}\frac{\partial\psi^*}{\partial x} \,+\, \text{c.c.} = \frac{\partial}{\partial x}\left(\psi\frac{\partial^2\psi^*}{\partial x^2} + \psi^*\frac{\partial^2\psi}{\partial x^2} - 2\frac{\partial\psi}{\partial x}\frac{\partial\psi^*}{\partial x}\right)$$

and integration by parts, we obtain

$$\frac{\mathrm{d}A}{\mathrm{d}t} = -\frac{\hbar^2}{2m^2}\int\left(\psi\,\frac{\partial^2\psi^*}{\partial x^2} + \psi^*\frac{\partial^2\psi}{\partial x^2}\right)\,\mathrm{d}x + \frac{\hbar^2}{m^2}\int\frac{\partial\psi}{\partial x}\frac{\partial\psi^*}{\partial x}\,\mathrm{d}x\,,$$

and a second integration by parts of the first term on the right-hand side yields the result for $B(t)$.

c. Using the Schrödinger equation again, we obtain

$$\frac{\mathrm{d}B}{\mathrm{d}t} = \frac{\mathrm{i}\hbar^3}{m^3}\int\left(\frac{\partial^3\psi}{\partial x^3}\frac{\partial\psi^*}{\partial x} - \frac{\partial\psi}{\partial x}\frac{\partial^3\psi^*}{\partial x^3}\right)\,\mathrm{d}x\,,$$

and integration by parts shows that this is zero. The coefficient B is a constant, and we take $B = 2v_1^2$ in the following, where v_1 has the dimensions of a velocity.

d. Integration of the equation of evolution for $A(t)$ yields $A(t) = 2v_1^2 t + \xi_0$ and $\langle x^2\rangle_t = \langle x^2\rangle_0 + \xi_0 t + v_1^2 t^2$.

e. Using $\langle x\rangle_t = \langle x_0\rangle_0 + v_0 t$, we obtain, for $\Delta x_t^2 = \langle x^2\rangle_t - \langle x\rangle_t^2$, the result (2.29).

Note that these results can be recovered in a simple way using Ehrenfest's theorem (Chap. 7).

2.3. The Gaussian wave packet.

a. For $t = 0$, we obtain

$$\psi(x,0) = \left(\sigma^2/\pi\right)^{1/4}\,\mathrm{e}^{\mathrm{i}p_0 x/\hbar}\,\mathrm{e}^{-x^2\sigma^2/2}$$

and

$$\langle x\rangle_0 = 0\,,\quad \langle p\rangle_0 = p_0\,,\quad \Delta x_0 = \frac{1}{\sigma\sqrt{2}}\,,\quad \Delta p_0 = \frac{\sigma\hbar}{\sqrt{2}}\,.$$

Hence the result $\Delta x_0\,\Delta p_0 = \hbar/2$. The Heisenberg inequality is saturated in the case of a Gaussian wave packet.

b. At time t, the wave function $\psi(x,t)$ is the Fourier transform of $e^{-ip^2t/(2m\hbar)}\,\varphi(p)$, which is still an exponential function of a second order polynomial in the variable p (with complex coefficients). The general results concerning the Fourier transform of Gaussian functions apply and one obtains, after a rather tedious calculation,

$$|\psi(x,t)|^2 = \frac{1}{\Delta x(t)\,\sqrt{2\pi}} \exp\left(-\left(x-\frac{p_0 t}{m}\right)^2 \frac{1}{2\Delta x^2(t)}\right),$$

where $\Delta x^2(t)$ is given by (2.46). We recover in this particular case the general results of Chap. 2 and of the previous exercise: propagation of the center of the wave packet at a velocity $\langle p\rangle_0/m$ and quadratic variation of the variance of the wave packet.

2.4. Characteristic size and energy in a linear or quadratic potential. For a quadratic potential, the relation (2.38) with $\gamma = 1$ yields the following typical energy E for the ground state:

$$E = \frac{\hbar^2}{2m\,\Delta x^2} + \frac{1}{2}m\omega^2\,\Delta x^2\,,$$

which is minimum for $\Delta x = \sqrt{\hbar/m\omega}$; $E = \hbar\omega$ in this case. This is indeed the correct order of magnitude for the spatial extension and the energy of the ground state: the exact result for the energy is $\hbar\omega/2$ (Chap. 4).
For a linear potential $\alpha|x|$, the order of magnitude of the energy of the ground state is obtained as the minimum of

$$E = \frac{\hbar^2}{2m\,\Delta x^2} + \alpha\Delta x$$

as the extension Δx varies. This minimum corresponds to $\Delta x = \left(\hbar^2/m\alpha\right)^{1/3}$, and the corresponding energy is $(3/2)\left(\hbar^2\alpha^2/m\right)^{1/3}$.

Chapter 3

3.1. Expectation values and variances. $\langle x\rangle = a/2$, $\Delta x = a\sqrt{1/12 - 1/2\pi^2}$, $\langle p\rangle = 0$, $\Delta p = \pi\hbar/a$ and therefore

$$\Delta x\,\Delta p = \hbar\pi\sqrt{1/12 - 1/2\pi^2} \sim 0.57\hbar\,.$$

3.2. The mean kinetic energy is positive. Using integration by parts, we find

$$\langle p_x^2\rangle = -\hbar^2 \int \psi^*(x)\,\frac{\partial^2\psi}{\partial x^2}\,\mathrm{d}x = \hbar^2 \int \left|\frac{\partial\psi}{\partial x}\right|^2 \mathrm{d}x \geq 0\,.$$

3.3. Real wave functions. If ψ is real, we have

$$\langle p \rangle = \frac{\hbar}{\mathrm{i}} \int_{-\infty}^{+\infty} \psi(x) \frac{\partial \psi}{\partial x} \mathrm{d}x = \frac{\hbar}{2i} \left[\psi^2\right]_{-\infty}^{+\infty} = 0$$

since ψ^2 vanishes at infinity for a physical state.

3.4. Translation in momentum space. Using the general properties of the Fourier transform, we obtain $\langle p \rangle = p_0 + q$ and $\Delta p = \sigma$.

3.5. The first Hermite function. We use

$$\frac{\mathrm{d}}{\mathrm{d}x}\left(\mathrm{e}^{-x^2/2}\right) = -x\mathrm{e}^{-x^2/2} \quad \text{and} \quad \frac{\mathrm{d}^2}{\mathrm{d}x^2}\left(\mathrm{e}^{-x^2/2}\right) = (x^2-1)\,\mathrm{e}^{-x^2/2}\,,$$

and hence obtain the result.

3.6. Ramsauer effect.

a. The continuity equations at $x = -a$ are

$$\mathrm{e}^{-\mathrm{i}ka} + A\mathrm{e}^{+\mathrm{i}ka} = B\mathrm{e}^{-\mathrm{i}qa} + C\mathrm{e}^{\mathrm{i}qa}\,,$$
$$\mathrm{i}k(\mathrm{e}^{-\mathrm{i}ka} - A\mathrm{e}^{\mathrm{i}ka}) = \mathrm{i}q(B\mathrm{e}^{-\mathrm{i}qa} - C\mathrm{e}^{\mathrm{i}qa})\,,$$

and at $x = +a$, they are

$$B\mathrm{e}^{\mathrm{i}qa} + C\mathrm{e}^{-\mathrm{i}qa} = D\mathrm{e}^{\mathrm{i}ka} \quad \text{and} \quad \mathrm{i}q(B\mathrm{e}^{\mathrm{i}qa} - C\mathrm{e}^{-\mathrm{i}qa}) = \mathrm{i}k\,D\mathrm{e}^{\mathrm{i}ka}\,.$$

b. Setting $\Delta = (q+k)^2 - \mathrm{e}^{4\mathrm{i}qa}(q-k)^2$, we obtain

$$D = \frac{4kq}{\Delta}\,\mathrm{e}^{-2\mathrm{i}(k-q)a}\,, \qquad A = \frac{(k^2-q^2)}{\Delta}\,\mathrm{e}^{-2\mathrm{i}ka}\,(1-\mathrm{e}^{4\mathrm{i}qa})\,.$$

We have $|\Delta|^2 = 16k^2q^2 + 4(k^2-q^2)^2 \sin^2 2qa$ and

$$R = |A|^2 = \frac{4(k^2-q^2)^2}{|\Delta|^2}\sin^2 2qa\,, \qquad T = |D|^2 = \frac{16k^2q^2}{|\Delta|^2}\,,$$

where $R + T = 1$.

c. For all values of q such that $\sin 2qa = 0$, i.e. $qa = n\pi/2$, the transmission probability is equal to 1, and there is no reflection; $T = 1$, $R = 0$. This happens when the size of the well $2a$ is a multiple of $\lambda/2$, where $\lambda = 2\pi/q$ is the de Broglie wavelength of the particle inside the potential well. All the reflected waves interfere destructively and the well becomes transparent to the incident wave (more precisely, the wave reflected at $x = -a$, which does not enter the well, interferes destructively in the backward direction with the sum of all the waves undergoing multiple reflections in the well).

d. The corresponding energies are

$$E_n = \frac{n^2\pi^2\hbar^2}{8ma^2} - V_0 \,.$$

Choosing $n = 1$ and $E = 0.7\,\mathrm{eV}$, we obtain $V_0 = 9.4 - 0.7 = 8.7\,\mathrm{eV}$.

e. When E tends to 0, k also tends to 0 and the transmission probability vanishes. The incident particle is reflected by the potential well. The sticking of hydrogen atoms on a liquid-helium surface occurs when the hydrogen atoms enter the potential well in the vicinity of the surface. In this well, a hydrogen atom may lose energy via the emission of a wave propagating on the surface of the liquid (a ripplon): after such a process, the energy of the hydrogen atom is too low to exit the well, and the atom is trapped at the surface of the liquid. At very low incident energies, incoming hydrogen atoms have a vanishing probability of entering the well, and hence the absorption probability tends to zero.

Chapter 4

4.1. Uncertainty relation for the harmonic oscillator. The functions $\psi_n(x)$ are either odd or even, so that the probability density $|\psi_n(x)|^2$ is even, and hence $\langle x\rangle = 0$. From this well-defined parity of the ψ_n's, we also deduce $\langle p\rangle = 0$, since the operator $\partial/\partial x$ changes the parity of the function. We have

$$\langle x^2\rangle = \Delta x^2 = \int_{-\infty}^{+\infty} \psi_n^*(x)\, x^2\, \psi_n(x)\,\mathrm{d}x \,.$$

Therefore, by applying (4.18), we obtain $\langle x^2\rangle = \Delta x^2 = (n + 1/2)\,\hbar/(m\omega)$. In order to calculate $\langle p^2\rangle$, we can just refer to the initial eigenvalue equation (4.10) and notice that $\langle p^2\rangle/2m + m\omega^2\langle x^2\rangle/2 = E_n$, and therefore $\langle p^2\rangle = \Delta p^2 = (n+1/2)\,m\hbar\omega$. Altogether, we find that in the eigenstate n, $\Delta x\Delta p = (n+1/2)\hbar$. For $n = 0$, $\Delta x\Delta p = \hbar/2$: the eigenfunction is a Gaussian and the Heisenberg inequality is saturated.

4.2. Time evolution of a one-dimensional harmonic oscillator. The initial wave function of the system is given as a linear combination of the eigenfunctions of the Hamiltonian, which makes the calculations quite simple.

a. The wave function at time t is

$$\psi(x,t) = \cos\theta\, \phi_0(x)\, \mathrm{e}^{-\mathrm{i}\omega t/2} + \sin\theta\, \phi_1(x)\, \mathrm{e}^{-3\mathrm{i}\omega t/2} \,.$$

b. We can deduce the expectation values

$$\langle E\rangle = (\cos^2\theta + 3\sin^2\theta)\,\hbar\omega/2 \,, \qquad \langle E^2\rangle = (\cos^2\theta + 9\sin^2\theta)\,\hbar^2\omega^2/4 \,,$$

and the variance $\Delta E^2 = \sin^2(2\theta)\hbar^2\omega^2/4$. The expectation values of functions of the energy are all time independent as a consequence of energy conservation (Ehrenfest's theorem; see Chap. 7).

c. For the position distribution, we obtain

$$\langle x \rangle = \sqrt{\frac{\hbar}{2m\omega}} \cos\omega t \, \sin 2\theta \,, \qquad \langle x^2 \rangle = \frac{\hbar}{2m\omega} \, (1 + 2\sin^2\theta) \,.$$

4.3. Three-dimensional harmonic oscillator.

a. The reasoning is similar to that for the three-dimensional square well, and we obtain

$$E_{n_1,n_2,n_3} = (n_1 + n_2 + n_3 + 3/2)\, \hbar\omega \,,$$

where the n_i's are nonnegative integers. The eigenenergies can therefore be written $E_n = (n + 3/2)\, \hbar\omega$, where n is a nonnegative integer. The degeneracy g_n of the level E_n is the number of triplets (n_1, n_2, n_3) whose sum $n_1 + n_2 + n_3$ is equal to n; we find that $g_n = (n + 1)(n + 2)/2$.
If the ϕ_n's are the eigenfunctions of the one-dimensional harmonic oscillator with angular frequency ω, we have

$$\Phi_{n_1,n_2,n_3}(\boldsymbol{r}) = \phi_{n_1}(x)\, \phi_{n_2}(y)\, \phi_{n_3}(z) \,.$$

b. If the oscillator is not isotropic, the energy levels read

$$E_{n_1,n_2,n_3} = (n_1 + 1/2)\, \hbar\omega_1 + (n_2 + 1/2)\, \hbar\omega_2 + (n_3 + 1/2)\hbar\omega_3 \,,$$

and the corresponding wave functions are

$$\Phi_{n_1,n_2,n_3}(\boldsymbol{r}) = \phi^{(1)}_{n_1}(x)\, \phi^{(2)}_{n_2}(y)\, \phi^{(3)}_{n_3}(z) \,,$$

where $\phi^{(i)}_n$ represents the nth eigenstate in a potential of frequency ω_i. These energy levels are generally not degenerate, except when the ratio of two frequencies ω_i and ω_j is a rational number.

4.4. One-dimensional infinite potential well.

a. For symmetry reasons, we obtain $\langle x \rangle = a/2$. For $\langle x^2 \rangle$, we find

$$\langle x^2 \rangle = \frac{2}{a} \int_0^a x^2 \, \sin \frac{n\pi x}{a} \, \mathrm{d}x = a^2 \left(\frac{1}{3} - \frac{1}{2n^2\pi^2} \right) ,$$

and therefore

$$\Delta x^2 = \frac{a^2}{12} \left(1 - \frac{6}{n^2\pi^2} \right) .$$

b. We first normalize the wave function $\psi(x) = Ax(a-x)$:

$$\int_0^a |\psi|^2 \,\mathrm{d}x = |A|^2 \int_0^a x^2(a-x)^2 \,\mathrm{d}x = |A|^2 \frac{a^5}{30},$$

and hence $|A|^2 = 30/a^5$. The probabilities are $p_n = |\alpha_n|^2$, where

$$\alpha_n = A\sqrt{\frac{2}{a}} \int_0^a x(a-x)\sin\frac{n\pi x}{a}\,\mathrm{d}x = \frac{4\sqrt{15}}{n^3\pi^3}\left[1-(-1)^n\right].$$

For symmetry reasons, $\alpha_n = 0$ if n is even. We find

$$\sum_{n=1}^{\infty} p_n = \frac{960}{\pi^6}\sum_{k=0}^{\infty}\frac{1}{(2k+1)^6} = 1,$$

$$\langle E\rangle = \sum E_n p_n = \frac{\hbar^2\pi^2}{2ma^2}\frac{960}{\pi^6}\sum\frac{1}{(2k+1)^4} = \frac{5\hbar^2}{ma^2},$$

$$\langle E^2\rangle = \sum E_n^2 p_n = \frac{\hbar^4\pi^4}{4m^2a^4}\frac{960}{\pi^6}\sum\frac{1}{(2k+1)^2} = \frac{30\hbar^4}{m^2a^4},$$

$$\Delta E = \frac{\sqrt{5}\hbar^2}{ma^2}.$$

c. We can calculate directly the average value and the variance of the kinetic-energy operator. For the average value, we recover the result obtained above:

$$\langle E\rangle = -\frac{\hbar^2}{2m}\int_0^a \psi(x)\,\frac{\mathrm{d}^2\psi}{\mathrm{d}x^2}\,\mathrm{d}x = \frac{\hbar^2 a}{m}|A|^2\int_0^a x(a-x)\,\mathrm{d}x = \frac{5\hbar^2}{ma^2}.$$

In calculating $\langle E^2\rangle$, one meets with a difficulty since, in the sense of functions (but not of distributions), $\mathrm{d}^4\psi/\mathrm{d}x^4 = 0$. This would lead to the absurd result $\langle E^2\rangle = 0$ and to a negative variance $\sigma^2 = \langle E^2\rangle - \langle E\rangle^2$.

In fact, in this problem, the Hilbert space is defined as the $\mathcal{C}_\infty$ periodic functions, of period a, which vanish at $x = 0$ and $x = a$. For such functions, we have

$$\int_0^a \phi_1(x)\,\frac{\mathrm{d}^4\phi_2}{\mathrm{d}x^4}\,\mathrm{d}x = \int_0^a \frac{\mathrm{d}^2\phi_1}{\mathrm{d}x^2}\,\frac{\mathrm{d}^2\phi_2}{\mathrm{d}x^2}\,\mathrm{d}x$$

for all pairs of functions $\phi_1(x), \phi_2(x)$. This relation can serve as a definition of the operator $\hat{H}^2$ in this problem. With this prescription, we obtain

$$\langle E^2\rangle = \frac{\hbar^4}{4m^2}\int_0^a\left(\frac{\partial^2\psi}{\partial x^2}\right)^2\mathrm{d}x = \frac{A^2\hbar^4 a}{m^2} = \frac{30\hbar^4}{m^2a^4}.$$

Actually, the correct general definition of the operator $\hat{E}_K^2$ uses the spectral representation $\hat{E}_K^2\psi(x) \equiv \int K(x,y)\,\psi(y)\,\mathrm{d}y$, where the kernel K is defined as

$$K(x,y) = \frac{2}{a} \sum_{n=1}^{\infty} E_n^2 \sin(n\pi x/a) \sin(n\pi y/a) .$$

This amounts to using the probabilities p_n as above in the calculation of the expectation values of powers of the energy $\langle E^q \rangle$.

4.5. Isotropic states of the hydrogen atom.

a. The Schrödinger equation is

$$-\frac{\hbar^2}{2m}\psi'' - \frac{A}{x}\psi = E\psi .$$

Therefore

$$-\frac{\hbar^2}{2m}\left(-\frac{2}{a} + \frac{x}{a^2}\right) \mathrm{e}^{-x/a} - A\mathrm{e}^{-x/a} = Ex\mathrm{e}^{-x/a} .$$

Equating the terms in $\mathrm{e}^{-x/a}$, we obtain

$$A = \hbar^2/ma , \quad \text{i.e.} \quad a = \hbar^2/mA = \hbar/mc\alpha .$$

The terms in $x\mathrm{e}^{-x/a}$ then give $E = -\hbar^2/2ma^2$, i.e. $E = -mc^2\alpha^2/2$.

b. $E = -13.6\,\text{eV}$, $a = 5.3 \times 10^{-2}\,\text{nm}$.

c. The condition $\int_0^\infty \psi^2 \,\mathrm{d}x = 1$ implies $C^2 a^3/4 = 1$, i.e. $C = 2/a^{3/2}$.

d. We find $\langle 1/x \rangle = C^2 \int_0^\infty x\mathrm{e}^{-2x/a} \,\mathrm{d}x = 1/a$, and therefore

$$\langle V \rangle = -A \left\langle \frac{1}{x} \right\rangle = -mc^2\alpha^2 = 2E .$$

The expectation value $\langle p^2/2m \rangle$ can be calculated directly by noticing that in the state under consideration,

$$E = \left\langle \frac{p^2}{2m} \right\rangle - \left\langle \frac{A}{x} \right\rangle \quad \text{and therefore} \quad \left\langle \frac{p^2}{2m} \right\rangle = \frac{1}{2} mc^2\alpha^2 = -E .$$

We obtain the relation $2\langle p^2/2m \rangle = +\langle A/x \rangle$, which is also true in classical mechanics if the averaging is performed over a closed orbit (the virial theorem).

4.6. δ-function potentials.

a. (i) We obtain $\lim_{\varepsilon \to 0} [\psi'(+\varepsilon) - \psi'(-\varepsilon)] = (2m\alpha/\hbar^2)\psi(0)$.

(ii) There is only one bound state, with $\psi(x) = \lambda_0^{-1/2}\,\mathrm{e}^{-|x|/\lambda_0}$. We have $K\lambda_0 = 1$ and $E = -\hbar^2/(2m\lambda_0^2)$.

b. The wave function can be written

$$\psi(x) = B\mathrm{e}^{K(x+d/2)} \qquad \text{for} \quad x < -d/2\,,$$
$$\psi(x) = C\mathrm{e}^{-K(x+d/2)} + C'\mathrm{e}^{K(x-d/2)} \quad \text{for} \; -d/2 \le x \le +d/2\,,$$
$$\psi(x) = B'\mathrm{e}^{-K(x-d/2)} \qquad \text{for} \quad x > d/2\,.$$

The solutions can be classified in terms of their symmetry with respect to $x = 0$:

- symmetric solution $\psi_S : B = B'\,, \quad C = C'$;
- antisymmetric solution $\psi_A : B = -B'\,, \quad C = -C'\,.$

The quantization condition is $(K\lambda_0 - 1)^2 = (\mathrm{e}^{-Kd})^2$.

The energy levels $E_\pm = -\hbar^2 K^2/2m$ are therefore obtained by solving the equation $K\lambda_0 = 1 \pm \mathrm{e}^{-Kd}$ (with $E_+ < E_- < 0$). E_+ corresponds to ψ_S and E_- to ψ_A. If $\lambda_0 > d$, there is only one bound state, ψ_S; if $\lambda_0 < d$, there are two bound states, ψ_S and ψ_A. One can compare this with the model of the ammonia molecule, in particular if $Kd \gg 1$.

4.7. Localization of internal atomic electrons. Following the same procedure as in Sect. 4.6, we find $K = \sqrt{2m|E_0|}/\hbar \sim 1.6 \times 10^{11}\,\mathrm{m}$, i.e. $\mathrm{e}^{-K\Delta} \sim 10^{-14}$ for $\Delta = 2\,\mathrm{Å}$. The tunneling time is $\sim 10^{-4}\,\mathrm{s}$, which is very long on the microscopic scale. We can therefore consider that the internal electrons are well localized, even for atoms inside condensed matter.

Chapter 5

5.1. Translation and rotation operators.

a. The proof is straightforward. We have

$$\hat{p} = \frac{\hbar}{\mathrm{i}}\frac{\mathrm{d}}{\mathrm{d}x} \qquad \Rightarrow \hat{T}(x_0) = \mathrm{e}^{-x_0(\mathrm{d}/\mathrm{d}x)}\,.$$

Therefore

$$\hat{T}(x_0)\,\psi(x) = \sum_{n=0}^{\infty} \frac{(-x_0)^n}{n!}\left(\frac{\mathrm{d}}{\mathrm{d}x}\right)^n \psi(x)\,,$$

which is simply the Taylor expansion of $\psi(x - x_0)$ around the point x.

b. Similarly, we find

$$\hat{R}(\varphi)\,\psi(r,\theta) = \sum_{n=0}^{\infty} \frac{(-\varphi)^n}{n!}\left(\frac{\partial}{\partial\theta}\right)^n \psi(r,\theta)\,,$$

where we recognize the Taylor expansion of $\psi(r, \theta - \varphi)$.

5.2. The evolution operator. The formula giving the derivative of the exponential of a function remains valid for the operator $\hat{U}$ since the Hamiltonian is time independent. Hence $\hat{U}(\tau)$ and $\hat{H}$ commute. We have, therefore,

$$\frac{\mathrm{d}}{\mathrm{d}t}\hat{U}(t-t_0) = -\frac{\mathrm{i}}{\hbar}\hat{H}\,\hat{U}(t-t_0)\,.$$

This implies that $|\psi(t)\rangle = \hat{U}(t-t_0)\,|\psi(t_0)\rangle$ is a solution of the Schrödinger equation with the appropriate initial condition since $\hat{U}(0) = \hat{1}$. The unitarity of $\hat{U}$ is a direct consequence of the fact that $\hat{H}$ is Hermitian:

$$\hat{U}^\dagger(\tau) = \mathrm{e}^{\mathrm{i}\hat{H}\tau/\hbar} = \hat{U}(-\tau) = \hat{U}^{-1}(\tau)\,.$$

5.3. Heisenberg representation. We have, by definition,

$$a(t) = \langle\psi(t)|\hat{A}|\psi(t)\rangle = \langle\psi(0)|\hat{U}^\dagger(t)\hat{A}\hat{U}(t)|\psi(0)\rangle\,.$$

We can define $\hat{A}(t) = \hat{U}^\dagger(t)\hat{A}\hat{U}(t)$, i.e.

$$\hat{A}(t) = \mathrm{e}^{\mathrm{i}\hat{H}t/\hbar}\,\hat{A}\mathrm{e}^{-\mathrm{i}\hat{H}t/\hbar}\,,$$

which is indeed a solution of the differential equation given in the exercise and is such that $a(t) = \langle\psi(0)|\hat{A}(t)|\psi(0)\rangle$.

5.4. Dirac formalism with a two-state problem.

a. $\langle\psi_2|\hat{H}|\psi_1\rangle = E_1\langle\psi_2|\psi_1\rangle = E_2\langle\psi_2|\psi_1\rangle$, and therefore $(E_1-E_2)\langle\psi_2|\psi_1\rangle = 0$ and $\langle\psi_2|\psi_1\rangle = 0$.
b. $\langle E\rangle = (E_1+E_2)/2$, $\quad \Delta E^2 = [(E_1-E_2)/2]^2$, $\quad \Delta E = \hbar\omega/2$.
c. $|\psi(t)\rangle = \left(\mathrm{e}^{-\mathrm{i}E_1t/\hbar}|\psi_1\rangle - \mathrm{e}^{-\mathrm{i}E_2t/\hbar}|\psi_2\rangle\right)/\sqrt{2}$.
d. $a = \pm 1$.
e. $|\psi_\pm\rangle = \left(|\psi_1\rangle \pm |\psi_2\rangle\right)/\sqrt{2}$.
f. $p = |\langle\psi_-|\psi(t)\rangle|^2 = \cos^2(\omega t/2)$.

5.5. Successive measurements and the principle of wave packet reduction.

1. a. The probability of finding α_i at time $t=0$ is $|\langle a_i|\psi_0\rangle|^2$ part (c) of the Second Principle.
   b. (i) Just after the measurement, the state of the system is $|\psi(0_+)\rangle = |a_i\rangle$ part (d) of the Second Principle.
   (ii) The Hamiltonian evolution between $t=0$ and t leads to the state vector $|\psi(t)\rangle = \mathrm{e}^{-\mathrm{i}\hat{H}_St/\hbar}|a_i\rangle$.
   (iii) The probability of finding β_j in a measurement of B at time t is:

$$|\langle b_j|\psi(t)\rangle|^2 = \left|\langle b_j|\mathrm{e}^{-\mathrm{i}\hat{H}_St/\hbar}|a_i\rangle\right|^2\,.$$

c. From the definition of a conditional probability, the desired probability is simply:

$$P(\alpha_i,0;\beta_j,t) = |\langle a_i|\psi_0\rangle|^2 \left|\langle b_j|\mathrm{e}^{-\mathrm{i}\hat{H}_S t/\hbar}|a_i\rangle\right|^2 .$$

2. a. We expand the vector $|\psi_0\rangle$ on the basis $|a_i\rangle$: $|\psi_0\rangle = \sum_i \mu_i|a_i\rangle$ with $\mu_i = \langle a_i|\psi_0\rangle$, i. e. $|\Psi_0\rangle = \sum_i \mu_i|a_i\rangle \otimes |\mathcal{A}_0\rangle \otimes |\mathcal{B}_0\rangle$. From linearity we then obtain $|\Psi(0_+)\rangle = \sum_i \mu_i|a_i\rangle \otimes |\mathcal{A}_i\rangle \otimes |\mathcal{B}_0\rangle$.

b. Since $\hat{H}_S$, $\hat{H}_\mathcal{A}$ and $\hat{H}_\mathcal{B}$ commute, we have $\mathrm{e}^{-\mathrm{i}\hat{H}t/\hbar} = \mathrm{e}^{-\mathrm{i}\hat{H}_S t/\hbar}\,\mathrm{e}^{-\mathrm{i}\hat{H}_\mathcal{A} t/\hbar}\,\mathrm{e}^{-\mathrm{i}\hat{H}_\mathcal{B} t/\hbar}$, which leads to:

$$|\Psi(t)\rangle = \sum_i \mu_i\,\mathrm{e}^{-\mathrm{i}(A_i+B_0)t/\hbar}\left(\mathrm{e}^{-\mathrm{i}\hat{H}_S t/\hbar}|a_i\rangle\right) \otimes |\mathcal{A}_i\rangle \otimes |\mathcal{B}_0\rangle .$$

c. We expand the vector $\mathrm{e}^{-\mathrm{i}\hat{H}_S t/\hbar}|a_i\rangle$ on the basis $\{|b_j\rangle\}$: $\mathrm{e}^{-\mathrm{i}\hat{H}_S t/\hbar}|a_i\rangle = \sum_j \gamma_{ij}|b_j\rangle$ with $\gamma_{ij} = \langle b_j|\mathrm{e}^{-\mathrm{i}\hat{H}_S t/\hbar}|a_i\rangle$, which leads to $|\Psi(t)\rangle = \sum_{i,j} \mu_i\,\mathrm{e}^{-\mathrm{i}(A_i+B_0)t/\hbar}\gamma_{ij}|b_j\rangle \otimes |\mathcal{A}_i\rangle \otimes |\mathcal{B}_0\rangle$. By linearity, we find:

$$|\Psi(t_+)\rangle = \sum_{i,j} \mu_i\,\mathrm{e}^{-\mathrm{i}(A_i+B_0)t/\hbar}\gamma_{ij}|b_j\rangle \otimes |\mathcal{A}_i\rangle \otimes |\mathcal{B}_j\rangle .$$

d. The probability of finding the detector $\mathcal{A}$ in the state $|\mathcal{A}_i\rangle$, corresponding to the result a_i at time $t = 0$, and the detector $\mathcal{B}$ in the state $|\mathcal{B}_j\rangle$, corresponding to the result b_j at time t, is obtained by squaring the modulus of the coefficient of $|\mathcal{A}_i\rangle\otimes|\mathcal{B}_j\rangle$ in this expansion part (c) of the Second Principle:

$$P(\alpha_i,0;\beta_j,t) = |\mu_i|^2|\gamma_{ij}|^2 = |\langle a_i|\psi_0\rangle|^2 \left|\langle b_j|\mathrm{e}^{-\mathrm{i}\hat{H}_S t/\hbar}|a_i\rangle\right|^2 .$$

This result coincides with what we found in the first question.

3. As a matter of principle, there is a great difference between the two approaches envisaged above. In the first approach, the application of part (d) of the Second Principle leads to an irreversible evolution during the first measurement, and it is impossible to recover the state $|\psi_0\rangle$ after the measurement has been performed. In the second approach, no irreversibility is introduced a priori and nothing forbids us to imagine a new interaction between S and the two detectors which could allow us to restore the state $|\psi_0\rangle$ we began with, by "erasing" the information registered on $\mathcal{A}$ and $\mathcal{B}$. However, from a practical point of view, the two descriptions are equivalent as soon as we deal with macroscopic detectors. Indeed, owing to the coupling of the two detectors with the outside environment, the coherent superposition $|\Psi(t_+)\rangle$ is very rapidly transformed into an incoherent mixture (see Appendix Appendix D for a precise definition of this terminology) and the reconstruction of the state $|\psi_0\rangle$ becomes impossible.

Chapter 6

6.1. Linear three-atom molecule.

a. The three energy levels and the corresponding eigenstates are

$$E_1 = E_0\,, \quad |\psi_0\rangle = \frac{1}{\sqrt{2}}\begin{pmatrix} 1 \\ 0 \\ -1 \end{pmatrix}$$

and

$$E_\pm = E_0 \pm a\sqrt{2}\,, \quad |\psi\pm\rangle = \frac{1}{2}\begin{pmatrix} 1 \\ \mp\sqrt{2} \\ 1 \end{pmatrix}.$$

b. The probabilities are $P_{\mathrm{L}} = P_{\mathrm{R}} = 1/4$ and $P_{\mathrm{C}} = 1/2$.

c. We have

$$|\psi_{\mathrm{L}}\rangle = \frac{1}{\sqrt{2}}|\psi_0\rangle + \frac{1}{2}|\psi_+\rangle + \frac{1}{2}|\psi_-\rangle$$

and hence $\langle E\rangle = E_0$ and $\Delta E = a$.

6.2. Crystallized violet and malachite green.

a. The Hamiltonian is

$$\hat{H} = \begin{pmatrix} 0 & -A & -A \\ -A & 0 & -A \\ -A & -A & 0 \end{pmatrix}.$$

As in the case of NH_3, the off-diagonal elements of the matrix H written in the basis of classical configurations represent the quantum effects, i. e., passage from one configuration to another one by tunneling.

b. $\langle E\rangle_1 = \langle\phi_1|\hat{H}|\phi_1\rangle = -2A$, $\langle E^2\rangle_1 = 4A^2$, and therefore $\Delta E^2 = 0$ in the state $|\phi_1\rangle$. Similarly, $\langle E\rangle_2 = \langle\phi_2|\hat{H}|\phi_2\rangle = +A$, $\langle E^2\rangle_2 = A^2$, and therefore $\Delta E^2 = 0$ in the state $|\phi_2\rangle$.

Consequently, these two states are energy eigenstates, with eigenvalues $-2A$ and $+A$, respectively. Naturally, this can be seen directly by letting $\hat{H}$ act on these two states.

c. Knowing two eigenvectors of $\hat{H}$, it suffices to look for a vector orthogonal to both of them. We find $|\phi_3\rangle = (2|1\rangle - |2\rangle - |3\rangle)/\sqrt{6}$. Altogether, we obtain
eigenvalue $\lambda = -2A$, eigenvector $|\phi_1\rangle = (|1\rangle + |2\rangle + |3\rangle)/\sqrt{3}$;
eigenvalue $\lambda = +A$, eigenvector $|\phi_2\rangle = (|2\rangle - |3\rangle)/\sqrt{2}$;
eigenvalue $\lambda = +A$, eigenvector $|\phi_3\rangle = (2|1\rangle - |2\rangle - |3\rangle)/\sqrt{6}$.
The eigenvalue $\lambda = A$ has a degeneracy equal to 2. This eigenbasis is not unique.

d. Absorption of light occurs at $\Delta E = 3A = 2.25$ eV, in the yellow part of the spectrum. The ion therefore has the complementary color, i. e. violet.
e. The Hamiltonian is now

$$\hat{H} = \begin{pmatrix} \Delta & -A & -A \\ -A & 0 & -A \\ -A & -A & 0 \end{pmatrix}$$

and the eigenvalue equation can be written

$$\det(\hat{H} - \lambda \hat{I}) = -(\lambda - A)\left(\lambda^2 + (A - \Delta)\lambda - A(\Delta + 2A)\right).$$

Hence the eigenvalues are

$$E_0 = A\,, \quad E_\pm = \left((\Delta - A) \pm \sqrt{(\Delta + A)^2 + 8A^2}\right)/2\,.$$

- If $\Delta \ll A$: $E_0 = A, \quad E_+ \sim A + 2\Delta/3, \quad E_- \sim -2A + \Delta/3.$
- If $\Delta \gg A$: $E_0 = A, \quad E_+ \sim \Delta, \quad E_- \sim -A.$

f. There are two possible transitions from the ground state E_-. One of them corresponds to

$$h\nu_1 = hc/\lambda_1 = hc/(450\,\text{nm}) \sim 2.75 \text{ eV} \;=\; \sqrt{(\Delta + A)^2 + 8A^2}\,,$$

which corresponds to $\Delta = 1\,\text{eV}$ and absorption in the violet part of the spectrum. With this value of Δ, we obtain

$$h\nu_2 = E_0 - E_- = (3A - \Delta)/2 + 1/2\sqrt{(\Delta + A)^2 + 8A^2} = 2 \text{ eV}\,.$$

This corresponds to absorption in the red–orange part of the spectrum and to a wavelength $\lambda_2 = hc/h\nu_2 = 620\,\text{nm}$, in good agreement with experimental observation.

Chapter 7

7.1. Commutator algebra. The first relation is immediate. To prove the second relation, one can start with

$$\begin{aligned} \hat{A}\hat{B}^n - \hat{B}^n\hat{A} = {} & \hat{A}\hat{B}^n - \hat{B}\hat{A}\hat{B}^{n-1} \\ & + \hat{B}\hat{A}\hat{B}^{n-1} - \hat{B}^2\hat{A}\hat{B}^{n-2} \\ & + \hat{B}^2\hat{A}\hat{B}^{n-2} - \ldots \\ & + \hat{B}^{n-1}\hat{A}\hat{B} - \hat{B}^n\hat{A}\,, \end{aligned}$$

and each line of the right-hand side can be written as $\hat{B}^s[\hat{A}, \hat{B}]\hat{B}^{n-s-1}$ with $s = 0, \ldots, n-1$, and hence the result. Finally, the Jacobi identity is derived by expanding all commutators, and checking that the twelve terms which appear cancel each other.

7.2. Glauber's formula. The derivative of $\hat{F} = \mathrm{e}^{t\hat{A}}\,\mathrm{e}^{t\hat{B}}$ with respect to t is

$$\frac{\mathrm{d}\hat{F}}{\mathrm{d}t} = \hat{A}\,\hat{F} + \hat{F}\,\hat{B}\,.$$

The second term on the right-hand side can be written

$$\mathrm{e}^{t\hat{A}}\mathrm{e}^{t\hat{B}}\hat{B} = \mathrm{e}^{t\hat{A}}\,\hat{B}\mathrm{e}^{t\hat{B}} = B\mathrm{e}^{t\hat{A}}\,\mathrm{e}^{t\hat{B}} + \sum_{n=0}^{\infty}\frac{t^n}{n!}[\hat{A}^n,\hat{B}]\mathrm{e}^{t\hat{B}}\,.$$

Since $\hat{A}$ commutes with the commutator $[\hat{A},\hat{B}]$, the result of the previous exercise gives

$$[\hat{A}^n,\hat{B}] = n\,[\hat{A},\hat{B}]\;\hat{A}^{n-1}\,,$$

from which we deduce the desired differential equation. This equation can be solved as a standard differential equation, and it yields

$$F(t) = \exp\left(t(\hat{A}+\hat{B}) + \frac{t^2}{2}[\hat{A},\hat{B}]\right)\,,$$

since the two operators $\hat{A}+\hat{B}$ and $[\hat{A},\hat{B}]$ commute. By choosing $t=1$, we obtain Glauber's formula, from which we deduce

$$\mathrm{e}^{\hat{x}/x_0}\mathrm{e}^{\hat{p}/p_0} = \mathrm{e}^{\hat{x}/x_0+\hat{p}/p_0}\mathrm{e}^{\mathrm{i}\hbar/(2x_0p_0)} \qquad \mathrm{e}^{\lambda\hat{a}}\mathrm{e}^{\mu\hat{a}^\dagger} = \mathrm{e}^{\lambda\hat{a}+\mu\hat{a}^\dagger}\mathrm{e}^{\lambda\mu/2}\,.$$

7.3. Classical equations of motion for the harmonic oscillator. For the harmonic oscillator, the Ehrenfest theorem gives

$$\frac{\mathrm{d}\langle x\rangle}{\mathrm{d}t} = \frac{\langle p\rangle}{m}\,, \qquad \frac{\mathrm{d}\langle p\rangle}{\mathrm{d}t} = -m\omega^2\langle x\rangle\,,$$

and hence the result $\mathrm{d}^2\langle x\rangle/\mathrm{d}t^2 + \omega^2\langle x\rangle = 0$.

7.4. Conservation law. For a system of n interacting particles, the total Hamiltonian reads

$$\hat{H} = \sum_{i=1}^{n}\frac{\hat{p}_i^2}{2m_i} + \frac{1}{2}\sum_i\sum_{j\neq i}V(\hat{\boldsymbol{r}}_i - \hat{\boldsymbol{r}}_j)\,.$$

The total momentum of the system is $\hat{\boldsymbol{P}} = \sum_i \hat{\boldsymbol{p}}_i$. This operator commutes both with the kinetic-energy term of the Hamiltonian and with the interaction term:

$$\begin{aligned}[\hat{P}_x, V(\hat{\boldsymbol{r}}_i - \hat{\boldsymbol{r}}_j)] &= [p_{x,i}, V(\hat{\boldsymbol{r}}_i - \hat{\boldsymbol{r}}_j)] + [p_{x,j}, V(\hat{\boldsymbol{r}}_i - \hat{\boldsymbol{r}}_j)]\\ &= \frac{\hbar}{\mathrm{i}}\frac{\partial V}{\partial x}(\hat{\boldsymbol{r}}_i - \hat{\boldsymbol{r}}_j) - \frac{\hbar}{\mathrm{i}}\frac{\partial V}{\partial x}(\hat{\boldsymbol{r}}_i - \hat{\boldsymbol{r}}_j) = 0\,.\end{aligned}$$

Therefore $\langle \boldsymbol{P}\rangle$ is a constant of motion.

7.5. Hermite functions. The expressions for the position and momentum operators in terms of the creation and annihilation operators are

$$\hat{x} = \sqrt{\frac{\hbar}{2m\omega}}\ (\hat{a} + \hat{a}^\dagger)\ , \qquad \hat{p} = \mathrm{i}\sqrt{m\hbar\omega/2}\ (\hat{a}^\dagger - \hat{a})\ ,$$

and hence we obtain the result.

7.6. Generalized uncertainty relations.

a. Applying the commutator to any function $\Psi(\boldsymbol{r})$, we obtain the following for all Ψ:

$$[\hat{p}_x, \hat{f}]\,\Psi(\boldsymbol{r}) = \frac{\hbar}{\mathrm{i}}\left(\frac{\partial}{\partial x}(f\Psi) - f\frac{\partial\Psi}{\partial x}\right) = -\mathrm{i}\hbar\frac{x}{r}f'(r)\,\Psi(\boldsymbol{r}),$$

and hence the relation $[\hat{p}_x, \hat{f}] = -\mathrm{i}\hbar(\hat{x}/\hat{r})\,f'(\hat{r})$.

b. • The square of the norm of $\hat{A}_x|\psi\rangle$ is

$$\begin{aligned}\|\,\hat{A}_x|\psi\rangle\,\|^2 &= \langle\psi|(\hat{p}_x + \mathrm{i}\lambda\hat{x}\hat{f})(\hat{p}_x - \mathrm{i}\lambda\hat{x}\hat{f}|\psi\rangle \\ &= \langle\psi|\hat{p}_x^2 + \lambda^2\hat{x}^2\hat{f}^2 - \mathrm{i}\lambda[\hat{p}_x, \hat{x}\hat{f}]|\psi\rangle\,.\end{aligned}$$

By direct calculation, we obtain $[\hat{p}_x, \hat{x}\hat{f}] = -\mathrm{i}\hbar(f + (x^2/r)f')$, and therefore

$$\|\,\hat{A}_x|\psi\rangle\,\|^2 = \langle p_x^2\rangle + \lambda^2\langle x^2 f^2\rangle - \hbar\lambda\left\langle f + \frac{x^2}{r}f'\right\rangle\,.$$

• Adding the analogous relations for $\hat{A}_y$ and $\hat{A}_z$, we obtain for any state $|\psi\rangle$,

$$\langle p^2\rangle + \lambda^2\langle r^2 f^2\rangle - \hbar\lambda\langle 3f + rf'\rangle \geq 0\,.$$

This second degree trinomial in λ must be nonnegative for all λ. Therefore, the discriminant is negative or zero, i. e.

$$4\langle p^2\rangle\langle r^2 f^2\rangle \geq \hbar^2\langle 3f + rf'\rangle^2\,.$$

c. For $f = 1$, we obtain $\langle p^2\rangle\,\langle r^2\rangle \geq (9/4)\hbar^2$; for $f = 1/r$, $\langle p^2\rangle \geq \hbar^2\langle r^{-1}\rangle^2$; for $f = 1/r^2$, $\langle p^2\rangle \geq (\hbar^2/4)\,\langle r^{-2}\rangle$.

d. *Harmonic oscillator.* For any state $|\psi\rangle$, we have $\langle E\rangle \geq 9\hbar^2/(8m\langle r^2\rangle) + m\omega^2\,\langle r^2\rangle/2$. Minimizing with respect to $\langle r^2\rangle$, we find the lower bound $\langle E\rangle \geq (3/2)\hbar\omega$ for the energy of the oscillator. Since there is a value of λ and a corresponding value of $\langle r^2\rangle$ for which the trinomial has a double root and vanishes, this means that there exists a state for which this lower bound is attained. This state is therefore the ground state since no state can have a lower energy than the ground state. We have $\hat{A}_x|\phi\rangle = 0 \to (\hat{p}_x + \mathrm{i}\lambda\hat{x})|\phi\rangle = 0$. In terms of wave functions, this corresponds to

$$\left(\hbar\frac{\partial}{\partial x} + \lambda x\right)\phi(\boldsymbol{r}) = 0\,,$$

and similar equations for y and z. The solution of this set of three equations is

$$\phi(r) = \mathcal{N} \exp(-\lambda r^2) = \mathcal{N} \exp(-3r^2/4r_0^2),$$

where $\langle r^2 \rangle = r_0^2$, $\mathcal{N}^{-1} = (2\pi r_0^2/3)^{3/4}$ and $\lambda = 3\hbar/4r_0^2$.

e. *Hydrogen atom.*
- Similarly, the lower bound for the energy of the hydrogen atom is $E_{\min} = -m_e e^4/(2\hbar^2)$ and this bound is attained; therefore it is the ground-state energy.
- The equation $\hat{A}_x|\psi\rangle = 0$ leads to the differential equation

$$\hbar \frac{\partial \psi}{\partial x} + \lambda \frac{x}{r} \psi(\boldsymbol{r}) = 0$$

and similar equations for y and z. The solution of this set of equations is $\psi(r) = \mathcal{N} \exp(-r/r_0)$, where $\mathcal{N} = 1/\sqrt{\pi r_0^3}$, $\langle r^{-1} \rangle = 1/r_0$ and $\lambda = 1/r_0$. This is indeed the ground state of the hydrogen atom.

7.7. Quasi-classical states of the harmonic oscillator.

a. By definition, we have

$$\hat{a}|\alpha\rangle = \sum_{n=1}^{\infty} C_n \sqrt{n}\, |n-1\rangle = \alpha \sum_{n=0}^{\infty} C_n\, |n\rangle .$$

The ensuing recursion relation $C_n\sqrt{n} = \alpha C_{n-1}$ allows us to calculate the coefficients C_n in terms of C_0 and α, whatever the value of the complex number α:

$$C_n = \frac{\alpha^n}{\sqrt{n!}} C_0 .$$

For any α, we therefore obtain

$$|\alpha\rangle = C_0 \sum_{n=0}^{\infty} \frac{\alpha^n}{\sqrt{n\,!}} |n\rangle$$

and, by normalizing the result,

$$\langle \alpha|\alpha\rangle = |C_0|^2 \sum_{n=0}^{\infty} \frac{|\alpha|^{2n}}{n!} = \mathrm{e}^{|\alpha|^2} |C_0|^2 \Rightarrow C_0 = \mathrm{e}^{-|\alpha|^2/2}$$

up to an arbitrary phase factor.

The probability $p(E_n)$ of finding E_n is

$$p(E_n) = |\langle n|\alpha\rangle|^2 = \mathrm{e}^{-|\alpha|^2} |\alpha|^{2n}/n!\,,$$

which is a Poisson distribution.

b. The expectation value of the energy is obtained by using

$$\langle E\rangle = \langle\alpha|\hat{H}|\alpha\rangle = \hbar\omega\langle\alpha|(\hat{a}^\dagger\hat{a}+1/2)|\alpha\rangle = (|\alpha|^2+1/2)\hbar\omega\,,$$
$$\langle E^2\rangle = \hbar^2\omega^2\langle\alpha|[(\hat{a}^\dagger\hat{a})^2+\hat{a}^\dagger\hat{a}+1/4]|\alpha\rangle\,.$$

We have $[\hat{a},\hat{a}^\dagger]=1$, and therefore

$$(\hat{a}^\dagger\hat{a})^2 = \hat{a}^\dagger\hat{a}\hat{a}^\dagger\hat{a} = \hat{a}^\dagger\hat{a}^\dagger\hat{a}\hat{a}+\hat{a}^\dagger\hat{a}\,.$$

Hence

$$\langle E^2\rangle = \hbar^2\omega^2(|\alpha|^4+2|\alpha|^2+1/4)$$

and the variance is given by

$$\Delta E^2 = \langle E^2\rangle - \langle E\rangle^2 = \hbar^2\omega^2|\alpha|^2\,,\qquad \Delta E = \hbar\omega|\alpha|\,,$$

or, equivalently, $\Delta E/\langle E\rangle = |\alpha|/(|\alpha|^2+1/2) \sim 1/|\alpha|$ for $|\alpha|\gg 1$. The *relative* dispersion of the energy $\Delta E/\langle E\rangle$ goes to zero as $|\alpha|$ increases.

c. The calculation of these expectation values yields

$$\langle X\rangle = \langle\alpha|(\hat{a}+\hat{a}^\dagger)/\sqrt{2}|\alpha\rangle = (\alpha+\alpha^*)/\sqrt{2}\,,$$
$$\langle P\rangle = \langle\alpha|(\hat{a}-\hat{a}^\dagger)/(\mathrm{i}\sqrt{2})|\alpha\rangle = \mathrm{i}(\alpha^*-\alpha)/\sqrt{2}\,,$$
$$\langle X^2\rangle = \langle\alpha|(\hat{a}^2+\hat{a}^{\dagger 2}+\hat{a}\hat{a}^\dagger+\hat{a}^\dagger\hat{a})|\alpha\rangle/2$$
$$= (\alpha^2+\alpha^{*2}+2|\alpha|^2+1)/2$$
$$\Delta X^2 = \langle X^2\rangle - \langle X\rangle^2 = 1/2\,,$$

and therefore

$$\Delta x = \sqrt{\hbar/2m\omega}\quad\text{and}\quad \langle x\rangle = (\alpha+\alpha^*)\sqrt{\hbar/2m\omega}\,.$$

Similarly, we obtain

$$\Delta p = \sqrt{m\hbar\omega/2}\quad\text{and}\quad \langle p\rangle = i(\alpha^*-\alpha)\sqrt{m\hbar\omega/2}\,,$$

and, for all α,

$$\Delta x\,\Delta p = \hbar/2\,.$$

Since the lower bound of the Heisenberg inequality is attained whatever the value of α, the X or P representation of $|\alpha\rangle$ is a Gaussian function of X or P. We shall check this explicitly in the following. We remark that $\langle x\rangle$ and $\langle p\rangle$ can be as large as we want if we increase $|\alpha|$, whereas Δx and Δp remain constant (and of course compatible with the uncertainty relations). Like the energy, the position and momentum become well defined in relative terms as $|\alpha|$ becomes large.

d. Time evolution: we start with $|\psi(0)\rangle = |\alpha\rangle$, and hence

$$|\psi(t)\rangle = \mathrm{e}^{-|\alpha|^2/2} \sum_{n=0}^{\infty} \frac{\alpha^n}{\sqrt{n!}} \mathrm{e}^{-\mathrm{i}(n+1/2)\omega t} |n\rangle = \mathrm{e}^{-\mathrm{i}\omega t/2} |\alpha \mathrm{e}^{-\mathrm{i}\omega t}\rangle\,.$$

Therefore $|\psi(t)\rangle$ is an eigenstate of $\hat{a}$ with the eigenvalue $\beta = \alpha \mathrm{e}^{-\mathrm{i}\omega t}$. From the result of exercise 7.3, we obtain

$$\langle X\rangle_t = (\alpha \mathrm{e}^{-\mathrm{i}\omega t} + \alpha^* \mathrm{e}^{\mathrm{i}\omega t})/\sqrt{2}\,, \qquad \langle P\rangle_t = i(\alpha^* \mathrm{e}^{\mathrm{i}\omega t} - \alpha \mathrm{e}^{-\mathrm{i}\omega t})/\sqrt{2}\,.$$

Setting $\alpha = \alpha_0 \mathrm{e}^{\mathrm{i}\varphi}$ with $\alpha_0 > 0$, we obtain

$$\langle x\rangle_t = \alpha_0 \sqrt{\frac{2\hbar}{m\omega}} \cos(\omega t - \varphi) = x_0 \cos(\omega t - \varphi)\,,$$
$$\langle p\rangle_t = -\alpha_0 \sqrt{2m\hbar\omega}\, \sin(\omega t - \varphi) = -p_0 \sin(\omega t - \varphi)\,,$$

and, naturally, $\Delta x_t\, \Delta p_t = \hbar/2$. The time evolution of $\langle x\rangle_t$ and $\langle p\rangle_t$ is the same as for a classical oscillator (exercise 7.3).

e. We have

$$\hat{P} = \frac{\hat{p}}{\sqrt{m\hbar\omega}} = -\mathrm{i}\sqrt{\frac{\hbar}{m\omega}} \frac{\partial}{\partial x} = -\mathrm{i}\frac{\partial}{\partial X}$$

and, similarly,

$$\hat{X} = \sqrt{\frac{m\omega}{\hbar}}\hat{x} = \mathrm{i}\sqrt{m\hbar\omega}\frac{\partial}{\partial p} = \mathrm{i}\frac{\partial}{\partial P}\,.$$

In terms of the variable X, we find

$$\frac{1}{\sqrt{2}}\left(X + \frac{\partial}{\partial X}\right)\psi_\alpha(X) = \alpha\psi_\alpha(X)\,,$$

whose solution is $\psi_\alpha(X) = C \exp\left[-(X - \alpha\sqrt{2})^2/2\right]$. In terms of the variable P, we have

$$\frac{\mathrm{i}}{\sqrt{2}}\left(P + \frac{\partial}{\partial P}\right)\varphi_\alpha(P) = \alpha\varphi_\alpha(P)$$

and the solution is $\varphi_\alpha(P) = C' \exp\left[-(P + i\alpha\sqrt{2})^2/2\right]$.

The wave function is a real Gaussian centered at $\langle X\rangle$, multiplied by a plane wave of wave vector $\langle P\rangle$. This wave function is called a minimal wave packet because the probability distribution is the same as for the ground state of the oscillator, except that it is shifted by $X_0 = \mathrm{Re}(\alpha\sqrt{2})$. In particular the Heisenberg inequality is saturated at all times. The time evolution consists in replacing α by $\alpha \exp(-\mathrm{i}\omega t)$. The oscillation of the center of the wave function is the same as that of a classical oscillator.

The states α are genuine quantum states in the sense that they satisfy all conditions of quantum mechanics. However, the physical properties of an oscillator prepared in the state $|\alpha\rangle$, with $|\alpha| \gg 1$, are very similar to those of a classical oscillator. Traditionally, these states are called "quasi-classical" or "coherent" states of the harmonic oscillator, and they play a key role in the quantum theory of radiation.

7.8. Time–energy uncertainty relation. The uncertainty relations give $\Delta a\, \Delta E \geq |\langle\psi|[\hat{A}, \hat{H}]|\psi\rangle|/2$. The Ehrenfest theorem gives

$$d\langle a\rangle/\mathrm{dt} = (1/i\hbar)\langle\psi|[\hat{A}, \hat{H}]|\psi\rangle\,,$$

and hence the inequality and the result.

7.9. Virial theorem.

a. We have

$$[\hat{H}, \hat{x}\hat{p}] = [\hat{H}, \hat{x}]\hat{p} + \hat{x}[\hat{H}, \hat{p}] = i\hbar\left(-\frac{\hat{p}^2}{m} + \hat{x}\frac{\partial V}{\partial x}\right)$$

and, for the potential under consideration

$$[\hat{H}, \hat{x}\hat{p}] = \mathrm{i}\hbar\left(-\frac{\hat{p}^2}{m} + nV(\hat{x})\right)\,.$$

b. For an eigenstate $|\psi_\alpha\rangle$ of $\hat{H}$, we have $\langle\psi_\alpha|[\hat{H}, \hat{x}\hat{p}]|\psi_\alpha\rangle = 0$. Hence

$$\langle\psi_\alpha| - \frac{\hat{p}^2}{m} + nV(\hat{x})|\psi_\alpha\rangle = 0 \quad \Rightarrow \quad 2\langle T\rangle = n\langle V\rangle\,.$$

The harmonic oscillator corresponds to the case $n = 2$: the average kinetic and potential energies are equal when the system is prepared in an eigenstate $|n\rangle$ of the Hamiltonian.

c. We obtain, in three dimensions,

$$[\hat{H}, \boldsymbol{r}\cdot\boldsymbol{p}] = i\hbar\left(-\frac{\boldsymbol{p}^2}{m} + \boldsymbol{r}\cdot\nabla V(\boldsymbol{r})\right)$$

and therefore $2\langle T\rangle = n\langle V\rangle$ as obtained previously. This applies to the Coulomb problem ($V(\boldsymbol{r}) = -e^2/r$). In this case, $n = -1$ and we have $2\langle T\rangle = -\langle V\rangle$. For the harmonic oscillator, we have $\langle T\rangle = \langle V\rangle$.

d. From the equation obtained above, we have, for a central potential $V(r)$,

$$\hat{r}\cdot\nabla V(r) = r\frac{\partial V}{\partial r}\,, \quad \text{and hence} \quad 2\langle T\rangle = \langle r\frac{\partial V}{\partial r}\rangle\,.$$

7.10. Benzene and cyclo-octatetraene molecules.

a. Obviously, $\hat{R}^6 = \hat{I}$. Therefore $\lambda_k^6 = 1$ and $\lambda_k = \mathrm{e}^{2\mathrm{i}k\pi/6}$, $k = 1, \ldots, 6$.

b. The definition $\hat{R}|\phi_k\rangle = \mathrm{e}^{2\mathrm{i}k\pi/6}|\phi_k\rangle$ gives the recursion relation

$$\mathrm{e}^{2\mathrm{i}k\pi/6} c_{k,p} = c_{k,p-1} \,.$$

Therefore we have, up to an arbitrary phase factor, $c_{k,p} = \mathrm{e}^{-2\mathrm{i}kp\pi/6}/\sqrt{6}$.

c. We have $\langle\phi_k|\phi_{k'}\rangle = (1/6)\sum_p \mathrm{e}^{-2\mathrm{i}(k'-k)p\pi/6} = \delta_{k,k'}$.

d. A direct calculation gives $\hat{R}^{-1}|\phi_k\rangle = \mathrm{e}^{-2\mathrm{i}k\pi/6}|\phi_k\rangle$. The eigenvalues of $\hat{R}^{-1} = \hat{R}^\dagger$ are $\lambda_k^{-1} = \lambda_k^*$.

e. A direct calculation gives $\hat{W}\hat{R}|\xi_k\rangle = |\xi_k\rangle + |\xi_{k+2}\rangle = \hat{R}\hat{W}|\xi_k\rangle$. Therefore $\hat{W}$ and $\hat{R}$, which are Hermitian and unitary operators, respectively, commute and possess a common eigenbasis.

f. We have $\hat{W} = -A(\hat{R} + \hat{R}^{-1})$. The eigenvectors of $\hat{W}$ are therefore

$$|\phi_k\rangle = \frac{1}{\sqrt{6}} \sum_{p=1}^{6} \mathrm{e}^{-2\mathrm{i}kp\pi/6} |\xi_p\rangle \,, \qquad k = 1, \ldots, 6 \,,$$

with eigenvalues $E_k = -2A\cos(2k\pi/6)$. The ground state $E_6 = -2A$ is nondegenerate, the levels $E_5 = E_1 = -A$ and $E_4 = E_2 = A$ are twofold degenerate and the level $E_3 = 2A$ is nondegenerate.

g. (i) Using a method similar to the preceding one, we obtain the following eight energy levels:
   - $E_8 = -2A$ is the ground state (nondegenerate),
   - $E_7 = E_1 = -A\sqrt{2}$ (twofold degenerate),
   - $E_6 = E_2 = 0$ (twofold degenerate),
   - $E_5 = E_3 = A\sqrt{2}$ (twofold degenerate),
   - $E_4 = 2A$ (nondegenerate).

   (ii) Using $\langle\phi_k|\xi_1\rangle = \mathrm{e}^{\mathrm{i}k\pi/4}/\sqrt{8}$, we write

$$|\psi(t=0)\rangle = |\xi_1\rangle = \frac{1}{\sqrt{8}} \sum_{k=1}^{8} \mathrm{e}^{\mathrm{i}k\pi/4} |\phi_k\rangle$$

and, therefore

$$|\psi(t)\rangle = \frac{1}{\sqrt{8}} \sum_{k=1}^{8} \mathrm{e}^{\mathrm{i}k\pi/4} \mathrm{e}^{-\mathrm{i}E_k t/\hbar} \, |\phi_k\rangle \,.$$

The probability of finding the electron again on the site $n = 1$ is $p_1(t) = |\langle\xi_1|\psi(t)\rangle|^2$, i.e. $p_1(t) = |(1/8)\sum_k \mathrm{e}^{-\mathrm{i}E_k t/\hbar}|^2$. This yields, putting $\omega = A/\hbar$,

$$p_1(t) = \left| \frac{1}{8} \sum_{k=1}^{8} \mathrm{e}^{-\mathrm{i}E_k t/\hbar} \right|^2 = \left| \frac{1}{4} \left(1 + 2\cos(\omega t\sqrt{2}) + \cos(2\omega t)\right) \right|^2 .$$

(iii) We find, of course, $p_1(0) = 1$. To obtain $p_1(t) = 1$ at a later time, one would need to find a time $t \neq 0$ such that $\cos(\omega t\sqrt{2}) = 1$ and $\cos(2\omega t) = 1$. This would mean that $\omega t\sqrt{2} = 2N\pi$ and $2\omega t = 2N'\pi$ with N and N' integer. Taking the ratio of these two quantities, we see that this would require $\sqrt{2} = N'/N$, i. e. that $\sqrt{2}$ is rational! Consequently, the particle never reaches its initial state on site 1 again, and the evolution of the state of the system is not periodic. However, it can be shown that the system comes back as close to the initial state as one wishes if one waits for a long enough time. This type of time evolution is said to be *quasi-periodic.*

h. These results can be readily extended to N centers. The levels are $E_n = -2A\cos(2n\pi/N)$ which are all twofold degenerate except the ground state $n = N$ and the highest level $n = N/2$ if N is even. These levels populate an energy *band* of fixed width $4A$, which becomes a continuum in the limit $N \to \infty$. The eigenstates are of the form $|\phi_n\rangle = (1/\sqrt{N})\sum_{p=1}^{N} \mathrm{e}^{-2\mathrm{i}np\pi/N}|\xi_p\rangle$.

Chapter 8

8.1. Determination of the magnetic state of a silver atom.

a. It is always possible to write the coefficients α and β as $\alpha = \cos(\theta/2)$, $\beta = \mathrm{e}^{\mathrm{i}\varphi}\sin(\theta/2)$ (if necessary, one can multiply the state (8.67) by a global phase factor in order to obtain a real value for α). Consider now a Stern–Gerlach magnet oriented along the unit vector $\boldsymbol{u}$, defined by the polar angles θ, φ:

$$\boldsymbol{u} = \sin\theta\cos\varphi\ \boldsymbol{e}_x + \sin\theta\sin\varphi\ \boldsymbol{e}_y + \cos\theta\ \boldsymbol{e}_z\,.$$

This gives

$$\boldsymbol{u}\cdot\hat{\boldsymbol{\mu}} = \mu_0\begin{pmatrix}\cos\theta & \mathrm{e}^{-\mathrm{i}\varphi}\sin\theta\\ \mathrm{e}^{\mathrm{i}\varphi}\sin\theta & \cos\theta\end{pmatrix}.$$

One can check easily that $\cos(\theta/2)|+\rangle + \mathrm{e}^{\mathrm{i}\varphi}\sin(\theta/2)|-\rangle$ is an eigenstate of $\boldsymbol{u}\cdot\hat{\boldsymbol{\mu}}$ with eigenvalue $+\mu_0$.

b. Bob has to choose an axis $\boldsymbol{u}'$ for his Stern–Gerlach measurement, but he does not know the value of $\boldsymbol{u}$. His measurement yields a binary answer $\pm\mu_0$, and the state of the system is then $|\pm\rangle_{\boldsymbol{u}'}$. This state is different from the initial state and subsequent measurements will not supply any new information on the state, sent by Alice. Consequently, Bob cannot determine the initial state from a measurement performed on a single magnetic moment. The only certainty that Bob can have when he obtains the result $+\mu_0$ or $-\mu_0$ in his measurement using the $\boldsymbol{u}'$ axis is that the initial state was not $|-\rangle_{\boldsymbol{u}'}$ or $|+\rangle_{\boldsymbol{u}'}$, respectively, which is very poor information.

c. If Alice sends a large number N of magnetic moments, all prepared in the same unknown state (8.67), Bob can split this ensemble into three sets. For the first $N/3$ magnetic moments, he measures μ_z. From the relative intensity of the two spots $\mu_z = \pm\mu_0$, he deduces $|\alpha|^2$ and $|\beta|^2$. For the second set of $N/3$ magnetic moments, he measures μ_x. The relative intensities of the two spots corresponding to $\mu_x = \pm\mu_0$ yield $|\alpha \pm \beta|^2$. Finally Bob measures μ_y for the last set of $N/3$ magnetic moments, which yields $|\alpha \pm \mathrm{i}\beta|^2$. From these three sets of results, Bob deduces α and β, within a global phase factor which is unimportant. Of course, this determination of α and β is only approximate and the relative statistical error is $N^{-1/2}$.

8.2. Results of repeated measurements: quantum Zeno paradox.

a. The energy levels of $\hat{H}$ are $E_\pm = \pm\hbar\omega/2$.
b. The state $|\psi(0)\rangle$ is an eigenstate of $\hat{\mu}_x$ with eigenvalues $+\mu_0$. Therefore, one obtains $+\mu_0$ with probability 1 in a measurement of μ_x.
c. The evolution of the state under consideration is

$$|\psi(T)\rangle = (|+\rangle \mathrm{e}^{-\mathrm{i}\omega T/2} + |-\rangle \mathrm{e}^{\mathrm{i}\omega T/2})/\sqrt{2}\,.$$

d. The corresponding probability is

$$P(T) = |_x\langle +|\psi(T)\rangle|^2 = |\langle\psi(0)|\psi(T)\rangle|^2 = \cos^2\frac{\omega T}{2}\,.$$

e. After a measurement giving the result $\mu_x = +\mu_0$, the system is again in the same state as initially, i. e. $|\psi(0)\rangle = (|+\rangle + |-\rangle)/\sqrt{2}$. The probability for all the N successive measurements to give the same result $+\mu_0$ is therefore $P_\mathrm{N}(T) = \cos^{2N}(\omega T/2N)$.
f. In the (mathematical) limit $N \to \infty$, we obtain

$$\begin{aligned} P_\mathrm{N} &= \exp\left\{N \ln\left[\cos^2\left(\frac{\omega T}{2N}\right)\right]\right\} \\ &\sim \exp\left[2N \ln\left(1 - \frac{\omega^2 T^2}{8N^2}\right)\right] \sim \exp\left(-\frac{\omega^2 T^2}{4N}\right) \to 1\,. \end{aligned}$$

This result may seem paradoxical: observing the system prevents it from evolving! Some people claim that watching water prevents it from boiling. However, the solution of the "quantum Zeno paradox" lies in the fact that any measurement has a finite extension both in space and in time. In practice, one cannot divide T into infinitely small parts except by interacting permanently with the system, which is another problem.

Chapter 9

9.1. Perturbed harmonic oscillator. The sum (9.21) involves two terms, those with $k = n \pm 2$. The matrix element $\langle k|\hat{x}^2|n\rangle$ can be calculated using the expression for the position operator as a function of $\hat{a}$ and $\hat{a}^\dagger$, $\hat{x} = [\hbar/(2m\omega)]^{1/2}(\hat{a} + \hat{a}^\dagger)$, so that

$$\langle n+2|\hat{x}^2|n\rangle = \frac{\hbar}{2m\omega}\sqrt{(n+1)(n+2)}\,, \qquad \langle n-2|\hat{x}^2|n\rangle = \frac{\hbar}{2m\omega}\sqrt{n(n-1)}\,.$$

The energy denominator is $\mp 2\hbar\omega$ for $k = n \pm 2$, and hence the result

$$\Delta E_n^{(2)} = -\frac{\lambda^2}{8}\,\hbar\omega(n+1/2)\,.$$

This result coincides with the second-order term of the expansion in λ of $\hbar\omega(n+1/2)\,\sqrt{1+\lambda}$.

9.2. Comparison of the ground states of two potentials. We denote by $|\psi_i\rangle$ $(i = 1, 2)$ the ground state of the Hamiltonian $\hat{H} = \hat{p}^2/2m + V_i(\boldsymbol{r})$, with energy E_i. Since $V_2(\boldsymbol{r}) > V_1(\boldsymbol{r})$ for any $\boldsymbol{r}$, we have

$$\begin{aligned} &\langle\psi_2|V_2(\hat{\boldsymbol{r}})|\psi_2\rangle > \langle\psi_2|V_1(\hat{\boldsymbol{r}})|\psi_2\rangle \\ \Rightarrow\; &E_2 = \langle\psi_2|\frac{\hat{p}^2}{2m} + V_2(\hat{\boldsymbol{r}})|\psi_2\rangle > \langle\psi_2|\frac{\hat{p}^2}{2m} + V_1(\hat{\boldsymbol{r}})|\psi_2\rangle\,. \end{aligned}$$

The second step of the reasoning consists of noticing that, because of the theorem on which the variational method is based, we have

$$\langle\psi_2|\frac{\hat{p}^2}{2m} + V_1(\hat{\boldsymbol{r}})|\psi_2\rangle \geq E_1 = \langle\psi_1|\frac{\hat{p}^2}{2m} + V_1(\hat{\boldsymbol{r}})|\psi_1\rangle\,,$$

and hence the result.

9.3. Existence of a bound state in a potential well. To show the existence of a bound state in a one-dimensional potential $V(x)$ which is negative everywhere and tends to zero at $\pm\infty$, we use the variational method with Gaussian trial functions,

$$\psi_\sigma(x) = [\sigma^2/(2\pi)]^{1/4}\,\exp(-\sigma^2 x^2/4)\,.$$

The mean kinetic energy is $T_\sigma = \hbar^2\sigma^2/(8m)$. This positive quantity tends to zero quadratically as σ tends to zero. The mean potential energy is

$$\langle V\rangle_\sigma = \frac{\sigma}{\sqrt{2\pi}}\int V(x)\,\mathrm{e}^{-\sigma^2 x^2/2}\,\mathrm{d}x\,.$$

By assumption this quantity is negative. As σ tends to zero, $\langle V\rangle_\sigma$ tends to zero linearly with σ if the integral $\int_{-\infty}^{\infty} V(x)\,\mathrm{d}x$ converges, or it may even diverge if the integral is itself divergent. In any case, there exists a range of value for the variable σ such that $|\langle V\rangle_\sigma| > T_\sigma$. For these values of σ, the total mean energy $E_\sigma = T_\sigma + \langle V\rangle_\sigma$ is negative. The ground-state energy is necessarily lower than E_σ and is also negative: the corresponding state is a bound state.

This demonstration cannot be extended to three dimensions. If we still take Gaussian functions $\mathrm{e}^{-\sigma^2 r^2/4}$ as a trial set, the kinetic energy varies as σ^2 as before, while the negative potential-energy term now scales as σ^3. As σ tends to zero, the potential energy tends to zero faster than the kinetic energy, and the total energy may be always positive, so that one cannot infer the existence of a bound state.

Indeed, there are simple three-dimensional potentials that are negative or zero everywhere and have no bound state. Consider for instance an isotropic square potential well, $V(\boldsymbol{r}) = -V_0$ for $r < r_0$ (with $V_0 > 0$) and $V(\boldsymbol{r}) = 0$ otherwise. We find that there is no bound state if $V_0 < \hbar^2\pi^2/(8mr_0^2)$. (To show this result, first consider states with zero angular momentum, and then generalize to arbitrary angular momenta using the result of the previous exercise).

9.4. Generalized Heisenberg inequalities. For any state $|\psi\rangle$, we have

$$\frac{1}{2m}\langle p^2\rangle + g\langle r^\alpha\rangle - \varepsilon_0\, g^{2/(\alpha+2)}\left(\frac{\hbar^2}{2m}\right)^{\alpha/(\alpha+2)} \geq 0\,.$$

Minimizing with respect to g, we obtain the result given in the exercise (it is safe to treat separately the case $\alpha > 0$, $g > 0$, in which case $\varepsilon_0 > 0$, and the case $\alpha < 0$, $g < 0$, in which case $\varepsilon_0 < 0$). We recover the usual results for $\alpha = 2$ and $\alpha = -1$. For the linear potential, $\alpha = 1$, we have $\varepsilon_0 = 2.33811$ and hence the uncertainty relation

$$\langle p^2\rangle\langle r\rangle^2 \geq \frac{4}{27}\,\varepsilon_0^3\,\hbar^2 \sim 1.894\,\hbar^2\,.$$

Chapter 10

10.1. Operator invariant under rotation. $[\hat{A}, \hat{L}_x] = 0$ and $[\hat{A}, \hat{L}_y] = 0 \Rightarrow [\hat{A}, [\hat{L}_x, \hat{L}_y]] = \mathrm{i}\hbar[\hat{A}, \hat{L}_z] = 0$.

10.2. Commutation relations for $\hat{\boldsymbol{r}}$ and $\hat{\boldsymbol{p}}$. A straightforward calculation gives

$$[\hat{L}_z, \hat{x}] = -\hat{y}[\hat{p}_x, \hat{x}] = \mathrm{i}\hbar\hat{y}\,, \quad [\hat{L}_z, \hat{y}] = \hat{x}[\hat{p}_y, \hat{y}] = -\mathrm{i}\hbar\hat{x}\,, \quad [\hat{L}_z, \hat{z}] = 0\,,$$

$$[\hat{L}_z, \hat{p}_x] = \hat{p}_y[\hat{x}, \hat{p}_x] = \mathrm{i}\hbar\hat{p}_y\,, \quad [\hat{L}_z, \hat{p}_y] = -\hat{p}_x[\hat{y}, \hat{p}_y] = -\mathrm{i}\hbar\hat{p}_x\,,$$

$$[\hat{L}_z, \hat{p}_z] = 0\,.$$

Therefore, since $[\hat{A}, \hat{B}^2] = [\hat{A}, \hat{B}]\hat{B} + \hat{B}[\hat{A}, \hat{B}]$, we obtain $[\hat{L}, \hat{p}^2] = [\hat{L}, \hat{r}^2] = 0$.

10.3. Rotation-invariant potential. If $V(\boldsymbol{r})$ depends only on $r = |\boldsymbol{r}|$, then $\hat{H} = \hat{p}^2/2m + V(\hat{r})$ (which is rotation invariant) commutes with the angular momentum $\hat{\boldsymbol{L}}$, which is a constant of the motion (Ehrenfest theorem).

$$[\hat{H}, \hat{\boldsymbol{L}}] = 0 \quad \Rightarrow \quad \frac{\mathrm{d}\langle \boldsymbol{L} \rangle}{\mathrm{d}t} = 0 \,.$$

This is not true if $V(\boldsymbol{r})$ depends not only on r but also on θ and φ.

10.4. Unit angular momentum.

a. The action of $\hat{L}_z$ on the basis set $|\ell = 1, m\rangle$ is given by $\hat{L}_z|1,m\rangle = m\hbar\,|1,m\rangle$, i.e.

$$\hat{L}_z = \hbar \begin{pmatrix} 1 & 0 & 0 \\ 0 & 0 & 0 \\ 0 & 0 & -1 \end{pmatrix} .$$

The actions of $\hat{L}_x$ and $\hat{L}_y$ are obtained using the operators $\hat{L}_+$ and $\hat{L}_-$, whose matrix elements are deduced from the recursion relation (10.17):

$$\begin{aligned} &\hat{L}_+|1,1\rangle = 0\,, \quad \hat{L}_+|1,0\rangle = \hbar\sqrt{2}\,|1,1\rangle\,, \\ &\hat{L}_+|1,-1\rangle = \hbar\sqrt{2}\,|1,0\rangle\,, \\ &\hat{L}_-|1,1\rangle = \sqrt{2}\,|1,0\rangle\,, \quad \hat{L}_-|1,0\rangle = \hbar\sqrt{2}\,|1,-1\rangle\,, \\ &\hat{L}_-|1,-1\rangle = 0 \end{aligned}$$

Hence the matrices of $\hat{L}_x = (\hat{L}_+ + \hat{L}_-)/2$ and $\hat{L}_y = \mathrm{i}(\hat{L}_- - \hat{L}_+)/2$ are

$$\hat{L}_x = \frac{\hbar}{\sqrt{2}} \begin{pmatrix} 0 & 1 & 0 \\ 1 & 0 & 1 \\ 0 & 1 & 0 \end{pmatrix} , \qquad \hat{L}_y = \frac{\hbar}{\sqrt{2}} \begin{pmatrix} 0 & -\mathrm{i} & 0 \\ \mathrm{i} & 0 & -\mathrm{i} \\ 0 & \mathrm{i} & 0 \end{pmatrix} .$$

b. The eigenvectors of $\hat{L}_x$ are

$$|1,\pm 1\rangle_x = \frac{1}{2}\left(|1,1\rangle \pm \sqrt{2}|1,0\rangle + |1,-1\rangle\right) , \qquad \text{eigenvalues } \pm\hbar$$

$$|1,0\rangle_x = \frac{1}{\sqrt{2}}\left(|1,1\rangle - |1,-1\rangle\right) , \qquad \text{eigenvalue } 0 \,.$$

The eigenfunction corresponding to $m_x = +1$ is

$$\begin{aligned} \psi(\theta,\varphi) &= \frac{1}{2}[Y_1^1(\theta,\varphi) + Y_1^{-1}(\theta,\varphi)] + \frac{1}{\sqrt{2}} Y_1^0(\theta,\varphi) \\ &= \sqrt{\frac{3}{8\pi}}\,(\cos\theta - \mathrm{i}\sin\theta\sin\varphi)\,. \end{aligned}$$

Therefore

$$I(\theta,\varphi) = |\psi(\theta,\varphi)|^2 = \frac{3}{8\pi}(1 - \sin^2\theta\cos^2\varphi)\,.$$

10.5. Commutation relations for $\hat{J}_x^2$, $\hat{J}_y^2$ and $\hat{J}_z^2$.

a. Since $[\hat{J}^2, \hat{J}_z] = 0$, we have also $[\hat{J}^2, \hat{J}_z^2] = 0$. Hence $[\hat{J}_x^2 + \hat{J}_y^2, \hat{J}_z^2] = 0$, and therefore $[\hat{J}_z^2, \hat{J}_x^2] = [\hat{J}_y^2, \hat{J}_z^2]$. The third equality is obtained using a circular permutation of x, y and z.

One can also calculate these commutators explicitly:

$$[\hat{J}_x^2, \hat{J}_y^2] = \mathrm{i}\hbar\{\hat{J}_x, \{\hat{J}_y, \hat{J}_z\}\}$$
$$= \frac{\mathrm{i}\hbar}{3}(\{\hat{J}_x, \{\hat{J}_y, \hat{J}_z\}\} + \{\hat{J}_y, \{\hat{J}_z, \hat{J}_x\}\} + \{\hat{J}_z, \{\hat{J}_x, \hat{J}_y\}\}).$$

where we have set $\{\hat{A}, \hat{B}\} = \hat{A}\hat{B} + \hat{B}\hat{A}$.

b. For $j = 0$ the result is obvious since $|0, 0\rangle$ is an eigenstate of all components with eigenvalue zero.

For $j = 1/2$, $\hat{J}_x^2$, $\hat{J}_y^2$ and $\hat{J}_z^2$ are proportional to the unit 2×2 matrix, with eigenvalue $+\hbar^2/4$. They obviously commute.

c. For $j = 1$, we consider the matrix elements

$$\langle 1, m_2 | [\hat{J}_x^2, \hat{J}_z^2] | 1, m_1 \rangle = (m_1^2 - m_2^2) \langle 1, m_2 | \hat{J}_x^2 | 1, m_1 \rangle .$$

For $m_1^2 = m_2^2$, this is obviously zero. We only have to consider the cases $m_1 = 0, m_2 = \pm 1$. Since $\hat{J}_x|1, 0\rangle \propto (|1, 1\rangle + |1, -1\rangle)$ and $\hat{J}_x|1, \pm 1\rangle \propto |1, 0\rangle$ the corresponding scalar products, i.e. the matrix elements under consideration, vanish.

Owing to the (x, y, z) symmetry, the common eigenbasis is $\{|j = 1, m_x = 0\rangle, |j = 1, m_y = 0\rangle, |j = 1, m_z = 0\rangle\}$, where $|j = 1, m_i = 0\rangle$ is the eigenvector of $\hat{J}_i$ $(i = x, y, z)$ associated with the eigenvalue 0:

$$|j = 1, m_x = 0\rangle = \frac{1}{\sqrt{2}} (|1, 1\rangle - |1, -1\rangle) ,$$
$$|j = 1, m_y = 0\rangle = \frac{1}{\sqrt{2}} (|1, 1\rangle + |1, -1\rangle) ,$$
$$|j = 1, m_z = 0\rangle = |1, 0\rangle .$$

In spherical coordinates, the corresponding (angular) wave functions are x/r, y/r and z/r, with a normalization coefficient $(4\pi/3)^{-1/2}$.

Chapter 11

11.1. Expectation value of r for the Coulomb problem.

a. Using $\int_0^\infty |u_{n,\ell}(\rho)|^2 \mathrm{d}\rho = 1$ and $\varepsilon = 1/n^2$, we obtain the first identity.

b. To show the second identity, we use

$$\int u'_{n,\ell}\, u_{n,\ell}\, \mathrm{d}\rho = 0 , \qquad \int \rho^2 u'_{n,\ell} u''_{n,\ell}\, \mathrm{d}\rho = \int \rho u_{n,\ell} u''_{n,\ell}\, \mathrm{d}\rho ,$$
$$\int \rho u'_{n,\ell} u_{n,\ell}\, \mathrm{d}\rho = -1/2 \int \rho^2 u'_{n,\ell} u_{n,\ell}\, \mathrm{d}\rho = -\langle \rho \rangle ,$$

which can be shown by integrating by parts.

c. Summing the two equations derived in the previous parts of the exercise, we obtain the desired expression for $\langle\rho\rangle$.

11.2. Three-dimensional harmonic oscillator in spherical coordinates.

a. The radial equation is

$$\left(-\frac{\hbar^2}{2m}\frac{1}{r}\frac{\mathrm{d}^2}{\mathrm{d}r^2}r + \frac{1}{2}m\omega^2 r^2 + \frac{\ell(\ell+1)}{2mr^2}\hbar^2 - E_{n,\ell}\right) R_{n,\ell}(r) = 0\,.$$

We set $\rho = r\sqrt{m\omega/\hbar}$ and $\epsilon = E/\hbar\omega$ and obtain (11.30).

b. The energy levels are indeed of the form

$$E_n = (n+3/2)\hbar\omega\,, \quad \text{where} \quad n = 2n' + \ell\,,$$

and the corresponding eigenstates can be labeled as $|n,\ell,m\rangle$. There exists a degeneracy with respect to ℓ, but it is different from the case of the hydrogen atom. For a given value of the energy, i. e. of n, ℓ has the same parity as n. Therefore the successive levels correspond alternately to even and odd angular momenta:

$$\begin{array}{llll} n=0\,, & E=3\hbar\omega/2\,, & \ell=0\,; & \\ n=1\,, & E=5\hbar\omega/2\,, & \ell=1\,; & \\ n=2\,, & E=7\hbar\omega/2\,, & \ell=0,2\,; & \\ n=3\,, & E=9\hbar\omega/2\,, & \ell=1,3\,; & \text{etc.} \end{array}$$

For a given value of n, the $(n+1)(n+2)/2$ states $|n,\,\ell,\,m\rangle$ with $\ell = 0,2\ldots n$ or $\ell = 1,3\ldots n$ are linear combinations of the $(n+1)(n+2)/2$ states $|n_1;n_2;n_3\rangle$ with $n_1+n_2+n_3 = n$.

c. For $n = 1$, three orthogonal wave functions are

$$\varphi_{100}(\boldsymbol{r}) = Cx\mathrm{e}^{-\alpha r^2/2}\,, \quad \varphi_{010}(\boldsymbol{r}) = Cy\mathrm{e}^{-\alpha r^2/2}\,, \quad \varphi_{001}(\boldsymbol{r}) = Cz\mathrm{e}^{-\alpha r^2/2}$$

where C is a normalization constant. We therefore obtain, by expressing x, y, z in terms of r and $Y_{1,m}(\theta,\varphi)$ (see (10.30)–(10.32))

$$|n=1,\ \ell=1,\ m=0\rangle = |n_1=0;\ n_2=0;\ n_3=1\rangle\,,$$

$$\begin{aligned} &|n=1,\ \ell=1,\ m=\pm1\rangle \\ &= \mp\frac{1}{\sqrt{2}}|n_1=1;\ n_2=0;\ n_3=0\rangle - \frac{\mathrm{i}}{\sqrt{2}}|n_1=0;\ n_2=1;\ n_3=0\rangle\,. \end{aligned}$$

11.3. Relation between the Coulomb problem and the harmonic oscillator.

a. The change of variable $\rho \to x$ and of the unknown function $u(\rho) \to f(x)$ leads to the following equation for the Coulomb problem:

$$\frac{\mathrm{d}^2 f}{\mathrm{d}x^2} + \frac{2\alpha - 1}{x}\frac{\mathrm{d}f}{\mathrm{d}x} + \left(\frac{\alpha(\alpha-2) - 4\ell(\ell+1)}{x^2} + 8Z + \frac{4E}{E_\mathrm{I}}x^2\right) f(x) = 0\,.$$

The choice $\alpha = 1/2$ eliminates the term $\mathrm{d}f/\mathrm{d}x$ and leads to an equation with the same structure as the radial equation for the harmonic oscillator:

$$\left(\frac{\mathrm{d}^2}{\mathrm{d}x^2} - \frac{(2\ell+1/2)(2\ell+3/2)}{x^2} + 8Z + \frac{4E}{E_\mathrm{I}}x^2\right) f(x) = 0\,.$$

b. The correspondence between the parameters of the harmonic oscillator and the Coulomb problem is

$$\ell_{\mathrm{harm.}} \leftrightarrow (2\ell_{\mathrm{coul.}} + 1/2)\,, \quad K^2 \leftrightarrow -4E_{\mathrm{coul.}}/E_\mathrm{I}\,, \quad E_{\mathrm{harm.}}/\hbar\omega \leftrightarrow 4Z\,.$$

In other words, the roles of the coupling constant and of the energy eigenvalue are interchanged! The shift in ℓ ensures the proper ℓ degeneracy of the hydrogen levels.

c. From the result of exercise 11.2, we know that the eigenvalues for the energy are

$$E_{\mathrm{harm}} = K\,(2n' + \ell_{\mathrm{harm}} + 3/2)\,\hbar\omega\,.$$

Using the correspondence that we have just found, this yields

$$4Z = \sqrt{\frac{-4E_{\mathrm{coul.}}}{E_\mathrm{I}}}\;(2n' + 2\ell_{\mathrm{coul.}} + 2)\,,$$

which can also be written $E_{\mathrm{coul.}} = -Z^2 E_\mathrm{I}/(n' + \ell_{\mathrm{coul.}} + 1)^2$. We recover indeed the energy levels of the Coulomb problem (cf. (11.28)). Notice that this provides an expression for the Laguerre polynomials in terms for the Hermite polynomials.

11.4. Confirm or invalidate the following assertions.

a. True. In fact, if $[\hat{H}, \hat{\boldsymbol{L}}] = 0$ we have $[\hat{H}, \hat{L}_z] = 0$ and $[\hat{H}, \hat{L}^2] = 0$. We consider an eigenbasis common to $\hat{H}$, $\hat{L}^2$ and $\hat{L}_z$, $|E_{\ell,m}, \ell, m\rangle$. Since $[\hat{H}, \hat{\boldsymbol{L}}] = 0$ implies that $[\hat{H}, \hat{L}_\pm] = 0$, we obtain

$$\hat{L}_\pm \hat{H} |E_{\ell,m}, \ell, m\rangle = E_{\ell,m} \hat{L}_\pm |E_{\ell,m}, \ell, m\rangle = \hat{H}\hat{L}_\pm |E_{\ell,m}, \ell, m\rangle\,.$$

We know that $\hat{L}_\pm |E_{\ell,m}, \ell, m\rangle$ is an eigenstate of $\hat{L}_z$ with eigenvalue $(m+1)\hbar$. Therefore the set of states $\{|E_{\ell,m}, \ell, m\rangle, m = -\ell \ldots \ell\}$ are eigenstates of $\hat{H}$ with the same eigenvalue $E_{\ell,m} \equiv E_\ell$.

b. Wrong. The square of the angular momentum commutes with the Hamiltonian and the energy levels can be labeled by ℓ. It is only the Coulomb and harmonic potentials, which have special symmetry properties, which produce degeneracies with respect to ℓ.

11.5. Centrifugal-barrier effects. Consider the Hamiltonians

$$\hat{H}_\ell = \frac{\hat{p}_r^2}{2m} + \frac{\ell(\ell+1)\hbar^2}{2mr^2} + V(r)\,, \quad \text{where} \quad \hat{p}_r^2 = -\hbar^2 \left(\frac{1}{r}\frac{\partial}{\partial r} r\right)^2 ,$$

which act only on the variable r. The state $|n' = 0, \ell, m\rangle \equiv |\psi_\ell\rangle$, which is an eigenstate of

$$\hat{H} = \frac{\hat{p}_r^2}{2m} + \frac{\hat{L}^2}{2mr^2} + V(r)\,,$$

is an eigenstate of $\hat{H}_\ell$ with eigenvalue E_ℓ, which is the smallest eigenvalue of $\hat{H}_\ell$. We obviously have

$$\hat{H}_{\ell+1} = H_\ell + \frac{(\ell+1)\hbar^2}{mr^2}\,.$$

Taking the expectation value of this expression for the state $|\psi_{\ell+1}\rangle = |n' = 0, \ell+1, m\rangle$, we obtain

$$E_{\ell+1} = \langle\psi_{\ell+1}|\hat{H}_\ell|\psi_{\ell+1}\rangle + \left\langle \frac{(\ell+1)\hbar^2}{mr^2} \right\rangle .$$

We have $\langle\psi|\hat{H}_\ell|\psi\rangle \geq E_\ell$ for all ψ, and $(\ell+1)\hbar^2/mr^2$ is a positive operator: its expectation value for any state is positive. Therefore $E_{\ell+1} \geq E_\ell$ and, more quantitatively,

$$E_{\ell+1} - E_\ell \geq \frac{(\ell+1)\hbar^2}{m} \langle\psi_{\ell+1}|\frac{1}{r^2}|\psi_{\ell+1}\rangle\,.$$

11.6. Algebraic method for the hydrogen atom.

a. The expression for $A_\ell^- A_\ell^+$ is

$$A_\ell^- A_\ell^+ = \frac{\mathrm{d}^2}{\mathrm{d}\rho^2} - \frac{\ell(\ell+1)}{\rho^2} + \frac{2}{\rho} - \frac{1}{(\ell+1)^2}\,.$$

Therefore the radial equation can be written as

$$\left(A_\ell^- A_\ell^+\right) u_\ell = \left(\epsilon - \frac{1}{(\ell+1)^2}\right) u_\ell\,.$$

b. We find, for $A_\ell^+ A_\ell^-$,

$$A_\ell^+ A_\ell^- = \frac{\mathrm{d}^2}{\mathrm{d}\rho^2} - \frac{(\ell+1)(\ell+2)}{\rho^2} + \frac{2}{\rho} - \frac{1}{(\ell+1)^2}\,.$$

By multiplying (11.33) by A_ℓ^+, we obtain

$$\left(A_\ell^+ A_\ell^-\right) A_\ell^+ u_\ell = \left(\epsilon - \frac{1}{(\ell+1)^2}\right) A_\ell^+ u_\ell\,,$$

which can also be written

$$\left(\frac{\mathrm{d}^2}{\mathrm{d}\rho^2} - \frac{(\ell+1)(\ell+2)}{\rho^2} + \frac{2}{\rho}\right) A_\ell^+ u_\ell(\rho) = \varepsilon\, A_\ell^+ u_\ell(\rho)\,.$$

Therefore, $A_\ell^+ u_\ell(\rho)$ satisfies the radial equation with the same eigenvalue ε but with an angular momentum $\ell' = \ell + 1$.

c. Similarly, we can write the equation for the angular momentum ℓ as

$$\left(A_{\ell-1}^+ A_{\ell-1}^-\right) u_\ell = \left(\epsilon - \frac{1}{\ell^2}\right) u_\ell\,.$$

We then obtain the result that $A_{\ell-1}^- u_\ell(\rho)$ satisfies the radial equation with the same eigenvalue ε but with an angular momentum $\ell' = \ell - 1$.

d. By multiplying (11.33) by u_ℓ^* and integrating over ρ, we find

$$\int_0^\infty u_\ell^*(\rho)\left(A_\ell^- A_\ell^+ u_\ell(\rho)\right)\,\mathrm{d}\rho = \left(\varepsilon_\ell - \frac{1}{(\ell+1)^2}\right)\int_0^\infty |u_\ell(\rho)|^2 \mathrm{d}\rho\,.$$

We integrate the left-hand side of this equation by parts and obtain

$$\int_0^\infty u_\ell^*(\rho)\left(A_\ell^- A_\ell^+ u_\ell(\rho)\right)\,\mathrm{d}\rho = -\int_0^\infty |A_\ell^+ u_\ell(\rho)|^2 \mathrm{d}\rho\,.$$

We deduce that the quantity $\varepsilon - 1/(\ell+1)^2$ is necessarily negative, and hence

$$\varepsilon \leq \frac{1}{(\ell+1)^2}\,.$$

e. The argument is then analogous to the case of the harmonic oscillator or to the quantization of angular momentum. By repeatedly applying A_ℓ^+, one can increase the value of ℓ by an integer. This is limited from above since $\varepsilon_\ell \leq 1/(\ell+1)^2$, and there is a maximum value ℓ_{max} of ℓ such that

$$\varepsilon = \frac{1}{(\ell_{\mathrm{max}}+1)^2} \equiv \frac{1}{n^2}\,.$$

The function $A_{\ell_{\mathrm{max}}}^+ u_{\ell_{\mathrm{max}}}(\rho)$ is identically zero. Therefore, $u_{\ell_{\mathrm{max}}}$ satisfies

$$\left(\frac{\mathrm{d}}{\mathrm{d}\rho} - \frac{n}{\rho} + \frac{1}{n}\right) u_{\ell_{\mathrm{max}}}(\rho) = 0\,.$$

f. The energy levels are $E_n = -E_{\rm I}/n^2$. The solution of the above equation is $u_{\ell_{\rm max}}(\rho) \propto \rho^n {\rm e}^{-\rho/n}$, up to a normalization factor. Coming back to the function R_ℓ, we recover (11.22) for $\ell = n-1$ (i. e. $n' = 0$): $R_{\ell_{\rm max}}(\rho) \propto \rho^{n-1}{\rm e}^{-\rho\sqrt{\varepsilon}}$. By repeatedly applying $A^-_{\ell-1}$, we obtain the other solutions of the same energy $\varepsilon = 1/n^2$ for $\ell = n-2, n-3, \dots, 0$.

11.7. Molecular potential.

a. The radial equation is

$$\left(-\frac{\hbar^2}{2m_{\rm e}}\frac{1}{r}\frac{{\rm d}^2}{{\rm d}r^2}r + \frac{\ell(\ell+1)\hbar^2 + 2m_{\rm e}A}{2m_{\rm e}r^2} - \frac{B}{r}\right)R(r) = ER(r)\,.$$

b. We define S as a positive real number such that $S(S+1) = \ell(\ell+1) + 2m_{\rm e}A/\hbar^2$, i. e.

$$S = -\frac{1}{2} + \frac{1}{2}\sqrt{(2\ell+1)^2 + 8m_{\rm e}A/\hbar^2}\,.$$

Note that S is generally not an integer. We set, as usual, $a_1 = \hbar^2/Bm_{\rm e}$, $E = -\varepsilon m B^2/2\hbar^2$ and $r = \rho a_1$. The radial equation becomes

$$\left(\frac{1}{\rho}\frac{{\rm d}^2}{{\rm d}\rho^2}\rho - \frac{S(S+1)}{\rho^2} + \frac{2}{\rho} - \varepsilon\right) R(\rho) = 0\,.$$

c. As in the case of hydrogen, the normalizable solutions can be labeled by an integer $n' \geq 0$ and are of the form $R(\rho) = {\rm e}^{-\rho\sqrt{\varepsilon}}\rho^S P_{n',\ell}(\rho)$, where $P_{n',\ell}(\rho)$ is a polynomial of degree n', and we must have $\varepsilon = (n'+S+1)^{-2}$. The energies are then

$$E_{n',\ell} = -\frac{B^2 m_{\rm e}}{2\hbar^2}\frac{1}{(n'+S+1)^2}\,.$$

Note that this potential is quantitatively very different from a molecular potential even though it has the same global features (attractive at long distances and repulsive at short distances). The long range attractive force in a molecule is not a Coulomb force and the repulsion at short distances is much stronger than an r^{-2} potential.

Chapter 12

12.1. Products of Pauli matrices. We first check that $\hat\sigma_j^2 = 1$. For $j \neq k$, a direct calculation yields the result (12.38).

12.2. Algebra with Pauli matrices. We first evaluate the products $(\boldsymbol{\sigma}\cdot\boldsymbol{A})(\boldsymbol{\sigma}\cdot\boldsymbol{B}) = \sum_{jk}\sigma_j\sigma_k A_j B_k$. The result (12.38) of the previous exercise then yields the desired formula.

12.3. Spin and orbital angular momentum.

a. We must have $\int_0^\infty |R(r)|^2 r^2 \,\mathrm{d}r = 1/2$, since the Y_ℓ^m's are orthonormal and $\int (|\psi_+|^2 + |\psi_-|^2)\,\mathrm{d}^3r = 1$.
b. We find $p(+\hbar/2) = 2/3$, $p(-\hbar/2) = 1/3$ for S_z, and $p(+\hbar/2) = 1/3$, $p(-\hbar/2) = 2/3$ for S_x.
c. $L_z = \hbar$ and $L_z = 0$ with $p(+\hbar) = 1/6$, $p(0) = 5/6$.

12.4. Geometric origin of the commutation relations of $\hat{\boldsymbol{J}}$.

a. (i) The second-order terms in $\mathrm{d}\alpha$ are, a priori, unchanged if $\beta = 0$. However, for $\mathrm{d}\beta = 0$, the composition of the four operations reduces obviously to the identity. Therefore the second-order terms in $\mathrm{d}\alpha$ vanish, and so do the terms in $\mathrm{d}\beta^2$ (and, of course the first order terms in $\mathrm{d}\alpha$ and $\mathrm{d}\beta$).

(ii) By applying successively the four infinitesimal rotations:

$$\begin{aligned} \boldsymbol{V} &\longrightarrow \boldsymbol{V}_1 = \boldsymbol{V} + \mathrm{d}\alpha\, \boldsymbol{e}_x \times \boldsymbol{V}\,, \\ \boldsymbol{V}_1 &\longrightarrow \boldsymbol{V}_2 = \boldsymbol{V}_1 + \mathrm{d}\beta\, \boldsymbol{e}_y \times \boldsymbol{V}_1\,, \\ \boldsymbol{V}_2 &\longrightarrow \boldsymbol{V}_3 = \boldsymbol{V}_2 - \mathrm{d}\alpha\, \boldsymbol{e}_x \times \boldsymbol{V}_2\,, \\ \boldsymbol{V}_3 &\longrightarrow \boldsymbol{V}_4 = \boldsymbol{V}_3 - \mathrm{d}\beta\, \boldsymbol{e}_y \times \boldsymbol{V}_3\,, \end{aligned}$$

we find with no difficulty that the resulting transformation is, to second order in $\mathrm{d}\alpha$ and $\mathrm{d}\beta$,

$$\boldsymbol{V} \longrightarrow \boldsymbol{V}_4 = \boldsymbol{V} - \mathrm{d}\alpha\,\mathrm{d}\beta\, \boldsymbol{u}_z \times \boldsymbol{V}\,.$$

(iii) This transformation $\boldsymbol{V} \to \boldsymbol{V}_4$ corresponds indeed to a rotation by an angle $-\mathrm{d}\alpha\,\mathrm{d}\beta$ around the z axis.

b. (i) The transformation of the state vector is $|\psi\rangle \to |\psi_4\rangle$, where

$$\begin{aligned} |\psi_4\rangle = &\left(1 + \frac{\mathrm{i}\mathrm{d}\beta}{\hbar}\hat{J}_y + \ldots\right)\left(1 + \frac{\mathrm{i}\mathrm{d}\alpha}{\hbar}\hat{J}_x + \ldots\right) \\ &\times \left(1 - \frac{\mathrm{i}\mathrm{d}\beta}{\hbar}\hat{J}_y + \ldots\right)\left(1 - \frac{\mathrm{i}\mathrm{d}\alpha}{\hbar}\hat{J}_x + \ldots\right)|\psi\rangle\,. \end{aligned}$$

To second order in $\mathrm{d}\alpha$ and $\mathrm{d}\beta$, we remark as above that the terms in $\mathrm{d}\alpha^2$ and $\mathrm{d}\beta^2$ do not contribute, and we obtain

$$|\psi_4\rangle = \left(1 + \frac{\mathrm{d}\alpha\,\mathrm{d}\beta}{\hbar^2}[\hat{J}_x, \hat{J}_y]\right)|\psi\rangle\,.$$

(ii) We know that the composition of the four rotations corresponds geometrically to a rotation by an angle $-\mathrm{d}\alpha\,\mathrm{d}\beta$ around the z axis. Therefore we are led to the set $[\hat{J}_x, \hat{J}_y] = \mathrm{i}\hbar\hat{J}_z$ in order to have

$$|\psi_4\rangle = \left(1 + \frac{\mathrm{i}\,\mathrm{d}\alpha\,\mathrm{d}\beta}{\hbar}\hat{J}_z\right)|\psi\rangle\,.$$

This commutation relation thus has a simple geometric origin, in the noncommutativity of the three-dimensional rotation group.

c. (i) For a spin-1/2 system, we obtain the following in a rotation of $\mathrm{d}\varphi$ around z:

$$\begin{pmatrix} \alpha \\ \beta \end{pmatrix} \rightarrow \begin{pmatrix} (1-\mathrm{i}\,\mathrm{d}\varphi/2)\,\alpha \\ (1+\mathrm{i}\,\mathrm{d}\varphi/2)\,\beta \end{pmatrix} .$$

(ii) This equation can be integrated for an arbitrary rotation of angle φ to obtain

$$\begin{pmatrix} \alpha \\ \beta \end{pmatrix} \rightarrow \begin{pmatrix} \mathrm{e}^{-\mathrm{i}\varphi/2}\,\alpha \\ \mathrm{e}^{\mathrm{i}\varphi/2}\,\beta \end{pmatrix} .$$

(iii) In the Larmor precession, the transformation of the state vector between times 0 and t corresponds to a rotation of angle $\omega_0 t$.

Chapter 13

13.1. Permutation operator. We use the eigenbasis of $\hat{S}_{1z}$ and $\hat{S}_{2z}$. In this basis, we find

$$\begin{aligned}
&(\hat{\sigma}_{1x}\hat{\sigma}_{2x} + \hat{\sigma}_{1y}\hat{\sigma}_{2y})\,|\sigma;\sigma\rangle = 0\,, \\
&(\hat{\sigma}_{1x}\hat{\sigma}_{2x} + \hat{\sigma}_{1y}\hat{\sigma}_{2y})\,|\sigma;-\sigma\rangle = 2|-\sigma;\sigma\rangle\,, \\
&(1 + \hat{\sigma}_{1z}\hat{\sigma}_{2z})\,|\sigma;\sigma\rangle = 2|\sigma;\sigma\rangle\,, \\
&(1 + \hat{\sigma}_{1z}\hat{\sigma}_{2z})\,|\sigma;-\sigma\rangle = 0\,,
\end{aligned}$$

where $\sigma = \pm 1$. Hence the result

$$\frac{1}{2}\,(1 + \hat{\boldsymbol{\sigma}}_1\hat{\boldsymbol{\sigma}}_2)\,|\sigma;\sigma\rangle = |\sigma;\sigma\rangle\,, \qquad \frac{1}{2}\,(1 + \hat{\boldsymbol{\sigma}}_1\hat{\boldsymbol{\sigma}}_2)\,|\sigma;-\sigma\rangle = |-\sigma;\sigma\rangle\,.$$

13.2. The singlet state. This invariance of the decomposition of the singlet state when $\boldsymbol{u}$ varies comes from the fact that the state is of angular momentum zero, and therefore that it is rotation invariant.

13.3. Spin and magnetic moment of the deuteron.

a. The eigenvalues of $\hat{K}^2$ are $K(K+1)\hbar^2$, where

$$K = J+I, J+I-1, J+I-2, \ldots, |J-I|\,.$$

b. We have $\hat{W} = A\hat{\boldsymbol{J}}\cdot\hat{\boldsymbol{I}}/\hbar^2$, where $A = a g_{\mathrm{I}} g_J \mu_{\mathrm{B}} \mu_{\mathrm{N}}$. We obtain $\hat{\boldsymbol{J}}\cdot\hat{\boldsymbol{I}} = (\hat{\boldsymbol{K}}^2 - \hat{\boldsymbol{J}}^2 - \hat{\boldsymbol{I}}^2)/2$, and hence the expression for W.
$E_{I,J,K} = E_{\mathrm{I}} + E_J + A\,[K(K+1) - J(J+1) - I(I+1)]\,/2$.

c. $E_{I,J,K} - E_{I,J,(K-1)} = AK$.

d. The electron has a total angular momentum of 1/2, its spin. The possible values of K are therefore $I \pm 1/2$. One of the levels is split into 4 sub-levels (angular momentum 3/2), while the other is split into two (angular momentum 1/2). Therefore $I = 1$.
e. Triplet state ($S = 1$).
f. We obtain the following value for the hyperfine splitting in deuterium:

$$\Delta E = \frac{3A}{2} = \frac{1.72}{1836} \alpha^4 m_e c^2 \sim 1.36\ 10^{-6}\,\text{eV}\,.$$

Hence the wavelength and frequency of the emitted radiation are

$$\lambda \sim 91\ \text{cm}\,, \qquad \nu \sim 328\ \text{MHz}\,.$$

The experimental value is $\lambda = 91.5720\,\text{cm}$. In a more accurate calculation, one must incorporate a more accurate value of g_{I} (0.8574) and take into account reduced mass effects and relativistic effects.

13.4. Determination of the Clebsch–Gordan coefficients.

a. Multiplying the relation $\hat{J}_z = \hat{J}_{1z} + \hat{J}_{2z}$ on the left by $\langle j_1, m_1; j_2, m_2|$ and on the right by $|j_1, j_2; j, m\rangle$, we find

$$\hbar\,(m_1 + m_2 - m)\; C^{j,m}_{j_1,m_1;j_2,m_2} = 0\,,$$

which means that the Clebsch–Gordan coefficient is zero if $m \neq m_1 + m_2$.
b. We have

$$\langle j_1, m_1; j_2, m_2|\hat{J}_{1+} =$$
$$\sqrt{j_1(j_1+1) - m_1(m_1-1)}\ \langle j_1, m_1 - 1; j_2, m_2|\,,$$
$$\langle j_1, m_1; j_2, m_2|\hat{J}_{2+} =$$
$$\sqrt{j_2(j_2+1) - m_2(m_2-1)}\ \langle j_1, m_1; j_2, m_2 - 1|\,,$$
$$\hat{J}_+|j_1, j_2; j, m\rangle =$$
$$\sqrt{j(j+1) - m(m+1)}\ |j_1, j_2; j, m+1\rangle\,.$$

Multiplying the relation $\hat{J}_+ = \hat{J}_{1+} + \hat{J}_{2+}$ on the left by $\langle j_1, m_1; j_2, m_2|$ and on the right by $|j_1, j_2; j, m\rangle$, we find the relation given in the exercise. For $m = j$, the recursion relation simplifies and reduces to

$$C^{j,j}_{j_1,m_1-1;j_2,m_2} = -\sqrt{\frac{j_2(j_2+1) - m_2(m_2-1)}{j_1(j_1+1) - m_1(m_1-1)}}\ C^{j,j}_{j_1,m_1;j_2,m_2-1}\,.$$

It therefore suffices to know one coefficient, for instance $C^{j,j}_{j_1,j_1;j_2,j-j_1}$, in order to determine all those of the form $C^{j,j}_{j_1,m_1;j_2,m_2}$. Since the vectors $|j_1, j_2; j, j\rangle$ and $|j_1, m_1; j_2, m_2\rangle$ are normalized, we obtain the relation

$$\sum_{m_1} \left| C^{j,j}_{j_1,m_1;j_2,j-m_1} \right|^2 = 1 ,$$

change

which allows us to determine these coefficients up to a global phase factor.

c. Imposing the requirement that the coefficient $C^{j,j}_{j_1,j_1;j_2,j-j_1}$ is real and positive removes the ambiguity in the phase and determines completely the coefficients $C^{j,j}_{j_1,m_1;j_2,m_2}$.

d. This recursion relation can be deduced by taking the matrix element of $\hat{J}_- = \hat{J}_{1-} + \hat{J}_{2-}$. Knowing all the coefficients of the type $C^{j,j}_{j_1,m_1;j_2,m_2}$, one can then deduce all the coefficients of the type $C^{j,j-1}_{j_1,m_1;j_2,m_2}$, then $C^{j,j-2}_{j_1,m_1;j_2,m_2}$ and so on.

13.5. Scalar operators.

a. Taking the matrix element of $[\hat{J}_z, \hat{O}]$ between $\langle \alpha, j, m|$ and $|\beta, j', m'\rangle$, we find

$$\hbar\,(m - m')\,\langle \alpha, j, m|\hat{O}|\beta, j', m'\rangle = 0 .$$

If $m \neq m'$, the matrix element therefore vanishes.

b. Taking the matrix element of $\hat{J}_+\hat{O} = \hat{O}\hat{J}_+$ between $\langle \alpha, j, m+1|$ and $|\beta, j', m\rangle$, we find

$$\begin{aligned}\sqrt{j(j+1) - m(m+1)}\;\langle \alpha, j, m|\hat{O}|\beta, j', m\rangle \\ = \sqrt{j'(j'+1) - m(m+1)}\;\langle \alpha, j, m+1|\hat{O}|\beta, j', m+1\rangle ,\end{aligned}$$

which is identical to the first relation in the exercise. Similarly, taking the matrix element of $\hat{J}_-\hat{O} = \hat{O}\hat{J}_-$ between $\langle \alpha, j, m|$ and $|\beta, j', m+1\rangle$, we find the second relation in the exercise.

c. If $\mathcal{O}_m$ and $\mathcal{O}_{m+1}$ are nonzero, the product of the two relations above yields $j = j'$ and their ratio gives $\mathcal{O}^2_m = \mathcal{O}^2_{m+1}$, i. e. $\mathcal{O}_m = \mathcal{O}_{m+1}$ since the two coefficients $\mathcal{O}_m$ and $\mathcal{O}_{m+1}$ have the same sign, given their recursion relation.

13.6. Vector operators and the Wigner–Eckart theorem.

a. It is straightforward to check the property. Consider for instance $\hat{\boldsymbol{V}} = \hat{\boldsymbol{r}}$, and $j = x$, $k = y$:

$$[\hat{L}_x, \hat{y}] = [\hat{y}\hat{p}_z - \hat{z}\hat{p}_y, \hat{y}] = \mathrm{i}\hbar\hat{z} .$$

Here, we have assumed $\hat{\boldsymbol{J}} = \hat{\boldsymbol{L}}$, but the incorporation of the spin of the particle would not change anything, since spin observables commute with $\hat{\boldsymbol{r}}$.

b. Again the verification is simple. Consider, for instance,

$$\left[\hat{J}_\pm, \hat{V}_{+1}\right] = \frac{-1}{\sqrt{2}}\left[\hat{J}_x \pm \mathrm{i}\hat{J}_y, \hat{V}_x + \mathrm{i}\hat{V}_y\right] = \frac{-1}{\sqrt{2}}\left(\mathrm{i}\left[\hat{J}_x, \hat{V}_y\right] \pm \mathrm{i}\left[\hat{J}_y, \hat{V}_x\right]\right)$$
$$= \frac{-\hbar}{\sqrt{2}}\left(-\hat{V}_z \pm \hat{V}_z\right).$$

This is equal to $\hbar\sqrt{2}\,\hat{V}_0$ for $\hat{J}_-$ and zero for $\hat{J}_+$.

c. (i) Multiplying the relation $[\hat{J}_z, \hat{V}_q] = \hbar q \hat{V}_q$ on the left by $\langle \alpha, j, m|$ and on the right by $|\beta, j', m'\rangle$, we find

$$\hbar\,(m - m' - q)\,\langle \alpha, j, m|\hat{V}_q|\beta, j', m'\rangle = 0\,,$$

which yields the result that $\langle \alpha, j, m|\hat{V}_q|\beta, j', m'\rangle$ vanishes if $m \neq m' + q$.

(ii) Multiplying the relation

$$\hat{J}_\pm \hat{V}_q = \hat{V}_q \hat{J}_\pm + \hbar\sqrt{2 - q(q \pm 1)}\;\hat{V}_q$$

on the left by $\langle \alpha, j, m|$ and on the right by $|\beta, j', m'\rangle$, we find

$$\sqrt{j(j+1) - m(m \mp 1)}\;\langle \alpha, j, m \mp 1|\hat{V}_q|\beta, j', m'\rangle$$
$$= \sqrt{j'(j'+1) - m(m \pm 1)}\;\langle \alpha, j, m|\hat{V}_q|\beta, j', m' \pm 1\rangle$$
$$+\sqrt{2 - q(q \pm 1)}\;\langle \alpha, j, m|\hat{V}_{q\pm 1}|\beta, j', m'\rangle\,.$$

If we make the identifications $(j_1, m_1) \leftrightarrow (j', m')$, $(j_2, m_2) \leftrightarrow (1, q)$ and

$$C^{j,m}_{j',m';1,q} \;\leftrightarrow\; \langle \alpha, j, m|\hat{V}_q|\beta, j', m'\rangle\,,$$

we observe that the required matrix elements satisfy the same recursion relations (13.34) and (13.35) as do the Clebsch–Gordan coefficients $C^{j,m}_{j',m';1,q}$.

(iii) The relations (13.34) and (13.35) completely define the Clebsch–Gordan coefficients up to a multiplicative factor, fixed by the normalization of the states $|j_1, j_2; j, m\rangle$. We therefore conclude that the matrix elements under consideration are proportional to those Clebsch–Gordan coefficients.

(iv) If $|j - j'| > 1$, the Clebsch–Gordan coefficients $C^{j,m}_{j',m';1,q}$ vanish whatever m, m' and q are. The same holds for the matrix elements of $\hat{V}_q$, as can be shown directly. Suppose, for instance, that $j > j' + 1$. Using the recursion relation of question c(i), we first deduce that the matrix elements $\langle \alpha, j, j|V_q|\beta, j', m'\rangle$ vanish for all values of m' and q between $-j'$ and j' and between -1 and 1, respectively. Using then the relation proven in c(ii), one can prove that the matrix elements $\langle \alpha, j, j-1|V_q|\beta, j', m'\rangle$ are also all zero, and so on.

Chapter 14

14.1. Bell measurement.

a. One can find $(+\hbar/2, +\hbar/2)$ with probability $|\alpha|^2$, $(+\hbar/2, -\hbar/2)$ with probability $|\beta|^2$, etc.

b. The operator associated with the occupation of a given state $|\Psi\rangle$ is the projector $P_\Psi = |\Psi\rangle\langle\Psi|$. The eigenvalues of this operator are 0 and 1. We have the following:

State $|\Psi_+\rangle$ occupied: probability $|\alpha + \delta|^2/2$;
State $|\Psi_-\rangle$ occupied: probability $|\alpha - \delta|^2/2$;
State $|\Phi_+\rangle$ occupied: probability $|\beta + \gamma|^2/2$;
State $|\Phi_-\rangle$ occupied: probability $|\beta - \gamma|^2/2$.

The Bell basis is an orthonormal basis and the sum of these four probabilities is equal to 1.

14.2. Quantum teleportation of a spin state.

a. The three-spin state is

$$|\Psi\rangle = \frac{\alpha}{\sqrt{2}}|+;+;-;\rangle + \frac{\beta}{\sqrt{2}}|-;+;-;\rangle - \frac{\alpha}{\sqrt{2}}|+;-;+;\rangle -! \frac{\beta}{\sqrt{2}}|-;-;+;\rangle .$$

This state can be written in the basis of the three-particle states formed by (i) the Bell basis for the subsystem $A+B$ multiplied by (ii) the basis $|\pm\rangle$ for C:

$$\begin{aligned}|\Psi\rangle &= \frac{1}{2}|\Psi_+\rangle \otimes (\alpha|-\rangle - \beta|+\rangle) + \frac{1}{2}|\Psi_-\rangle \otimes (\alpha|-\rangle + \beta|+\rangle) \\ &\quad - \frac{1}{2}|\Phi_+\rangle \otimes (\alpha|+\rangle - \beta|-\rangle) - \frac{1}{2}|\Phi_-\rangle \otimes (\alpha|+\rangle + \beta|-\rangle) .\end{aligned}$$

The probability of finding the pair AB in any particular Bell state is thus $(|\alpha|^2 + |\beta|^2)/4 = 1/4$.

b. After a measurement which yields the result "the pair AB is in the state $|\Phi_-\rangle$", the state of the spin of C is $\alpha|+\rangle + \beta|-\rangle$.

c. In order to teleport the unknown state $\alpha|+\rangle + \beta|-\rangle$ from particle A to particle C, Alice must not try to measure this state. She has "only" to perform a Bell measurement on the pair AB and then transmit her result to Bob. If Alice finds the pair AB in the state $|\Phi_-\rangle$ (this happens with a probability 1/4), Bob has nothing to do: the state of the spin of C after Alice's measurement is equal to the state of the spin of A before this measurement. In the remaining cases, Bob can reconstruct the initial spin state of A using a simple transformation. For instance, if Alice finds the pair AB in the Bell state $|\Phi_+\rangle$, the state of the spin of C after the measurement is $\alpha|+\rangle - \beta|-\rangle$, and Bob can convert it back to the initial state $\alpha|+\rangle + \beta|-\rangle$ using a rotation by an angle π around the z axis of the spin of C.

d. One cannot use this idea to transmit information faster than with a classical channel. As long as Alice has not told Bob the result of her Bell measurement, Bob has no valuable information. The density operator associated with his spin is simply $(1/2)\hat{1}$ (see Appendix Appendix D). It is only after he has learned about Alice's result, and after he has rejected or reconstructed the fraction $(3/4)$ of the experimental runs which did not yield $|\Phi_-\rangle$, that Bob can benefit from this teleportation of the quantum state of particle A.

Chapter 15

15.1. The Lorentz force in quantum mechanics.

a. One can check easily that $\boldsymbol{\nabla} \times \boldsymbol{A} = \boldsymbol{B}$.

b. The classical equations of motion $m\ddot{\boldsymbol{r}} = \boldsymbol{f}$ and $\dot{\boldsymbol{r}} = \boldsymbol{v}$ give $\ddot{\boldsymbol{r}} = (q/m)\dot{\boldsymbol{r}} \times \boldsymbol{B}$. The Lorentz force does not do any work. The energy boils down to the kinetic energy $E = mv^2/2$, which is a constant of the motion.

The longitudinal velocity, parallel to $\boldsymbol{B}$, is constant. The transverse velocity, perpendicular to $\boldsymbol{B}$, rotates around $\boldsymbol{B}$ with an angular velocity $\omega = -qB/m$. The trajectory is a helix with its axis: parallel to the z axis: the motion along z is uniform and the motion in the xy plane is circular.

c. We have $\hat{A}_x = -B\hat{y}/2$, $\hat{A}_y = B\hat{x}/2$, $\hat{A}_z = 0$. Therefore

$$[\hat{p}_x, \hat{A}_x] = [\hat{p}_y, \hat{A}_y] = [\hat{p}_z, \hat{A}_z] = 0 \quad \Rightarrow \quad \hat{\boldsymbol{p}} \cdot \hat{\boldsymbol{A}} = \hat{\boldsymbol{A}} \cdot \hat{\boldsymbol{p}} .$$

The commutation relations between the components of $\hat{\boldsymbol{u}}$ are

$$[\hat{u}_x, \hat{u}_y] = [\hat{p}_x - q\hat{A}_x, \hat{p}_y - q\hat{A}_y] = -q[\hat{p}_x, \hat{A}_y] - q[\hat{A}_x, \hat{p}_y] = \mathrm{i}\hbar qB$$

and $[\hat{u}_x, \hat{u}_z] = [\hat{u}_y, \hat{u}_z] = 0$.

d. We have $[\hat{B}, \hat{C}^2] = [\hat{B}, \hat{C}]\hat{C} + \hat{C}[\hat{B}, \hat{C}]$. Therefore

$$[\hat{\boldsymbol{r}}, \hat{H}] = \frac{\mathrm{i}\hbar}{m}(\hat{\boldsymbol{p}} - q\hat{\boldsymbol{A}}) = \frac{\mathrm{i}\hbar}{m}\hat{\boldsymbol{u}} ,$$

and, by applying the Ehrenfest theorem,

$$\frac{\mathrm{d}}{\mathrm{d}t}\langle \boldsymbol{r} \rangle = \frac{1}{\mathrm{i}\hbar}\langle \psi | [\hat{\boldsymbol{r}}, \hat{H}] | \psi \rangle = \frac{\langle \boldsymbol{u} \rangle}{m} .$$

On the other hand, we calculate the following commutators:

$$[\hat{u}_x, \hat{H}] = \frac{1}{2m}[\hat{u}_x, \hat{u}_y^2] = \mathrm{i}\hbar\frac{qB}{m}\hat{u}_y ,$$

$$[\hat{u}_y, \hat{H}] = \frac{1}{2m}[\hat{u}_y, \hat{u}_x^2] = -\mathrm{i}\hbar\frac{qB}{m}\hat{u}_x ,$$

and $[\hat{u}_z, \hat{H}] = 0$. In other words,

$$[\hat{\boldsymbol{u}}, \hat{H}] = \mathrm{i}\frac{\hbar q}{m}\,\hat{\boldsymbol{u}} \times \boldsymbol{B} \quad \Rightarrow \quad \frac{\mathrm{d}}{\mathrm{d}t}\langle \boldsymbol{u} \rangle = \frac{q}{m}\langle \boldsymbol{u} \rangle \times \boldsymbol{B}\,.$$

Altogether, we obtain

$$\frac{\mathrm{d}^2}{\mathrm{d}t^2}\langle \boldsymbol{r} \rangle = \frac{q}{m}\frac{\mathrm{d}}{\mathrm{d}t}\langle \boldsymbol{r} \rangle \times \boldsymbol{B}\,,$$

which is identical to the classical equations of motion. Note that we recover the classical equations of motion identically, not approximately, because the Hamiltonian is a second-degree polynomial in the dynamical variables.

e. If we set $\hat{\boldsymbol{v}} = \hat{\boldsymbol{u}}/m$, the observable $\hat{\boldsymbol{v}}$ corresponds to the velocity operator. In terms of that observable, the Hamiltonian can be written as $\hat{H} = m\hat{\boldsymbol{v}}^2/2$, which is the indeed the kinetic energy. Therefore, in a magnetic field, the linear momentum $m\hat{\boldsymbol{v}}$ does not coincide with the conjugate momentum $\hat{\boldsymbol{p}}$. These quantities are related by $m\hat{\boldsymbol{v}} = \hat{\boldsymbol{p}} - q\hat{\boldsymbol{A}}$, which corresponds to the classical relation (15.28).

f. The commutation relations between the components of $\hat{\boldsymbol{u}}$ show that in the presence of a magnetic field, the various components of the velocity cannot be defined simultaneously. One can define simultaneously the longitudinal component and one of the transverse components. The two transverse components satisfy the uncertainty relation

$$\Delta v_x\, \Delta v_y \geq \frac{\hbar}{2}\frac{|qB|}{m^2}\,.$$

15.2. Landau levels.

a. The eigenvalue equation for the energy reads

$$\frac{-\hbar^2}{2m}\left(\frac{\partial^2}{\partial x^2} + \left(\frac{\partial}{\partial y} - \mathrm{i}\frac{qB}{\hbar}x\right)^2 + \frac{\partial^2}{\partial z^2}\right)\Psi(x,y,z) = E_{\mathrm{tot}}\,\Psi(x,y,z)\,.$$

b. This eigenvalue equation is separable and one can check immediately that the functions $\Psi(x,y,z) = \mathrm{e}^{\mathrm{i}k_z z}\,\psi(x,y)$, where $\psi(x,y)$ satisfies (15.36) with $E = E_{\mathrm{tot}} - \hbar^2 k_z^2/2m$, are eigenfunctions of the energy. As in the classical case, the motion along z is linear and uniform.

c. (i) The substitution $\psi(x,y) = \mathrm{e}^{\mathrm{i}k_y y}\,\chi(x)$ leads to the following equation for $\chi(x)$:

$$-\frac{\hbar^2}{2m}\frac{\mathrm{d}^2\chi}{\mathrm{d}x^2} + \frac{1}{2}m\omega_{\mathrm{c}}^2\,(x - x_{\mathrm{c}})^2\,\chi = E\chi\,,$$

where we have set $x_{\mathrm{c}} = \hbar k_y/(qB)$. This is the Schrödinger equation for a one-dimensional harmonic oscillator of frequency $\omega_{\mathrm{c}}/2\pi$, centered at x_{c}.

(ii) The energy eigenvalues are $E = (n + 1/2)\,\hbar\omega_c$, where n is a nonnegative integer. These eigenvalues do not depend on k_y.

d. (i) The periodic boundary conditions for the y axis imply $e^{ik_y Y} = 1$, i.e. $k_y = 2\pi j/Y$, where j is an integer, which can a priori be positive or negative.

(ii) In order for us to have a wave function localized in the desired rectangle, the center of the oscillator corresponding to the motion along the x axis has to be between 0 and X:

$$0 < x_c < X \quad \Rightarrow \quad 0 < j < j_{\max} = \frac{qBXY}{2\pi\hbar} \,.$$

Since the extension of the wave function along x is of the order of a few times a_0 for the first Landau levels, the hypothesis $a_0 \ll X$ implies that the particle is indeed localized, with a probability close to 1, in the rectangle $X \times Y$.

(iii) The number of independent states corresponding to a given Landau level is $j_{\max} = \Phi/\Phi_0$, where we have set $\Phi = BXY$ and $\Phi_0 = 2\pi\hbar/q$. This is the degeneracy of the level. The functions $\psi_n(x, y) = e^{ik_y y}\chi_n(x)$ constitute a basis for this level. Another possible basis is obtained by exchanging the roles of x and y, using the gauge $\boldsymbol{A}(\boldsymbol{r}) = -By\,\boldsymbol{u}_x$. A third possible gauge choice is the symmetric gauge, studied in the next exercise, which leads to a third possible basis for each of the Landau levels (for simplicity, the next exercise is actually restricted to the lowest Landau level).

15.3. The lowest Landau level (LLL).

a. For the gauge chosen in the exercise, the eigenvalue equation for the motion in the xy plane is

$$\left[\frac{-\hbar^2}{2m}\left(\frac{\partial^2}{\partial x^2} + \frac{\partial^2}{\partial y^2}\right) - \frac{\omega_c}{2}\hat{L}_z + \frac{1}{8}m\omega_c^2\left(x^2 + y^2\right)\right]\Psi(x, y)$$
$$= E\,\Psi(x, y)\,.$$

b. We introduce the polar coordinates ρ, θ in the xy plane. The functions $\psi_\ell(x, y) = (x + iy)^\ell e^{-(x^2+y^2)/(2\,a_0^2)} = \rho^\ell e^{i\ell\theta} e^{-\rho^2/(2a_0^2)}$ are eigenstates of $\hat{L}_z = -i\hbar\,\partial/\partial\theta$, with eigenvalue $\ell\hbar$. Inserting this result into the above eigenvalue equation, we reach the desired result, after a relatively long but straightforward calculation.

c. The state $\psi_\ell(x, y)$ is relevant if this wave function is essentially localized inside the disk of radius R. The probability density $|\psi_\ell(x, y)|^2 \propto \rho^{2\ell} e^{-\rho^2/a_0^2}$ has a maximum located at a distance $\ell^{1/2}\,a_0$ from the origin, with a width $\ell^{1/4} a_0$ (for $\ell \gg 1$). Therefore the quantum numbers ℓ must be between 0 and $\ell_{\max} = R^2/a_0^2$, which can also be written $\ell_{\max} = \Phi/\Phi_0$, where $\Phi_0 = 2\pi\hbar/q$ and $\Phi = \pi R^2 B$ represents the field flux across the accessible surface. We indeed recover the degeneracy found in the previous exercise.

15.4. The Aharonov–Bohm effect.

a. We denote the distances OBC and $OB'C$ by D and D', respectively. If the vector potential is zero, the classical action corresponds to a uniform motion at velocity $D/\Delta t$ (or $D'/\Delta t$) between O and B (or B') and then between B (or B') and C:

$$S_0 = \frac{mD^2}{2\,\Delta t}\,, \qquad S_0' = \frac{mD'^2}{2\,\Delta t}\,,$$

where $\Delta t = t_2 - t_1$. For a point C located at a distance x from the center of the screen O', we find $D^2 - D'^2 \simeq 2D_0 ax/L$, where D_0 represents the distance $OBO' = OB'O'$, a is the distance between the holes and L is the distance between the plane pierced by the two holes and the detection screen (we assume $x \ll L$). The quantity $D_0/\Delta t$ represents the average velocity v of the particles, and we set $\lambda = h/(mv)$. We find

$$\begin{aligned} |A(C)|^2 \propto \left| \mathrm{e}^{\mathrm{i}S_0/\hbar} + \mathrm{e}^{\mathrm{i}S_0'/\hbar} \right|^2 &\propto 1 + \cos\left[(S_0 - S_0')/\hbar\right] \\ &= 1 + \cos(2\pi x/x_{\mathrm{s}})\,, \end{aligned}$$

which corresponds to the usual signal found in a Young double slit experiment, with a fringe spacing $x_{\mathrm{s}} = \lambda L/a$.

b. When a current flows in the solenoid, the vector potential is not zero anymore, and the classical action is changed because of the term $q\dot{\boldsymbol{r}}\cdot\boldsymbol{A}(\boldsymbol{r})$ in the Lagrangian (15.27). The classical trajectories are not modified, since no force acts on the particle. We have thus

$$S = S_0 + \int_{OBC} q\,\dot{\boldsymbol{r}}\cdot\boldsymbol{A}(\boldsymbol{r})\,\mathrm{d}t\,, \qquad S' = S_0' + \int_{OB'C} q\,\dot{\boldsymbol{r}}\cdot\boldsymbol{A}(\boldsymbol{r})\,\mathrm{d}t\,,$$

and the intensity at C is

$$|A(C)|^2 \propto 1 + \cos\left[(S_0 - S_0')/\hbar + \varphi\right]\,,$$

where the phase φ reads

$$\begin{aligned} \varphi &= \frac{q}{\hbar}\left(\int_{OBC} \dot{\boldsymbol{r}}\cdot\boldsymbol{A}(\boldsymbol{r})\,\mathrm{d}t - \int_{OB'C} \dot{\boldsymbol{r}}\cdot\boldsymbol{A}(\boldsymbol{r})\,\mathrm{d}t\right) \\ &= \frac{q}{\hbar}\left(\int_{OBC} \boldsymbol{A}(\boldsymbol{r})\cdot\mathbf{dr} - \int_{OB'C} \boldsymbol{A}(\boldsymbol{r})\cdot\mathbf{dr}\right) = \frac{q}{\hbar}\oint \boldsymbol{A}(\boldsymbol{r})\cdot\mathbf{dr}\,. \end{aligned}$$

The last integral is calculated along the closed contour $OBCB'O$. Its value does not depend on the position of C. This integral is equal to the flux of the magnetic field inside the contour, i. e. $\Phi = \pi r^2 B$. Therefore the current induces a global shift of the interference pattern, corresponding to a phase change $\varphi = 2\pi\Phi/\Phi_0$.

Chapter 16

16.1. Identical particles incident on a beam splitter.

a. The final wave packet must be normalized: $|\alpha|^2 + |\beta|^2 = 1$. In addition the two final states $[\phi_3(\boldsymbol{r}) + \phi_4(\boldsymbol{r})]/\sqrt{2}$ and $\alpha\phi_3(\boldsymbol{r}) + \beta\phi_4(\boldsymbol{r})$ must be orthogonal since the two initial states $\phi_1(\boldsymbol{r})$ and $\phi_2(\boldsymbol{r})$ are orthogonal. This implies that $\alpha + \beta = 0$. We shall take $\alpha = -\beta = 1/\sqrt{2}$ in the following.

b. The initial state for two fermions reads

$$|\Psi(t_\mathrm{i})\rangle = \frac{1}{\sqrt{2}}\left(|1:\phi_1\,;2:\phi_2\rangle - |1:\phi_2\,;2:\phi_1\rangle\right) .$$

We neglect the interaction between the fermions when they pass through the beam splitter. The final state is then obtained by linearity:

$$\begin{aligned}|\Psi(t_\mathrm{f})\rangle = &\frac{1}{2\sqrt{2}}\left(|1:\phi_3\rangle + |1:\phi_4\rangle\right) \otimes \left(|2:\phi_3\rangle - |2:\phi_4\rangle\right) \\ &- \frac{1}{2\sqrt{2}}\left(|1:\phi_3\rangle - |1:\phi_4\rangle\right) \otimes \left(|2:\phi_3\rangle + |2:\phi_4\rangle\right) .\end{aligned}$$

This state can also be written

$$|\Psi(t_\mathrm{f})\rangle = \frac{1}{\sqrt{2}}\left(|1:\phi_4\,;2:\phi_3\rangle - |1:\phi_3\,;2:\phi_4\rangle\right) .$$

The two fermions never come out on the same side of the beam splitter, this is a direct consequence of the exclusion principle.

c. The initial state for two bosons is

$$|\Psi(t_\mathrm{i})\rangle = \frac{1}{\sqrt{2}}\left(|1:\phi_1\,;2:\phi_2\rangle + |1:\phi_2\,;2:\phi_1\rangle\right) ,$$

and hence the final state is

$$\begin{aligned}|\Psi(t_\mathrm{f})\rangle = &\frac{1}{2\sqrt{2}}\left(|1:\phi_3\rangle + |1:\phi_4\rangle\right) \otimes \left(|2:\phi_3\rangle - |2:\phi_4\rangle\right) \\ &+ \frac{1}{2\sqrt{2}}\left(|1:\phi_3\rangle - |1:\phi_4\rangle\right) \otimes \left(|2:\phi_3\rangle + |2:\phi_4\rangle\right) \\ = &\frac{1}{\sqrt{2}}\left(|1:\phi_3\,;2:\phi_3\rangle - |1:\phi_4\,;2:\phi_4\rangle\right) .\end{aligned}$$

The two bosons are always detected on the same side of the beam splitter. This surprising conclusion results from the destructive interference between the two quantum paths

$$\begin{cases}\phi_1 \to \phi_3 \\ \phi_2 \to \phi_4\end{cases} \quad \text{and} \quad \begin{cases}\phi_1 \to \phi_4 \\ \phi_2 \to \phi_3\end{cases} ,$$

which would both lead to the final state $(|1:\phi_3\,;2:\phi_4\rangle + |1:\phi_4\,;2:\phi_3\rangle)/\sqrt{2}$, corresponding to one boson in each output port.

16.2. Bose–Einstein condensation in a harmonic trap.

a. The population of each energy level must be positive, in particular that of the ground state, of energy E_0. This implies $\mu < E_0 = (3/2)\,\hbar\omega$.

b. The number of particles not in the ground state is

$$N' = \sum_{n=1}^{\infty} \frac{g_n}{e^{(E_n-\mu)/k_B T} - 1} \, .$$

Since μ is smaller than E_0, we obtain

$$N' < \sum_{n=1}^{\infty} \frac{g_n}{e^{(E_n-E_0)/k_B T} - 1} = F(\xi) \, .$$

c. For $\xi \ll 1$, we can replace the sum defining $F(\xi)$ by an integral, and we find

$$F(\xi) = \frac{1}{2\xi^3} \int_0^{\infty} \frac{(x+\xi)(x+2\xi)}{e^{\xi} - 1} \, dx \, .$$

To lowest order in ξ, we can replace the numerator by x^2. In addition, using:

$$\frac{1}{e^{\xi} - 1} = \frac{e^{-\xi}}{1 - e^{-\xi}} = e^{-\xi} \sum_{n=0}^{\infty} e^{-n\xi}$$

and $\int_0^{\infty} x^2 e^{-nx} \, dx = 2n^{-3}$, we reach the desired result for N'_{max}.

d. At a given temperature, for a small number N of particles, the distribution of the bosons in the various energy levels is close to the Boltzmann distribution. When N increases and becomes of the order of N'_{max}, the distribution becomes quite different from Boltzmann's law. When N becomes larger than N'_{max}, the population of the excited states ($n > 0$) saturates at its maximal value N'_{max}, and the $N - N'_{max}$ remaining atoms accumulate in the ground state $n = 0$. This is the Bose–Einstein condensation phenomenon. For the values given in the text of the exercise, Bose–Einstein condensation occurs for $T \simeq 0.45\,\mu$K.

16.3. Fermions in a square well.

a. The energy levels for the one-body Hamiltonian are $E_n = n^2 E_1$, where $E_1 = \pi^2\hbar^2/(2mL^2)$. The four lowest levels correspond to
   - the state $|1+, 1-\rangle$, of energy $2E_1$;
   - the four states $|1\pm, 2\pm\rangle$, of energy $5E_1$;
   - the state $|2+, 2-\rangle$, of energy $8E_1$;
   - the four states $|1\pm, 3\pm\rangle$, of energy $10E_1$.

b. We must diagonalize the restriction of the potential to each of the eigensubspaces found above. For the nondegenerate subspaces, we just need to calculate the matrix element $V_1 = \langle \alpha+, \alpha- | \hat{V} | \alpha+, \alpha- \rangle$, which gives, after simple algebra, $V_1 = 3g/(2L)$. Each level $|\alpha+, \alpha-\rangle$ is displaced by this amount.

The case of the eigensubspaces $\mathcal{E}_{\alpha,\beta}$, spanned by the four vectors $|\alpha\pm, \beta\pm\rangle$ with $\alpha \neq \beta$, is more subtle. The problem can be simplified by noticing that $\hat{V}$ does not affect the spin variables. The diagonalization of $\hat{V}$ inside $\mathcal{E}_{\alpha,\beta}$ then amounts to three distinct eigenvalue problems:

- The dimension-1 subspace corresponding to $|\alpha+, \beta+\rangle$ is not coupled to the other vectors of $\mathcal{E}_{\alpha,\beta}$. A simple calculation yields

$$\langle \alpha+, \beta+ | \hat{V} | \alpha+, \beta+ \rangle = 0 .$$

This energy level is not displaced by $\hat{V}$.
- The same conclusion is reached for $|\alpha-, \beta-\rangle$.
- The restriction of $\hat{V}$ to the two dimensional subspace spanned by $|\alpha+, \beta-\rangle$ and $|\alpha-, \beta+\rangle$ reads:

$$\frac{g}{L}\begin{pmatrix} 1 & -1 \\ -1 & 1 \end{pmatrix},$$

whose eigenvalues are $2g/L$ and 0.

To summarize, an energy level corresponding to a four-dimensional eigensubspace is split into two sublevels: one sublevel is threefold degenerate and is not shifted by $\hat{V}$, and the other level is not degenerate and its shift is $2g/L$.

16.4. Heisenberg–Pauli inequality (continuation).

a. Let k be the quantum number of the Fermi level. We have

$$N = \sum_{n=1}^{k} (2s+1) n^2 , \qquad E_0 = -\sum_{n=1}^{k} (2s+1) n^2 \frac{E_{\mathrm{I}}}{n^2} ,$$

i. e., replacing the sum over n by an integral (for large N),

$$N \simeq (2s+1)\frac{k^3}{3} , \qquad E_0 \simeq -(2s+1) E_{\mathrm{I}} k .$$

b. One can eliminate k in these two equations to obtain the desired relation between N and E_0.

c. For any state $|\Psi\rangle$, we have $E_0 \leq \langle \Psi | \hat{H} | \Psi \rangle$. This inequality leads to a trinomial with respect to the variable e^2, which must always be positive. The discriminant of the trinomial has to be negative, which leads to the desired result.

Chapter 17

17.1. Excitation of an atom with broadband light.

a. The state vector of an atom reads, at time t, $\alpha(t)|a\rangle + \beta(t)\mathrm{e}^{-\mathrm{i}\omega_0 t}\,|b\rangle$. The evolutions of α and β are given by

$$\dot{\alpha} = -\mathrm{i}\Omega_1\, f(t)\, \cos(\omega t)\, \mathrm{e}^{-\mathrm{i}\omega_0 t}\, \beta(t) \simeq -i\frac{\Omega_1}{2}\, f(t)\, \mathrm{e}^{\mathrm{i}\delta t}\, \beta(t)\,,$$

$$\dot{\beta} = -\mathrm{i}\Omega_1\, f(t)\, \cos(\omega t)\, \mathrm{e}^{\mathrm{i}\omega_0 t}\, \alpha(t) \simeq -i\frac{\Omega_1}{2}\, f(t)\, \mathrm{e}^{-\mathrm{i}\delta t}\, \alpha(t)\,,$$

where we have neglected the nonresonant terms and have defined $\delta = \omega - \omega_0$. The initial conditions are $\alpha(0) = 1$, $\beta(0) = 0$. The solution of the second equation is, to lowest order in Ω_1,

$$\beta(\tau) = -\mathrm{i}\frac{\Omega_1}{2}\int_{-\tau}^{\tau} f(t)\, \mathrm{e}^{-\mathrm{i}\delta t}\, \mathrm{d}t\,.$$

The bounds of the integral can be extended to $\pm\infty$ since $f(t)$ is zero outside the interval $[-\tau, \tau]$. This gives

$$n_b = n|\beta(\tau)|^2 = n\frac{\pi}{2}\Omega_1^2\, |g(-\delta)|^2\,.$$

b. We generalize the preceding calculation and obtain

$$n_b = n\frac{\pi}{2}\Omega_1^2\, |g(-\delta)|^2 \left|\sum_{p=1}^{\ell} \mathrm{e}^{\mathrm{i}\omega_0 t_p}\right|^2 .$$

c. We calculate the statistical average of $\left|\sum_{p=1}^{\ell} \mathrm{e}^{\mathrm{i}\omega_0 t_p}\right|^2$ as follows. First,

$$\left|\sum_{p=1}^{\ell} \mathrm{e}^{\mathrm{i}\omega_0 t_p}\right|^2 = \sum_{p=1}^{\ell}\sum_{p'=1}^{\ell} \mathrm{e}^{\mathrm{i}\omega_0 (t_p - t'_p)}\,.$$

Since the various times t_p are uncorrelated, the statistical average of $\mathrm{e}^{\mathrm{i}\omega_0 (t_p - t'_p)}$ is zero unless $p = p'$, in which case this term is equal to 1. Therefore there are $\ell \sim \gamma T$ nonzero terms in this sum, and we obtain

$$\bar{n}_b(T) = n\frac{\pi}{2}\Omega_1^2\, |g(-\delta)|^2\, \gamma T\,.$$

The average number of atoms in state b increases linearly with time and we can define a transition rate from a to b,

$$\Gamma_{a\to b} = \frac{\pi}{2}\Omega_1^2\, |g(-\delta)|^2\, \gamma\,.$$

d. The energy contained in a wave packet is $(\epsilon_0 c/2)\, E_0^2 \int f^2(t)\, \mathrm{d}t$, and the mean energy flux reads

$$\Phi = \frac{\epsilon_0 c}{2} E_0^2 \gamma \int f^2(t)\, \mathrm{d}t = \frac{\epsilon_0 c}{2} E_0^2 \gamma \int |g(\Omega)|^2\, d\Omega = \int w(\omega+\Omega)\, d\Omega\,,$$

where we have used the Parseval–Plancherel equality. The function $w(\omega)$ is the spectral density of the energy. This function is related to $\Gamma_{a\to b}$ by

$$\Gamma_{a\to b} = \frac{\pi \mathrm{d}^2}{\hbar^2 \epsilon_0 c^3}\, w(\omega_0)\,.$$

e. The preceding reasoning can be transposed to the case where the atoms are initially in the state b. In this way we can define a transition rate from b to a, and obtain $\Gamma_{b\to a} = \Gamma_{a\to b}$.

f. For this incoherent excitation, the evolution of the atom numbers n_a and n_b is obtained by simply adding the two transition rates that we have just found (this can be proven rigorously using the density operator formalism in Appendix Appendix D):

$$\dot{n}_a = -\Gamma_{a\to b} n_a + \Gamma_{b\to a} n_b\,, \qquad \dot{n}_b = \Gamma_{a\to b} n_a - \Gamma_{b\to a} n_b\,.$$

The steady state is simply $n_a = n_b = n/2$, since $\Gamma_{b\to a} = \Gamma_{a\to b}$.

17.2. Atoms in equilibrium with black-body radiation. Within the framework of the previous exercise, we predict that the atom numbers n_a and n_b are equal. In contrast, statistical physics requires the result $n_b/n_a = \exp(-\hbar\omega_0/k_\mathrm{B}T)$. Indeed, we know that a system (here the atomic assembly) in contact with a heat reservoir at temperature T (the black-body radiation) must reach a thermodynamic equilibrium characterized by the same temperature T.

Einstein's hypothesis consists of adding a second decay process from level b to level a, which creates an asymmetry between the populations of these levels. Suppose that this second process is characterized by the rate $\Gamma'_{b\to a}$. The evolutions of the atom numbers n_a and n_b are now

$$\begin{aligned}\dot{n}_a &= -\Gamma_{a\to b} n_a + (\Gamma_{b\to a} + \Gamma'_{b\to a}) n_b\,,\\ \dot{n}_b &= \Gamma_{a\to b} n_a - (\Gamma_{b\to a} + \Gamma'_{b\to a}) n_b\,,\end{aligned}$$

and the equilibrium state is

$$\frac{n_b}{n_a} = \frac{\Gamma_{a\to b}}{\Gamma_{b\to a} + \Gamma'_{b\to a}}\,, \qquad \text{where} \qquad \Gamma_{a\to b} = \Gamma_{b\to a}\,.$$

Let us impose the value $\exp(-\hbar\omega_0/k_\mathrm{B}T)$ on this ratio. This leads to the value of the rate $\Gamma'_{b\to a}$:

$$\Gamma'_{b\to a} = \Gamma_{b\to a}\left(\mathrm{e}^{\hbar\omega_0/k_\mathrm{B}T} - 1\right)\,.$$

We have obtained in the previous exercise the relation between $\Gamma_{b\to a}$ and the spectral density of the energy $w(\omega)$. In the case of black body radiation, this relation implies that

$$\Gamma'_{b\to a} = \frac{\pi d^2}{\hbar^2 \epsilon_0 c^3} w(\omega_0) \left(e^{\hbar\omega_0/k_B T} - 1 \right) = \mu \frac{\pi d^2}{\hbar^2 \epsilon_0 c^3} .$$

We find that the rate $\Gamma'_{b\to a}$ is independent of temperature, and hence of the state of the electromagnetic field. This rate corresponds to the spontaneous emission process described qualitatively in Sect. 17.3. Einstein's reasoning gives an account of two important characteristics of the spontaneous emission rate. This rate is proportional to the square of the average dipole moment d of the transition, and it varies as the cube of the Bohr frequency ω_0 of the transition (see (17.24)).

17.3. Ramsey fringes. The state vector of the neutron is $\gamma_+(t)\, e^{-i\omega_0 t/2}\,|+\rangle + \gamma_-(t)\, e^{i\omega_0 t/2}\,|-\rangle$. Inside the cavities, the evolutions of the coefficients $\gamma_\pm$ are given by

$$i\dot{\gamma}_+ = \frac{\omega_1}{2} e^{i(\omega_0-\omega)t}\, \gamma_-(t) , \qquad i\dot{\gamma}_- = \frac{\omega_1}{2} e^{i(\omega-\omega_0)t}\, \gamma_+(t) .$$

At the entrance of the first cavity, we have $\gamma_+(0) = 1$ and $\gamma_-(0) = 0$. At the exit of the first cavity ($t_1 = L/v$), $\gamma_-(t_1)$ is given by

$$\gamma_-(t_1) = \frac{\omega_1}{2(\omega-\omega_0)} \left(1 - e^{i(\omega-\omega_0)t_1} \right) ,$$

where we restrict ourselves to the first-order term in B_1. The coefficient γ_- does not evolve anymore until the neutron enters the second cavity at time $T = D/v$. The evolution equation of γ_- during the crossing of the second cavity can be integrated similarly, and we finally obtain:

$$\gamma_-(T+t_1) = \frac{\omega_1}{2(\omega-\omega_0)} \left(1 - e^{i(\omega-\omega_0)t_1} \right) \left(1 + e^{i(\omega-\omega_0)T} \right) ,$$

and hence the spin flip probability is

$$P_{+\to -} = \frac{\omega_1^2}{(\omega-\omega_0)^2} \sin^2[(\omega-\omega_0)t_1/2] \cos^2[(\omega-\omega_0)T/2] .$$

When one varies ω around the resonance frequency ω_0, one obtains a resonance with a width $\sim \pi/T$. This is much narrower than the width that one would obtain with a single cavity ($\sim \pi/t_1$). This setup, which allows one to accurately measure the resonance frequency ω_0, is currently used in metrology and high-resolution spectroscopy.

17.4. Damping of a quantum oscillator.

a. We obtain

$$\frac{\mathrm{d}\hat{a}}{\mathrm{d}t} = -\mathrm{i}\omega\hat{a} - \sum_\lambda \frac{\mathrm{i}g_\lambda}{\hbar}\hat{b}_\lambda\,, \qquad \frac{\mathrm{d}\hat{b}_\lambda}{\mathrm{d}t} = -\mathrm{i}\omega_\lambda\hat{b}_\lambda - \frac{\mathrm{i}g_\lambda}{\hbar}\hat{a}\,.$$

b. The equation for $\hat{b}_\lambda$ can be integrated into

$$\hat{b}_\lambda(t) = \hat{b}_\lambda(0)\,\mathrm{e}^{-\mathrm{i}\omega_\lambda t} - \frac{\mathrm{i}g_\lambda}{\hbar}\int_0^t \mathrm{e}^{-\mathrm{i}\omega_\lambda(t-t')}\,\hat{a}_\lambda(t')\,\mathrm{d}t'\,.$$

We insert this result in the equation of evolution for $\hat{a}$ and obtain

$$\frac{\mathrm{d}\hat{a}}{\mathrm{d}t} = -\mathrm{i}\omega\hat{a} - \mathrm{e}^{-\mathrm{i}\omega t}\int_0^t \mathcal{N}(t'')\,\mathrm{e}^{\mathrm{i}\omega(t-t'')}\,\hat{a}(t-t'')\,\mathrm{d}t'' + \hat{F}(t)\,,$$

where $\hat{F}(t) = -\mathrm{i}\sum_\lambda g_\lambda\hat{b}_\lambda(0)\,\mathrm{e}^{-\mathrm{i}\omega_\lambda t}$.

c. If the bath of oscillators is initially in its ground state, the term $F(t)$ plays the role of a "fluctuating force" whose average value is zero, like the Langevin force in the theory of Brownian motion. Using the fact that the times t'' which contribute significantly to this integral are close to 0 and that the quantity $\langle \mathrm{e}^{\mathrm{i}\omega(t-t'')}\,\hat{a}(t-t'')\rangle$ does not evolve much on this timescale, we obtain the following approximate evolution equations for $\langle\hat{a}\rangle$ and $\langle\hat{a}^\dagger\rangle$:

$$\frac{\mathrm{d}\langle\hat{a}\rangle}{\mathrm{d}t} = -\left(i(\omega+\delta\omega) + \frac{1}{2\tau}\right)\langle\hat{a}\rangle\,,$$
$$\frac{\mathrm{d}\langle\hat{a}^\dagger\rangle}{\mathrm{d}t} = \left(i(\omega+\delta\omega) - \frac{1}{2\tau}\right)\langle\hat{a}^\dagger\rangle\,,$$

where we set, as in (17.41),

$$\int_0^\infty \mathcal{N}(t'')\,\mathrm{d}t'' = \mathrm{i}\delta\omega + \frac{1}{2\tau}\,.$$

This shows that the average position of the oscillator is damped in a time of the order of τ, whatever its initial state is.

This model, and its generalization to nonharmonic systems, is frequently used for the study of dissipation in quantum mechanics (see, e. g., R.P. Feynman and F.L. Vernon, Ann. Phys. **24**, 118 (1963); A.O. Caldeira and A.J. Leggett, Ann. Phys. **149**, 374 (1983)).

Chapter 18

18.1. Scattering length for a hard-sphere potential.

a. We solve (18.53) for a square barrier. The boundary condition is $u(0) = 0$, and the solution between 0 and b reads $u(r) = C' \sinh(Kr)$ where $K = \sqrt{2mV_0}/\hbar$. For $r > b$, we look for a solution varying as $u(r) = C\,(r - a_s)$, as indicated in Sect. 18.5.2. The continuity of u and u' at $r = b$ implies

$$a_s = b - \frac{\tanh Kb}{K} .$$

The scattering length is always positive and smaller than b. If V_0 is small, more precisely if $Kb \ll 1$, we have $\tanh(Kb) \simeq Kb - (Kb)^3/3$, and the scattering length is given by $a_s = K^2b^3/3$. This result coincides with the result derived in the Born approximation (18.52):

$$Kb \ll 1 \quad \Rightarrow \quad a_s = \frac{m}{2\pi\hbar^2}\frac{4\pi}{3}V_0 b^3 = \frac{K^2b^3}{3} .$$

b. The scattering length is an increasing function of V_0. When V_0 tends to infinity, a_s tends to b. In the limiting case of a hard-sphere potential, the scattering length is equal to the radius of the sphere. The collision cross section is then $4\pi b^2$.

c. If we let V_0 tend to $+\infty$ and b tend to 0 while keeping a constant product $V_0 b^3$, we necessarily leave the range of validity of the Born approximation since Kb tends to infinity. We have to use the exact result, which gives $a_s \to 0$: there is no scattering for a $\delta(\boldsymbol{r})$ potential, contrary to what the Born approximation suggests.

18.2. Scattering length for a square well.

a. The mathematical formalism presented in the previous exercise remains valid, even though the conclusions are very different. We find

$$a_s = b - \frac{\tan Kb}{K} ,$$

where $K = \sqrt{2m\,|V_0|}/\hbar$. The scattering length exhibits a series of resonances for $Kb = \pi/2 + n\pi$ (n integer), where it becomes infinite. Each of these resonances corresponds to the appearance of a new bound state in the potential well. This can be checked using the argument of Sect. 4.3.2, and restricting consideration to the eigenstates such that $\psi(0) = 0$ (corresponding to the boundary condition $u(0) = 0$). These peaks in the variation of a_s and of the total cross section $\sigma = 4\pi a_s^2$ are zero-energy bound states (they are also called *zero-energy resonances*). Such states have wave functions which extend to infinity (the limit of

$e^{-\kappa r}$ when $\kappa = \sqrt{2m|E|}/\hbar \to 0$). In other words, although the range of the potential is finite, its depth can be such that, owing to constructive interference, its effect remains important even at very large distances. The corresponding scattering length is anomalously large and negative if $Kb = \pi/2 - \epsilon$ (modulo π), and large and positive if $Kb = \pi/2 + \epsilon$ (modulo π).

These zero-energy bound states or resonances also appear for potentials other than the square well considered here.[2] When one studies the variation of a_s as a function of a parameter defining the potential (depth, size, etc.), these resonances always occur at values of the parameter corresponding to the appearance of a new bound state in the potential well (Levinson's theorem[3]).

b. For small values of V_0 ($Kb \ll 1$), we recover the result of the Born approximation:

$$Kb \ll 1 \quad \Rightarrow \quad a_s = \frac{m}{2\pi\hbar^2}\frac{4\pi}{3}V_0 b^3 = -\frac{K^2 b^3}{3} \qquad (a_s < 0)\,.$$

For a purely attractive potential treated in the Born approximation, the scattering length is negative. This is not true anymore outside the range of validity of the Born approximation (for $Kb = \pi$, the exact result is $a_s = b > 0$). Notice that the validity criterion for the Born approximation ($Kb \ll 1$) requires that no bound state exists in the potential well.

18.3. The pseudopotential.

a. We calculate the action of $\hat{H} = \hat{p}^2/2m + \hat{V}$ on the wave function $\psi(\boldsymbol{r})$. For a plane wave $e^{i\boldsymbol{k}\cdot\boldsymbol{r}}$, the calculation is very simple and yields

$$\hat{H} e^{i\boldsymbol{k}\cdot\boldsymbol{r}} = \frac{\hbar^2 k^2}{2m}\, e^{i\boldsymbol{k}\cdot\boldsymbol{r}} + g\delta(\boldsymbol{r})\,,$$

where we have used the fact that $\delta(\boldsymbol{r})f(\boldsymbol{r}) = \delta(\boldsymbol{r})f(0)$ if $f(\boldsymbol{r})$ is a regular function of $\boldsymbol{r}$. The action of $\hat{H}$ on the spherical wave e^{ikr}/r is more subtle since the expression (10.22) for the Laplacian operator is only valid for regular functions and cannot be applied in this case. To progress further, we write

$$\frac{e^{ikr}}{r} = \frac{1}{r} + \frac{e^{ikr} - 1}{r}\,.$$

We know that $\Delta(1/r) = -4\pi\delta(\boldsymbol{r})$ (see (B.14) in Appendix B). The function $(e^{ikr} - 1)/r$ is regular at $r = 0$ and we can use (10.22). The action of

[2] Such a zero-energy resonance occurs, for instance, in collisions of ultracold cesium atoms (M. Arndt et al., Phys. Rev. Lett. **79**, 625 (1997)).

[3] N. Levinson, Danske Videnskab. Selskab. Mat.-Fys. Medd. **25**, 9 (1949).

$\hat{L}^2$ is simple in this case since we are dealing with an isotropic function, corresponding to zero angular momentum. We thus obtain

$$\hat{H}\psi(\boldsymbol{r}) = \frac{\hbar^2 k^2}{2m}\,\psi(\boldsymbol{r}) + \delta(\boldsymbol{r})\,\frac{1}{1+ika}\left(g - \frac{2\pi\hbar^2 a}{m}\right).$$

It is then straightforward to see that, if we set $a = mg/(2\pi\hbar^2)$, the states (18.50) are eigenstates of Hamiltonian with eigenvalue $\hbar^2 k^2/(2m)$. Consequently, for each wave vector $\boldsymbol{k}$, we have found a scattering state $\psi_{\boldsymbol{k}}(\boldsymbol{r})$.

b. The limit $k \to 0$ is simple and yields $a_s = a = mg/(2\pi\hbar^2)$.
c. The scattering is always isotropic since the scattering amplitude $f(k) = -a/(1+ika)$ does not depend on the angles θ and φ.
d. The total cross section is $\sigma = 4\pi a^2/(1+k^2a^2)$. It tends to $4\pi a^2$ at low energy ($ka \ll 1$) and varies as $4\pi/k^2$ at high energy.

Appendix A

A.1. Distribution of impacts. Going to polar coordinates gives

$$f(\rho) = \frac{1}{\sigma^2}\,\rho\,\mathrm{e}^{-\rho^2/2\sigma^2} \qquad (\rho \geq 0)\,.$$

One can check that $\int_0^{+\infty} f(\rho)\,\mathrm{d}\rho = 1$.

A.2. Is is a fair game? The probability that 6 does not appear is $(5/6)^3$; that it appears once, $3\times(5/6)^2\times(1/6)$; that it appears twice, $3\times(5/6)\times(1/6)^2$; and it appears three times, $(1/6)^3$. The respective gains are $-1, 1, 2, 5$. The expectation value of the gain is therefore

$$\frac{1}{6^3}\,\left(-5^3 + 3\times 5^2 + 2\times 3\times 5 + 5\right) = -\frac{15}{216}\,.$$

Don't play! In order for the game to be fair, the third gain would need to be 20 euros!

A.3. Spatial distribution of the molecules in a gas. We start with the simple alternative: either a molecule is inside v ($p = v/V$) or it is not ($q = 1 - v/V$). The probability of finding k molecules in v is therefore

$$P_{\mathrm{N}}(k) = \binom{N}{k}\left(\frac{v}{V}\right)^k\left(1 - \frac{v}{V}\right)^{N-k}.$$

The expectation value is $\langle k\rangle = Np = Nv/V$, and the dispersion is $\sigma = \sqrt{Npq} \sim \sqrt{Nv/V} = \sqrt{\langle k\rangle}$. In our numerical example $\langle k\rangle \sim 3\times 10^{16}$, and therefore $\sigma \sim 1.7\times 10^8$. The relative dispersion $\sigma/\langle k\rangle$ is very small ($\sim 10^{-8}$). The probability of finding a number of molecules outside the interval $\langle k\rangle \pm 2\sigma$ is 5% (Gaussian law).

Appendix B

B.1. Laplacian operator in three dimensions. We notice that for $r \neq 0$, $\Delta(1/r) = 0$. Using integration by parts, we find, for any φ in S,

$$\begin{aligned}\int \Delta\left(\frac{1}{r}\right)\varphi(\boldsymbol{r})\,\mathrm{d}^3r &= -\int \nabla\left(\frac{1}{r}\right)\cdot\nabla\varphi(\boldsymbol{r})\,r^2\,\mathrm{d}r\,\mathrm{d}^2\Omega \\ &= +\int \mathrm{d}^2\Omega\int_0^\infty \frac{1}{r^2}\left(\frac{\partial}{\partial r}\varphi(\boldsymbol{r})\right)r^2\,\mathrm{d}r \\ &= -\int \mathrm{d}^2\Omega\,\varphi(0) = -4\pi\,\varphi(0)\,,\end{aligned}$$

which demonstrates the very useful identity $\Delta(1/r) = -4\pi\delta(\boldsymbol{r})$.

B.2. Fourier transform and complex conjugation. The Fourier transform of $f^*(k)$ is $g^*(-x)$. If $f(k)$ is real, then $g^*(-x) = g(x)$. If $f(k)$ is even, then $g(x)$ is also even. We can then conclude:

$$f(k)\text{ real and even} \quad\longleftrightarrow\quad g(x)\text{ real and even.}$$

Appendix D

D.1. Trace of $\hat{\rho}^2$. The eigenvalues of $\hat{\rho}^2$ are the Π_i^2 and we have

$$\mathrm{Tr}\,\hat{\rho}^2 = \sum_\mathrm{i}\Pi_\mathrm{i}^2 \leq \left(\sum_\mathrm{i}\Pi_\mathrm{i}\right)^2 = 1\,.$$

This inequality turns into an equality if and only if all the eigenvalues Π_i are equal to 0 but one, which is equal to 1. In this case the system is in a pure state.

D.2. Evolution of a pure state. We have

$$\frac{\mathrm{d}}{\mathrm{d}t}\mathrm{Tr}\,\hat{\rho}^2 = \mathrm{Tr}\left(\hat{\rho}\frac{\mathrm{d}\hat{\rho}}{\mathrm{d}t} + \frac{\mathrm{d}\hat{\rho}}{\mathrm{d}t}\hat{\rho}\right) = \frac{1}{\mathrm{i}\hbar}\mathrm{Tr}\left(\hat{\rho}\,[\hat{H},\hat{\rho}] + [\hat{H},\hat{\rho}]\,\hat{\rho}\right) = 0\,,$$

where we have used the invariance of the trace in a circular permutation. A system which is initially in a pure state ($\mathrm{Tr}(\hat{\rho}^2) = 1$) will thus remain in a pure state in a Hamiltonian evolution.

D.3. Inequalities related to ρ. We set $\langle\phi|\hat{\rho}|\psi\rangle = \langle\psi|\hat{\rho}|\phi\rangle^* = \alpha\,\mathrm{e}^{\mathrm{i}\beta}$ with α positive, and we introduce the vector $|\chi_\theta\rangle = \cos\theta\,|\psi\rangle + \mathrm{e}^{\mathrm{i}\beta}\sin\theta\,|\phi\rangle$. The probability of finding the system in the state $|\chi_\theta\rangle$ is $P(\theta)=\mathrm{Tr}(\hat{\rho}\,|\chi_\theta\rangle\langle\chi_\theta|)$, which can also be written

$$P(\theta) = \langle\chi_\theta|\hat{\rho}|\chi_\theta\rangle = \cos^2\theta\,\left(t^2\langle\phi|\hat{\rho}|\phi\rangle + 2t\alpha + \langle\psi|\hat{\rho}|\psi\rangle\right)\,,$$

where we have set $t = \tan\theta$. This quantity must be positive for any θ, which means that the discriminant of the trinomial in the variable t must be negative. This provides the desired inequality.

D.4. Density operator of a spin-1/2 system. The density operator of a spin-1/2 system is a 2×2 Hermitian matrix, which can be written as a linear combination of the three Pauli matrices and the identity operator, with real coefficients. We can thus set $\hat{\rho} = \left(a_0 \hat{1} + \boldsymbol{a} \cdot \hat{\boldsymbol{\sigma}}\right)/2$, where the a_i's $(i = x, y, z)$ and a_0 are real.

We now have to express the fact that the eigenvalues of $\hat{\rho}$ are two positive numbers whose sum is 1. This can be expressed simply in terms of the trace and of the determinant of $\hat{\rho}$. We have $\mathrm{Tr}(\hat{\rho}) = a_0 = 1$ and

$$\begin{aligned}\det(\hat{\rho}) &= \frac{1}{4}\left[(a_0 + a_z)(a_0 - a_z) - (a_x + ia_y)(a_x - ia_y)\right] \\ &= \frac{1}{4}\left(a_0^2 - \mathbf{a}^2\right) \geq 0\,,\end{aligned}$$

i.e. $\boldsymbol{a}^2 \leq 1$. The expectation value of the spin operator is

$$\langle \boldsymbol{S} \rangle = \mathrm{Tr}\left(\hat{\rho}\,\hat{\boldsymbol{S}}\right) = \frac{\hbar}{4}\,\mathrm{Tr}\left((\hat{1} + \boldsymbol{a} \cdot \hat{\boldsymbol{\sigma}})\,\hat{\boldsymbol{\sigma}}\right) = \frac{\hbar}{2}\boldsymbol{a}\,.$$

The spin is completely polarized if $\boldsymbol{a}^2 = 1$. In this case the spin is in a pure state corresponding to the eigenstate of $\boldsymbol{a} \cdot \hat{\boldsymbol{S}}$ associated with the eigenvalue $+\hbar/2$. The opposite case $\boldsymbol{a} = 0$ corresponds to an unpolarized spin.

Index